Sustainable Cropping Systems

Sustainable Cropping Systems

Special Issue Editor
Jeffrey A. Coulter

MDPI • Basel • Beijing • Wuhan • Barcelona • Belgrade • Manchester • Tokyo • Cluj • Tianjin

Special Issue Editor
Jeffrey A. Coulter
University of Minnesota
USA

Editorial Office
MDPI
St. Alban-Anlage 66
4052 Basel, Switzerland

This is a reprint of articles from the Special Issue published online in the open access journal *Agronomy* (ISSN 2073-4395) (available at: https://www.mdpi.com/journal/agronomy/special_issues/sustainable_cropping_systems).

For citation purposes, cite each article independently as indicated on the article page online and as indicated below:

LastName, A.A.; LastName, B.B.; LastName, C.C. Article Title. *Journal Name* **Year**, *Article Number*, Page Range.

ISBN 978-3-03928-907-3 (Pbk)
ISBN 978-3-03928-908-0 (PDF)

Cover image courtesy of the United States Department of Agriculture-Natural Resources Conservation Service. The photographer was Lynn Betts.

Contents

About the Special Issue Editor

Jeffrey A. Coulter is a Professor and Extension Specialist of Maize-Based Cropping Systems in the Department of Agronomy and Plant Genetics at the University of Minnesota. He received a B.S. degree in Agronomy from South Dakota State University and M.S. and Ph.D. degrees in Crop Sciences from the University of Illinois. Dr. Coulter's research is focused on understanding agronomic practices that increase crop productivity, environmental stewardship, and profitability of maize-based cropping systems. This is centered on field research conducted on farms and at university research and outreach centers, and serves as the foundation for educational programs for agricultural professionals and farmers.

 *agronomy*

Editorial

Sustainable Cropping Systems

Jeffrey A. Coulter

Department of Agronomy and Plant Genetics, University of Minnesota, St. Paul, MN 55108, USA;
jeffcoulter@umn.edu

Received: 25 March 2020; Accepted: 30 March 2020; Published: 1 April 2020

Abstract: Crop production must increase substantially to meet the needs of a rapidly growing human population, but this is constrained by the availability of resources such as nutrients, water, and land. There is also an urgent need to reduce negative environmental impacts from crop production. Collectively, these issues represent one of the greatest challenges of the twenty-first century. Sustainable cropping systems based on ecological principles, appropriate use of inputs, and soil improvement are the core for integrated approaches to solve this grand challenge. This special issue includes several review and original research articles on these topics for an array of cropping systems, which can advise implementation of best management practices and lead to advances in agronomics for sustainable intensification of crop production.

Keywords: cropping systems; sustainable crop production; agroecology; nutrient use efficiency; water use efficiency; environmental quality

---

1. Introduction

The global human population reached 7.7 billion in 2019 and is predicted to be 8.5 billion in 2030 and 9.7 billion in 2050 [1]. Increases in crop production will be needed to meet the requirements of the growing human population. Worldwide, total demand for all agricultural products is anticipated to increase by 1.1% per year until 2050, while that for cereals is expecded to grow by 0.9% per year until 2050, compared to the demand in 2005 to 2007 [2]. However, many of the resources needed for crop production are limited, including land, water, and nutrients, making it essential that they be used responsibly.

Since the 1960s, increases in global crop production have been associated with expansion of land in crop production, increased cropland under irrigation, and greater use of chemical fertilizers [2], along with reliance on chemical pesticides [3]. However, these factors have been linked to negative impacts on the environment. Expansion of cropland can result in loss of soil organic carbon [4], biodiversity [5,6], and ecosystem services [7], and cultivation of marginal land that is highly susceptible to soil erosion [8]. Large-scale irrigation can result in decreased downstream river flow [9], declining aquifer levels [10], and soil salinization and desertification [11]. High application rates of chemical fertilizers, especially nitrogen (N) and phosphorus fertilizers, promote nutrient losses to the ambient environment, which can lead to contamination of surface and ground waters [12–14], eutrophication of non-saline surface waters and coastal waters [13], and greenhouse gas emissions [15]. Additionally, chemical pesticide use can cause a reduction in beneficial predators and parasites, pesticide resistance in pests, a decline in pollinators, injury to target and non-target crops, reduced biodiversity, and pollution of ground and surface waters [16]. Therefore, it is imperative to increase crop production while decreasing its negative effects on the environment. Sustainable cropping systems rooted in agroecological principles, coupled with judicious use of external inputs and soil enhancement, are key to accomplishing this task.

2. Special Issue Overview

This special issue provides an international base for revealing the underlying mechanisms of sustainable cropping systems to drive agronomic innovations and guide the application of best management practices. It includes two review and 16 original research articles reporting novel scientific findings on the development of cropping systems for improved crop yields with greater resistance to abiotic and biotic stressors, enhanced resource use efficiency and profitability, reduced risk of negative environmental impacts, improved soil conditions, and enriched ecosystem diversity. These papers are broadly focused on farming system design, crop rotation, N management, residue management, cover crops, organic management, and crop management for efficient use of irrigation water, and are introduced below.

2.1. Farming System Design

Adaptation of farming systems for improved sustainability requires an understanding of the multifaceted interactions that they are affected by [17], along with the effects of altered agronomic practices on system performance. The article by Merot and Beohouchette [18] proposes a method for applying ecologically-based hierarchical patch dynamics theory to farming systems analysis, which considers spatiotemporal heterogeneity and variation in crop management and fields. The authors applied this method in a case study of a French vineyard undergoing transition to organic production. The results showed that it was useful for hierarchical characterization of the farming system from the farm to field scale and for understanding interactions between farming practices and biological processes. This revealed new biophysical indicators that should be considered in the development of an organic farming system, along with fields that were good candidates to remove from production prior to organic transition due to their greater management requirements.

To control stemborer damage in maize (*Zea mays* L.), improve soil fertility, and increase maize yield on smallholder farms in Ethiopia, a push–pull cropping system was developed [19], by which the drought-tolerant legume desmodium [*Desmodium intortum* (Mill.) Urb.] is intercropped with maize and used to repel (i.e., push) stemborers to a trap (i.e., pull) crop of Bracharia hybrid grass planted in the field borders [20]. The paper by Kumela et al. [21] evaluated this cropping system with monoculture maize in on-farm trials across two neighborhoods in 2016 and four neighborhoods in 2017. Maize grain yield with the push–pull cropping system was significantly greater than that of monoculture maize in all cases except for in one neighborhood in 2017. In a survey of farmers following the 2016 growing season, the majority of respondents who tested the push–pull cropping system rated it as better than monoculture maize based on its ability to provide livestock feed and control stemborer damage, and 96% reported that they were interested in utilizing the push–pull system in the next growing season.

Intercropping shade-tolerant crops with trees is a viable strategy for increasing the utilization efficiency of arable land to meet the needs of an ever-growing human population. Potato (*Solanum tuberosum* L.) is a shade-tolerant food crop [22] that is commonly intercropped with trees in tropical and sub-tropical regions where there are high levels of solar irradiance. In these regions, shading has been shown to have minimal effect on potato tuber yield [22–25]. However, little is known about the performance of potato under shade at higher latitudes where solar irradiance is less. To provide a basis for the development of potato/tree intercropping at higher latitudes, the article by Schulz et al. [26] reports on the effects of artificial shading on potato in southwestern Germany. Compared to the non-shaded control during the three study years, tuber starch content was not significantly affected by shading, while tuber dry matter yield was reduced in zero, one, and two, years with shade levels of 12%, 26%, and 50%, respectively. Since the yield reduction with 26% shading occurred in the year with low total solar irradiance, the authors concluded that potato is a suitable intercrop for agroforestry systems with shading up to 26% under normal levels of solar irradiance in this region.

The review by Sellami et al. [27] synthesizes data from agronomic research on protein crops in Europe. It shows that sowing date and density, fertilization, and deficit irrigation had the greatest effect on seed yield, that faba bean (*Vicia faba* Roth), pea (*Pisum sativum* L.), and lupin (*Lupinus albus* L.)

produced greater seed yield than quinoa (*Chenopodium quinoa* Willd.), amaranth (*Amaranthus* spp.), chickpea (*Cicer arietinum* L.), and lentil (*Lens culinaris* Medik.), and that highest seed yield occurred in central European growing environments, providing insight for sustainable intensification of protein crop production.

2.2. Crop Rotation

Crop rotation has long been recognized as a key component of sustainable cropping systems. The paper by Pagnani et al. [28] reports on the effects of crop rotation and tillage system (conventional tillage and no-tillage) on durum wheat (*Triticum turgidum* L. subsp. *durum* (Desf.) Husn.) over two growing seasons in a Mediterranean region. Compared to monocropped wheat, a two-year faba bean-wheat rotation increased wheat grain yield by 8%, across years and tillage systems. Crop rotation also promoted remobilization of N to grain, and crop rotation coupled with no-tillage improved multiple indices of grain quality when compared to monocropped wheat under conventional tillage.

The article by Qaswar et al. [29] summarizes a long-term study on green manure rotations on an acidic paddy soil in a double rice (*Oriza sativa* L.) cropping system. Compared to a rice–rice–winter fallow cropping system, replacing winter fallow with a green manure crop improved grain yield and the sustainable yield index of early and late rice, soil organic matter, soil total and available N and phosphorus, and phosphatase and urease activities, and reduced the apparent N and phosphorus balances. The authors also quantified the effect of soil biochemical properties on grain yield, providing understanding into the mechanisms driving the rotation effect.

2.3. Nitrogen Management

Nitrogen is typically the most limiting nutrient for the production of cereal crops [30], but the recovery of applied N in harvested grain is only about 35% of the amount applied worldwide [31], rendering the excess N susceptible to loss, which can lead to environmental degradation [32,33]. To advance N use efficiency in maize production through improved synchrony between N supply and crop N requirements, the review by Asibi et al. [34] summarizes recent research and provides new understanding on N assimilation, utilization, and remobilization in maize.

Compost, or decomposed organic material, can improve soil fertility [35] and serve as a slow-release source of N to crops [36,37]. The paper by Maucieri et al. [38] evaluated different types of compost as a substitute for mineral N fertilizer in a three-year maize-wheat (*Triticum aestivum* L.)-sunflower (*Helianthus annus* L.) sequence in Italy. Treatments included a non-N-fertilized control and N applied as 100% mineral fertilizer or compost, and 50% compost plus 50% mineral fertilizer. Highest maize grain yield occurred with 100% mineral fertilizer or 50% compost plus 50% mineral fertilizer, and was lowest with 100% compost. In the subsequent years, grain yields of wheat and sunflower were not significantly different among treatments, but total aboveground biomass of all crops in the three-year study was greatest with 100% mineral fertilizer or 50% compost plus 50% mineral fertilizer. These findings indicate that compost can be used to offset synthetic N fertilizer application in crop production, and that it can be a key component of sustainable cropping systems.

2.4. Residue Management

Crop residues are an important source of nutrients, and their return to soils is considered fundamental for sustaining crop production and soil quality [39]. The article by Tian et al. [40] reports on soil carbon and N storage as affected by the method of maize residue retention under simulated tillage in northeastern China. Treatments included mixing maize residue with soil (to simulate rotary tillage) to a soil depth of 10, 30, or 50 cm, or burying maize residue (to simulate moldboard plow tillage) at 10, 30, and 50 cm soil depths. When residue was mixed with soil, its decomposition and the release of carbon and N decreased as the depth of mixing increased in both study years, but the results differed between years when residue was buried. Greater soil organic carbon content occurred when residue was buried in soil compared to when it was mixed with soil. There was a positive correlation between residue decomposition and the release of carbon and N in the 0–20 cm soil layers, while a negative correlation was present for the 20–60 cm soil layers. Hence, the authors concluded that burying maize residue at a depth of 30 cm was the most suitable method for residue retention among the treatments that they tested.

2.5. Cover Crops

Integrating cover crops into annual cropping systems can provide many benefits, including protection against soil erosion, improved physical, chemical, and biological properties of soil, uptake of water and N to reduce nitrate-N leaching, pest suppression, and enhanced crop yields [41]. The paper by Halwani et al. [42] assessed the performance of soybean [*Glycine max* (L.) Merr.] in northeastern Germany following a winter rye (*Secale cereal* L.) cover crop among systems where rye was harvested as silage followed by plowing and planting of soybean, or terminated by herbicide or crimping followed by no-tillage planting of soybean. The system with rye crimping did not use herbicides for weed control in soybean, while the other systems did. Additionally, the planting date of soybean in this system averaged 11 days later than that of the other systems. Weed control was effective in all systems. Soybean yield was greatest with herbicide termination of rye and no-tillage in all three study years, but was not significantly greater than that from the plow-based system in two of the years. Averaged across years, crimping of rye and no-tillage produced the lowest soybean yield. Dates for termination of rye and soybean planting varied among systems, and earlier dates of these operations were associated with higher soybean yield. This paper also reports on cover crop biomass, soybean plant density, and net economic return, and provides a basis for future research and development of no-tillage cover crop systems for soybean.

The article by Everett et al. [43] summarizes a study across 19 sites in the upper Midwest USA that evaluated the effects of a winter rye cover crop on soil nitrate-N and maize yield when sown before autumn injection of liquid manure in fields where the previous crop was silage maize or soybean. Across sites, use of a rye cover crop reduced nitrate-N concentration in the 0–60 cm soil layer in the spring by 36% without affecting maize silage or grain yield, demonstrating its utility for reducing the risk of N losses without restricting maize productivity in these cropping systems.

The paper by Andersen et al. [44] evaluated five faba bean cultivars, one forage pea cultivar, and one field pea cultivar for their potential as cover and forage crops when sown after wheat in the USA Northern Great Plains. On average, ground coverage prior to the first killing frost was greatest with forage pea (44%), followed by field pea (27%) and the faba bean cultivars (17%), and was least with no cover crop (6%). Forage yield was not significantly different among cover crops (mean = 450 Mg ha^{-1} of dry matter), but crude protein in forage was higher for faba bean and forage pea compared to field pea. Grain yield of the subsequent unfertilized maize crop was not significantly affected by cover cropping. These results show that faba bean and pea can be suitable options for dual-purpose cover and forage in a wheat–maize rotation.

The effects of a winter cover crop of Chinese milk vetch (*Astragalus sinicus* L.) on subsequent double-cropped rice in southern China is discussed in the article by Nie et al. [45]. Under moderate and high N input, they found that a Chinese milk vetch cover crop increased grain yield of the early and late rice crops compared to no cover crop, largely due to increased tillering. Additionally, annual grain yield was not significantly different between rice following Chinese milk vetch with a moderate N rate and rice following no cover crop with a high N rate, confirming the value of Chinese milk vetch for improving the sustainability of double-crop rice production.

Living mulch systems are an innovative approach to cover cropping, in which a perennial cover crop is intercropped with a cash crop. The paper by Alexander et al. [46] reports on the yield and economic responses to N fertilization for first- and second-year maize grown in kura clover (*Trifolium ambiguum* M. bieb) living mulch in the upper Midwest USA. Nitrogen fertilization did not increase grain or stover yields of first-year maize, and the economically optimum N rate for grain yield of second-year maize was similar to the local recommendation for maize following soybean.

Across the two study years, net economic return from grain and stover of first- and second-year maize grown in kura clover living mulch averaged \$138 ha^{-1} greater than net economic return from grain of conventionally managed maize following soybean. These findings reveal that kura clover living mulch can reduce N fertilizer requirements for maize while enhancing profitability for farmers, and contribute to a growing body of literature indicating that use of kura clover living mulch is a viable tactic for sustainable maize-based cropping systems [47–51].

2.6. Organic Management

Organic crop production uses non-genetically modified crops and relies on ecologically based practices such as diversified crop rotations including forage legumes, cover cropping, use of manure and other organic soil amendments, and mechanical weed control in place of synthetic fertilizers and pesticides. Crops produced using certified organic practices are eligible for price premiums, but there is a three-year transition period following the conversion from conventional to organic production in the USA when crops must be produced organically before organic certification is approved [52]. During this transition period, organically produced crops are not eligible for price premiums, thereby complicating decisions on whether to convert to organic production and what the optimal agronomic practices are during the transition period. To address these issues, the article by Cox et al. [53] assesses the agronomic and economic performance of red clover (*Trifolium pretense* L.)–maize, maize–soybean, and soybean–wheat/red clover rotations under organic and conventional management with recommend or high levels of inputs during the transition period in the northeastern USA. With recommended inputs, organic maize had significantly lower grain yield, higher production cost, and lower net economic return in the maize–soybean rotation, but equivalent grain yield, production cost, and net economic return in the red clover–maize rotation, compared to conventional maize. For soybean with recommended inputs, grain yield, production cost, and net economic return were comparable between organic and conventional production in the soybean-wheat/red clover and maize–soybean rotations. With recommended inputs, organic wheat in the soybean–wheat/clover rotation had equivalent grain yield, significantly higher production cost, and lower net economic return compared to conventional wheat. Across all crops, net economic return was greatest for the conventional maize–soybean rotation with either level of inputs. High levels of inputs did not enhance the agronomic and economic performance of organic compared to conventional management. With recommended inputs, net economic return across all crops was similar among the three rotations. These results provide a basis for best practices during the transition from conventional to organic production and serve a foundation to advance the development of ecologically based cropping systems.

Organic cropping systems can vary widely, with some intensive organic systems lacking an overarching ecological approach to crop production and simply replacing conventional inputs with organic inputs. Although such systems can produce high crop yields while meeting organic certification standards in some countries, they can have negative environmental impacts that are similar to those from their conventional counterparts [54]. The paper by Ciaccia et al. [55] compares organic soil fertility management systems during a two-year vegetable rotation under un-heated tunnel greenhouses in a Mediterranean environment, and provides a comprehensive assessment of crop performance, soil fertility, and the abundance of soil arthropods. The results showed that the input substitution system, where commercial organic fertilizers were substituted for synthetic fertilizers, had slightly higher crop productivity than systems utilizing agroecological service crops (i.e., crops grown for cover and green manure) combined with cattle manure or compost, but no improvement in long-term soil fertility parameters or soil arthropods. Meanwhile, the two systems with agroecological service crops exhibited improvement in long-term soil fertility parameters that were associated with changes in the community structure of soil arthropods. This confirms the importance of concurrently evaluating multiple agronomic, soil fertility, and soil biodiversity indices when assessing agroecosystem performance, and shows that soil arthropods can be used as bioindicators for comprehensive evaluation of cropping systems.

2.7. Crop Management for Efficient Use of Irrigation Water

Declining aquifer levels are a serious threat to crop production in semi-arid and arid areas that rely on them for irrigation. The article by Leghari et al. [56] uses data from the North China Plain and simulation modeling to assess crop productivity and water and N use efficiencies for the predominant winter wheat–summer maize double-crop system under standard and optimized rates of irrigation and N, and for monocropped spring maize and a two-year winter wheat–summer maize–spring maize rotation under optimized irrigation and N rates. Across the two study years, the double-cropped system under optimized irrigation and N rates produced the greatest total grain yield, which was 6% greater than that of the same system when 22% more water and 162% more N were applied. Monocropped spring maize and the two-year rotation produced the lowest total grain yield over the two study years, which was 23% less, on average, than that of the double-cropped system under optimized irrigation and N rates. However, monocropped spring maize received 47% less water and 51% less N than the double-cropped system under optimized irrigation and N rates, and, therefore, exhibited the highest water and N use efficiencies in this study. This article also reveals opportunities to improve water and N use efficiencies in these cropping systems based on simulation modeling for a range of irrigation and N rates, and suggests that tradeoff in crop productivity may be needed for responsible stewardship of water and N resources in some areas.

The paper by Machicek et al. [57] reports on forage yield and quality, and water use efficiency of irrigated brown midrib sorghum–sudangrass (*Sorghum bicolor* (L.) Moench ssp. *Drummondii*) and brown midrib pearl millet (*Pennisetum glaucum* (L.) Leeke)) under three, two, and one harvests on 30-, 45-, and 90-day intervals, respectively. Averaged across harvest intervals, sorghum–sudangrass had 35%–52% greater forage dry matter yield than pearl millet in the two study years, but 26% lower water use efficiency in the one year when it was measured. For both crops, one harvest per year produced greater total forage dry matter yield and water use efficiency, but lower quality forage compared to multiple harvests per year. These results demonstrate the importance of forage crop and harvest schedule on forage production and water use efficiency.

3. Conclusions

Increasing global crop production to keep pace with rising demand on a limited supply of resources, while reducing its negative environmental effects, is one of the greatest challenges facing humanity. Sustainable cropping systems based on agroecology, rational use of inputs, and soil improvement are key to meeting this challenge. This special issue contains several articles that synthesize previous research and present original research on various aspects of these topics from a wide range cropping systems around the world. This information can guide on-farm adoption of best management practices and it serves as a base for the agronomic innovations needed for widescale sustainable intensification of crop production.

Funding: This research received no external funding.

Acknowledgments: Thank you to all authors who submitted papers to the special issue of Agronomy entitled "Sustainable Cropping Systems", to the reviewers of these papers for their constructive comments and thoughtful suggestions, and to the editorial staff of Agronomy.

Conflicts of Interest: The author declares no conflict of interest.

References

1. United Nations-Department of Economic and Social Affairs. World Population Prospects 2019: Highlights. 2019. Available online: https://population.un.org/wpp/Publications/Files/WPP2019_Highlights.pdf (accessed on 24 March 2020).

2. Alexandratos, N.; Bruinsma, J. *World Agriculture towards 2030/2050: The 2012 Revision*; ESA Working Paper No. 12-03; FAO: Rome, Italy, 2012; Available online: http://www.fao.org/3/ap106e/ap106e.pdf (accessed on 24 March 2020).

3. Zhang, W. Global pesticide use: Profile, trend, cost/benefit and more. *Proc. Int. Acad. Ecology Environ. Sci.* **2018**, *8*, 1–27.

4. Davidson, E.A.; Ackerman, I.L. Changes in soil carbon inventories following cultivation of previously untilled soils. *Biogeochemistry* **1993**, *20*, 161–193. [CrossRef]

5. Green, R.E.; Cornell, S.J.; Scharlemann, J.P.W.; Balmford, A. Farming and the fate of wild nature. *Science* **2005**, *307*, 550–555. [CrossRef] [PubMed]

6. Stoate, C.; Báldi, A.; Beja, P.; Boatman, N.D.; Herzon, I.; van Doorn, A.; de Snoo, G.R.; Rakosy, L.; Ramwell, C. Ecological impact of early 21st century agricultural change in Europe—A review. *J. Environ. Manag.* **2009**, *91*, 22–46. [CrossRef] [PubMed]

7. Nelson, E.; Sander, H.; Hawthorne, P.; Conte, M.; Ennaanay, D.; Wolny, S.; Manson, S.; Polasky, S. Projecting global land-use change and its effect on ecosystem service provision and biodiversity with simple models. *PLoS ONE* **2010**, *5*, e14327. [CrossRef]

8. Lark, T.J.; Salmon, J.M.; Gibbs, H.K. Cropland expansion outpaces agricultural and biofuel policies in the United States. *Environ. Res. Lett.* **2015**, *10*, 044003. [CrossRef]

9. Cai, X.M.; McKinney, D.C.; Rosegrant, M.W. Sustainability analysis for irrigation water management in the Aral Sea region. *Agric. Syst.* **2003**, *76*, 1043–1066. [CrossRef]

10. McGuire, V.L. *Water-Level Changes in the High Plains Aquifer, Predevelopment to 2001, 1999 to 2000, and 2000 to 2001*; USGS Fact Sheet: FS-078-03; U.S. Geological Survey: Denver, CO, USA, 2003. Available online: https://pubs.usgs.gov/fs/FS078-03/pdf/FS078-03.pdf (accessed on 24 March 2020).

11. Ji, X.B.; Kang, E.S.; Chen, R.S.; Zhao, W.Z.; Zhang, Z.H. The impact of the development of water resources on environment in arid inland river basins of Hexi region, Northwestern China. *Environ. Geol.* **2006**, *50*, 793–801. [CrossRef]

12. Spalding, R.F.; Exner, M.E. Occurrence of nitrate in groundwater—A review. *J. Environ. Qual.* **1993**, *22*, 392–402. [CrossRef]

13. Sharpley, A.N.; Chapra, S.C.; Wedepohl, R.; Sims, J.T.; Daniel, T.C.; Reddy, K.R. Managing agricultural phosphorus for protection of surface waters: Issues and options. *J. Environ. Qual.* **1994**, *23*, 437–451. [CrossRef]

14. Addiscott, T.M. Fertilizers and nitrate leaching. In *Agricultural Chemicals and the Environment. Issues in Environmental Science Technology*; Hester, R.E., Harrison, R.M., Eds.; The Royal Soc. Chem.: Cambridge, UK, 1996; Volume 5, pp. 1–26.

15. Shcherbak, I.; Millar, N.; Robertson, G.P. Global meta-analysis of the nonlinear response of soil nitrous oxide emissions to fertilizer nitrogen. *Proc. Natl. Acad. Sci. USA* **2014**, *111*, 9199–9204. [CrossRef] [PubMed]

16. Pimentel, D.; Burgess, M. Environmental and economic costs of the application of pesticides primarily in the United States. In *Integrated Pest Management*; Pimentel, D., Peshin, R., Eds.; Springer: Dordrecht, The Netherlands, 2014; pp. 47–71.

17. Merot, A.; Wery, J. Converting to organic viticulture increases cropping system structure and management complexity. *Agron. Sustain. Dev.* **2017**, *37*, 19. [CrossRef]

18. Merot, A.; Belhouchette, H. Hierarchical patch dynamics perspective in farming system design. *Agronomy* **2019**, *9*, 604. [CrossRef]

19. Khan, Z.R.; Amudavi, D.M.; Midega, C.A.O.; Wanyama, J.M.; Pickett, J.A. Farmers' perceptions of a 'push–pull' technology for control of cereal stemborers and striga weed in western Kenya. *Crop Prot.* **2008**, *27*, 976–987. [CrossRef]

20. Cook, S.M.; Khan, Z.R.; Pickett, J.A. The use of "Push-pull" strategies in integrated pest management. *Ann. Rev. Entomol.* **2007**, *52*, 375–400. [CrossRef]

21. Kumela, T.; Mendesil, E.; Enchalew, B.; Kassie, M.; Tefera, T. Effect of the push-pull cropping system on maize yield, stem borer infestation and farmers' perception. *Agronomy* **2019**, *9*, 452. [CrossRef]

22. Mariana, M.; Hamdani, J.S. Growth and yield of *Solanum tuberosum* at medium plain with application of paclobutrazol and paranet shade. *Agric. Agric. Sci. Procedia* **2016**, *9*, 26–30. [CrossRef]

23. Pleijel, H.; Danielsson, H.; Vandermeiren, K.; Blum, C.; Colls, J.; Ojanperä, K. Stomatal conductance and ozone exposure in relation to potato tuber yield—Results from the European CHIP Programme. *Eur. J. Agron.* **2002**, *17*, 303–317. [CrossRef]

24. Abdrabbo, M.A.; Farag, A.A.; Abul-Soud, M. The intercropping effect on potato under net house as adaption procedure of climate change impacts. *Appl. Res.* **2013**, *5*, 48–60.

25. Nadir, S.W.; Ng'etich, W.K.; Kebeney, S.J. Performance of crops under Eucalyptus tree-crop mixtures and its potential for adoption in agroforestry systems. *Aust. J. Crop Sci.* **2018**, *12*, 1231. [CrossRef]

26. Schulz, V.S.; Munz, S.; Stolzenburg, K.; Hartung, J.; Weisenburger, S.; Graeff-Hönninger, S. Impact of different shading levels on growth, yield and quality of potato (*Solanum tuberosum* L.). *Agronomy* **2019**, *9*, 330. [CrossRef]

27. Sellami, M.H.; Pulvento, C.; Aria, M.; Stellacci, A.M.; Lavini, A. A systematic review of field trials to synthesize existing knowledge and agronomic practices on protein crops in Europe. *Agronomy* **2019**, *9*, 292. [CrossRef]

28. Pagnani, G.; Galieni, A.; D'Egidio, S.; Visioli, G.; Stagnari, F.; Pisante, M. Effect of soil tillage and crop sequence on grain yield and quality of durum wheat in Mediterranean areas. *Agronomy* **2019**, *9*, 488. [CrossRef]

29. Qaswar, M.; Huang, J.; Ahmed, W.; Li, D.; Liu, S.; Ali, S.; Liu, K.; Xu, Y.; Zhang, L.; Liu, L.; et al. Long-term green manure rotations improve soil biochemical properties, yield sustainability and nutrient balances in acidic paddy soil under a rice-based cropping system. *Agronomy* **2019**, *9*, 780. [CrossRef]

30. Fageria, N.K.; Baligar, V.C. Enhancing N use efficiency in crop plants. *Adv. Agron.* **2005**, *88*, 97–185. [CrossRef]

31. Omara, P.; Aula, L.; Oyebiyi, F.; Raun, W.R. World cereal nitrogen use efficiency trends: Review and current knowledge. *Agrosyst. Geosci. Environ.* **2019**, *2*, 180045. [CrossRef]

32. Galloway, J.N. The nitrogen cycle: Changes and consequences. *Environ. Pollut.* **1998**, *102*, 15–24. [CrossRef]

33. Galloway, J.N.; Dentener, F.J.; Capone, D.G.; Boyer, E.W.; Howarth, R.W.; Seitzinger, S.P.; Asner, G.P.; Cleveland, C.C.; Green, P.A.; Holland, E.A.; et al. Nitrogen cycles: Past, present, and future. *Biogeochemistry* **2004**, *70*, 153–226. [CrossRef]

34. Asibi, A.E.; Chai, Q.; Coulter, J.A. Mechanisms of nitrogen use in maize. *Agronomy* **2019**, *9*, 775. [CrossRef]

35. Fecondo, G.; Guastadisegni, G.; D'Ercole, M.; Del Bianco, M.; Buda, P.A. Utilizzo. Use of quality compost on arboreus cultivation to improve soil fertility. *Ital. J. Agron.* **2008**, *1*, 31–36. [CrossRef]

36. Sullivan, D.; Bary, A.; Thomas, D.; Fransen, S.; Cogger, C. Food waste compost effects on fertilizer nitrogen efficiency, available nitrogen, and tall fescue yield. *Soil Sci. Soc. Am. J.* **2002**, *66*, 154–161. [CrossRef]

37. Sullivan, D.; Bary, A.; Nartea, T.; Myrhe, E.; Cogger, C.; Fransen, S. Nitrogen availability seven years after a high-rate food waste compost application. *Compos. Sci. Util.* **2003**, *11*, 265–275. [CrossRef]

38. Maucieri, C.; Barco, A.; Borin, M. Compost as a substitute for mineral N fertilization? Effects on crops, soil and N leaching. *Agronomy* **2019**, *9*, 193. [CrossRef]

39. Kumar, K.; Goh, K.M. Crop residues and management practices: Effects on soil quality, soil nitrogen dynamics, crop yield, and nitrogen recovery. *Adv. Agron.* **1999**, *68*, 197–319. [CrossRef]

40. Tian, P.; Sui, P.; Lian, H.; Wang, Z.; Meng, G.; Sun, Y.; Wang, Y.; Su, Y.; Ma, Z.; Qi, H.; et al. Maize straw returning approaches affected straw decomposition and soil carbon and nitrogen storage in northeast China. *Agronomy* **2019**, *9*, 818. [CrossRef]

41. Snapp, S.S.; Swinton, S.M.; Labarta, R.; Mutch, D.; Black, J.R.; Leep, R.; Nyiraneza, J.; O'Neil, K. Evaluating cover crops for benefits, costs and performance within cropping system niches. *Agron. J.* **2005**, *97*, 322–332. [CrossRef]

42. Halwani, M.; Reckling, M.; Schuler, J.; Bloch, R.; Bachinger, J. Soybean in no-till cover-crop systems. *Agronomy* **2019**, *9*, 883. [CrossRef]

43. Everett, L.A.; Wilson, M.L.; Pepin, R.J.; Coulter, J.A. Winter rye cover crop with liquid manure injection reduces spring soil nitrate but not maize yield. *Agronomy* **2019**, *9*, 852. [CrossRef]

44. Andersen, B.J.; Samarappuli, D.P.; Wick, A.; Berti, M.T. Faba bean and pea can provide late-fall forage grazing without affecting maize yield the following season. *Agronomy* **2020**, *10*, 80. [CrossRef]

45. Nie, J.; Yi, L.; Xu, H.; Liu, Z.; Zeng, Z.; Dijkstra, P.; Koch, G.W.; Hungate, B.A.; Zhu, B. Leguminous cover crop *Astragalus sinicus* enhances grain yields and nitrogen use efficiency through increased tillering in an intensive double-cropping rice system in southern China. *Agronomy* **2019**, *9*, 554. [CrossRef]

46. Alexander, J.R.; Baker, J.M.; Venterea, R.T.; Coulter, J.A. Kura clover living mulch reduces fertilizer N requirements and increases profitability of maize. *Agronomy* **2019**, *9*, 432. [CrossRef]

47. Prasifka, N.P.; Schmidt, J.R.; Kohler, K.A.; O'Neal, M.E.; Hellmich, R.L.; Singer, J.W. Effects of living mulches on predator abundance and sentinel prey in a corn-soybean-forage rotation. *Environ. Entomol.* **2006**, *35*, 1423–1431. [CrossRef]

48. Berkevich, R.J. Kura Clover Used as a Living Mulch in a Mixed Cropping System. Master's Thesis, University of Wisconsin, Madison, WI, USA, 2008.

49. Singer, J.W.; Kohler, K.A.; Moore, K.J.; Meek, D.W. Living mulch forage yield and botanical composition in a corn-soybean-forage rotation. *Agron. J.* **2009**, *101*, 1249–1257. [CrossRef]

50. Ochsner, T.E.; Albrecht, K.A.; Schumacher, T.W.; Baker, J.M.; Berkevich, R.J. Water balance and nitrate leaching under corn in kura clover living mulch. *Agron. J.* **2010**, *102*, 1169–1178. [CrossRef]
51. Siller, A.R.S.; Albrecht, K.A.; Jokela, W.E. Soil erosion and nutrient runoff in corn silage production with kura clover living mulch and winter rye. *Agron. J.* **2016**, *108*, 989–999. [CrossRef]
52. USDA-Agricultural Marketing Service. What is Organic Certification? 2012. Available online: http://www.ams.usda.gov/sites/default/files/media/What%20is%20Organic%20Certification.pdf (accessed on 24 March 2020).
53. Cox, W.J.; Hanchar, J.J.; Cherney, J. Agronomic and economic performance of maize, soybean, and wheat in different rotations during the transition to an organic cropping system. *Agronomy* **2018**, *8*, 192. [CrossRef]
54. Frison, E.A. IPES-Food. In *From Uniformity to Diversity: A Paradigm Shift from Industrial Agriculture to Diversified Agroecological Systems*; IPES: Louvain-la-Neuve, Belgium, 2016.
55. Ciaccia, C.; Ceglie, F.G.; Burgio, G.; Madžaric, S.; Testani, E.; Muzzi, E.; Mimiola, G.; Tittarelli, F. Impact of agroecological practices on greenhouse vegetable production: Comparison among organic production systems. *Agronomy* **2019**, *9*, 372. [CrossRef]
56. Leghari, S.J.; Hu, K.; Liang, H.; Wei, Y. Modeling water and nitrogen balance of different cropping systems in the North China Plain. *Agronomy* **2019**, *9*, 696. [CrossRef]
57. Machicek, J.A.; Blaser, B.C.; Darapuneni, M.; Rhoades, M.B. Harvesting regimes affect brown midrib sorghum-sudangrass and brown midrib pearl millet forage production and quality. *Agronomy* **2019**, *9*, 416. [CrossRef]

 *agronomy*

Article

Hierarchical Patch Dynamics Perspective in Farming System Design

Anne Merot [1],* and Hatem Belhouchette [2]

[1] INRA, UMR System, 34060 Montpellier, France
[2] CIHEAM-IAMM, UMR System, 34060 Montpellier, France
* Correspondence: anne.merot@inra.fr; Tel.: +0033-0499613049

Received: 15 July 2019; Accepted: 26 September 2019; Published: 2 October 2019

Abstract: Farming systems are complex and include a variety of interacting biophysical and technical components. This complexity must be taken into account when designing farming systems to improve sustainability, but more methods are needed to be able to do so. This article seeks to apply the Hierarchical Patch Dynamics theory (HPD) to farming systems to understand farming system complexity and be better able to support farming system re-design. A six-step framework is proposed to adapt the HPD theory to farming system analysis by taking into account (i) spatial and temporal interactions and (ii) field and management diversity. This framework was applied to a vineyard case study. The result was a hierarchical formalization of the farming system. The HPD framework improved understanding and enabled the formalization of (i) the hierarchical structure of the farming system, (ii) the interactions between structure and processes and (iii) scaling up and down from field to farm scale. HPD theory proved to be successful in analyzing farming system complexity at the farm scale. The framework can help with specific aspects of farming system design, such as how to change the scale of study or determining which scale should be used when choosing indicators for crop management and integrating multi-scale constraints and processes.

Keywords: hierarchical patch dynamics; cropping system design; up-scaling; vineyard system; complexity; organization

---

1. Introduction

In the current context of socioeconomic, climate and environmental changes, farmers have to adapt their farming systems to reduce environmental impacts while ensuring agronomic performances and farm profitability. The challenge for agricultural research is to provide knowledge, tools and methods to redesign sustainable farming systems in various production contexts [1] and analyze induced changes. For certain farms, a simple adjustment of management practices without additional organizational changes may suffice, but for others a redesign of the whole farming system is required [2], which makes the task more complicated. In fact, farmers must strike the right balance between production, economic considerations and environmental protection. The first key step in any such redesign is to understand the farming system organization, which is a challenging undertaking because the farming system is so complex [2]: indeed, crop management sequences are numerous; farmers generally manage several heterogeneous fields; and multiple interactions and different feedback over space and time between the various farming system components must be taken into account. With regard to each management practice, for practical reasons and resource constraints (e.g., access to labor and equipment [3]), farmers do not manage an isolated field but rather a group of fields with the same attributes and crops for greater efficiency [4].

Two main approaches are often used to characterize the organization of farming systems. The first is a field/crop-centered approach with a focus on the crop management systems. This focus can either be on the biophysical components (soil, pest and disease components, etc.) to link the components of the system under the influence of a few management practices and their performance [5], or on the technical components to formalize farmers' knowledge and management practice interactions [3,6]. Two main limitations are identified for this type of approach. First, the link between the technical components and the biophysical components is often limited or simplified in many studies; thus, not all practices are studied but only those relating to one process such as irrigation and soil water dynamics [6]. Second, the farm organization based on resource constraints, the socioeconomic context and field diversity are often not taken into account when characterizing farming system organization [7].

The second approach is farm-centered and describes the farm system organization as a combination of a set of agricultural activities [8]. Each agricultural activity is defined as an archetype of a unique combination of soil type, crop, previous crop and type of management [9]. Each activity is then associated with agricultural, environmental and socio-economic performances. This second approach is particularly suited to assessing the impact of policies and economic drivers on farm performances on a larger scale [9]. However, it is limited when designing innovative farming systems because the temporal and spatial interactions between the activities are often not considered, such as the order of when certain practices on the farm are performed, the priorities for certain fields compared to others or the phenological delay between fields that imposes temporal practices to be managed differently.

The hierarchical patch dynamics (HPD) theory [10] was developed in landscape ecology as a way to break down spatial and temporal complexity. HPD theory combines two approaches: the hierarchical theory of systems [11] and the patch dynamics theory [12]. It describes the links between patterns, processes and scales in a system through a vertical hierarchical organization of interacting patches [10,13,14]. HPD theory seemed to be a good approach to better handle the three issues presented above for understanding the complex farming system and redesign: (i) characterizing the spatial heterogeneity of the farming system components, (ii) identifying interactions and feedback over space and time of the different system components, and (iii) describing biophysical, economic, social, technical and even ecological processes from field to farm scales.

To overcome the limits of the above two approaches, this article proposes applying HPD theory to farming systems to understand farming system complexity and their functional organization. By mobilizing the HPD theory, the farming systems will be represented as interacting patches with a functional "logical" spatial and temporal organization. These patches must be considered by taking into account the links between patterns, processes and scales in a system through a vertical hierarchical organization. Moreover, the HPD concept seems to be a suitable approach towards the re-design of sustainable farming systems through the characterization of their socio-economic and environmental performance by taking into consideration their biophysical structure and crop/patch management practices.

2. Materials and Methods

2.1. Applying the HPD Theory to Farming Systems

2.1.1. Conceptual Framework to Assess Farming Systems Hierarchy and Organization

According to Wu and David [10], "Hierarchy theory assumes that complex systems have a vertical structure that is composed of levels of organization and a horizontal structure that consists of holons." In ecology, patch dynamics theory [12] deals explicitly with patch spatial heterogeneity and changes and assumes that patches are both structural and functional homogeneous units. By combining both hierarchical and dynamic patch theories, we can assume that a holon is likened to a dynamic patch (Figure 1). Therefore, the term "patch" is used in this article only to mean a type of holon. Dynamic patches are defined here as specific components of a spatial organization at a specific level of a given system that have the same spatial-temporal dynamic and are influenced by the same decision-making processes [15]. Using the patch theory, the farming system spatial representation could then be organized across several patches at each level of organization: fields, groups of fields and the farm, to be structured in a hierarchy [13] according to: (i) geomorphological and climatic processes (topography, climate, soil structure and texture), (ii) biological processes including crop characteristics (species, rootstock, structure and spatial arrangement of the plantation, weeds), insect outbreaks, disease dispersal and soil processes, and iii) human disturbances such as management practices grouped under technical tasks and resource competition (labor, water, equipment).

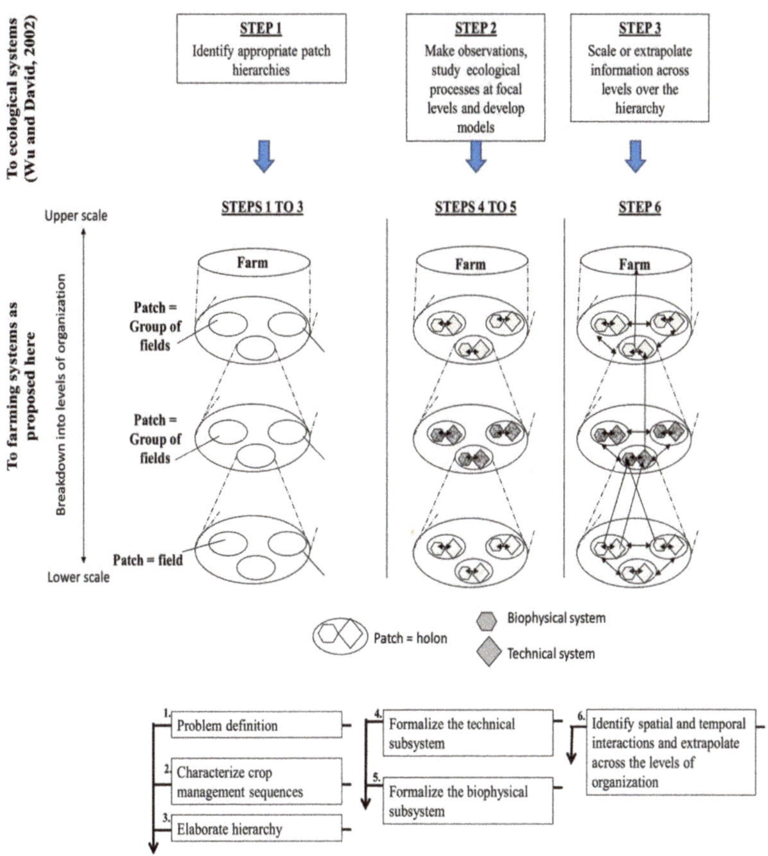

Figure 1. Hierarchical patch dynamics theory in farming systems derived from ecological systems [11]: framework of analysis and six-step procedure to break the farming system down into a patch hierarchy.

Every patch is characterized by two subsystems: the biophysical subsystem represented by different components interacting through various processes, such as transpiration, nutrient absorption, resource competition (water, nitrogen, etc.), plant growth or pest infestation; and the technical subsystem, which is composed of the whole set of management practices during a production season [15]. Both subsystems are mainly affected by the decisions made by the farmer and access to resources (water, land, labor, etc.). The farm level is more closely associated with strategic decisions or certain operational decisions applied to the whole farm. Fields are characterized by operational decision-making processes that result in numerous technical operations. Both subsystems are related to processes that affect spatial elements of the farm such as fields, hedges, etc., which shape the spatial structure of the farming system.

For a given patch, there are numerous and frequent interactions between biophysical and technical subsystems; for example, irrigating a field will change the water status. With regard to the HPD theory, patterns and processes in structural and functional patches only interact when both operate on the same or similar spatial-temporal scales. When a spatial pattern is more or less static relative to the process under study, only the effect of pattern on process, and not process on pattern, must be considered. For example, considering the lower level as a mosaic of pests and diseases for a group of fields and the higher level as a topographic group of fields, the processes related to pest and disease dynamics can be impacted by topographical aspects, but topography dynamics are not affected by pest and disease dynamics. At a given level, three types of interactions could be observed between patches [16]:

Interactions related to transport-based ecological processes (e.g., fungal spore dispersal, runoff, etc.) that are not specific to a patch but applicable to several patches

Constraints on access to resources (labor, equipment, financial resources, etc.)

Patch configuration in space and accessibility from roads—Choosing a preferential pathway to perform management practices, farmers initiate interactions between patches as priority rules for certain management practices

2.1.2. An Innovative Six-Step Procedure to Assess Farming System Organization

This study proposes a six-step procedure adapted from Wu and David [10] to characterize farming system organization using the HPD theory (Figure 1).

Step 1: Defining the Problem and Delineating the Farming System

Step 1 aims to define the problem to be addressed and delineate the farming system in terms of components of interest related to the problem [5]. Thus, in this step, a "systemic representation of the question to be addressed concerning a tangible object" [5] is established. This means that the system environment should be specified and the system should be named and explained as a whole [17], organized around biophysical and technical subsystems. This representation should also include the spatial extent and scales of observation, the time of analysis including duration, and the resolution if the system dynamics are part of the problem to be tackled. This first step is essential to guide data acquisition and make adequate choices in the formalization.

Step 2: Data Acquisition

Step 2 consists in collecting the data required to characterize both farming system functioning and structure in space and time. Based on the definition of a farming system [15], three types of data must be considered: biophysical subsystem data, technical subsystem data, and data on the spatial structure of the farming system (mainly fields).

For the biophysical subsystem, the data to be collected relate to the biophysical components of the farming system such as plant, soil or pest and disease components. The phenological stages are identified and the variables associated with these components (e.g., for the soil component, an associated variable might be soil water status).

To characterize the technical subsystem, the different sequences of management practices used by a farmer on every field during the production season must then be provided. Crop management sequences are analyzed to identify the different management practices and their interactions. Decision rules must be explained, particularly the descriptive variables associated with these rules. Descriptive variables are based on the agronomic characteristics of the soil, climate, crops, a farmer's strategic objectives or farm resources.

Lastly, the spatial elements of the farming system (fields, pathways, ecological infrastructures, etc.) related to the problem definition must be described. Factors of heterogeneity of spatial elements are also analyzed; these might include climate (frost, wind, temperature), topography (slope), soil characteristics (depth, stoniness, soil water retention, waterlogging), accessibility, pest and disease pressure, weed pressure, access to water (irrigation equipment), field location, product labels (certified origin label, protected geographical indication, etc.) and crop characteristics (varieties and cultivars). The heterogeneity aspects taken into account by farmers in their decisions are identified among the various characteristics of these spatial elements. The level of detail for these data depends on the issue defined in Step 1. While there are considerable data collected at the end of this step, they do not show the spatial-temporal interactions between the various components of the farming system.

Step 3: Elaboration of the Appropriate Hierarchy of Patch Mosaics

Step 3 consists in establishing an appropriate patch hierarchy by organizing the farming system into homogeneous structural and functional patches. The hierarchy is built so as to limit the number of levels in the hierarchy following a principle of parsimony in the farming system breakdown.

As specified in [10] and [16], there are a number of methods in spatial pattern analysis for identifying patch hierarchies. For example, in landscape analysis, a land cover approach or ecosystem functional type concept can be used [10]. These methods are based on recognizing scale-dependent heterogeneity among spatial elements resulting from biophysical and technical processes. The detailed specification of the biophysical and technical subsystems as described in Step 2 is useful for several quantitative methods to break down the farming system into hierarchical patch levels following a progressive top-down approach. At the end of Step 3, the depth and breadth of the hierarchy is known, as are the number and size of the different patches at each level.

Step 4: Formalization of the Technical Subsystem in the Patch Hierarchy

Step 4 aims to formalize the technical subsystem throughout the patch hierarchy and to describe the technical components. The technical components are chiefly the management practices identified in Step 2. While management practices are performed at field level, they are often decided and designed for a group of fields. Thus, management practices are associated with the level at which management practices are decided, which often corresponds to the labor or resources organization level.

Step 5: Formalization of the Biophysical Subsystem in the Patch Hierarchy

Step 5 consists in formalizing the biophysical subsystem throughout the patch hierarchy. The biophysical components should be defined according to their horizontal interactions with technical components. In other words, for each interaction identified between a technical component and a biophysical component, a biophysical component must be positioned at the hierarchy level corresponding to the technical component concerned.

Step 6: Spatial and Temporal Interactions between Patches and Across Levels

Step 6 aims to formalize the interactions between patches and across levels and introduce dynamics into the hierarchy. With regard to vertical interactions for technical components, the extrapolation led to scaling down the technical subsystem from higher levels where practices are decided (activated/started/stopped/reported) to the levels where they are performed. This extrapolation mainly concerns constraints imposed by the upper levels on the lower levels. The last level is the field or the infra-field pattern corresponding to the level where management practices are carried out.

Extrapolation across levels of the biophysical components can be seen as supplying initial conditions for the activation, deactivation, starting and stopping of management practices from the lowest levels aggregated to the upper levels. It corresponds to scaling up information across the hierarchy. The aggregation process could vary depending on the information involved, such as means, totals, maximum, minimum, reference patch or ratios involving the various biophysical variables (e.g. soil nitrogen content).

Information collected on temporal crop management sequences leading to complete horizontal interactions between the different management practices are also reported in the different hierarchy levels in each patch. Horizontal interactions at a given level between patches, such as spatial elements, are also indicated in Step 6. Information on interactions between patches could be provided in Step 2 from interviews with farmers, such as when they mention specific pathways for management practices or specific spatial order for processes (erosion, pest and disease development). Finally, the lowest hierarchy level includes simple technical components that directly impact the biophysical components.

2.2. Implementation of the HPD Approach to Vineyard Farming Systems

2.2.1. Presentation of the Test Case: Vineyard Farming System

The framework was applied to a French vineyard farm (Beziers plain, Languedoc region, southern France). It was used to understand and formalize the vineyard system management before its conversion from conventional to organic production. Due to the increasing number of management practices to perform and associated indicators to cope with a limited set of inputs, conversion to organic farming would likely introduce more complexity into the vineyard system [2]. The selected vineyard farm studied here has a total area of 28.5 ha, which comprises the entire area of vines in production on the farm.

2.2.2. Data Collection

The data needed to apply the HPD theory to this case study were collected through a two-hour semi-comprehensive interview with one farmer held on one farm. We met the winegrower in person, taking notes on the content of the semi-directive discussion. The interview was organized into four parts:
Introduction: presentation of the farm
Spatial description: description of the vineyard area, groups of fields and field heterogeneity
Technical subsystem description: description of practices over time and space
Biophysical subsystem description: explanation of decision rules and elicitation of biophysical indicators
In the technical subsystem description, an exhaustive collection of the management practices between budburst and harvest was performed for each field: winter pruning, bud pruning, fertilization, tillage, weeding within and between the rows, the entire pest and disease management process, topping and harvest. When information on the crop management sequences for each field was not available, the farmer was asked if he had to adjust it to the characteristics of each field (e.g., whether mechanical weeding was done differently in a stony field). The farmer was also asked to explain how he decided when and how to carry out a given management practice.

2.2.3. Statistical Approach to Build the Patch Hierarchy

The data were then analyzed and formalized using traditional procedures. A schematic map of the vineyard was built so as to highlight geographical groups of fields and factors of heterogeneity that were used (or not) by the farmer in the crop management. We organized the practices temporally into crop management sequences [3]. The decision-making processes linking the field, decision-rules and practices were also listed. A table was then created to show the different fields with the conditions selected for the different field characteristics. Statistical analysis was performed on these data.

Different methods can be used to break the vineyard system down into a patch hierarchy. We chose to avoid qualitative methods requiring significant expertise and propose basing the breakdown on a more reproducible and statistical approach that takes the different impacts of the various field characteristics into account to explain the heterogeneity observed in the vineyard. The chosen method must enable the farming system to be divided into subgroups of fields via a dissimilarity criterion between fields. Contrary to hierarchical classification methods, which base division on the distance matrix between fields, it was important for each breakdown level to be associated with one meaningful criteria. Accordingly, we used a multiple correspondence analysis (MCA, R®-2.12.0 software, package FactomineR, Agrocampus Rennes, Rennes, France). MCA made it possible to extract the most important and structuring variables characterizing a set of individuals according to criteria of dissimilarity, followed by an analysis of the contribution of these variables on a projection plan. One output of MCA is the ordered contribution on the first projection axes. The higher the contribution on the first axe, the more the structuring the variable.

The impact of the selected field characteristics on heterogeneity in the farming system was analyzed on the first projection axis. The order of the field characteristics contributions was used for the farming system breakdown. Thus, the farm area (first hierarchy level) was first divided according to the field characteristic that best explained its heterogeneity. A second level was then determined for a certain number of patches based on the diversity observed for this first field characteristic. For example, if the first field characteristic is soil, and if two types of soils were observed in the vineyard, then the second hierarchy level is composed of two patches according to soil type. The other levels were determined in the same way for all heterogeneity characteristics. The principle of parsimony was ensured by considering a new level only when the characteristics analyzed led to a separation of a patch into several patches.

When geographical groups of fields could be observed on the farm, a hierarchy of geographical groups of fields followed by a hierarchy of fields was created and two successive MCAs were performed. It was assumed that geographical field accessibility (mainly in terms of distances between fields via pathways and roads) was a strong determinant of vineyard patchiness.

3. Results: Applying the HPD Theory to the Vineyard System

3.1. Step 1: Problem and System Definition

The objective of this step is to define the target of the study, which is to support the conversion to organic farming, and to delimit the studied farm. The selected farm is composed only of a grape production system. All productive grapevine fields (more than five years old) are considered. All the fields produce grapes that are sold to a local cooperative, which makes and sells wine. With regard to management practices, long-term practices such as the planting of perennial crops were not taken into account in our study. Only annual management practices performed just before the conversion to organic farming were included in the analysis. Particular focus was given to all practices impacted by the organic conversion such as pest and disease management, fertilization, weeding, soil management and bud pruning. Field hedges, pathways, roads and ecological infrastructures were not considered in this study because organic certification does not include this type of landscape planning. Fields were considered to be the smallest patches in the farming system because the winegrower did not take into account infra-field heterogeneity. Climate and production factors were part of the environment because they act on the system and are not considered to be affected by the system [5].

All the fields are used to produce the same type of wine and the yield is high (around 60 to 80 hL per ha). The climate is Mediterranean. The farm is located on a windy plain prone to waterlogging. The main pests and diseases are vine moths, downy mildew and powdery mildew. With regard to productive resources, the vineyard is not irrigated and the farmer is the only permanent worker.

3.2. Step 2: Characterization of Specific Crop Management Sequences and Field Diversity

The objective of this step is to collect data to characterize the biophysical and technical subsystems and to describe the spatial structure of the studied farm.

Data Collection

An exhaustive inventory of the management practices between budburst and harvest was drawn up for each field: winter pruning, bud pruning, fertilization, tillage, weeding within and between the rows, pest and disease management, topping and harvest. When information on the crop management sequences for each field was unavailable, the farmer was asked if he had to make adjustments according to the characteristics of each field (e.g., whether mechanical weeding was done differently in a stony field). The farmer was also asked to explain his decisions on when and how to carry out a given management practice.

A schematic map of the vineyard was created to highlight the groups of fields and factors of heterogeneity used by the farmer for crop management. We organized the practices temporally into crop management sequences [5]. The biophysical indicators and decision-making process linking these biophysical indicators, rules and practices were also listed.

Data Analysis: Identification and Description of Groups of Fields

The vineyard studied was made up of 23 fields, arranged in 12 geographical groups of fields ranging from 0.5 ha to 6.5 ha, located around the farm at distances of 0.5 km to 5 km (Figure 2). The geographical groups of fields are lettered from A to L. Certain groups had only two fields while others were composed of five fields. They differed in terms of soil, frost risk and temperature, and powdery mildew susceptibility. No variations in wind, slope, weed pressure or wine production objectives between these geographical groups were identified. However, individual fields, numbered from 1 to 23, varied greatly in their characteristics. Seven characteristics that varied between fields were identified: weed pressure; distance between rows; powdery mildew susceptibility; vine moth pressure; leaf branch and bud development; rabbit presence and grape color. The grapevine cultivar was not considered as a whole; we preferred instead to consider certain cultivar characteristics to explain heterogeneity: bud development, leaf and branch development, and wine color.

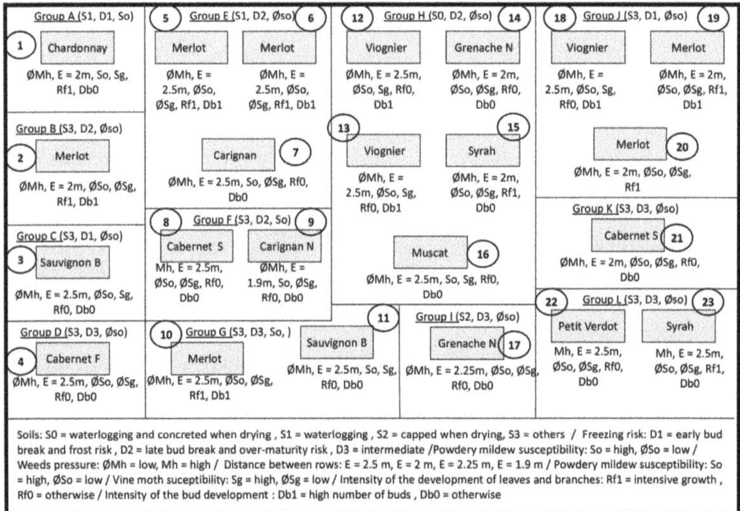

Figure 2. Farm area for the case study. The geographical groups of fields spatially separated, lettered from A to L and delimited by black lines, are characterized by three variables: soils, frost risk and powdery mildew susceptibility. Each field, numbered from 1 to 23, is characterized by the other variables. The various values taken from these variables among the fields illustrate the biophysical heterogeneity throughout the farm area. Variables in italics were not taken into account in the analysis.

The various management practices performed during the year were identified. These could be grouped into four types: those which affect (i) the soil and weeds (e.g., weeding, tillage, nutrient supply), (ii) pests and diseases (e.g., pesticide applications and preventative practices), (iii) the vine architecture and vegetative biomass (e.g., pruning, topping, etc.) or iv) the vine reproductive biomass (e.g., cluster thinning, harvest). Each management practice was associated with biophysical characteristics used to determine when to start a task (Table 1). These characteristics are used by the winegrower as indicators for decision-making: for example, mechanical weeding is done when weeds reach 0.15 m high.

Table 1. Characteristics that are taken into account to adjust the general crop management sequence for each field and their consequences on crop management at the geographical group of fields scale and at field scale.

	Selected Variables	Crop Management Operations Adjusted	Modalities Observed on the Farm	Adjustments for Each Modality
Geographical Group of Field Scale	Soil type	Tillage, and every technical operation performed with a tractor	S0 = waterlogging after rain and concreting when drying	First fields where the technical operations are performed if drying and delay if waterlogging—If drying change in the equipment
			S1 = waterlogging	Delay in the technical operations if waterlogging
			S2 = soils concreted when drying	First fields where the technical operations are performed if drying, change of the equipment
			S3 = none of these characteristics	Technical operations performed when it is not possible in the other fields
	Powdery mildew Susceptibility	Pesticide application	So = high susceptibility	Patches with a higher susceptibility first treated
			Øso = lower susceptibility	Patches treated after the others
	Root disease	Fertilization (N, P, K supply)	Mb = Root disease presence	No nitrogen supply
			ØMb = no root disease	Annual organic fertilization
	Frost risk	Pruning	D1 = early bud break and freezing risk	Last fields pruned
			D2 = late bud break with a risk of over-maturity	First fields pruned
			D3 = none of these characteristics	Intermediate
	Selected variables	Crop management operations adjusted	Modalities observed on the farm	Adjustments for each modality
	Weeds	Mechanical weeding in the row and in the inter-row	Mh: Fields with a lot of weeds	2 mechanical weeding's completed with a manual weeding in the row and a third intervention in the inter-row
			ØMh: Fields with not too many weeds	2 mechanical weeding's in the row- and in the inter-row

Table 1. *Cont.*

	Selected Variables	Crop Management Operations Adjusted	Modalities Observed on the Farm	Adjustments for Each Modality
Field Scale	Width of the inter-row	Mechanical weeding, pesticides application, natural inter-cropping	Distance = 2.5m	1/4 row intercropped and 3/4 tilled-mechanical weeding with 10 teeth—Pesticide application every two rows
			Distance = 2m	1/6 row intercropped and 5/6 tilled-mechanical weeding with 7 teeth—Pesticide application every three rows
			Distance = 2.25m	1/4 row intercropped and 3/4 tilled-mechanical weeding with 10 teeth—Pesticide application every two rows
			Distance = 1.9m	1/6 row intercropped and 5/6 tilled—Manual weeding of the row—Pesticide application every three rows
	Susceptibility to powdery mildew (field)	Pesticide application	So = high susceptibility	Much longer pesticide application for high susceptibility fields, manual application if needed, application $\frac{1}{2}$ rows when the distance between two rows is 2m otherwise 1/3 row
			Øso = lower susceptibility	The treatment is stopped early, no manual treatment
	Susceptibility to vine moth	Pesticide application	Sg = high susceptibility	The treatment is performed in only certain years
			Øsg = lower susceptibility	No treatment
	Leaves and branches development	Trimming	Rf1 = intensive vegetative growth	Three trimmings
			Rf0 = not intensive vegetative growth	Two trimmings
	Bud development	Bud pruning	Db1 = Intensive primary bud development	Two bud prunings
			Db0 = not intensive primary bud development	One bud pruning

Three characteristics of the geographical groups of fields were taken into account by the farmer: soil type, frost risk and powdery mildew susceptibility. The links between these characteristics and their effects on management practices are presented in Table 1. Furthermore, five field characteristics identified during the interview resulted in management practices being adjusted from one field to another (Table 1). For example, the number of toppings in a field with prolific vegetative development (three toppings) differed from those in less prolific fields (two toppings). This means that potential variables of interest were identified during the interviews but they were not necessarily used by the farmer to adjust the crop management sequences to the field heterogeneity. In fact, the farmer did not take the heterogeneity related to grape color and rabbit presence into account in his management.

3.3. Step 3: Hierarchy of Patch Mosaics

This step consists in establishing an appropriate patch hierarchy. To do this, eight factors of heterogeneity were used to construct the farming system hierarchy: three characterizing the groups of fields and five characterizing the fields mentioned in Table 1. Two successive MCAs on the characteristics of the groups of fields and then on the individual fields helped explain the variable weight on field diversity. The vineyard was first broken down into four patches (Figure 3) by the first characteristic (e.g., according to the four soil types). Each of the four soil patches was then divided according to the three modalities of management related to frost risk. As a result, the vineyard was broken down into seven patches. They were further broken down based on the following characteristics: powdery mildew susceptibility, leaf and branch development, bud development and equal distance between rows, vine moth susceptibility and weed pressure (Figure 3). The vineyard farming system was thus divided into ten organizational and structural levels according to the order of these characteristics (Table 2).

These levels and the way the fields and patches of fields were spread throughout the hierarchy are presented in Table 2.

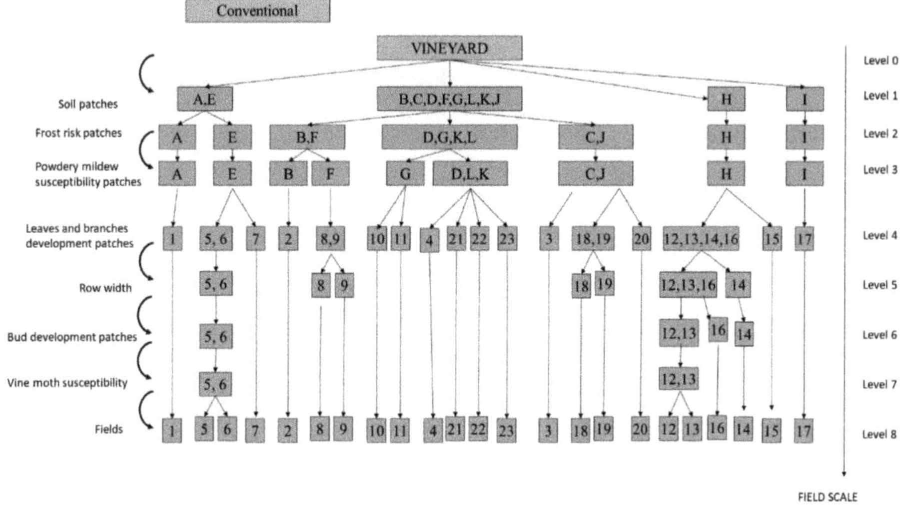

Figure 3. Patch hierarchy established for the studied vineyard (Step 3). Each level was associated with a characteristic of the fields or geographical group of fields. (Figure 3a). The patches at a given level were created by breaking down the patches at the upper level according to the characteristics given on the left side of the figure. The letters represent the geographical groups of fields (see Figure 2) and the field numbers (see Figure 2). Each patch is composed of a technical subsystem and a biophysical subsystem.

Table 2. Results of the MCA analysis-contribution of the selected variables to the first axe of MCA.

	Selected Variables	Contribution to the First Axe of MCA
Geographical Group of Fields Scale	Soil type	0.779
	Powdery mildew susceptibility	0.734
	Frost risk	0.688
Field Scale	Leaves and branches development	0.8111
	Primary bud development	0.339
	Width of the inter-row	0.429
	Vine moth susceptibility	0.109

3.4. Step 4: Formalization of the Technical Subsystem

To build the technical subsystem hierarchy, the associated management practices at each level were reported along with information on decision-making processes as presented in Table 1. For example, because harvesting methods were the same for the whole vineyard, harvest is reported at vineyard level. Because the starting date for pruning was adjusted according to frost risk, pruning was positioned at the frost risk level. The management practices were then reported as extrapolations across the hierarchy in the lower levels and until the field level. For example, mechanical weeding was adjusted at soil type level

3.5. Step 5: Formalization of the Biophysical Subsystem

The various indicators used to manage management practices were positioned and linked to the management practices with which they were associated (Figure 4). These indicators have been identified in Step 2. They are part of decision rules and therefore formalize interactions between management practices and biophysical components. For example, the grapevine phenological stage was needed at the frost risk level because the phenological stage was used by the farmer to manage pruning at frost risk level. Another example, mechanical weeding was done when weeds were 0.15 m high. A biophysical component 'weed' was thus positioned in each patch at the level defined by the variable 'weed' and was characterized by the variable 'height'.

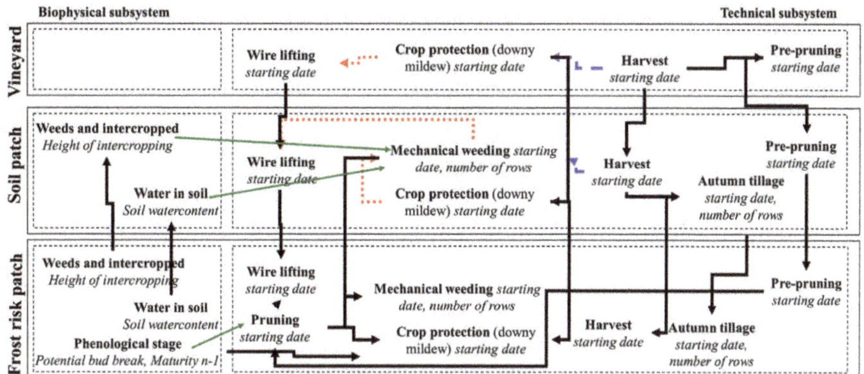

Figure 4. Figure 4 shows the hierarchical breakdown of the technical and biophysical subsystems on the three first levels of organization at the end of Step 6. The interactions between the two subsystems are represented for the three first levels in the hierarchy. Each interaction could be associated with a rule linking a value or the state of the biophysical variable. Finally, Figure 4 shows also the establishment of horizontal and vertical interactions among the technical subsystem and the biophysical subsystem for the three first levels. Each arrow represents an interaction (which can be described by a decision rule) between two crop management practices: continuous thin arrow = start; thin dotted arrow = report; long dotted arrow = stop; continuous thick arrow = activate.

3.6. Step 6: Spatial and Temporal Interactions between Patches and Scaling Up and Down Across Levels

All interactions between patches and across organization levels were positioned in this step. With regard to management practices, vertical interactions from higher levels to the lower levels were reported (Figure 4 for the three first levels—Figure S1). For example, if mechanical weeding at soil type level starts/stops, mechanical weeding starts/stops at frost risk level as well, with consequences at the lower levels. This means that at levels lower than soil type (levels 2 to 9), a component (here, mechanical weeding) and associated interactions were also reported.

For the biophysical subsystem, information on biophysical states of the vineyard system was up-scaled from field level to higher levels. For example, at each level from the field to frost risk levels, a phenological stage component was positioned (see Figure 4)—It aggregated information on the phenological stage of the lower levels and was aggregated at the upper level. The aggregation process from field to lower levels depends on each variable. For example, pruning was activated when one of the fields—the earliest—reached a phenological stage threshold identified in the interview. Figure 4 shows that the height of the intercropped plant was the indicator used by the farmer to determine the mechanical weeding starting date. At frost risk level, the phenological stage of the vine was used to activate pruning. The HPD representation is only qualitative at this stage but if quantitative data is needed, the aggregation method should be indicated.

Horizontal interactions between management practices at a given level of the hierarchy were also formalized in this step (Figure 4). For example, mechanical weeding was activated by pruning at frost risk level. Harvest deactivated or stopped mechanical weeding. Pre-pruning activated pruning. Information on biophysical states of the vineyard system was up-scaled from field level to higher levels.

Finally, interactions between patches are strongly related to spatial pathways of performed practices. In fact, the farmer used preferential spatial pathways to carry out management practices. For example, for pesticide application, the winegrower begins with the group of fields E, followed by groups A, B, C, D, F, G, H, J, K, L and ends with I (Figure 2). This means that regardless of the hierarchical level, when management practices had to be performed on several patches at a given level, the priority between patches was assigned according to the spatial position of patches. This was all the truer at field level. This example did not consider the ecological interactions related to pest and disease dispersion, but this could be possible for another issue.

4. Discussion

4.1. Using the HPD Conceptualization to Support Conversion towards Organic Farming

The HPD theory was successfully applied to our case study to understand and formalize the complexity of interactions within the farming system. In our case study, the farming system was broken down from the farm to the field scale, with numerous levels and associated field characteristics reflecting the complexity of farming system [2].

Considering our objective to support conversion towards organic farming, applying HPD theory made it possible to identify specific elements to take into account for conversion. Converting to organic farming implies no longer using synthetic chemicals. For vineyard systems, farmers must replace (i) herbicides in the inter-row and the row with mechanical or manual weeding, (ii) mineral fertilization with voluminous organic fertilization based on mineralization processes, (iii) chemical bud pruning with manual or mechanical bud pruning, and finally, they must stop using (iv) chemicals against powdery mildew, downy mildew and pests such as vine moths [18]. In our case study, the survey revealed that only weeding in the inter-row was already performed following organic rules. The other management practices as well as the biophysical indicators used to manage these practices must also evolve [2]. Whereas chemical weeding under the row does not require weed growth to be taken into consideration in a precise way, mechanical or manual weeding are started based on weed growth several times during the production season. Similarly, shifting from chemicals to biocontrol or sulfur/copper-based products is required to better take into account the various pest and disease susceptibilities between fields to control pest and disease risk. In the hierarchy before conversion, weed pressure under the row between fields or powdery mildew susceptibility are not used as management indicators, even when there are known differences between fields (Figure 2). This is not the case for vine moth susceptibility. Variations in bud development between fields was already taken into account, which can facilitate changes for bud pruning. Thus, the conversion to organic farming will require following at least two new biophysical indicators, weed pressure and powdery mildew susceptibility, which can result in a higher number of levels in the hierarchy. HPD representation led to planning for a more complex farming system due to the increased number of indicators for technical practice management [2].

Applying HPD theory highlighted the specific position of certain fields in the farming system. In this case study, two fields (field 1, group A and field 17, group I) differed from the others with regard to several biophysical characteristics and associated management practices. In fact, they were isolated quite high in the hierarchy (fields 1 at the third level of the hierarchy and field 17 at the second level). These fields are isolated in terms of management and consequently are time-consuming. Conversion towards organic farming in vineyard systems is a period of readjustment of the area in production with regard to available labor [18]. Farmers often decrease the area in production and vines are frequently removed in one or even two fields on the farm. In the case study, field 17 and potentially field 1 are good candidates to reduce labor requirements by decreasing the area in production. Moreover, the removal of field 17 or field 1 could limit the number of technical management sequences and the complexity of the farming system. Removing field 17 could therefore reduce the number of soil patches in the farming system from four to three.

4.2. What the HPD Framework Brings to Farming System Redesign

Applying HPD theory to the farming system redesign, which ensured more sustainability, was shown to be of significant added value because other technical system representations [3,9] do not analyze farming system organization. In several studies, the design of innovative/sustainable production systems first covers an evaluation of the performance of the current systems. This assessment often involves biophysical variables such as soil moisture content, organic matter evolution, or yield [6,7] when the assessment is at field scale. These variables are frequently of a socio-economic nature when the assessment is at the scale of the farm [8]. In both cases, the proposed alternatives in the design of new systems aiming to improve the performance of those studied often have two important limitations: (i) the variables used to select alternative activities for the newly designed innovative systems are rarely multi-criteria (including production, biophysical, socio-economic criteria) [19] or multi-scale combining variables at the level of cropping systems (yield, organic matter, etc.) and variables at farm level (gross margin, labor, farm water consumption, etc.) [8,17], and ii) the selection and feasibility of these alternatives are often limited to their agronomic and/or socio-economic performance without accounting for organizational constraints over time in line with the management of current activities [19,20]. HPD as applied here will allow us to overcome these two limits. The design of innovative systems using HPD and considering the spatial organization of the studied systems will make it possible to prioritize the systems not only according to biophysical and socio-economic criteria, but also with regard to the logic of crop management intervention (grouped in patches) in relation to the performance (e.g., yield) of these systems. On this basis, these criteria will be accounted for during the implementation of the innovative systems [20].

5. Conclusions

The adaptation of HPD theory to farming systems from ecological theory is a novel way of breaking down farming system complexity while maintaining the essential traits responsible for the overall system performance and constraints at farm level. The HPD framework requires a detailed analysis of the farming system and thorough interviews. However, it can improve the implementation of alternative farming systems that take into account the farming system complexity and the analysis of organizational changes. It is a novel way of integrating farm scale processes (e.g., organization and resources), while keeping the essence of the biophysical field diversity. In this way, it allows up- and down-scaling across farm and fields and between the socioeconomic and biophysical dimensions of the farming systems. Furthermore, the ecological background of the method will also ensure that the representation of the farming system proposed here will be compatible with the ecosystem structuration and natural resources in interaction with farming systems.

Supplementary Materials: The following are available online at http://www.mdpi.com/2073-4395/9/10/604/s1, Figure S1: The whole hierarchical breakdown of the technical subsystem of the patches at the different levels of organization of the cropping system.

Author Contributions: Conceptualization, A.M.; formal analysis, A.M. and H.B.; Methodology, A.M.; Writing—original draft, A.M. and H.B.; Writing—review & editing, A.M. and H.B.

Acknowledgments: The work presented in this publication is funded by INRA – CiAB Agribio3 project AIDY (Integrated Analysis of the Conversion to Organic Farming). The author is grateful to farmers involved in this study for their fruitful collaboration, time and commitment. The authors thank Teri Jones-Villeneuve for the English revision and the anonymous reviewers for their helpful comments.

Conflicts of Interest: The authors declare no conflicts of interest.

References

1. Rodriguez, D.; Cox, H.; de Voilk, P.; Power, B. A participatory whole farm modelling approach to understand impacts and increase preparedness to climate change in Australia. *Agric. Syst.* **2014**, *126*, 50–61. [CrossRef]

2. Merot, A.; Wery, J. Converting to organic viticulture increases cropping system structure and management complexity. *Agron. Sustain. Dev.* **2017**, *37*. [CrossRef]

3. Aubry, C.; Papy, F.; Capillon, A. Modelling decision-making processes for annual crop management. *Agric. Syst.* **1998**, *56*, 45–65. [CrossRef]

4. Schaller, N.; Lazrak, E.G.; Martin, P.; Mari, J.-F.; Aubry, C.; Benoit, M. Combining farmers' decision rules and landscape stochastic regularities for landscape modeling. *Landsc. Ecol.* **2012**, *27*, 433–446. [CrossRef]

5. Lamanda, N.; Roux, S.; Delmotte, S.; Merot, A.; Rapidel, B.; Adam, M.; Wery, J. A protocol for the conceptualisation of an agro-ecosystem to guide data acquisition and analysis and expert knowledge integration. *Eur. J. Agron.* **2012**, *34*, 104–116. [CrossRef]

6. Merot, A.; Bergez, J.E. IRRIGATE: A dynamic integrated model combining a knowledge-based model and mechanistic biophysical models for border irrigation management. *Environ. Model. Soft.* **2010**, *25*, 421–432. [CrossRef]

7. Ripoche, A.; Rellier, J.P.; Martin-Clouaire, R.; Paré, N.; Biarnès, A.; Gary, C. Modelling adaptive management of intercropping in vineyards to satisfy agronomic and environmental performances under Mediterranean climate. *Environ. Model. Soft.* **2011**, *26*, 1467–1480. [CrossRef]

8. Hammouda, M.; Wery, J.; Darbin, T.; Belhouchette, H. Agricultural Activity concept for simulating strategic agricultural production decisions: Case study of weed resistance to herbicide treatments in South-West France. *Comp. Elect. Agric.* **2018**, *155*, 167–179. [CrossRef]

9. Belhouchette, H.; Louhichi, K.; Therond, O.; Mouratiadou, I.; Wery, J.; van Ittersum, M.K.; Lichman, G. Assessing the impact of the Nitrate Directive on farming systems using a bio-economic modelling chain. *Agr. Syst.* **2011**, *104*, 135–145. [CrossRef]

10. Wu, J.G.; David, J.L. A spatially explicit hierarchical approach to modelling complex ecological systems: Theory and applications. *Ecol. Model.* **2002**, *153*, 7–26. [CrossRef]

11. Pattee, H. *Hierarchy Theory: The Challenge or Complex Systems*, 1st ed.; Braziller: New York, NY, USA, 1973; 156p.

12. Meurant, G. *The Ecology of Natural Disturbance and Patch Dynamics*; Pickett, S.T.A., White, P.S., Eds.; Academic Press: London, UK; Orlando, FL, USA, 1985; 472p.

13. Wu, J.; Loucks, O.L. From balance to hierarchical patch dynamics: A paradigm shift in ecology. *Q. Rev. Biol.* **1995**, *70*, 439–466. [CrossRef]

14. Ratze, C.; Gillet, F.; Muller, J.P.; Stoffel, K. Simulation modelling of ecological hierarchies in constructive dynamical systems. *Ecol. Complex.* **2006**, *4*, 13–25. [CrossRef]

15. Le Gal, P.-Y.; Merot, A.; Moulin, C.H.; Navarrete, M.; Wery, J. A modelling framework to support farmer in designing innovative agricultural production systems. *Environ. Model. Soft.* **2010**, *25*, 258–268. [CrossRef]

16. Turner, M.G. Landscape Ecology: The Effect of Pattern on Process. *Annu. Rev. Ecol. Syst.* **1989**, *20*, 171–197. [CrossRef]

17. Lafond, D.; Coulon, T.; Métral, R.; Merot, A.; Wery, J. EcoViti: A systemic approach to design low pesticide vineyards. *Integ. Protect. Prod. Viticulture* **2013**, *85*, 77–86.

18. Merot, A.; Belhouchette, H.; Saj, S.; Wery, J. Implementing organic farming in vineyards. *Agroecol. Sustain. Food Syst.* **2019**. [CrossRef]

19. Saddique, Q.; Cai, H.; Ishaque, W.; Chen, H.; Wai Chau, H.; Umer Chattha, M.; Umair Hassan, M.; Imran Khan, M.; He, J. Optimizing the Sowing Date and Irrigation Strategy to Improve Maize Yield by Using CERES (Crop Estimation through Resource and Environment Synthesis)-Maize Model. *Agronomy* **2019**, *9*, 109. [CrossRef]

20. Prost, L.; Reau, R.; Paravanoc, L.; Cerf, M.; Jeuffroy, M.-H. Designing agricultural systems from invention to implementation: The contribution of agronomy. Lessons from a case study. *Agric. Syst.* **2018**, *164*, 122–132. [CrossRef]

 *agronomy*

Article

Effect of the Push-Pull Cropping System on Maize Yield, Stem Borer Infestation and Farmers' Perception

Teshome Kumela [1,†], Esayas Mendesil [2,†], Bayu Enchalew [1], Menale Kassie [3] and Tadele Tefera [1,*]

[1] International Center of Insect Physiology & Ecology (ICIPE), PO Box 5689 Addis Ababa, Ethiopia
[2] Department of Horticulture & Plant Sciences, Jimma University College of Agriculture & Veterinary Medicine, PO Box 307 Jimma, Ethiopia
[3] International Center of Insect Physiology & Ecology (ICIPE), PO Box 30772-00100 Nairobi, Kenya
* Correspondence: ttefera@icipe.org; Tel.: +251-116172000
† These authors contributed equally to this work.

Received: 3 July 2019; Accepted: 7 August 2019; Published: 15 August 2019

Abstract: The productivity of maize in Ethiopia has remained lower than the world average because of several biotic and abiotic factors. Stemborers and poor soil fertility are among the main factors that contribute to this poor maize productivity. A novel cropping strategy, such as the use of push-pull technology, is one of the methods known to solve both challenges at once. A push-pull technology targeting the management of maize stemborers was implemented in the Hawassa district of Ethiopia with the ultimate goal of increased food security among smallholder farmers. This study evaluated farmers' perception of push-pull technology based on their experiences and observations of the demonstration plots that were established on-farm in Dore Bafano, Jara Gelelcha and Lebu Koremo village of the Hawasa district in 2016 and 2017. This study examined farmers' perception of the importance of push-pull technology in controlling stemborers and improving soil fertility and access to livestock feed. In both cropping seasons, except for Jara Gelelcha, the maize grain yields were significantly higher in the climate-adapted push-pull plots compared to the maize monocrop plots. The majority (89%) of push-pull technology-practising farmers rated the technology better than their maize production methods on attributes such as access to new livestock feed and the control of stemborer damage. As a result, approximately 96% of the interviewed farmers were interested in adopting the technology starting in the upcoming crop season. Awareness through training and effective dissemination strategies should be strengthened among stakeholders and policymakers for the sustainable use and scaling-up of push-pull technology.

Keywords: farmer's perception; maize; push-pull technology; stemborer

---

1. Introduction

In Ethiopia, agriculture is the dominant economic sector that contributes 42% of the total gross domestic product (GDP), employs approximately 83% of the population, is a source for more than 90% of export revenues, and provides raw materials for more than 70% of the country's industries [1]. According to the World Bank [2] report, within the Ethiopian agricultural sector, 60% of income comes from crop production, while 30% comes from livestock, and the rest (10%) comes from forestry products. Smallholder farmers account for the largest share of agricultural production. The country produces more maize than all other crops, accounting for more than 27% of production. Maize is widely produced in almost all agro-ecologies with both rain-based and artificial irrigation systems. Although the crop plays a leading role in maintaining food security with a high population growth, productivity remains low with an average yield of 3.24 ton/ha compared to the world average of 4.5 ton/ha [3]. Maize productivity is limited by biotic and abiotic factors. Abiotic factors include inefficient production methods, low soil fertility, drought, and small landholdings, while biotic factors include diseases, weeds and insect pests [4]. The stemborer insect pest can cause an average yield loss of 20–50% and, in some cases, a complete loss of maize and sorghum crops in Ethiopia [5].

Stemborer moths lay eggs on maize that hatch into larvae that eat the maize leaves and burrow into the stems as they grow. Hence, the stem borer eats the food that the maize plants would use to produce grains. The economically important species of stem borers in Ethiopia include *Busseola fusca* Fuller (Lepidoptera: Noctuidae), *Chilo partellus* (Swinhoe) (Lepidoptera: Crambidae) and *Sesamia calamistis* Hampson (Lepidoptera: Noctuidae) [6]. A study that was conducted on the ecology of the African maize stemborers also indicated that *B. fusca* occurs in all agro-ecological zones from the lowland semi-arid and arid savannahs to the highland wet mountain forests in Africa and is one of the most economically important species of the stemborers [7]. Cultural methods, such as intercropping maize with beans and crop residue disposal, biological control methods, and host plant resistance and insecticidal methods, have been recommended for the control of stem borers. These control methods are either not effective or not affordable by farmers, hence, alternative stem borer control strategies are necessary. As a response to this, the International Centre of Insect Physiology and Ecology (ICIPE) and its partners have developed a novel cropping strategy, the push-pull technology, which can address such efficacy and affordability constraints at once [8]. Push-pull technology is a system of biological intensification that involves attracting gravid female stem borer moths with a trap plant, either Napier grass or Brachiaria, along the border (pull) while driving them away (push) from the main crop, either maize or sorghum, using a repellent intercrop of Desmodium [9].

Previous studies have shown that companion grasses (Brachiaria and Desmodium) can also provide high-value animal fodder, thereby improving milk and meat production and diversifying farmers' income sources [10]. Moreover, Desmodium is an efficient nitrogen-fixing legume that enhances soil fertility, conserves soil moisture, and prevents soil degradation. Push-pull technology is appropriate, environmentally friendly and fits well with resource-poor smallholder farmers of traditional crop-livestock mixed farming systems as it uses locally available and adapted bio-resources [8]. Furthermore, recent studies in East Africa have demonstrated the effectiveness of push-pull technology in controlling the invasive fall armyworm, *Spodoptera frugiperda* (J E Smith), in maize [11]. Push-pull technology was introduced in early 2016 on 32 smallholder farmers' fields in the area of the Hawassa district as demonstration plots targeting the control of stem borers. All other practicing neighborhood farmers in the villages have exercised and shared the knowledge on the implementation and importance of the technology. Since the technology is new to the farmers in the district, they could have different insights into its practical implementations and attributes. In this study, therefore, farmers' perception of push-pull technology was assessed and analyzed to evaluate the level of understanding of practicing (demonstrating) and non-practicing (visiting) farmers based on observations and day-to-day follow-up of the demonstration plots against their local farming practices. Therefore, this paper reports on the assessments of farmers' perception of push-pull technology, their awareness of the levels of yield reduction caused by stem borers, and the control methods that the farmers were exercising.

2. Materials and Methods

2.1. Study Area

The experiment was conducted during the 2016 and 2017 cropping seasons. In the 2016 cropping season, the study was conducted at two kebeles (kebele is the lowest administrative unit in Ethiopia), Dore Bafano and Jara Gelelcha, in the Hawassa district of the Sidama zone in the South Nation, Nationality and People Regional (SNNPR) State of Ethiopia. In 2017, the study was conducted in four kebeles: Dore Bafano, Jara Gelelcha, Wudo Wotatie, and Lebu Koremo. The Hawassa district is located 250 km from the capital, Addis Ababa, to the south following the main road to Kenya (Figure 1). The soil type of the study areas is clay-loam with an average pH of 5.67; the elevation is 1709 masl with a longitude of 38°21.535′ E and latitude of 07°01.921′ N. The annual total rainfall was 654 mm, while the average annual temperature was 11.6 °C, calculated as averages from the data from 2000–2012.

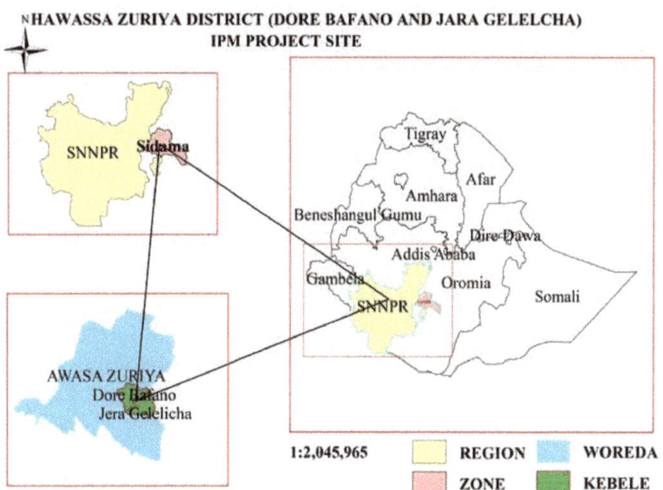

Figure 1. A map showing the Hawassa district and the study sites in Ethiopia.

2.2. Push-Pull Technology Demonstration Plots

As per the information from the agricultural office of the district, the Hawasa district ranks first in the zone for its maize production area of approximately 11,000 hectares. Unfortunately, maize is susceptible to production constraints such as insect pests, particularly stemborers, and loss of soil fertility. This calls for easily accessible and appropriate integrated pest management (IPM) technologies such as push-pull technology. Accordingly, through the IPM project in partnership with the Ministry of Agriculture and Natural Resource (MoANR), the ICIPE established push-pull technology demonstration plots in the two selected kebeles of the district beginning in May 2016. The sites were selected in consultation with the zone and district agricultural experts and the development agents of the two kebeles based on the potential productivity of maize and the presence of major constraints, such as maize stem borers and low soil fertility. From all 23 kebeles of the district, Dore Bafeno and Jara Gelelcha were selected as the implementation sites for the demonstration plots. A total of 32 model farmers, 16 from each kebele, were selected for the establishment of the demonstration and control plots. The farmers were selected based on their willingness and readiness to adopt the new technologies on their lands by setting selection criteria of having their land, experience in maize cultivation and livestock ownership. All the selected farmers with their spouses (husband or wife), the development agents and the agricultural experts were trained on the strategy and implementation methods of push-pull technology.

After acquiring enough levels of awareness regarding the push-pull technology, the farmers planted the demonstration and control plots with an average area of 900 square meters on their lands. In the demonstration plots, maize was planted on average areas of 30 m by 30 m with 28 total rows, 40 cm spacing between maize plants and 80 cm spacing between the rows. The companion plants used for the control of maize stem borers were the grass Brachiaria (Mulato-II) and the legume Greenleaf Desmodium, *Desmodium intortum* (Mill.) Urb. Desmodium seeds were planted by broadcasting between the maize rows. Desmodium produces a smell (semio-chemical) that stem borer moths do not like, hence, it pushes the stem borers away from the maize and enhances soil fertility. Brachiaria was planted in 3 rows surrounding the maize plots (40 cm between plants and by 40 cm between rows) as a trap plant to attract stem borer moths, hence, the term pull [12,13]. Due to the incorporation of drought-tolerant companion plants (Brachiaria cv mulato II and *D. intortum*), the push-pull system is termed the climate–adapted push-pull system [13]. All the maize, Desmodium and Brachiaria seeds were planted from 5–10 June 2016 on the same date at each plot with the support and close supervision of the development agents and ICIPE field technicians. Similarly, the control plots were also planted on the same land area as the push-pull plots following the farmers' conventional practices (without using the companion plants). The control plots were used for checking and comparing the effectiveness and importance of the technology to the farms' conventional practices. In both the demonstration and the control plots, the maize seed hybrid, fertilizer rate, and type and additional agronomic practices were each applied as per the research recommendations for the area.

2.3. Data Collection

2.3.1. Maize Grain Yields

At physiological maturity, all the maize plants in each plot were harvested and the cobs sun-dried separately for each plot (see Section 2.2 for description). Then, the cobs were shelled manually and the maize grain sun-dried to 12% moisture content, and the grain weights were individually taken for each plot and farmer. The grain weights were calculated per plot area harvested, and the yield data converted to kg/hectare.

2.3.2. Survey Data Collection

The household survey data were conducted in the project kebeles from September to October 2016. The data were collected from 71 (29 push-pull practising and 42 randomly selected non-practising) farmers. The structured questionnaires were used to obtain information from the sampled farmers. The questionnaires comprised questions including demographic details such as gender, age, education level, and household size, the owner's land size and the proportion of its area used for maize production. The farmers' awareness and perception of stem borer damage, cultural control methods and push-pull technology were also included in the questionnaires. To promote their understanding of the levels of stem borer damage, the farmers were asked to estimate the potential yield of their maize crops (if no stem borer damage) compared with their actual harvested yield in the 2015–2016 crop season. The questionnaires were pre-tested before starting the survey to check the consistency of the questions. The data collection was conducted by trained enumerators using local languages under the close supervision of researchers and supervisors.

2.4. Statistical Analysis

The data on the maize grain yields were averaged for each plot and farmer (each farmer being a replicate), and comparisons between the climate-adapted push-pull technology and the farmers' practice plots were analysed using t-tests. All statistical analyses were performed using MINITAB 16 statistical software.

The survey data were summarized, and descriptive statistics (means and percentages) were calculated using the Statistical Package for Social Sciences (SPSS). The percentage of farmers who gave similar responses to each question was calculated for each site. The surveys that did not contain responses to certain questions were excluded from the calculations. In instances where a farmer indicated more than one answer to a given question, the percentages were calculated for each group of similar responses. The comparative statistical tools, such as chi-square and t-tests, were conducted to assess the differences regarding the socio-demographics and farm characteristics and the knowledge and perceptions of stemborers and their management practices. The level of significance was set at 0.05, and the means were separated by Tukey's honestly significant difference (HSD) test.

3. Results

3.1. Maize Grain Yields

In all kebeles except for Jara Gelelcha in the 2017 cropping season, the maize grain yields were significantly higher in the climate-adapted push-pull plots than in the maize monocrop plots ($p < 0.05$).

In the 2016 cropping season, the maize yields ranged from 3359 to 3983 kg/ha in the climate-adapted push-pull plots and from 2641 to 2960 kg/ha in the maize monocrop plots. In 2017, the season yields ranged from 4761 kg/ha in Jara Gelecha to 6451 kg/ha in Dore Bafano in the climate-adapted push-pull plots and from 4360 kg/ha in Lebu Koromo to 4721 kg/ha in Wudo Wotatie in the maize monocrop plots (Figure 2).

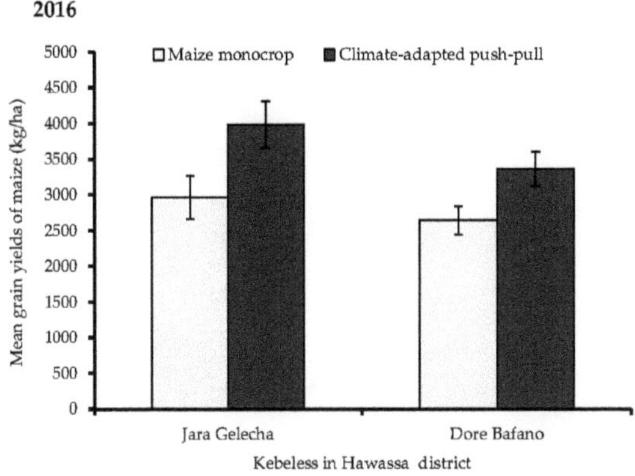

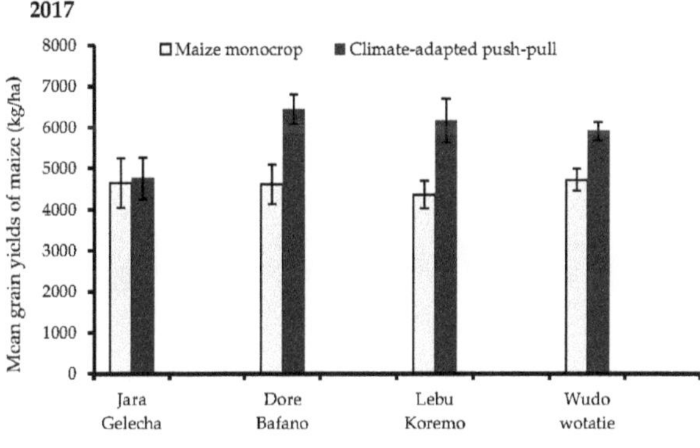

Figure 2. The mean (±S.E.) grain yields of maize (kg/ha) planted in the maize monocrop or climate-adapted push-pull stands in the Hawassa district in the 2016 and 2017 cropping season.

3.2. Socio-Economic and Farm Characteristics

The majority (80%) of both the push-pull technology (PPT)-practising and the non-practising farmers were male. The age of the household head ranged from 24 to 73 years with an average of 42.5. The family size of the households differed significantly between PPT-practising and non-practising farmers with an average of 6.6 persons. Most farmers (94%) had some formal education with an average of 5.15 years of education. The average farm size of the households was 1.15 hectares, while the area used for maize production was 0.71 hectares (63.5%) (Table 1). The survey results indicated that approximately 15% of the interviewed farmers grew fodder crops, such as Napier grass and Rhodes, on their small plots, while the remaining majority (85%) of them collected feed from other sources, such as crop residues from their farms (64%) or natural pastures (24%), or they purchased from other farmers in the village (12%) (Table 1).

The survey results indicated that the majority (approximately 97%) of PPT-practising and non-practising farmers practised mixed, maize-based, crop-livestock farming systems, while 3.4% of the farmers cultivated only crops, whereas the remaining households practised crop-livestock production (Figure 3A). Maize ranked first both in its use as food (94.4%) and as a cash resource (80.3%) (Figure 3B,C), whereas enset (*Ensete ventricosum*) was the second most important food crop in supporting household food security (Figure 3B). Most PPT-practising and non-practising farmers gave high ratings for farming constraints, such as maize stemborer damage (62.5%), the loss of soil fertility (42.75%) and the lack of quality livestock fodder (34.85%), in the study areas (Table 2).

Table 1. Socio-economic characteristics of the respondents in the Hawassa district, Ethiopia.

Variable	PPT-Practicing	Non-Practicing	Mean	χ^2	*t*-Test
	$N = 29$	$N = 42$	$N = 71$		
Gender					
Male	86.2	73.8	80	1.582 ns	
Female	13.8	26.2	20		
Age	43.2	41.8	42.5		0.649 ns
Level of education (years)	4.6	5.7	5.15		−1.441 ns
Family size	7.2	6	6.6		2.841 **
Total land size	1.1	1.2	1.15		−1.030 ns
Total area of maize	0.62	0.80	0.71		−1.369 ns
Source of livestock feed (%)					
Own fodder	62.1	66.7	64.4	2.6919 ns	
Buy fodder	6.9	16.7	11.8		
Free grazing fields	31.0	16.7	23.85		

Statistically significant at * $p < 0.05$, ** $p < 0.01$; ns = not significant.

Table 2. Farmers' perception of farming constraints in the Hawassa district, Ethiopia.

Farming Constraints Response	PPT-Practicing	Non-Practicing	Mean	χ^2
	$N = 29$	$N = 42$	$N = 71$	
Stemborer damage				
Very high	31.0	38.1	34.55	0.412 ns
High	65.5	59.5	62.5	
Low	3.4	2.4	2.9	
Soil fertility problem				
Very high	20.7	21.4	21.05	1.606 ns
High	37.9	47.6	42.75	
Low	34.5	21.4	27.95	
Not a problem	6.9	9.5	8.2	
Shortage of livestock feed				
Very high	34.5	54.8	44.65	5.950 ns
High	48.3	21.4	34.85	
Low	13.8	21.4	17.6	
Not a problem	3.4	2.4	2.9	

ns = not significant.

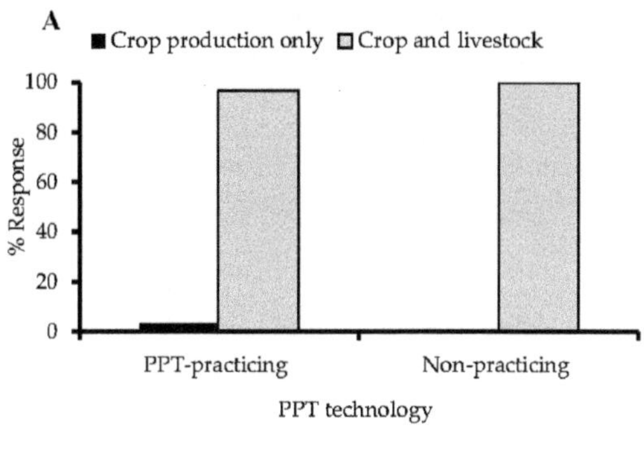

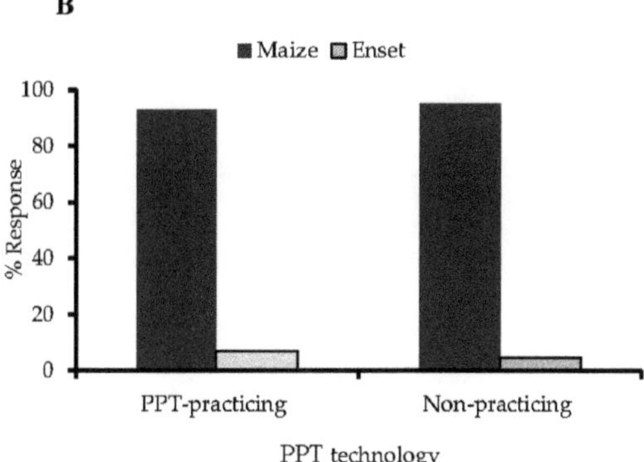

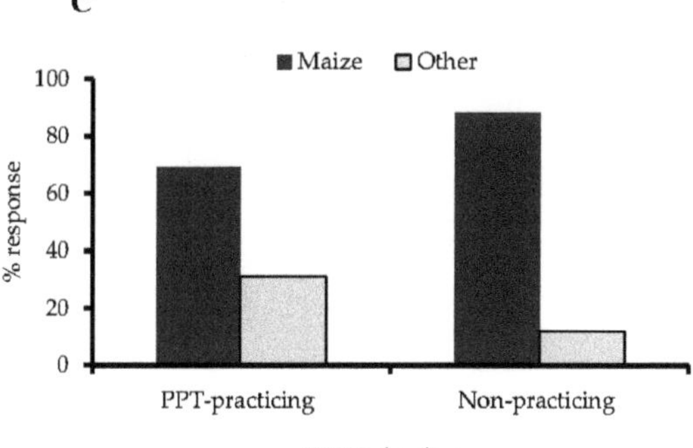

Figure 3. (**A**) Farming systems, (**B**) major food crops grown and (**C**) major cash crops grown in push-pull technology (PPT)-practicing and non-practicing plots in the Hawassa district, Ethiopia.

3.3. Maize Stemborer Knowledge, Damage and Control Methods

All of the interviewed farmers knew stemborers by their local name, Santoo, and approximately 98.1% of them cited maize damage caused by stemborers in the last cropping season. All of the farmers produced maize during the main/rainy season from June to September. The majority (80.8%) of PPT-practising and non-practising farmers perceived that stemborer damage to their crops was serious during the rain shortage. However, 17.5% of the interviewed farmers revealed that neither rain shortages nor long-term rain had any effect on the occurrence of stem borer damage (Table 3). The survey results indicated that stemborers reduced the average maize yield by approximately 29.3%, comparing the estimated yield with the actual harvested maize yield in the 2015 cropping season. This was computed from 2.15 ton/hectare of actual harvested maize with 3.0 ton/hectare of potential maize yield estimated by farmers if there were no stemborer damage. Approximately 86% of PPT-practising and non-practising farmers practised different stemborer control methods. The methods used by the farmers included timely planting (34.05%), insecticide applications (20.4%), the application of wood ash (16.8%), up-rooting damaged stems (14.55%) and intercropping with other legume crops (1.7%) (Table 3). Farmers mentioned that of all the methods used were neither affordable nor consistent in controlling stemborer damage.

Table 3. Farmers' knowledge and perceptions of stemborer in the Hawassa district, Ethiopia.

Variables	PPT-Practicing	Non-Practicing	Mean	χ^2	*t*-Test
	$N = 29$	$N = 42$	$N = 71$		
Know stemborer (Yes)	100	98.0	99.3	2.013 ^{ns}	
Encountered stemborer damage (Yes %)	98.2	98.0	98.1	6.003 *	
A season where stemborer is serious					
Long rain	3.4	0.0	1.7	2.068 ^{ns}	
Short rain	75.9	85.7	80.8		
All-season	20.7	14.3	17.5		
Expected maize yield (kg/ha)					
If no damage by stemborer	2670	3230	2950		4.969 ^{ns}
Infested by stemborer	1910	930	1420		4.462 ^{ns}
The severity of stemborer damage					
Very high	31	38.1	34.55	0.412 ^{ns}	
High	65.5	59.5	62.5		
Low	3.4	2.4	2.9		
Pest control method					
Insecticide spray	24.1	16.7	20.4	9.393 ^{ns}	
Timely planting	27.6	40.5	34.05		
Intercropping	3.4	0.0	1.7		
Wood ash	24.1	9.5	16.8		
Uprooting damaged stem	17.2	11.9	14.55		
No control	3.4	21.4	12.4		
Have you heard/know as PPT control stemborer damage? (Yes %)	100	100	100		
Interested to adopt PPT (Yes %)	79.3	95.2	87.3	4.253 *	

Statistically significant at * $p < 0.05$; ns = not significant.

3.4. Farmers' Perception of Push-Pull Technology

The majority of the PPT-practising farmers gave high ratings to the PPT technology compared to their maize production practices (Table 4). These perceptions were based on the major attributes of the technology, such as access to new livestock feed (Desmodium and Brachiaria) (51%), the control of stemborer damage (by observing holes on the maize stems and or leaves) (41%), the improved soil fertility (by observing changes in soil colour and moisture) (6%) and erosion control (through observations of reduced run-off) (3%) (Figure 4A). Although the push-pull technology is new in the district, an average of 87% of the interviewed farmers were interested and ready to practice the push-pull technology on their land starting in the upcoming crop season based on their perception of the multiple benefits of the push-pull technology. On the other hand, a few (13%) farmers were still hesitating to adopt push-pull technology mainly because of the shortage of cultivable land (7%) and fear of taking risks of using an unknown technology (4%) (Table 3). Most (45%) interviewed farmers mentioned that the implementation of the push-pull technology decreased workload, while other farmers reported no change (28.2%) or an increased (26.8%) workload during field management, including land preparation and weeding, compared to their maize cultivation methods (Figure 4B). Some farmers also suggested that the technology demanded more labour for the construction of fences to protect the perennial companion grasses from damage caused by free-grazing livestock after the crops were harvested.

Table 4. Push-pull technology practicing farmer's perceptions on PPT in the Hawassa district, Ethiopia.

Technology Attributes Observed	Frequency	Percent
Controls stemborer		
Very high	32	45.1
High	32	45.1
Lower	6	8.5
No	1	1.4
Increased maize grain yield		
Very high	20	28.2
High	45	63.4
Lower	4	5.6
No	2	2.8
Increases fertility		
Very high	21	29.6
High	38	53.5
No change	9	12.7
Reduced fertility	3	4.2
Provides fodder		
Very high	44	62
High	23	32.4
Lower	4	5.6
Increased milk production		
Very high	25	35.2
High	9	12.7
Lower	2	2.8
No change	2	2.8

A

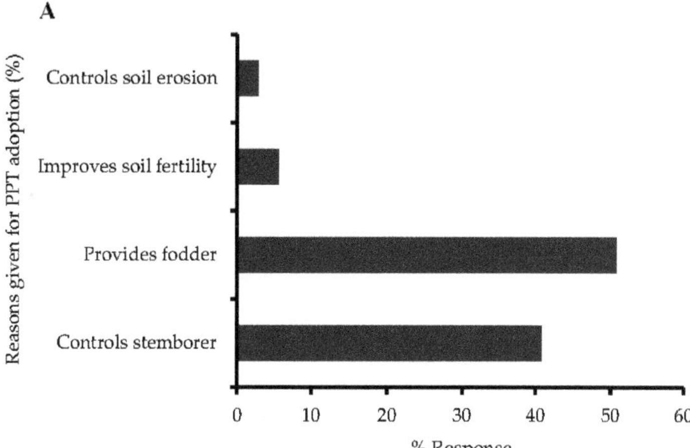

Figure 4. *Cont.*

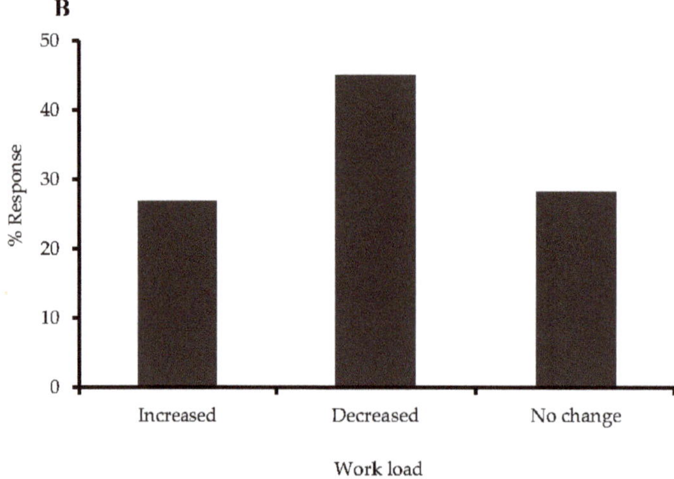

Figure 4. Farmers' perception of (**A**) motivating factors for adoption of the push-pull technology and (**B**) workload in implementing push-pull technology in the Hawassa district, Ethiopia.

4. Discussion

The average family size of the households in the study area was 6.6, which is greater than the national average of 5 persons [1]. Previous research findings indicate that family and landholding size have a significant and positive influence on a households' decisions to employ new agricultural technologies that are particularly targeted towards smallholder farmers because of the requirement for more family labour. In this study, a larger family size, therefore, indicates a potential for the sustainability of the push-pull technology because of access to more family labour. The average landholding size of the households was 1.15 hectares per household, where the majority (63.5%) of farmers allocated a larger portion of the average area, 0.71 hectares, to maize production. Maize has vital importance as both a food (94%) and cash (80%) crop sold either as green cobs or grains for household use. This is consistent with the study by Below et al. [14], which stated that maize producers in Ethiopia are mostly smallholder subsistence farmers in both land size and production volume. The reasons for the maize dependency of the households was not only from their farming preferences but also from their limited land size to cultivate other crops. Usually, households with a larger area of maize are relatively more food secure and are more concerned with adopting and using proper technologies.

The study area was characterized by rain-fed (bimodal, with short rain from February to April and long rain from June to September), crop-livestock, mixed farming systems that were mainly recognized by mono-cropping patterns with dominant maize cultivation. Crop production is the main source of food and income for households. The farmers cited that livestock provided not only milk, meat and draft power for the household, but also sources for manure that can improve soil fertility. This can be considered a better option for strategies that enhance maize productivity, as push-pull companion plants (Desmodium and Brachiaria) could be an important source of animal feed [15].

Farmers rated stemborers as their most important constraint limiting maize production. However, other challenges, such as the lack of quality livestock feed, the loss of soil fertility and run-off, were also reported by farmers. All these constraints can be solved by the novel, push-pull technology at once [9]. These attributes of the technology-initiated farmers to adopt the technology on their farms starting with the next cropping season. The farmers estimated that stemborers can reduce the estimated maize yield by approximately 29.3% compared with actual harvested yields in the 2015–2016 cropping season. This result is consistent with Getu et al. [5] study on the level of damage caused by stemborers, which was an average of 20–50% maize yield loss in Ethiopia. Similarly, a study by Groote et al. [16] also showed that the average maize yield loss caused by stemborer damage was estimated to be 20–40% in Africa.

In the study area, only approximately 20% of farmers applied chemicals for stemborer control, whereas 80% used chemical-free cultural methods, including timely planting, the application of wood ash, the up-rooting of damaged maize stems and intercropping with other legume crops. Similarly, a study conducted by Oben et al. [17] reported the use of different cultural methods, such as wood ash and botanicals, for the control of maize stem borers in Cameroon. In addition, Tefera [18] also indicated that intercropping and crop rotation of cereals with legumes decreased the incidence of pests, which resulted in increased yields. As a result, farmers held a positive attitude that the push-pull technology was the best fit for their maize cultivation and livestock management practices through their observations of the demonstration plots on field days and their frequent visits. A study conducted by Khan et al. [8] also showed that such attitudes were found to positively and significantly influence the likelihood of farmers adopting the technology.

Labour is an important factor of production in crop cultivation and is one of the main constraints in the adaptation of new technologies [19]. In this study, about 27% of farmers reported that the technology required more labour during land preparation, plantation and weeding activities compared to their conventional practices. Women and children were often subjected to weeding the companion grasses. In this regard, a study by Khan et al. [20] revealed that push-pull technology increased the cost of labour during the initial cropping season due to the extra labour required for the planting and hand weeding of Desmodium and Brachiaria compared to the farmers' conventional practices. However, 45% of farmers appreciated that the technology decreased the farm labour required, particularly after the first weeding of the maize field since Desmodium covers the soil and reduces/suppresses the presence of weeds. This result corroborates earlier findings that push-pull technology begins to yield benefits in terms of increased production and decreased labour demand in the second and third years after its establishment [12,21].

5. Conclusions

The findings of the present study demonstrate that stemborers and poor soil fertility are among the main factors that contribute to poor maize productivity in the study areas. In both cropping seasons, except for Jara Gelelcha, better maize grain yield was obtained from the climate-adapted push-pull plots than in the maize monocrop plots. Furthermore, most push-pull technology-practicing farmers perceived that push-pull technology is better than their practices due to the different attributes of climate-adapted push–pull technology. Given its multiple benefits, the climate adapted push-pull technology has great potential to positively impact the livelihoods of smallholder farmers in the region [12,21]. However, the dissemination of the climate adapted push-pull technology to reach more and more farmers requires strong partnerships among different practitioners/actors such as farmers, researchers, non-government organizations (NGOs), the Ministry of Agriculture, agricultural universities and seed companies.

Author Contributions: Conceptualization, T.T., T.K., E.M.; methodology, T.K., M.K., T.T., B.E., E.M.; data analysis, E.M., T.K.; investigation, T.K., T.T., B.E., E.M.; writing—original draft preparation, T.K., T.T., E.M.; writing—review and editing, E.M., M.K., T.T.; funding acquisition, T.T.

Funding: This research was funded by the USAID Feed the Future IPM Innovation Lab through Virginia Tech., Cooperative Agreement No. AID-OAA-L-15-00001.

Acknowledgments: We gratefully acknowledge the financial support provided by the UK's Department for International Development (DFID); the Swedish International Development Cooperation Agency (Sida); the Swiss Agency for Development and Cooperation (SDC); the Kenyan Government and Ethiopian Government. The views expressed herein do not necessarily reflect the official opinion of the donors. The authors acknowledge the farmers who participated in this study.

Conflicts of Interest: The authors declare no conflict of interest.

References

1. CSA (Central Statistical Authority). *Federal Democratic Republic of Ethiopia Central Statistical Agency Agricultural Sample Survey (2009/10), Report on Area and Production of Crops*; CSA: Addis Ababa, Ethiopia, 2010; Volume I.
2. World Bank. Ethiopia-Accelerating equitable growth—country economic memorandum (Vol. 2): Thematic chapters (English). World Bank: Washington, DC, USA, 2007. Available online: http://documents.worldbank.org/curated/en/949951468030574203/Thematic-chapters (accessed on 10 February 2019).
3. CSA (Central Statistical Authority). *Federal Democratic Republic of Ethiopia Central Statistical Agency Agricultural Sample Survey (September–December, 2011), Report on Land Utilization (Private Peasant Holding, Meher Season*; CSA: Addis Ababa, Ethiopia, 2012; Volume IV.
4. Geta, E.; Bogale, A.; Kassa, B.; Elias, E. Productivity and efficiency analysis of smallholder maize producers in southern Ethiopia. *J. Hum. Ecol.* **2013**, *41*, 67–75. [CrossRef]
5. Getu, E.; Overholt, W.A.; Kairu, E. Status of stemborers and their management in Ethiopia. In Proceedings of the Integrated Pest Management Conference, Kampala, Uganda, 8–12 September 2002.
6. Getu, E.; Overholt, W.A.; Kairo, E. Distribution and species composition of stem borers and their natural enemies in maize and sorghum in Ethiopia. *Insect Sci. Appl.* **2001**, *21*, 353–359.
7. Calatayud, P.A.; Le Ru, B.P.; Van den Berg, J.; Schulthess, F. Ecology of the African maize stalk borer, *Busseola fusca* (Lepidoptera: Noctuidae) with special reference to insect-plant interactions. *Insects* **2014**, *5*, 539–563. [CrossRef] [PubMed]
8. Khan, Z.R.; Amudavi, D.M.; Midega, C.A.O.; Wanyama, J.M.; Pickett, J.A. Farmers' perceptions of a 'push–pull' technology for control of cereal stemborers and striga weed in western Kenya. *Crop Prot.* **2008**, *27*, 976–987. [CrossRef]
9. Cook, S.M.; Khan, Z.R.; Pickett, J.A. The use of "Push-pull" strategies in integrated pest management. *Ann. Rev. Entomol.* **2007**, *52*, 375–400. [CrossRef] [PubMed]
10. Midega, C.A.O.; Khan, Z.R.; Berg, J.V.D.; Ogol, K.P.O.; Pickett, J.A.; Wadhams, L.J. Maize stemborer predator activity under 'push-pull' system and Bt-maize: A potential component in managing Bt resistance. *Int. J. Pest Manag.* **2006**, *52*, 1–10. [CrossRef]
11. Midega, C.A.O.; Pittchar, J.O.; Pickett, J.A.; Hailu, G.W.; Khan, Z.R. A climate-adapted push-pull system effectively controls fall armyworm, *Spodoptera frugiperda* (J E Smith), in maize in East Africa. *Crop Prot.* **2018**, *105*, 10–15. [CrossRef]
12. Khan, Z.R.; Midega, C.A.O.; Amudavi, D.M.; Hassanali, A.; Pickett, J.A. On-farm evaluation of the 'push–pull' technology for the control of stemborers and striga weed on maize in western Kenya. *Field Crops Res.* **2008**, *106*, 224–233. [CrossRef]
13. Midega, C.A.O.; Bruce, T.J.A.; Pickett, J.A.; Pittchar, J.O.; Murage, A.; Khan, Z.R. Climate-adapted companion cropping increases agricultural productivity in East Africa. *Field Crops Res.* **2015**, *180*, 118–125. [CrossRef]
14. Below, T.; Artner, A.; Siebert, R.; Sieber, S. *Micro-level Practices to Adapt to Climate Change for African Small-scale Farmers*; IFPRI Discussion Paper No. 953; International Food Policy Research Institute: Washington, DC, USA, 2010.
15. Khan, Z.R.; Pickett, J.A. The 'push–pull' strategy for stemborer management: A case study in exploiting biodiversity and chemical ecology. In *Ecological Engineering for Pest Management: Advances in Habitat Manipulation for Arthropods*; Gurr, G.M., Wratten, S.D., Altieri, M.A., Eds.; CABI Publishing: Wallingford, UK, 2004; pp. 155–164.
16. De Groote, H.; Overholt, W.; Ouma, J.O.; Mugo, S. Assessing the potential impact of Bt maize in Kenya using a GIS model. In Proceedings of the International Agricultural Economics Conference, Durban, South Africa, 16–22 August 2003.
17. Oben, E.O.; Ntonifor, N.N.; Kekeunou, S.; Abbeytakor, M. Farmers knowledge and perception on maize stem borers and their indigenous control methods in south western region of Cameroon. *J. Ethnobiol. Ethnomed.* **2015**, *11*, 77. [CrossRef] [PubMed]
18. Tefera, T. Farmers' perceptions of sorghum stem-borer and farm management practices in eastern Ethiopia. *Int. J. Pest Manag.* **2004**, *50*, 35–40. [CrossRef]
19. White, D.S.; Labarta, R.A.; Leguía, E.J. Technology adoption by resource-poor farmers: Considering the implications of peak-season labor costs. *Agric. Syst.* **2005**, *85*, 183–201. [CrossRef]

20. Khan, Z.R.; Midega, C.A.O.; Bruce, T.J.A.; Hooper, A.M.; Pickett, J.A. Exploiting Phyto-chemicals for developing a 'push–pull' crop protection strategy for cereal farmers in Africa. *J. Exp. Bot.* **2010**, *61*, 4185–4196. [CrossRef] [PubMed]
21. Kassie, M.; Stage, J.; Diiro, G.; Muriithi, B.; Muricho, G.; Ledermann, S.T.; Pittchar, J.; Midega, C.; Khan, Z. Push–pull farming system in Kenya: Implications for economic and social welfare. *Land Use Policy* **2018**, *77*, 186–198. [CrossRef]

Article

Impact of Different Shading Levels on Growth, Yield and Quality of Potato (*Solanum tuberosum* L.)

Vanessa S. Schulz [1,*], Sebastian Munz [1], Kerstin Stolzenburg [2], Jens Hartung [1], Sebastian Weisenburger [2] and Simone Graeff-Hönninger [1]

[1] Department of Agronomy (340), Institute for Crop Science, University of Hohenheim, 70599 Stuttgart, Germany; s.munz@uni-hohenheim.de (S.M.); moehring@uni-hohenheim.de (J.H.); simone.graeff@uni-hohenheim.de (S.G.-H.)

[2] Centre for Agricultural Technology Augustenberg (LTZ), 76287 Rheinstetten-Forchheim, Germany; Kerstin.Stolzenburg@ltz.bwl.de (K.S.); Sebastian.Weisenburger@ltz.bwl.de (S.W.)

* Correspondence: V.Schulz@uni-hohenheim.de; Tel.: +49-721-9518-216

Received: 17 April 2019; Accepted: 18 June 2019; Published: 21 June 2019

Abstract: In agroforestry systems (AFS), trees shade the understory crop to a certain extent. Potato is considered a shade-tolerant crop and was thus tested under the given total solar irradiance and climatic conditions of Southwestern Germany for its potential suitability in an AFS. To gain a better understanding of the effects of shade on growth, yield and quality; a three-year field experiment with different artificial shading levels (12%, 26% and 50%) was established. Significant changes in growth occurred at 50% shading. While plant emergence was not affected by shade, flowering was slightly delayed by about three days. Days until senescence also showed a delay under 50% shade. The number of tubers per plant and tuber mass per plant were reduced by about 53% and 69% under 50% shade. Depending on the year, tuber dry matter yield showed a decrease of 19–44% at 50% shade, while starch content showed no significant differences under shade compared to unshaded treatment. The number of stems per plant, plant height and foliage mass per plant as well as tuber fraction, black spot bruise and macronutrient content were unaffected. Overall, potato seems to tolerate shading and can therefore be integrated in an AFS, and can cope with a reduced total irradiance up to 26%.

Keywords: potato (*Solanum tuberosum*); shade; light; yield; growth; quality

---

1. Introduction

Due to increasing pressure on cultivated land, intercropping systems may provide an alternative option of economic and environmental interest in temperate regions. Research on temperate intercropping peaked in the 1980s, and was focused on the promotion of sustainable agricultural management strategies [1,2]. These past studies presented intercropping systems as ecologically advantageous when compared to monocultures. Intercropping allows more efficient use of land area, changes the microclimate, improves the biodiversity, offers economic diversity, creates wildlife habitats, and minimizes climate variabilities [3–5]. Within the past decade, research on temperate intercropping has increased because it is considered as an effective strategy to mitigate food insecurities and agriculture-related environmental degradation of land and water. This increased interest is partially associated with recent technological advancements, which improve the labor efficiency potential of the practice [6].

A special form of intercropping is the agroforestry system (AFS). These systems combine an annual agricultural component (crop or livestock production) with a perennial woody component (trees, hedgerows) at the same time on the same area of land [7–9]. The advantages of AFS include increased carbon sequestration, improved water regulation, better soil fertility, reduced erosion, and additional aesthetic value [10–13]. However, in most silvoarable agroforestry systems (a combination of annual crop production with woody perennials), competition not only exists aboveground (competition for light), but also comes from belowground (competition for soil moisture and nutrients), both of which may lead to lower crop yields.

Worldwide, there are numerous options for combining trees and crops in AFS (e.g., alley cropping, forest farming, riparian buffer, silvopasture or windbreaks) [14]. However, most of these systems show a reduction in crop yields due to tree competition, especially when the plantation design is too dense. An example of an AFS is apple trees (*Malus pumila* Mill.) with soybean (*Glycine max* L. Merr.) and peanut (*Arachis hypogaea* L.) in the Loess Plateau region of China. The yields were reduced by about 3–4% in 2.5 m distances to the tree trunk, respectively [15]. An AFS of jujube trees (*Ziziphus jujube* Mill.) and wheat (*Triticum aestivum* L.) in northwest China showed a grain yield reduction of 18% under 4-year-old trees planted with a row distance of 6 m, and a yield reduction of 30% under 6-year-old trees planted with a 3 m row distance compared with the unshaded control [16]. Other experiments with maize (*Zea mays* L.) and beans (*Phaseolus* spp. L.) grown between 15 m wide rows of Paulownia trees (*Paulownia elongate* S. Y. Hu) showed reduced grain yields of 32% and 37%, respectively [17]. Rice (*Oryza sativa* L.) or wheat grown in a 20 m x 20 m field in Western Himalaya together with one row of *Grewia optiva* (J.R. Drumm. ex Burret), *Morus alba* (L.) or *Eucalyptus* spp. hybrids (L'Hér.) in the center of the field, reduced yields of rice by 28–34% and of wheat by 28–29% compared with the control without trees [18]. Beans (*Phaseolus vulgaris* L.) grown under Timor Mountain Gum (*Eucalypthus urophylla* S.T. Blake) in Brazil showed significantly reduced bean yields of almost 50% [19].

Most of these studies examined the reduction of incident radiation as the main factor for reduced yields [15,18,20,21], thus studying the use of shade tolerant crops in an AFS could be advantageous. Such crops are able to reach their light saturation point at lower total solar irradiance, have a better yield performance under shade, and therefore, can be grown in an AFS.

Potato (*Solanum tuberosum* L.) is known to be a shade-tolerant crop. As a C3 plant, potato needs moderate irradiance conditions [22]. Its light saturation point for photosynthetically active radiation (PAR) is considered to be around 400 μmol m^{-2} s^{-1}, which corresponds to 14.86 MJ m^{-2} day^{-1} [23]. Especially in tropical and subtropical zones (0–23.5° N/S and 23.5–40° N/S latitude) where potato can be grown throughout the year and radiation is up to 30 MJ m^{-2} day^{-1}, potato is quite often integrated in an AFS. Studies from Nigeria, Kenya and South Asia show only minor effects on yield by tree shading in AFS.

An experiment in Nigeria showed that growing potato (*Solanum tuberosum* L.) between rows of rattle trees (*Albizia lebbeck* L. Enth.) increased the tuber yield and the number of tubers [24]. Under unfertilized, open field conditions in Kenya, potatoes also obtained higher yields in an AFS with *Eucalyptus grandis* (W. Hill ex Maiden) [25]. An Indonesian experiment that used artificial shading showed that plant height and tuber yield increased under 50% light reduction, compared with full sunlight. The height of some potato cultivars was affected by artificial shade [22]. Such changes in plant height represent a shade avoidance response, with plant height increasing under shade to reach more light. This stimulates the plants and leads to height growth and elongation to obtain more irradiation [26]. In Egypt, taller plants were obtained under colored nets in comparison to the open field [27]. Earlier experiments in Egypt on potatoes found that potatoes grown under low irradiance were taller, but the tubers were smaller and irregularly shaped. Furthermore, the tuber dry weight was reduced under low light conditions [28].

It has been proven that the duration of each potato growth phase determines the later yield [29]. In the tropics and subtropics, there is still enough radiation (even under shady AFS conditions) available to reach the light saturation point of potato. However, it might not be reached at higher latitudes. In the temperate zone of Europe where the growing season lasts from March to October, the amount of radiation available is between 10–20 MJ m^{-2} day^{-1} [30]. Since light has a decisive influence on plant growth, yield is reduced by shade and lower total solar irradiance in higher latitudes, while in lower latitudes competition for water and nutrients has a major effect. So far, little research is available on the impact of shady conditions at higher latitudes on the growth, yield and quality of potato in an

AFS under non-tropical conditions. In the few studies on AFS with potatoes in temperate (potatoes and hazel (*Corylus avellane* (L.)) and subarctic zones (potatoes and willow (*Salix* sp. (L.)), experiments have mainly focused on potato cultivation beside windbreaks [31–33]. Beside these windbreaks, other abiotic factors such as wind reduction, reduced soil evaporation, reduction of mechanical stimulus (e.g., twisting of plants) have an influence on growth and yield, and water and nutrients are also affected. In an AFS, these interactions make it difficult to determine the influence of shade. Therefore, the influence of shade has to be determined by artificial shading.

The objectives of this study were to evaluate the impact of four different shade levels (0%, 12%, 26% and 50%) on potato growth, tuber yield and quality parameters under the given total solar irradiance of Southwestern Germany. The determined threshold could be an indicator for farmers as to which level of shade potato cultivation might be profitable. Fertilization or irrigation can compensate for some limitations, but a reduction in light cannot be mitigated.

2. Materials and Methods

2.1. Site Conditions and Experimental Design

The field experiment was carried out from 2015 to 2017 in Southwest Germany at the Centre for Agricultural Technology Augustenberg (LTZ) in Rheinstetten-Forchheim (48°58′ N, 8°18′ E, 117 m above sea level). The site is located in the lower Rhine valley on a Luvisol (60.2% sand, 13.7% clay and 26.1% silt) soil. The mean long-term annual precipitation was 742 mm and the average temperature was 10.1 °C (1981–1990). During the main growing season at this site (April to October), the mean average total solar irradiance from 2009 to 2017 amounted to 17 MJ m^{-2} day^{-1}. Weather data were collected in a linear distance of 270 m from the experimental site. Total solar irradiance was measured by a SCAPP (scanning pyrheliometer and pyranometer, Fa. Siggelkow Gerätebau, Hamburg). The monthly air temperature averages, cumulative precipitation and average total solar irradiance for the experimental years are given in Figure 1. In all of the experimental years, the previous crop was winter barley. Different green manure crops were incorporated in the potato experimental plots during the winter months of each experimental year. Green manure crops included 25 kg ha^{-1} *Sinapsis alba* L. in 2014/2015, 18 kg ha^{-1} flower mixture (FAKT M2, BSV Saaten; 20.0% leguminosae, 6.0% rough leguminosae, 27.5% herbs, 46.5% others [34]) in 2015/2016 and 25 kg ha^{-1} *Raphanus sativus* L. cv. 'Denfender' in 2016/2017.

On 20 September 2014 (day of the year (DOY) 263), primary tillage was done with a moldboard plough (25 cm depth). Potatoes were planted on 16 April 2015 (DOY 106), 13 April 2016 (DOY 104) and 13 April 2017 (DOY 103) after secondary tillage with a chisel plow (15 cm depth). The mid-early potato variety 'Selma' (*Solanum tuberosum* L., Bavaria Saat) was planted with a row distance of 0.75 m and an intra-row distance of 0.35 m, which resulted in four plants per m^2. The experimental design was a randomized complete block design with three replicates. Plots were 10 m long and 6 m wide, consisting of a total of 8 rows per plot. Core plots for tuber harvest were 8 m long and 1.5 m wide, including two rows and leaving three rows on the left and right as a border. Planting depth was 5 cm. Hoeing and earthing up was done prior to pre-emergence herbicide application. Amount of fertilizer was calculated based on nutrient removal. The date, amount and type of fertilizer is shown in Table 1. Fertilization was done by a pneumatic centrifugal spreader (RAUCH AERO 2212, Sinzheim, Germany). Plant protection was done based on the risk assessment of the online tool 'ISIP' [35]. The amount and type of pesticides are given in Table A1 in the Appendix A. Plant protection was conducted according to the codes of "Good Agricultural Practice in Plant Protection and Fertilization" [36]. Irrigation was done by an overhead irrigation-gun on 29 May 2015 (DOY 149), 29 June 2015 (DOY 180), 7 July 2015 (DOY 188), 16 July 2015 (DOY 197), 3 August 2015 (DOY 215), 7 July 2016 (DOY 189), 13 July 2016 (DOY 195), 29 July 2016 (DOY 211), 12 August 2016 (DOY 225), 31 August 2016 (DOY 244), 31 May 2017 (DOY 151), 20 June 2017 (DOY 171) and 4 July 2017 (DOY 185), with 30 mm of water at each irrigation event. The irrigation was based on the recommendations of the online irrigation tool, 'Agrowetter' [37]. Harvest was conducted using a one-row potato elevator-digger (Niewöhner

Wühlmaus, Weimar, Germany) on 8 September 2015 (DOY 251), 6 September 2016 (DOY 250) and 6 September 2017 (DOY 249).

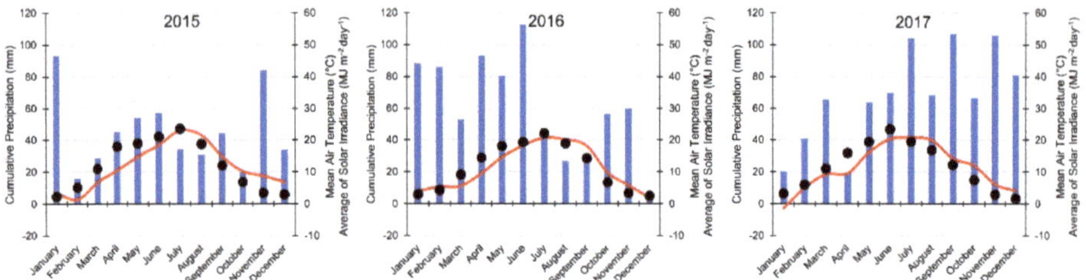

Figure 1. The monthly cumulative precipitation (mm, blue bars), mean air temperature (°C, solid, red line) and average total solar irradiance (MJ m^{-2} day^{-1}, filled, black circles) during the experimental years 2015 to 2017 at Rheinstetten-Forchheim.

Table 1. Date, amount, active ingredient and pure nutrient amount of the applied fertilizer. The day of the year (DOY) is given in parentheses beneath the corresponding date.

Date	Fertilizer	Active Ingredient	Pure Nutrient
16 April 2015 (DOY 106)	130 kg ha^{-1} lime-nitrogen	20% N, 50% CaO	26 kg N, 46 kg Ca
	300 kg ha^{-1} ALZON46	46% N	138 kg N
	600 kg ha^{-1} potassium sulfate with magnesium	23% P$_2$O$_5$, 9% S	60 kg P, 54 kg S
	200 kg ha^{-1} superphosphate 18	18% P$_2$O$_5$, 12% S	16 kg P, 24 kg S
11 April 2016 (DOY 102)	350 kg ha^{-1} lime-nitrogen	20% N, 50% CaO	70 kg N, 125 kg Ca
12 April 2016 (DOY 103)	260 kg ha^{-1} calcium ammonium nitrate	27% N	70 kg N
	450 kg ha^{-1} superphosphate 18	18% P$_2$O$_5$, 12% S	35 kg P, 54 kg S
	1110 kg ha^{-1} sulphate of potash containing magnesium salt	30% K$_2$O, 10% MgO, 17% S	276 kg K, 67 kg Mg, 189 kg S
13 April 2017 (DOY 103)	260 kg ha^{-1} ALZON46	46% N	120 kg N
	970 kg ha^{-1} sulphate of potash containing magnesium salt	30% K$_2$O, 10% MgO, 17% S	242 kg K, 58 kg Mg, 165 kg S
27 April 2017 (DOY 117)	390 kg ha^{-1} superphosphate 18	18% P$_2$O$_5$, 12% S	31 kg P 47 kg S

2.2. Shading Levels

Shading was created by nets which reduced the incoming solar radiation by 12%, 26% and 50%. The different shading levels were compared with full sunlight (0% shade). The nets were made of polyethylene and had different mesh sizes to create the different shading levels. The 12% net had a mesh size of 3 × 8 mm and was black; the 26% net had a mesh size of 12 × 12 mm and was green, and the 50% net had a mesh size of 3 × 3 mm and was green (AGROFLOR Kunststoff GmbH, Wolfurt, Austria). Nets were installed at the time of potato emergence (growth stage (GS) 009 according to [38]), on 20 May 2015 (DOY 140), 10 May 2016 (DOY 131) and 9 May 2017 (DOY 129). Nets were clipped on to steel wires, which were connected between wooden posts. The height of the nets could be adapted to the plant growth, and to 1 or 2 m in height. A distance of 0.5 m between the nets and canopy surface was guaranteed. Further information about the experiment layout can be found in Schulz et al. [39]. Table 2 shows the total incoming daily solar irradiance at the experimental site from the time of the potato crop emergence (Growth Stage (GS) 009) to the tuber harvest (GS 909) for each experimental year and the theoretically reduced incoming total solar irradiance under the shading nets.

Table 2. The calculated total solar irradiance for the shading treatments during the period without shading (-S, planting growth stage (GS) 000 to emergence GS 009), the period with shading (+S, emergence GS 009 to harvest GS 909) and the whole growing period (GP, planting GS 000 to harvest GS 909) (MJ m^{-2} day^{-1}), the duration of these time periods (days) is given in parentheses.

		Total Solar Irradiance (MJ m^{-2} day^{-1})								
	Year	2015			2016			2017		
	Time Period	-S (26)	+S (112)	GP (138)	-S (26)	+S (121)	GP (147)	-S (32)	+S (115)	GP (147)
	0%		20.22	19.90		19.15	18.90		20.13	18.96
Shading level	12% ‡	18.52	17.80	17.93	17.70	16.86	17.00	14.87	17.72	17.08
	26% ‡		14.97	15.64		14.17	14.80		14.90	14.89
	50% ‡		10.11	11.70		9.58	10.01		10.07	11.14

‡ values for +S were calculated by subtracting the light reduction by nets from the measured total irradiance at 0% shade.

2.3. Data Collection and Analysis

2.3.1. Growth Parameters

In 2015, no growth parameters were determined; only the tuber dry matter yield and quality were determined. During the vegetation periods 2016 and 2017, destructive and non-destructive measurements were done. Growth stages according to the BBCH-scale were determined twice a week [40]. Potato plant height measurements were obtained every week during the emergence stage (GS 009) through to tuber formation (GS 405) on four plants per plot. Plant height was determined using a meter stick to measure the highest point of the soil surface to the highest point of the plant canopy. When the potato plant flowers, the stem and leaves have reached their maximum growth (GS 405), and tubers have reached 50% of their final mass (GS 625) [33–35]. Due to the high workload at GS 405/625, two plants per plot were randomly selected from the 3rd or 6th row and harvested for further observations. The observed parameters were stems per plant, tubers per plant, tuber mass per plant, total foliage mass per plant (including all above ground biomass; leaves, stem, flowers, berries), the ratio between foliage and tuber mass, total mass per plant and the harvest index (HI). Leaf area (LA) was determined using Equation (1):

$$LA = LL \cdot LW \cdot 0.55,\tag{1}$$

where LL is the leaf length from leaf tip to leaf attachment at stem, LW is the maximum leaf width and 0.55 is a constant [41]. Leaf length and the width of a leaf from the middle leaf layer were measured with a meter-stick. The leaf was dried for three days at 60 °C and the specific leaf area (SLA) was calculated. LA and SLA were only determined in 2017. Growing degree days (GDD) were calculated using Equation (2), where i is the day between planting (P) and harvest (H):

$$GDD = \sum_{i=P}^{H} \left(\frac{T_{max_i} + T_{min_i}}{2} - T_{base} \right).\tag{2}$$

For potato, a base temperature (T_{base}) of 6 °C was assumed since no sprout growth is expected at lower temperatures [42–45]. If T_{max} or T_{min} at day i were smaller than T_{base} they were set to T_{base} [46].

2.3.2. Yield Parameters

In all years, all harvested tubers from the center rows of each plot were weighed to calculate yield on a hectare basis. Then, a sub-sample of 2 kg per plot were fresh weighed, oven-dried (1 week, 105 °C) and the dry weight was determined to calculate the dry mass and substance. In 2016 and 2017, all fresh-harvested tubers per plot were sorted according to the size classes: <30 mm (undersized fraction), 30–60 mm (table fraction), and >60 mm (oversized fraction) [47]. Selma is listed in the German variety list as a variety that has long oval tubers [48].

2.3.3. Quality Parameters

An additional sub-sample of 2 kg from the harvested tubers per plot was used to determine nitrogen (N) via the combustion method after Dumas, and phosphorus (P), potassium (K), calcium (Ca), magnesium (Mg) and sulfur (S) via spectrometry [49–51]. Analysis of starch content was done according to the polarimetry method [52]. Sub-sample of 30 tubers per plot between 30–60 mm were analyzed for black spot bruise [47]. The black spot bruise index (BSB) was calculated from the number of light, middle and strong discolored tubers ($tuber_{light}$, $tuber_{middle}$, and $tuber_{strong}$, respectively):

$$BSB = \frac{\left(0.3 \cdot tuber_{light}\right) + \left(0.5 \cdot tuber_{middle}\right) + tuber_{strong}}{tuber_{total}} \times 100. \tag{3}$$

A tuber is counted as light discolored when 1/4 of the circumference is discolored to a 5 mm depth. A tuber is counted as middle discolored when 1/4 of the circumference is discolored and this discoloration is deeper than 5 mm and/or when half of the circumference is discolored to 5 mm. A strong discoloration occurs when tubers are discolored up to half of the circumference and are discolored deeper than 5 mm and/or more than 1/2 is discolored up to 5 mm depth. To measure BSB, samples were spun in a washing machine for 45–90 s (determination of the time took place every year with a standard potato variety). Afterwards, samples were stored for 4–5 days at room temperature. Then the tubers were cut at the greatest diameter and the number of tubers with discoloration (blue, grey or black) was determined [53].

2.3.4. Data Analysis and Statistics

Analysis of the yield data was performed for each year by using the following fitted model:

$$y_{ij} = \mu + r_i + s_j + e_{ij}, \tag{4}$$

where y_{ij} is the tuber dry matter yield, μ the general effect, r_i is the fixed effect of the i-th replicate, s_j is the fixed effect of the j-th shading level and e_{ij} is the residual error of y_{ijk}.

For the analysis of repeated measurements (duration of growing phases, number of stems per plant, number of tubers per plant, tuber mass per plant, foliage mass per plant, foliage:tuber mass ratio, total mass per plant and HI) on two plants per plot at GS 405/625 the model was as follows:

$$y_{ijk} = \mu + r_i + s_j + (rs)_{ij} + e_{ijk}, \tag{5}$$

where y_{ikj} is the response, μ the general effect, r_i is the fixed effect of the i-th replicate, s_j is the fixed effect of the j-th shading level, $(rs)_{ij}$ is the random plot effect where the j-th shading level is used in the i-th replicate, and e_{ijk} is the residual error of y_{ijk} which corresponds to the kth plant effect in the ij^{th} plot. For both models the PROC MIXED procedure of Statistical Analysis Software SAS, version 9.4 (SAS Institute Inc., Cary, NC, USA) was used.

The multi-year analysis of quality data (macronutrients: nitrogen, phosphorus, potassium, calcium, magnesium, and sulfur) was done by using the Residual Maximum Likelihood of the PROC MIXED procedure of SAS. The following linear mixed model was fitted:

$$y_{ijl} = \mu + a_l + s_j + (ra)_{il} + (as)_{lj} + e_{ijl}, \tag{6}$$

where y_{ijl} is the response, μ the general effect, a_l is the fixed effect of the l-th year, s_j is the fixed effect of the j-th shading level, $(ra)_{il}$ is the fixed effect of the i-th replicate in the l-th year, $(as)_{lj}$ is the random interaction effect between the l-th year and the j-th shading level, and e_{ijl} is the residual error of y_{ijl}. For all models, the assumptions of normality and homogenous variances of residuals were checked graphically. If necessary, that is, if the AIC decreases, year-specific error variances were fitted. In all cases, after finding significant differences via the F-test, differences between treatments were compared at $\alpha = 5\%$ using Fisher's least significant difference test (LSD). More information on the statistics used can be found in Schulz et al. [39].

The growth parameters for plant height were fitted for each plot with the function 'nls' of the R packages 'nlstools' and 'car' [54,55]. The non-linear regression matched the following equation:

$$y = \frac{\theta_1}{1 + e^{-(\theta_2 + \theta_3 \cdot GDD)}}, \tag{7}$$

where y is the dependent variable for height in the single years 2016 and 2017, θ_1 is the asymptote of the dependent variable, θ_2 is the parallel shift, θ_3 the slope of the function; and GDD are the growing degree days, calculated after Equation (2). Estimates for θ_1, θ_2 and θ_3 from each plot were then submitted to multi-year analysis via model (6).

3. Results and Discussion

3.1. Growth and Development

In 2016 and 2107, artificial shading started after emergence (GS 009), therefore, shading had no influence on the emergence of the potatoes (Table 3). These results agree with an experiment with diverse potato cultivars in the Philippines, where uniform plant emergence was observed at 54% shading and at full light [56]. Because potatoes do not have photosynthetically active biomass until emergence, a change in total solar irradiance has no direct effect on the emergence of plants by influencing their radiation use. However, an indirect influence due to changing soil temperature and moisture might occur. Our study revealed that flowering initiation (GS 601) was prolonged at shading levels >12% shade. In 2017, there was only a significant prolongation under 50%, from 440 GDD under 0% to 467 GDD under 50%. The time from flowering initiation to senescence initiation (GS 901) was prolonged from 973 GDD under 0% and 12% shade to 1211 GDD under 26% and 50% shade. In 2017, no change was observable between 12% and 26% shade compared with 0%. This can be explained by differing climatic conditions in 2016 and 2017. In 2016, the 26% and 50% shade treatment needed a higher amount of GDD to reach senescence due to the cooler and rainy growing period. The light saturation of 14.86 MJ m^{-2} day^{-1} could not be reached. The rainy period lasted from April to June (Figure 1). During these months the total solar irradiance was lower (14.34, 18.02 and 19.28 MJ m^{-2} day^{-1}) than in 2015 (17.94, 18.93 and 21.08 MJ m^{-2} day^{-1}) and 2017 (15.9, 19.38 and 23.39 MJ m^{-2} day^{-1}). Table 2 showed that in 2016 the light saturation point of potatoes could not be reached at levels of 26% and 50% shade, while in 2015 and 2017 this was only observable under 50% shade. The time from senescence initiation until harvest day (GS 909) in both 2016 and 2017, did not show any significant changes by shade. The harvestable tuber yield was determined by the duration of the growing season. This was also shown in a Dutch experiment. The authors observed that the growth of potato plants and the dry matter production of tubers were mainly determined by the duration of its growth cycle [29], that is, the duration of each single growth phase is important for the later yield.

The authors of the study concluded that the development depends on temperature and daylength. At higher latitudes (e.g., >55° N) growth limitations could occur due to cooler temperatures, which do not fit the optimum values for the single growing phases.

Table 3. Duration of growing phases in Growing Degree Days (°Cd) and the range of days from planting to emergence (P-E), emergence to flowering initiation (E-F), flowering initiation to senescence initiation (F-S) and senescence initiation to harvest day (S-H) in 2016 and 2017 for the four shading levels (0%, 12%, 26% and 50%). From planting to emergence is a phase without shading (-S), from emergence to harvesting potatoes were shaded (+S; see also Table 2). Phases correspond to the GS 000 to 009 (P-E), 009 to 601 (E-F), 601 to 901 (F-S) and 901 to 909 (S-H). SEM gives the standard error of means.

		Duration of Growing Phases							
		-S		+S					
Year	Shade	P-E		E-F		F-S		S-H	
		GDD	days	GDD	day	GDD	days	GDD	days
	0%	132	26	559 c [†]	42	973 b	30	1689	48
2016	12%	132	26	573 b	43	973 b	29	1685	48
	26%	132	26	580 b	44	1211 a	43	1685	33
	50%	137	26	598 a	45	1211 a	42	1685	33
	SEM	2.24		3.62		0.00 [‡]		2.03	
		p-values [$]							
	Replicate	0.422		0.422		1.000		0.422	
	Shade	0.455		0.002		<0.0001		0.455	
	0%	169	32	440 b	21	1003	39	1755	54
2017	12%	173	32	447 b	21	1016	39	1768	54
	26%	173	32	444 b	21	1011	39	1764	54
	50%	169	32	467 a	24	1007	36	1760	54
	SEM	2.93		2.79		4.17		4.02	
		p-values [$]							
	Replicate	0.670		1.000		0.823		0.708	
	Shade	0.654		0.002		0.249		0.243	

[†] Means with identical letters within each column and year show non-significant differences between the shade levels of the single years (LSD test, $\alpha \leq 0.05$). [‡] Note: The SEM was between 0 and 0.005, so rounding to two decimal places resulted in a SEM of zero. [$] *p*-value for the *F*-test of the corresponding factor.

An experiment conducted in the Philippines showed no significant change in plant height at different light intensities for potatoes grown in December (long-day), while potatoes grown in March (short-day) showed differences [56]. Under short-day conditions potatoes develop a canopy, which causes faster senescence and low tuber yields. Since the plants do not receive enough irradiation, they get into a stress situation and start to relocate their nutrients from the leaves to the generative organs, which causes senescence of the leaves. Under long-day conditions the above-ground organs do not die off as quickly and can use the solar irradiance longer and generate higher yields. An additional shade under short-day conditions can delay development and so, the potato growth phase is prolonged. Additionally, high temperatures reduce the above-ground biomass. Potatoes grown under temperatures of 17 °C showed dry matter production of 22.8 g m^{-2} day^{-1} [57], while under higher temperature, biomass is reduced. To detect if artificial shade affects plant height at higher latitudes, plant height obtained from our experiment was fitted using a sigmoid growth curve. Results indicated that the year-specific and/or shading level-specific curve determining parameters, θ_1, θ_2 and θ_3 for the trait plant height (Equation (7)) were not significantly different from each other (the test for year-specific parameters showed $p = 0.607$, $p = 0.076$ and $p = 0.826$ for θ_1, θ_2 and θ_3, respectively; the test for shade-specific parameters showed $p = 0.649$, $p = 0.282$ and $p = 0.837$ for θ_1, θ_2 and θ_3, respectively). Thus, a single curve across both years can be fitted. This indicates that there were no significant effects of shading and year on plant height. The observed values and the fitted curve are shown in Figure 2. Note that year-by-shade interactions were assumed as random in Equation (6).

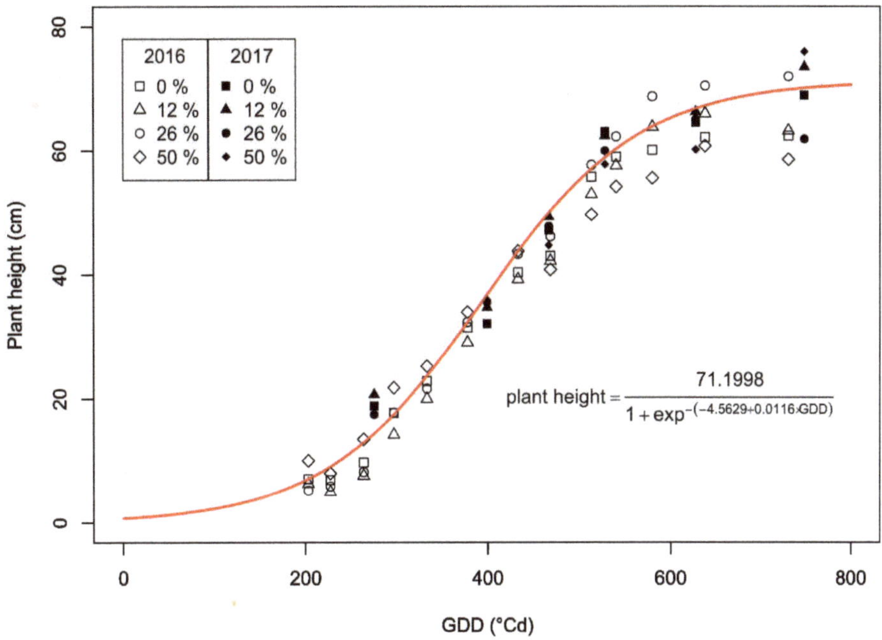

Figure 2. Average values for the observed (symbols) plant height (cm) depending on GDD (°Cd) for the four shade levels and the two years; 2016 (0% open square, 12% open triangle, 26% open circle and 50% open diamond) and 2017 (0% filled square, 12% filled triangle, 26% filled circle and 50% filled diamond) and the fitted growth function (solid, red line) for plant height over all shade levels and years.

In 2016, the control and 26% shade plants reached their maximum height after 730 GDD (62.3 and 72.2 cm). Plants in the 12% and 50% shading treatments reached their maximum heights after 637 GDD with 65.9 and 60.7 cm, respectively. In the second year, all treatments, with the exception of the 26% treatment, reached their maximum height after 747 GDD (68.8, 73.5 and 75.9 cm). Plants in the 26% shading treatment reached their maximum after 627 GDD at 65.0 cm. In Sri Lanka, potatoes in an AFS with Leucaena leucocephala ((LAM.) DE WIT) showed no changes in plant height [58]. No change in plant height was observed when potatoes were intercropped with maize in a tropical experiment in Uganda [59]. An experiment in temperature-controlled cabinets showed an increase in plant height only at radiations below 7.7 MJ m^{-2} day^{-1} [60]. The authors related the increase in plant height to increased gibberellin activity under shade and a reduced assimilation of CO_2. Table 2 shows that in the current study, the total solar irradiance never fell below 7.7 MJ m^{-2} day^{-1}, resulting in no difference in plant height as shown in Figure 2. In addition, the cultivar was a strong influence on the growth of the potato [61]. At lower latitudes, two out of four shaded potato cultivars showed no changes in height. One cultivar showed an increase in height at 30% shade, and the other cultivar at 50% shade [22]. The authors ascribed this to a higher auxin level while the gibberellin level also increased, which promoted stem growth. These results suggest that the cultivar plays a crucial role in height growth under shade. Abu-Zinada and Mousa generally attributed height changes to genetic differences in different potato cultivars [62]. A study of a shade-effect on different phytohormones showed that due to the total irradiance reduction and the associated change in the wavelength spectrum, changes in phytochrome B occurred, which led to growth expansion [63].

3.2. Yield Determining Parameters and Yield

The tuber yield of potatoes is influenced by various factors such as nitrogen, cultivar, planting density and spacing of planting tubers, climatic conditions and geographic location [64]. The four main tuber yield determining growth parameters are the number of plants per hectare, number of stems per plant, number of tubers per plant and average tuber weight per plant [65]. The experiment revealed no changes in the number of plants per hectare under the different shade levels. This is due to the fact that the shade was only established after potato emergence. As discussed above, a change in solar total irradiance does not affect plant emergence directly; so, all planted potatoes were able to emerge (Table 2).

Table 4. Mean growth parameters for two potato plants under four different shade levels (0%, 12%, 26% and 50%) evaluated in 2016 and 2017 at GS 405/625 (maximum foliage growth was reached); number of stems per plant, number of tuber per plant, tuber mass per plant (g), foliage mass per plant (g), foliage:tuber mass ratio (%), total mass per plant (g) and the harvest index (HI). SEM gives the standard error of means.

Shade	Number of Stems per Plant	Number of Tubers per Plant	Tuber Mass per Plant	Foliage Mass per Plant	Foliage:Tuber Mass Ratio	Total Mass per Plant	HI
Year			**2016**				
0%	2.50	10.50	44.47	48.21	1.32	92.68	0.45
12%	3.67	12.44 [⊥]	56.70	57.06	1.25	113.76	0.48
26%	4.17	12.83	49.81	75.45	2.45	125.25	0.36
50%	4.17	13.67	36.28	57.05	1.81	93.33	0.39
SEM	0.85	2.79	10.97	9.98	0.53	19.66	0.04
			***p*-values** [$]				
Replicate	0.248	0.144	0.104	0.509	0.092	0.488	0.002
Shade	0.479	0.873	0.612	0.300	0.417	0.585	0.213
Year			**2017**				
0%	4.83	19.00 a [†]	103.60 a	79.37	0.95 b	182.97 a	0.54
12%	3.50	17.83 a	51.28 b	64.70	1.58 b	115.98 b	0.43
26%	3.17	13.17 ab	66.32 ab	62.15	1.00 b	128.47 ab	0.52
50%	3.33	9.00 b	32.68 b [‖]	60.07	4.67 a [∇]	86.12 b	0.25
SEM	0.51	2.47	14.05	10.59	0.92	21.15	0.06
			***p*-values** [$]				
Replicate	0.835	0.605	0.510	0.259	0.573	0.473	0.788
Shade	0.186	0.038	0.020	0.575	0.051	0.032	0.064

[†] Means with identical letters within each column and year show non-significant differences between the shade levels of the single years (LSD test, $\alpha \leq 0.05$). [⊥] SEM for 12% shade ± 3.08 due to missing value. [‖] SEM for 50% shade ± 15.53 due to missing value. [∇] SEM for 50% shade ± 1.02 due to missing value. [$] p-value for the F-test of the corresponding factor.

The different shading treatments had no significant impact on the number of stems per plant in either year (2016 $p = 0.479$ and 2017 $p = 0.186$) (Table 4). Studies showed that the number of stems depended on the size of the seed tubers or potato variety, but not on the given environmental factors [59]. Genotype is also an influence on the number of produced stems [66]. Other sources also show that the age of the planted tubers influences the number of stems. Young tubers produced one stem and older tubers more stems [67]. A study conducted in the Philippines showed that the number of stems per plant was not affected by a shade level of 54% [56]. Another study showed that number of stems was determined by the number of sprouts, which is influenced by moisture, temperature and structure of soil, and the number of plants per hectare [66]. Since the development of sprouts into stems takes place below-ground, the shade only impacts soil temperature and moisture. In our experiment, shade nets were installed after emergence. By this time sprouts were already developed. In a real AFS where the distance between single trees is wide enough, and trees are pruned and/or varieties with thinner crowns are used, their influence on soil temperature and moisture will be quite small, therefore, this potential influence on sprouts and emergence can be neglected.

Every stem produces leaves, which are photosynthetically active. As described above no change in the number of stems per plant was determined, therefore no effect on the foliage mass per plant was observable. Even in a rather overcast year like 2016, plants did not compensate for the reduced total solar irradiance with increased photosynthetically active biomass. However, shading is often accompanied by a changed in the partitioning of dry matter between the source and sink organs. In 2017, the number of tubers per plant were significantly reduced at a shade level of 50%. An experiment in the United Kingdom with different potato cultivars showed that the cultivar 'Estima' showed no change in time of tuber initiation up until to an artificial shading of 75%, while 'Maris Piper' showed delayed tuber initiation in shading of 50% or more [68]. Since the cultivar remained the same every year, it is suspected that the reduction in 2017 was caused by environmental factors (e.g., soil temperature or moisture) other than irradiance reduction. On average, the number of tubers was reduced by ten tubers per plant compared with the control. The studies of Sun and de Luca et al. showed a decrease in the number of tubers per plant under 54% shade and attributed this to a shade induced increase in the gibberellin (GA) content [69,70]. Studies with peas (*Pisum sativum* L.), lotus (*Nelumbo* spp. Adans.) and *Brassica* spp. (L.) at different shade levels also showed a higher GA, therefore, the change of GA under shade seems to be important for plant development, especially for tuber formation [71,72]. In potatoes, higher content of GA has been shown to inhibit tuber formation [20,47,50,58]. Wurr et al. found a reduced number of tubers under field conditions at a shade level of 70% in experimental sites in the United Kingdom [73]. The authors attributed this to a reduced number of stolons, which was caused by lower temperatures slowing down growth. The number of stolons formed indicate the final tuber number. The number of tubers per plant is initiated in a very short time of ten days, the maximum number is reached when shoot dry matter starts to decrease [74]. Ewing et al. observed that tuber formation is promoted by soil moisture [75]. It is possible that in 2016, the naturally occurring low total solar irradiance in combination with the shading provided more moisture than in 2017, leading to a significant reduction in the number of tubers formed in 2017. The results show that less tubers with lower weight were observed in 2017 under 50% shade compared with 0%. Pohjakalli stated that tuber weight decreased about 80% at light intensities of 67% to 33% of full sunlight (which corresponds to 33% to 67% shade) [76]. A Philippine experiment showed that depending on the cultivar, under 54% shade a reduction in dry matter weight of tubers can be determined between 0% and 80 % compared with potatoes grown under full sunlight [77]. Under 74% light (corresponds to 26% shade) most of the used cultivars showed a reduction of up to 29%. Under 30% shade, 3% more tubers were formed, while under 50% shade there was an increase of about 55% [78]. In tomatoes, it has been observed that during the bulking period, the radiation use efficiency is highly related to fruit development because at this time the canopy is fully developed [79]. In our experiment, we observed that the onset of bulking occurred even under shaded conditions. During this time, in 2016 only the 0% and 12% shade, and in 2017 all treatments except for the 50% received adequate total solar irradiance for light saturation.

Our results showed no significant changes in the foliage mass per plant under the different shade levels in any experimental year. Mean values ranged from 48.21 g (0%) to 75.45 g (26%) in 2016 and 60.07 g (50%) to 79.37 g (0%) in 2017 (Table 4). This corresponds well with the results for plant height (Figure 2). The literature shows that the rate of foliage development is highly dependent on the cultivar. Some cultivars grow faster than others. Also, the age of the seed tubers affects the foliage, while older tubers enhance the foliage production [56]. Data on leaf area (LA) and specific leaf area (SLA) were only available for 2017. However, no significant differences between the shade treatments (LA $p = 0.772$

and SLA $p = 0.963$) were observed. Another experiment with 50% and 90% shaded potato leaves showed that shading up to 50% also did not influence LA. However, shading levels of 90% showed a decline in green leaf area. The authors postulated that shading reduces the transpiration, and so, the distribution of cytokinins. Parts which are exposed to more light or less shade have a higher amount of cytokinins, which can promote cell division, branching and leaf growth at shading levels >50% [57].

Foliage mass showed no change based on the tested shading treatment; however, in combination with a decreased tuber mass in 2017, a shift to the above-ground biomass occurred. The ratio changed from 0.95 in the control to 4.67 at the 50% shade level. This shift has been documented in the literature [42]. Under low irradiance (2000 to 3000 lux or lower) a shift to the aboveground biomass occurred, which was not observed under high irradiance (8000 to 16,000 lux). An increase in above-ground biomass growth and an increase in below-ground biomass was also observed in maize plants under 69% artificial shade [80]. In 2016, the potato plants received more total solar irradiance during the phase without shading (until emergence) than in 2017 (17.70 and 14.87 MJ m^{-2} day^{-1}, Table 2). Until onset of tuber initiation, most dry mass was partitioned in leaves and stems and after this time in tubers. If light is reduced, the plant will use more assimilates for leaf mass than for tubers to provide an adequate level of photosynthesis. This can be seen in Table 4. Leaf mass showed no change under reduced light as the plant tried to provide an adequate amount of photosynthetically active biomass, while the tuber mass was reduced.

The lower number of tubers in 2017 and the constant starch content is in line with results found in the literature. Under shade, more sugar is needed to provide photosynthetically active leaf mass. The large amount of sugar that is translocated in the tubers to form starch (because tubers are not photosynthetically active) cannot be covered [81].

As mentioned above, potatoes grown under optimum conditions are able to form 22.8 g of biomass per m^2 and day. Total mass per plant only showed significant changes in 2017. Biomass in the 50% shading treatment was reduced by almost 100 g compared to the control. In 2017, the 0% shade had a radiation use efficiency (RUE) of 2.41 g MJ^{-1} which fits well with the values mentioned in the literature [82]. The 50% shade had a RUE of 1.14 g MJ^{-1}.

The harvest index (HI) could not be determined for the core plot due to defoliation for facilitated harvest. Therefore, the HI was determined at GS 405/625. Table 4 shows that there was no influence from shade prior to defoliation, neither in 2016 nor in 2017. Therefore, the trend of a decreasing HI with increasing shade was observed.

Dry matter tuber yield (DMY, Figure 3) was significantly reduced by shade in 2016 ($p = 0.040$) and 2017 ($p = 0.004$), but not in 2015 ($p = 0.467$). Under 26% shade DMY was significantly reduced by 44% in 2016, while in 2017 a significant reduction of 44% occurred at 50% shade. This is related to the total solar irradiance values in Table 2. The light saturation point of potatoes was reached in 2015 and 2017 in up to 26% shade. Since the plants were able to cover their need for total solar irradiance of 14.86 MJ m^{-2} day^{-1} during the shaded time, no significant changes were observable (Table 2). Both 2015 and 2017 were rather sunny years, while in contrast, spring 2016 had comparatively low total solar irradiance. In June 2016, hot and dry phases alternated with rainfall events. Light saturation was reached up until 12% shade (Table 2). These observations in combination with the already discussed changes, suggest that the phase after emergence is crucial for yield formation, especially since the plant has no photosynthetically active biomass before emergence that can use the light. Figure 3 and Table 2 suggest that the light saturation point does not necessarily have to be met to generate adequate yields. In 2015, even a total solar irradiance of 10.11 MJ m^{-2} day^{-1} from emergence to harvest showed no yield changes. In 2017, the 50% shading received 10.07 MJ m^{-2} day^{-1} after emergence. This indicates that after emergence, potatoes need a total solar irradiance >10.11 MJ. In 2016, the weather was very unsteady, and yield was probably more influenced by temperature, which led to a cooling of the dam (soil piled up to 30 cm). Under air temperatures near optimum, more tubers than shoots are built. When air temperature increases, there is a shift to more shoot biomass than tuber mass [83]. If the air temperature is below the base temperature there will be no growth, neither above-ground nor below-ground.

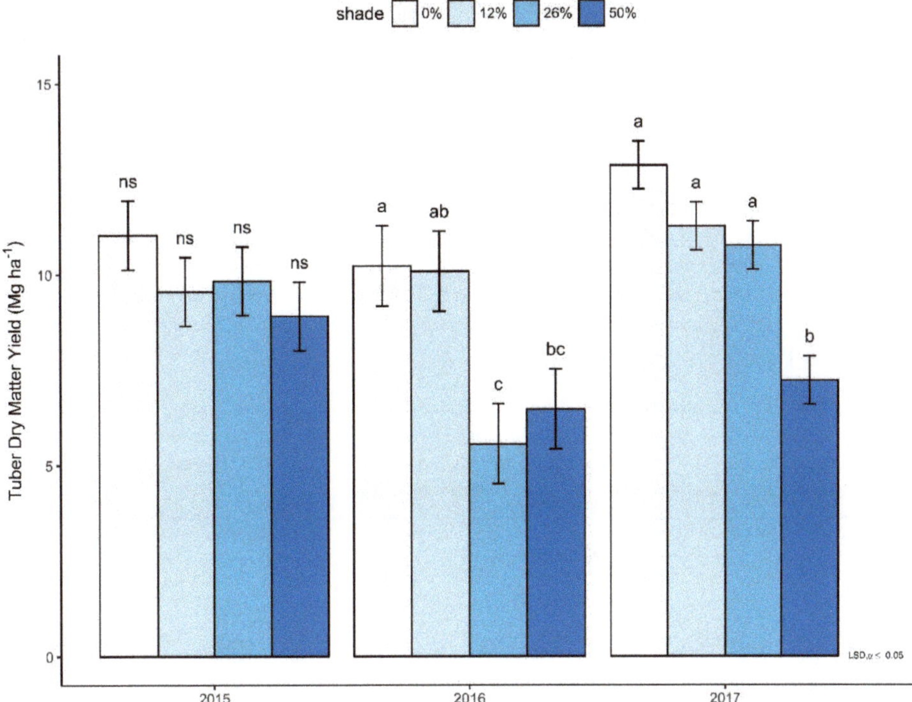

Figure 3. Tuber Dry Matter Yield (Mg ha^{-1}) for the different shade levels (0%, 12%, 26% and 50%) in the single experiment years. Black bars represent the standard error of mean. Means with identical letters within one year show non-significant differences between the shade levels (LSD, $\alpha \leq 0.05$).

Demagante and Vander Zaag indicated that shading of 54% led to total dry matter yields similar to those under full sunlight in the Philippines [56]. A series of experiments in The Netherlands, Rwanda and Tunisia revealed that the tuber dry matter production is highly dependent on growth duration, which is determined by temperature and daylength [29]. The Netherlands is located in a zone with temperate climate and long-day conditions which fits best to the long-day requirement of potatoes, Rwanda is located under short-day conditions with high temperatures, and Tunisia is located in an interface zone between long- and short-day conditions with adequate temperatures from October to April. Hence, as potato is a long-day plant requiring a maximum temperature of >20 °C, Rwanda with its short daylength and high temperature could be unfavorable, while in The Netherlands and Tunis the day-length during the growing period is adequate. However, shading can lower the temperature unfavorably and a short day-length hastens tuber initiation, which reduces the final tuber yield [84]. Kuruppuarachchi showed that shading potatoes at a level of 50% by suspended coconut leaves during the whole cropping season reduced tuber yield significantly by about 56% in Sri Lanka [58]. He concluded that permanent shading compared with shade in the first four weeks resulted in variation in the day/night temperatures of the soil, which may be unfavorable for tuber growth. A study by Sale with potatoes shaded at a level of 34% throughout the growing period showed a 26–42% decrease in yield [85]. Cultivation of potato beneath stone pines (*Pinus pinea* L.) reached tuber yields of 60-86% yield when compared with the national average yields [86].

Experiments with 30% and 50% shade have showed reduced yields by approximately 2–56% [78]. In 2016 and 2017, we also observed a 50% reduction in yield under 50% shade.

Overall, the yield reductions in our experiment are comparable with the results of other experiments, mostly from tropical countries where irradiance is in general much higher. Therefore, it can be concluded that potatoes tolerate shade up to 26% even in the temperate zone and are able to reach adequate yields.

3.3. Quality Parameters of Tubers

With regard to the tuber fraction, an increased proportion of undersized tubers was found up until 26% shade (Table 5). Under 50% shade the share of undersized potatoes (<30 mm) decreased insignificantly. The table fraction (30–60 mm) also showed an insignificant increase at higher shade levels. The 50% treatment had a share of 83.90%, while the control only had 74.83%. An insignificant decreasing share with increasing shade was observed for the oversized fraction (>60 mm). The literature indicates that tuber fractions are generally determined by numerous factors, but these do not include light or shade [66].

Table 5. Mean of starch content (% DM), fractions of undersized (<30 mm), table sized (30–60 mm) and oversized tubers (>60 mm) (%), the black spot bruise index (BSB, %) and the macronutrient content of N, P, K, Ca, Mg and S (% DM) for the different shade levels (0%, 12%, 26% and 50%) averaged over the three experiment years. SEM gives the standard error of means.

Shade	Starch	Fraction [O]			BSB	N	P	K	Ca	Mg	S
		Undersized	Table	Oversized							
0%	70.45	4.39	74.83	19.18	16.80	1.30	0.21	2.62	0.03	0.13	0.19
12%	71.04	4.69	77.27	17.06	19.81	1.31	0.22	2.65	0.03	0.13	0.19
26%	70.06	7.84	76.48	12.62	19.70 [□]	1.36	0.22	2.70	0.03	0.13	0.18
50%	68.43	5.62	83.90	8.86	26.65	1.42	0.23	2.69	0.03	0.13	0.19
SEM	0.67	2.60	4.997	4.02	3.82 [□]	0.05	0.01	0.07	0.00 [‡]	0.00 [‡]	0.01
					p-values [$]						
Year	0.043	0.034	0.071	0.011	0.157	0.066	<0.0001	0.011	<0.0001	0.026	0.055
Shade	0.063	0.806	0.642	0.415	0.386	0.339	0.448	0.864	0.808	0.921	0.853
Year x Replicate	0.424	0.705	0.234	0.028	0.011	0.206	0.299	0.287	0.104	0.043	0.138

[O] Data available for 2016 and 2017 only. [□] SEM for 26% shade ± 3.89% due to missing value. [‡] Note: The SEM was between 0 and 0.005, so rounding to two decimal places resulted in a SEM of zero. [$] *p*-value for the global F-test of the corresponding factor.

The tuber size is mainly influenced by the size of the seed tubers and the growing conditions during the growth of the seed tubers. The number of tubers m^{-2} will be determined by the number of formed stolons per stem. It has been shown that irradiance has no effect on this parameter. It is more sensitive to seed size, number of stems, temperature and drought. Studies by Tekalign and Hammes showed that the cultivar also has an influence on the number of tubers. They showed that the fruit or berry development affects the total and marketable tuber mass and the final tuber yield [87,88]. Berries have an influence on the sink-distribution, leading to yield decreases at higher berry numbers.

Hence, tuber size distribution can be influenced by total tuber yield, seeding rate and size of seed tubers, and the number of stems per plant [66]. As mentioned above, older tubers produce more stems than younger tubers. The tuber size distribution is mainly determined by the date of initiation, position and size of the stolon [61]. This shows that shade has no influence on tuber fraction. A study by Knowles and Knowles showed that under the climate conditions of higher northern latitudes, less tubers are formed, but the number of formed tubers of marketable size are higher than for potatoes grown at lower northern latitudes [89]. More potatoes per plant were formed; however, they are smaller, which ultimately led to lower yields. Other experiments have shown that a late harvest results in a larger range of tuber sizes. No additional tubers will grow, but small tubers continue to grow in the later stages of the growing season, resulting in the larger fraction for tuber size [61].

For most potato cultivars (being determinate), the vegetative plant growth ends with flowering when maximum above-ground biomass has formed [32–34]. During flowering, the tuber formation is completed and the potato plant begins to reallocate the sugars from the above-ground parts to the tubers, where starch is formed. After this, only the tuber mass increases and the quality of the tuber changes. The maximum starch yield can be found when half of the leaves are dead and stems begin to die [74]. Due to the simultaneous harvesting in all shade treatments (date determined after the 0% shade treatment) and the delay in ripening (days from senescence initiation to harvest, Table 3) in 2016, the potatoes had less time to reallocate their sugars from leaves to tubers and build up starch. While the effect of the year ($p = 0.043$) was significant, the effect of the shading treatment was not ($p = 0.063$). So, the year should show a statistical difference. A weather-induced delay in development increased the share of smaller tubers in comparison to larger tubers. Smaller tubers have a lower sink demand for sugars that are reallocated from leaves and stored as starch in tubers. This explains the year effect on the starch content. Across years, the starch content showed no significant differences between the shade levels and the unshaded control. An experiment with 34% and 57% shaded tomatoes (*Solanum lycopersicum* L.) showed that there was no influence on glucose by different levels of irradiance [88]. Other studies with shaded tomatoes showed that shade had no influence on final sugar content [89]. This suggests that the starch content in potatoes is also not affected by shade.

Across years, no effect on black spot bruise (BSB) was detectable. None of the macronutrients showed significant treatment effects across years (Table 5) and values were in the given range of values reported in the literature [90].

The above results showed that the influence of shade on plant growth and tuber yield depends on total solar irradiance but also on other factors (e.g., cultivar, soil temperature, and soil moisture). To minimize the shade, which is a controllable effect, different management techniques can be used. If the trees are still small in the first years of an AFS and need grow first, there will be little or no shade influence on the understory crop in the first years. To obtain high yields, potatoes can be integrated in an AFS in the first years without yield reduction. In addition, a large distance between the single trees, the pruning of the trees, the direction of tree strips from north-to-south and the choice of trees with thinner crowns can keep the shade influence at a minimum. Additionally, shade does not remain static on the field during the whole growing period (as in our experimental setup). In a real AFS, the shading varies during the day and moves on a parabolic shape over the crop as the solar position changes, so, the influence of shade in a real AFS can be regarded as smaller than in our experimental setup.

3.4. Prospects for AFS: Potential Total Solar Irradiance in the Temperate Zone of North-European Latitudes

Theoretically, potatoes need an average total irradiation of 14.86 MJ m^{-2} day^{-1} to reach the given light saturation point of 400 µmol m^{-2} s^{-1} PAR to maximize yields. To reach this light saturation under shading, the required total irradiance would amount to 16.89 MJ m^{-2} day^{-1} at 12% shading, 20.08 MJ m^{-2} day^{-1} under 26% shade and 29.72 MJ m^{-2} day^{-1} under 50% shade. Figure 4 shows the hypothetical growing regions in Europe with an assumed limited available irradiance, taking mean total solar irradiance data from 1984–2013 into account [30]. Under a generalized, assumed potato growing season in Europe (30° N, 20° W to 75° N, 40° E) from 1 March to 31 October (DOY 60–304) and without taking any other climatic growth factors except for irradiance into account, potato cultivation under 50% shade would be possible up to 35° N without yield losses (Figure 4). For 26% shade, cultivation would theoretically be possible from 35° to 45° N, for 12% shade from 45° to 55° N, and from 55° to the northern polar circle at 66° N, which is the geographical limit of potato cultivation. In years with high total solar irradiance, the borders for cultivation under shade will shift to the north, while in years with lower irradiance levels the borders will shift to the south. Possible reasons for this shift include less clouds, low variation in the inclination of the earth's axis, high solar activity, low air pollution or weather phenomena (e.g., fog) or depending on the elevation of the potato cultivation site (in higher elevations, a greater amount of total solar irradiance reaches the surface).

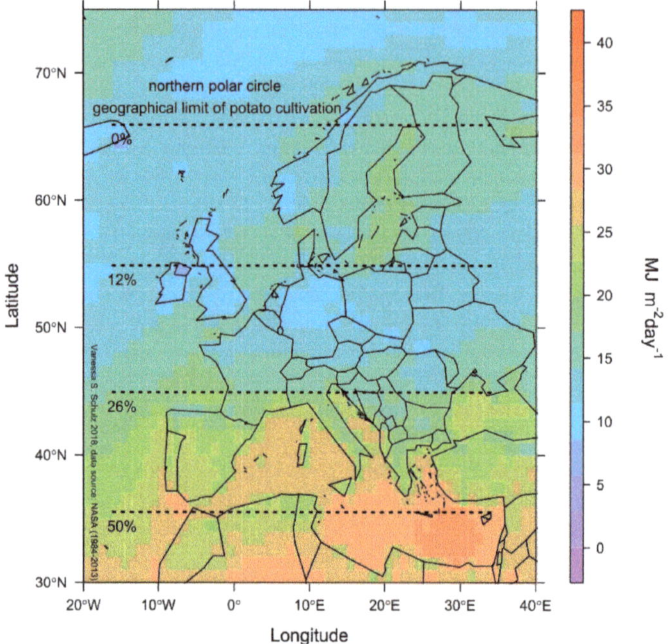

Figure 4. Total solar irradiance (MJ m^{-2} day^{-1}) during the potential potato growing season in Europe (01 March–31 October, 1984–2013) and the theoretical limits of cultivation under shade values of 0%, 12%, 26% and 50% (0.5 × 0.5 m grid, data source NASA [91]).

4. Conclusions

Potatoes are known as being a shade tolerant crop. The results of this study indicated that the DMY was only significantly reduced in 50% shade in years with high irradiance, while a significant reduction at a shade level of 26% only occurred in years with low irradiance. Shading had no significant influence on starch content. Other quality parameters were also not significantly influenced by shade. Yield determining factors like the number of plants per hectare, number of stems per plant, number of tubers per plant and tuber mass per plant were slightly affected by shade. As long as shade is the only influencing factor and no below-ground factors, such as competition for water and nutrients occur, potatoes can be cultivated at latitudes lower than 35° N under 50% shade, while with every increase of 10° N the accepted shade levels have to be halved. Therefore, potatoes can be recommended as an understory crop in AFS up to a shading level of 26% without significant yield and quality reductions under the given total solar irradiance in Southwestern Germany. However, depending on the year (low-irradiance or high-irradiance), this can shift latitudinally in one direction or the other.

Author Contributions: Conceptualization, K.S.; methodology, V.S.S., S.M. and K.S.; software, V.S.S and J.H.; validation, V.S.S. S.M., J.H. and S.G.-H.; formal analysis, V.S.S., S.M. and S.G.H.; investigation, V.S.S.; data curation, V.S.S; writing—original draft preparation, V.S.S., S.M. and S.G.-H.; writing—review and editing, V.S.S., S.M., J.H. and S.G.-H.; supervision, S.G.-H.; project administration, V.S.S, S.M. and S.W.; funding acquisition, K.S.

Funding: This research was funded by the Federal Ministry of Food and Agriculture (BMEL) through the project agency Fachagentur Nachwachsende Rohstoffe (FNR) e.V. within the project "Agro-Wertholz: Agroforstsysteme mit Mehrwert für Mensch und Umwelt", grant number 2201514.

Acknowledgments: The authors would like to thank the staff of the LTZ Augustenberg for making this joint research project possible and Melanie Hinderer for data collection. We also thank Cameron Anderson and Willem Molenaar for English proofreading.

Conflicts of Interest: The authors declare no conflict of interest.

Appendix A

Table A1. Date, product, trade name, amount and active ingredients of plant protection agent and also the Mode of Action (MoA) after HRAC (Herbicide Resistance Action Committee), FRAC (Fungicide Resistance Action Committee) and IRAC (Insecticide Resistance Action Committee) for all three years.

Date	Product	Trade Name	Amount and Active Ingredient	MoA
		2015		
18 May	H	2.0 kg ha^{-1} Artist (Bayer AG)	240 g kg^{-1} *flufenacet*, 175 g kg^{-1} *metribuzin*	K3 C1
10 June	F	2.0 kg ha^{-1} Ridomil Gold (Syngenta AG)	40 g kg^{-1} *metalaxyl-M*, 640 g kg^{-1} *mancozeb*	A1 M3
10 June	I	0.3 L ha^{-1} Biscaya (Bayer AG)	240 g L^{-1} *thiacloprid*	4A
25 June	F	2 kg ha^{-1} Acrobat Plus WG (BASF SE)	90 g kg^{-1} *dimethomorph*, 600 g kg^{-1} *mancozeb*	H5 M3
10 July	F	2 kg ha^{-1} Acrobat Plus WG (BASF SE)	90 g kg^{-1} *dimethomorph*, 600 g kg^{-1} *mancozeb*	H5 M3
10 July	I	0.3 L ha^{-1} Biscaya (Bayer AG)	240 g L^{-1} *thiacloprid*	4A
24 July	F	2 kg ha^{-1} Acrobat Plus WG (BASF SE)	90 g kg^{-1} *dimethomorph*, 600 g kg^{-1} *mancozeb*	H5 M3
6 August	F	2 kg ha^{-1} Acrobat Plus WG (BASF SE)	90 g kg^{-1} *dimethomorph*, 600 g kg^{-1} *mancozeb*	H5 M3
		2016		
6 May	H	3 L ha^{-1} Boxer (Syngenta AG)	800 g L^{-1} *prosulfocarb*	N
6 May	H	0.3 kg ha^{-1} Sencor WG (Syngenta AG)	700 g kg^{-1} *metribuzin*	C1
2 June	I	0.3 L ha^{-1} Biscaya (Bayer AG)	240 g L^{-1} *thiacloprid*	4A
20 June	F	2 kg ha^{-1} Acrobat Plus WG (BASF SE)	90 g kg^{-1} *dimethomorph*, 600 g kg^{-1} *mancozeb*	H5 M3
28 June	I	0.3 L ha^{-1} Biscaya (Bayer AG)	240 g L^{-1} *thiacloprid*	4A
28 June	F	2 kg ha^{-1} Acrobat Plus WG (BASF SE)	90 g kg^{-1} *dimethomorph*, 600 g kg^{-1} *mancozeb*	H5 M3
8 July	F	1.5 L ha^{-1} Infinito (Bayer SE)	62.5 g L^{-1} *fluopicolide*, 625,0 g L^{-1} *propamocarb*-HCl	B5 F4
15 July	F	1.6 L ha^{-1} Infinito (Bayer SE)	62.5 g L^{-1} *fluopicolide*, 625.0 g L^{-1} *propamocarb*-HCl	B5 F4
15 July	I	0.3 L ha^{-1} Biscaya (Bayer AG)	240 g L^{-1} *thiacloprid*	4A
3 August	H	0.8 L ha^{-1} Quickdown (Ceminova Deutschland GmbH & Co. KG)	24.2 g L^{-1} *pyraflufen*	E14
3 August	H	2 L ha^{-1} Toil (Ceminova Deutschland GmbH & Co. KG)	836 g L^{-1} rapeseed oil methyl ester	
		2017		
5 May	H	2.0 kg ha^{-1} Artist (Bayer AG)	240 g kg^{-1} *flufenacet*, 175 g kg^{-1} *metribuzin*	K3 C1
2 June	F	2.0 kg ha^{-1} Ridomil Gold (Syngenta AG)	40 g kg^{-1} *metalaxyl-M*, 640 g kg^{-1} *mancozeb*	A1 M3
2 June	I	0.3 L ha^{-1} Biscaya (Bayer AG)	240 g L^{-1} *thiacloprid*	4A
16 June	F	2 kg ha^{-1} Acrobat Plus WG (BASF SE)	90 g kg^{-1} *dimethomorph*, 600 g kg^{-1} *mancozeb*	H5 M3
16 June	I	0.3 L ha^{-1} Biscaya (Bayer AG)	240 g L^{-1} *thiacloprid*	4A
5 July	I	0.06 L ha^{-1} Coragen (DuPont)	200 g L^{-1} *chlorantraniliprole*	28
5 July	F	2 kg ha^{-1} Acrobat Plus WG (BASF SE)	90 g kg^{-1} *dimethomorph*, 600 g kg^{-1} *mancozeb*	H5 M3
9 August	H	2.5 L ha^{-1} Reglone (Syngenta AG)	374 g L^{-1} *diquat dibromide*	D

H herbicide, F fungicide, I insecticide.

References

1. Horwith, B. A role for intercropping in modern agriculture. *BioScience* **1985**, *35*, 286–291. [CrossRef]
2. Power, J.F.; Follett, R.F. Monoculture. *Sci. Am.* **1987**, *256*, 78–87. [CrossRef]
3. Gliessman, S.R. Multiple cropping systems: A basis for developing an alternative agriculture. In *US Congress Off Ice of Technology Assessment. Innovative Biological Technologies for Lesser Developed Countries: Workshop Proceedings*; Congress of the USA: Washington, DC, USA, 1985; pp. 69–83.

4. Thevathasan, N.V.; Gordon, A.M.; Simpson, J.A.; Reynolds, P.E.; Price, G.; Zhang, P. Biophysical and ecological interactions in a temperate tree-based intercropping system. *J. Crop Improv.* **2004**, *12*, 339–363. [CrossRef]

5. Kremen, C.; Miles, A. Ecosystem services in biologically diversified versus conventional farming systems: Benefits, externalities, and trade-offs. *Ecol. Soc.* **2012**, *17*. [CrossRef]

6. Nair, P.R. The coming of age of agroforestry. *J. Sci. Food Agric.* **2007**, *87*, 1613–1619. [CrossRef]

7. Nair, P.R. Classification of agroforestry systems. *Agrofor. Syst.* **1985**, *3*, 97–128. [CrossRef]

8. Raintree, J.B. Agroforestry pathways: Land tenure, shifting cultivation and sustainable agriculture. *Unasylva* **1986**, *38*, 2–15.

9. Nair, P.R. *An Introduction to Agroforestry*; Springer Science & Business Media: Berlin, Germany, 1993.

10. Vandermeer, J.H. *The Ecology of Intercropping*; Cambridge University Press: Cambridge, UK, 1992.

11. Montagnini, F.; Nair, P.K.R. Carbon sequestration: An underexploited environmental benefit of agroforestry systems. *Agrofor. Syst.* **2004**, *61*, 281–295.

12. Rigueiro-Rodríguez, A.; Fernández-Núñez, E.; González-Hernández, P.; McAdam, J.H.; Mosquera-Losada, M.R. Agroforestry systems in Europe: Productive, ecological and social perspectives. In *Agroforestry in Europe*; Springer: Berlin/Heidelberg, Germany, 2009; pp. 43–65.

13. Roces-Diaz, J.V.; Rolo, V.; Kay, S.; Moreno, G.; Szerencsits, E.; Fagerholm, N.; Plieninger, T.; Torralba, M.; Graves, A.; Giannitsopoulos, M.; et al. Exploring the Relationships among Bio-Physical and Socio-Cultural Ecosystem Services of Agroforestry Systems across Europe. In Proceedings of the Agroforestry as Sustainable Land Use, Nijmegen, The Netherlands, 28–30 May 2018; EURAF: Nijmegen, The Netherlands, 2018.

14. Elevitch, C.; Mazaroli, D.; Ragone, D. Agroforestry standards for regenerative agriculture. *Sustainability* **2018**, *10*, 3337. [CrossRef]

15. Gao, L.; Xu, H.; Bi, H.; Xi, W.; Bao, B.; Wang, X.; Bi, C.; Chang, Y. Intercropping competition between apple trees and crops in agroforestry systems on the Loess Plateau of China. *PLoS ONE* **2013**, *8*, e70739. [CrossRef]

16. Zhang, W.; Ahanbieke, P.; Wang, B.J.; Xu, W.L.; Li, L.H.; Christie, P.; Li, L. Root distribution and interactions in jujube tree/wheat agroforestry system. *Agrofor. Syst.* **2013**, *87*, 929–939. [CrossRef]

17. Newman, S.M.; Bennett, K.; Wu, Y. Performance of maize, beans and ginger as intercrops in Paulownia plantations in China. *Agrofor. Syst.* **1997**, *39*, 23–30. [CrossRef]

18. Khybri, M.L.; Gupta, R.K.; Ram, S.; Tomar, H.P.S. Crop yields of rice and wheat grown in rotation as intercrops with three tree species in the outer hills of Western Himalaya. *Agrofor. Syst.* **1992**, *17*, 193–204. [CrossRef]

19. Ceccon, E. Production of bioenergy on small farms: A two-year agroforestry experiment using Eucalyptus urophylla intercropped with rice and beans in Minas Gerais, Brazil. *New For.* **2008**, *35*, 285–298. [CrossRef]

20. Friday, J.B.; Fownes, J.H. Competition for light between hedgerows and maize in an alley cropping system in Hawaii, USA. *Agrofor. Syst.* **2002**, *55*, 125–137. [CrossRef]

21. Rahman, M.; Bari, M.; Rahman, M.; Ginnah, M.; Rahman, M. Screening of Potato Varieties under Litchi Based Agroforestry System. *Am. J. Exp. Agric.* **2016**, *14*, 1–10. [CrossRef]

22. Mariana, M.; Hamdani, J.S. Growth and yield of Solanum tuberosum at medium plain with application of paclobutrazol and paranet shade. *Agric. Agric. Sci. Procedia* **2016**, *9*, 26–30. [CrossRef]

23. Pleijel, H.; Danielsson, H.; Vandermeiren, K.; Blum, C.; Colls, J.; Ojanperä, K. Stomatal conductance and ozone exposure in relation to potato tuber yield—Results from the European CHIP Programme. *Eur. J. Agron.* **2002**, *17*, 303–317. [CrossRef]

24. Kareem, I.A. Rattle Tree (ALBIZIA LEBBECK) Effects on Soil Properties and Productivity of Irish Potato (SOLANUM TOBEROSUM) on The Jos Plateau, Nigeria. Ph.D. Thesis, Environmental Resource Planning, University of Jos, Jos, Nigeria, 2007.

25. Nadir, S.W.; Ng'etich, W.K.; Kebeney, S.J. Performance of crops under Eucalyptus tree-crop mixtures and its potential for adoption in agroforestry systems. *Aust. J. Crop Sci.* **2018**, *12*, 1231. [CrossRef]

26. Li, L.; Ljung, K.; Breton, G.; Schmitz, R.J.; Pruneda-Paz, J.; Cowing-Zitron, C.; Cole, B.J.; Ivans, L.J.; Pedmale, U.V.; Jung, H.-S.; et al. Linking photoreceptor excitation to changes in plant architecture. *Genes Dev.* **2012**, *26*, 785–790. [CrossRef]

27. Abdrabbo, M.A.; Farag, A.A.; Abul-Soud, M. The intercropping effect on potato under net house as adaption procedure of climate change impacts. *Appl. Res.* **2013**, *5*, 48–60.

28. Gawronska, H.; Dwelle, R.B.; Pavek, J.J. Partitioning of photoassimilates by potato plants (*Solanum tuberosum* L.) as influenced by irradiance: II. Partitioning patterns by four clones grown under high and low irradiance1. *Am. J. Potato Res.* **1990**, *67*, 163–176. [CrossRef]
29. Kooman, P.L.; Fahem, M.; Tegera, P.; Haverkort, A.J. Effects of climate on different potato genotypes 2. Dry matter allocation and duration of the growth cycle. *Eur. J. Agron.* **1996**, *5*, 207–217. [CrossRef]
30. NASA Atmospheric Science Data Center. Available online: https://eosweb.larc.nasa.gov/sse/global/text/lat_tilt_radiation (accessed on 24 December 2017).
31. Smith, J.; Pearce, B.D.; Wolfe, M.S. Reconciling productivity with protection of the environment: Is temperate agroforestry the answer? *Renew. Agric. Food Syst.* **2013**, *28*, 80–92. [CrossRef]
32. Barbeau, C.D.; Wilton, M.J.; Oelbermann, M.; Karagatzides, J.D.; Tsuji, L.J. Local food production in a subarctic Indigenous community: The use of willow (Salix spp.) windbreaks to increase the yield of intercropped potatoes (Solanum tuberosum) and bush beans (Phaseolus vulgaris). *Int. J. Agric. Sustain.* **2018**, *16*, 29–39. [CrossRef]
33. Wilton, M.J.; Karagatzides, J.D.; Tsuji, L.J. Nutrient Concentrations of Bush Bean (*Phaseolus vulgaris* L.) and Potato (*Solanum tuberosum* L.) Cultivated in Subarctic Soils Managed with Intercropping and Willow (*Salix* spp.) Agroforestry. *Sustainability* **2017**, *9*, 2294. [CrossRef]
34. BSV Saaten–Blühmischung M2 FAKT Maßnahme E2.1 und E2.2. Available online: https://bsv-saaten.de/bluehmischungen-und-bienenweiden/foerderprogramme-1/fakt-1/naturplus-fakt-m2-zr-100-bluehmischung-m2-fuer-fakt-massnahme-e2-1-und-e2-2.html (accessed on 24 November 2018).
35. Röhrig, M.; Sander, R.; eV Geschäftsstelle, I. Interaktive Online-Beratung mit dem Informationssystem Integrierte Pflanzenproduktion (ISIP). In Proceedings of the GIL Jahrestagung, Potsdam, Germany, 6–8 March 2006; pp. 221–224.
36. Bundesministerium für Ernährung. *Landwirtschaft und Verbraucherschutz (BMELV). Gute fachliche Praxis im Pflanzenschutz-Grundsätze für die Durchführung*; BMELV: Bonn, Germany, 2010.
37. Janssen, W. Online irrigation service for fruit and vegetable crops at farmers site. In Proceedings of the 9th EMS Annual Meeting, 9th European Conference on Applications of Meteorology (ECAM) Abstracts, Toulouse, France, 28 September–2 October 2009; Available online: http://meetings.copernicus.org/ems2009/ (accessed on 5 September 2017).
38. Hack, H.; Gall, H.; Klemke, T.H.; Klose, R.; Meier, U.; Stauss, R.; Witzenberger, A. The BBCH scale for phenological growth stages of potato (*Solanum tuberosum* L.). In Proceedings of the 12th Annual Congress of the European Association for Potato Research, Paris, France, 1993; pp. 153–154.
39. Schulz, V.S.; Munz, S.; Stolzenburg, K.; Hartung, J.; Weisenburger, S.; Mastel, K.; Möller, K.; Claupein, W.; Graeff-Hönninger, S. Biomass and Biogas Yield of Maize (*Zea mays* L.) Grown under Artificial Shading. *Agriculture* **2018**, *8*, 178. [CrossRef]
40. Mushagalusa, G.N.; Ledent, J.-F.; Draye, X. Shoot and root competition in potato/maize intercropping: Effects on growth and yield. *Environ. Exp. Bot.* **2008**, *64*, 180–188. [CrossRef]
41. Bodlaender, K.B.A. *Influence of Temperature, Radiation and Photoperiod on Development and Yield*; Wageningen University: Gelderland, The Netherlands, 1963.
42. Cao, W.; Tibbitts, T.W. Leaf emergence on potato stems in relation to thermal time. *Agron. J.* **1995**, *87*, 474–477. [CrossRef]
43. Sprenger, H.; Rudack, K.; Schudoma, C.; Neumann, A.; Seddig, S.; Peters, R.; Zuther, E.; Kopka, J.; Hincha, D.K.; Walther, D.; et al. Assessment of drought tolerance and its potential yield penalty in potato. *Funct. Plant Biol.* **2015**, *42*, 655–667. [CrossRef]
44. Fernández, S.D.M.; Aguilar, R.M.; Moreno, Y.S.; Pérez, J.E.R.; León, M.T.C.; Saldaña, H.L. Growth and sugar content of potato tubers in four maturity stages under greenhouse conditions. *Rev. Chapingo Ser. Hortic.* **2018**, *24*, 53–67. [CrossRef]
45. McMaster, G.S.; Wilhelm, W.W. Growing degree-days: One equation, two interpretations. *Agric. For. Meteorol.* **1997**, *87*, 291–300. [CrossRef]
46. Bundessortenamt. *Richtlinien für die Durchführung von Landwirtschaftlichen Sortenversuchen*; Bundessortenamt: Hannover, Germany, 2000.
47. Bundessortenamt. *Beschreibende Sortenliste Kartoffel—2017*; Bundessortenamt: Hannover, Germany, 2017.
48. Hoffmann, G. *VDLUFA-Methodenbuch Band I: Die Untersuchung von Boeden*; Loseblattsammlung; VDLUFA-Verlag: Darmstadt, Germany, 1991; ISBN 3-922712-42-8.

49. Janßen, E. *VDLUFA-Methodenbuch Band VII: Umweltanalytik*, 3rd ed.; VDLUFA: Darmstadt, Germany, 2003.
50. Bassler, R. *VDLUFA-Methodenbuch, Band III: Die Chemische Untersuchung von Futtermitteln*; 3. Aufl., 2. Ergänzungslieferung 1988 und 3. Ergänzungslieferung 1993; VDLUFA-Verlag: Darmstadt, Germany, 1988.
51. Amtsblatt, D.E.U. Verordnung (EG) Nr. 152/2009 der Kommission vom 27.01. 2009 zur Festlegung der Probenahmeverfahren und Analysemethoden für die amtliche Untersuchung von Futtermitteln. *L* **2009**, *54*, 130.
52. Heinecke, A. *Beitrag zur Ermittlung der Biochemischen Ursachen der Schwarzfleckigkeit bei Kartoffeln*; Georg-August-Universität Göttingen: Gottingen, Germany, 2007.
53. Baty, F.; Ritz, C.; Charles, S.; Brutsche, M.; Flandrois, J.-P.; Delignette-Muller, M.-L. A toolbox for nonlinear regression in R: The package nlstools. *J. Stat. Softw.* **2015**, *66*, 1–21. [CrossRef]
54. Fox, J. *Nonlinear Regression and Nonlinear Least Squares*; Wiley: New York, NY, USA, 2002.
55. Demagante, A.L.; Zaag, P.V. The response of potato (*Solanum* spp.) to photoperiod and light intensity under high temperatures. *Potato Res.* **1988**, *31*, 73–83. [CrossRef]
56. Sale, P.J.M. Productivity of vegetable crops in a region of high solar input. II. Yields and efficiencies of water use and energy. *Aust. J. Agric. Res.* **1973**, *24*, 751–762. [CrossRef]
57. Kuruppuarachchi, D.S.P. Intercropped potato (*Solanum* spp.): Effect of shade on growth and tuber yield in the northwestern regosol belt of Sri Lanka. *Field Crops Res.* **1990**, *25*, 61–72. [CrossRef]
58. Ebwongu, M.; Adipala, E.; Ssekabembe, C.K.; Kyamanywa, S.; Bhagsari, A.S. Effect of intercropping maize and Solanum potato on yield of the component crops in central Uganda. *Afr. Crop Sci. J.* **2001**, *9*, 83–96. [CrossRef]
59. Menzel, C.M. Tuberization in potato at high temperatures: Interaction between temperature and irradiance. *Ann. Bot.* **1985**, *55*, 35–39. [CrossRef]
60. Struik, P.C.; Vreugdenhil, D.; Haverkort, A.J.; Bus, C.B.; Dankert, R. Possible mechanisms of size hierarchy among tubers on one stem of a potato (*Solanum tuberosum* L.) plant. *Potato Res.* **1991**, *34*, 187–203. [CrossRef]
61. Abu-Zinada, I.A.; Mousa, W.A. Growth and productivity of different potato varieties under Gaza Strip conditions. *Int. J. Agric. Crop Sci.* **2015**, *8*, 433.
62. Stamm, P.; Kumar, P.P. The phytohormone signal network regulating elongation growth during shade avoidance. *J. Exp. Bot.* **2010**, *61*, 2889–2903. [CrossRef] [PubMed]
63. Arsenault, W.J.; LeBlanc, D.A.; Tai, G.C.; Boswall, P. Effects of nitrogen application and seedpiece spacing on yield and tuber size distribution in eight potato cultivars. *Am. J. Potato Res.* **2001**, *78*, 301–309. [CrossRef]
64. Putz, B. *Kartoffeln: Züchtung, Anbau, Verwertung*; Behr's Verlag: Hamburg, Germany, 1989.
65. Struik, P.C.; Haverkort, A.J.; Vreugdenhil, D.; Bus, C.B.; Dankert, R. Manipulation of tuber-size distribution of a potato crop. *Potato Res.* **1990**, *33*, 417–432. [CrossRef]
66. Delaplace, P.; Brostaux, Y.; Fauconnier, M.-L.; du Jardin, P. Potato (*Solanum tuberosum* L.) tuber physiological age index is a valid reference frame in postharvest ageing studies. *Postharvest Biol. Technol.* **2008**, *50*, 103–106. [CrossRef]
67. O'Brien, P.J.; Firman, D.M.; Allen, E.J. Effects of shading and seed tuber spacing on initiation and number of tubers in potato crops (Solanum tuberosum). *J. Agric. Sci.* **1998**, *130*, 431–449. [CrossRef]
68. de Lucas, M.; Davière, J.-M.; Rodríguez-Falcón, M.; Pontin, M.; Iglesias-Pedraz, J.M.; Lorrain, S.; Fankhauser, C.; Blázquez, M.A.; Titarenko, E.; Prat, S. A molecular framework for light and gibberellin control of cell elongation. *Nature* **2008**, *451*, 480. [CrossRef]
69. Sun, T. Gibberellin-GID1-DELLA: A Pivotal Regulatory Module for Plant Growth and Development. *Plant Physiol.* **2010**, *154*, 567–570. [CrossRef]
70. García-Martinez, J.L.; Gil, J. Light regulation of gibberellin biosynthesis and mode of action. *J. Plant Growth Regul.* **2001**, *20*, 354–368. [CrossRef]
71. Kamiya, Y.; García-Martínez, J.L. Regulation of gibberellin biosynthesis by light. *Curr. Opin. Plant Biol.* **1999**, *2*, 398–403. [CrossRef]
72. Wurr, D.C.E.; Hole, C.C.; Fellows, J.R.; Milling, J.; Lynn, J.R.; O'Brien, P.J. The effect of some environmental factors on potato tuber numbers. *Potato Res.* **1997**, *40*, 297–306. [CrossRef]
73. Kolbe, H.; Stephan-Beckmann, S. Development, growth and chemical composition of the potato crop (*Solanum tuberosum* L.). II. Tuber and whole plant. *Potato Res.* **1997**, *40*, 135–153. [CrossRef]
74. Ewing, E.E.; Wareing, P.F. Shoot, Stolon, and Tuber Formation on Potato (*Solanum tuberosum* L.) Cuttings in Response to Photoperiod. *Plant Physiol.* **1978**, *61*, 348–353. [CrossRef] [PubMed]

75. Pohjakalli, O. On the effect of the intensity of light and length of day on the energy economy of certain cultivated plants. *Acta Agric. Scand.* **1950**, *1*, 153–175. [CrossRef]

76. Saha, R.R.; Sarker, A.Z.; Talukder, A.H.M.M.R.; Akter, S.; Golder, P.C. Variability in Growth and Yield of Potato Varieties at Different Locations of Bangladesh. *Bangladesh Soc. Hortic. Sci.* **2016**, *2*, 101–110.

77. Hébert, Y.; Guingo, E.; Loudet, O. The Response of Root/Shoot Partitioning and Root Morphology to Light Reduction in Maize Genotypes. *Crop Sci.* **2001**, *41*, 363–371. [CrossRef]

78. Geigenberger, P.; Kolbe, A.; Tiessen, A. Redox regulation of carbon storage and partitioning in response to light and sugars. *J. Exp. Bot.* **2005**, *56*, 1469–1479. [CrossRef]

79. Manrique, L.A.; Kinry, J.R.; Hodges, T.; Axness, D.S. Dry Matter Production and Radiation Interception of Potato. *Crop Sci.* **1991**, *31*, 1044–1049. [CrossRef]

80. Dam, J.V.; Kooman, P.L.; Struik, P.C. Effects of temperature and photoperiod on early growth and final number of tubers in potato (*Solanum tuberosum* L.). *Potato Res.* **1996**, *39*, 51–62.

81. Streck, N.A.; de Paula, F.L.M.; Bisognin, D.A.; Heldwein, A.B.; Dellai, J. Simulating the development of field grown potato (*Solanum tuberosum* L.). *Agric. For. Meteorol.* **2007**, *142*, 1–11. [CrossRef]

82. Sale, P.J.M. Effect of shading at different times on the growth and yield of the potato. *Aust. J. Agric. Res.* **1976**, *27*, 557–566. [CrossRef]

83. Loewe, V.; Delard, C. Stone pine (Pinus pinea L.): An interesting species for agroforestry in Chile. *Agrofor. Syst.* **2019**, *93*, 703–713. [CrossRef]

84. Tekalign, T.; Hammes, P.S. Growth and productivity of potato as influenced by cultivar and reproductive growth—I. Stomatal conductance, rate of transpiration, net photosynthesis, and dry matter production and allocation. *Sci. Hortic.* **2005**, *105*, 13–27. [CrossRef]

85. Tekalign, T.; Hammes, P.S. Growth and productivity of potato as influenced by cultivar and reproductive growth—II. Growth analysis, tuber yield and quality. *Sci. Hortic.* **2005**, *105*, 29–44. [CrossRef]

86. Knowles, N.R.; Knowles, L.O. Manipulating stem number, tuber set, and yield relationships for northern-and southern-grown potato seed lots. *Crop Sci.* **2006**, *46*, 284–296. [CrossRef]

87. Kolbe, H.; Stephan-Beckmann, S. Development, growth and chemical composition of the potato crop (*Solanum tuberosum* L.). I. leaf and stem. *Potato Res.* **1997**, *40*, 111–129. [CrossRef]

88. Kläring, H.-P.; Krumbein, A. The effect of constraining the intensity of solar radiation on the photosynthesis, growth, yield and product quality of tomato. *J. Agron. Crop Sci.* **2013**, *199*, 351–359. [CrossRef]

89. Gautier, H.; Diakou-Verdin, V.; Bénard, C.; Reich, M.; Buret, M.; Bourgaud, F.; Poëssel, J.L.; Caris-Veyrat, C.; Génard, M. How does tomato quality (sugar, acid, and nutritional quality) vary with ripening stage, temperature, and irradiance? *J. Agric. Food Chem.* **2008**, *56*, 1241–1250. [CrossRef] [PubMed]

90. Burlingame, B.; Mouillé, B.; Charrondière, R. Nutrients, bioactive non-nutrients and anti-nutrients in potatoes. *J. Food Compos. Anal.* **2009**, *22*, 494–502. [CrossRef]

91. NASA POWER Data Access Viewer. Available online: https://power.larc.nasa.gov/data-access-viewer/ (accessed on 1 November 2018).

Review

A Systematic Review of Field Trials to Synthesize Existing Knowledge and Agronomic Practices on Protein Crops in Europe

Mohamed Houssemeddine Sellami [1,*], Cataldo Pulvento [1], Massimo Aria [2], Anna Maria Stellacci [3] and Antonella Lavini [1]

1. National Research Council of Italy (CNR), Institute for Agricultural and Forestry Systems in the Mediterranean (ISAFOM), 80056 Ercolano, Italy; cataldo.pulvento@cnr.it (C.P.); antonella.lavini@cnr.it (A.L.)
2. Department of Economic and Statistics, University of Naples Federico II, 80126 Naples, Italy; aria@unina.it
3. Department of Soil, Plant and Food Sciences, University of Bari Aldo Moro; 70126 Bari, Italy; annamaria.stellacci@uniba.it
* Correspondence: hsellemi@yahoo.fr; Tel.: +39-081-788-6704

Received: 13 May 2019; Accepted: 1 June 2019; Published: 6 June 2019

Abstract: Protein crops can represent a sustainable answer to growing demand for high quality, protein-rich food in Europe. To better understand the state of scientific studies on protein crops, a systematic review of field trials results to collect existing knowledge and agronomic practices on protein crops in European countries was conducted using published data from the literature (1985–2017). A total of 42 publications was identified. The following seven protein crops were considered: quinoa, amaranth, pea, faba bean, lupin, chickpea, and lentil. Observations within the studies were related to one or more of eight wide categories of agronomic managements: deficit irrigation (n = 130), salinity (n = 6), tillage (n = 211), fertilizers (n = 146), sowing density (n = 32), sowing date (n = 92), weed control (n = 71), and multiple interventions (n = 129). In 86% of the studies, measures of variability for yield mean values are missing. Through a multiple correspondence analysis (MCA) based on protein crops, European environments, and agronomic management factors, we provide a state of art of studies carried out in Europe on protein crops over the 32-year period; this study will allow us to understand the aspects that can still be developed in the topic. Most investigated studies refer to southern Europe and showed some trends: (i) faba bean, pea, and lupin provide highest seed yields; (ii) sowing date, sowing density, fertilization, and deficit irrigation are the agronomic practices that most influence crop yield; (iii) studies conducted in Central Europe show highest seed yields. The output from this study can be used to guide policies for sustainable crop management.

Keywords: protein crops; systematic review; Europe; multiple correspondence analysis (MCA)

1. Introduction

Meeting the globally growing demand for high quality, protein-rich food, that can satisfy the need of a growing world population while considering environmental sustainability, adapted land-use practices, and food security are special challenges [1]. Europe has a large consumption of animal-based proteins for food, i.e., meat and dairy products whereas most plant protein in the European market is used as feed (for instance more than 95% in the Netherlands; [2]).

The European Union has a 70% deficit in protein-rich grains that is met primarily by imports of GMO soya for feed from USA and South America [3–5]. The deficiency in locally grown protein sources also creates price volatility and trade distortions [6].

Increasing the proportion of vegetable proteins and decreasing those of animal origin in the human diet is a win-win situation from both environmental and nutritional standpoints [6], but also for bio-diversity and sustainability in crop production. However, there are currently few options to encourage this. Vegetable protein sources should have high protein content, high digestibility, low anti-nutritional factor levels, high amount of essential amino acids, and be comparable in price to animal protein sources [7].

Arable land for dry pulses in the Europe, from 2007 to 2017, ranged between 1.7 and 4.2 million hectares. In recent years, this area increased considerably. In that period, an increase of 53.15% of the area harvested at Europe-level was recorded [8]. In 2017, the total harvest of dry pulses in Europe was 9.47 million tons (Figure 1). The production in 2016 was 7.43 million tons with an increase of 17.62% in comparison to 2015. The dry pulse harvested in 2017 was 101.1% higher than the average production of 4.7 million tons registered in 2007–2016. Other high-quality protein crops like quinoa were recently introduced in Europe; the area under quinoa cultivation in Europe increased from 0 in 2008 to 5000 ha in 2015, mainly in France, Spain, and the UK [9].

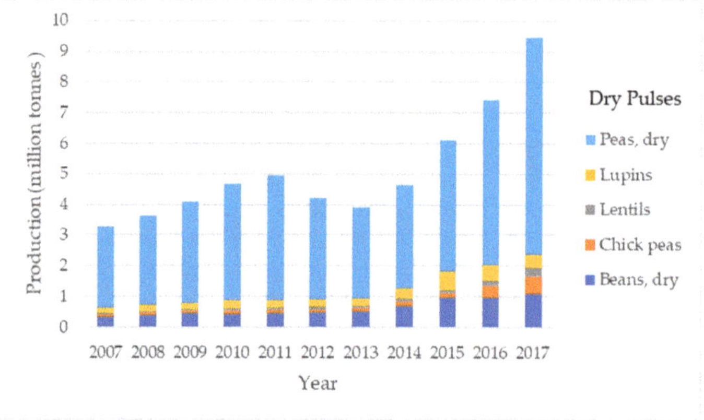

Figure 1. Evolution of harvest production (million tons) of dry pulses by species on Europe.

Producing protein crops as grain legumes is often viewed as risky by European farmers, who tend to prefer cultivating non-legume species such as cereals, oilseeds, and tubers [10,11]. Several authors have in fact hypothesized that a large adoption of legume crops by European farmers is hampered by frequent losses due to high inter-annual yield variability [3,10,12].

The objective of this study was to evaluate, using a systematic review, the agronomic management practices in different European environments on the basis of the literature on protein crops, in order to establish the weight and the significance of the three factors (agronomic management, European environment, and protein crop) for the improvement of production and diffusion of these crops.

A systematic review (SR) of the literature provides a replicable, transparent, and reliable method of identifying, assessing, and summarizing available evidence on a research question, with reduced bias [13,14].

This work is part of a European project called "Development of high-quality food protein from multi-purpose crops through optimized, sustainable production and processing methods" (PROTEIN2FOOD, AMD-635727). This project aims to provide a platform for a diverse production of protein crops throughout various EU-climate zones, supporting reduction of soybean import, and generating novel protein food products by combination of nutritious protein sources, i.e., the quality and quantity of protein from selected high protein quality seed crops (quinoa, amaranth) and legumes with high protein content (lupin, faba beans, pea, chickpea, lentil), to support the reduction of meat consumption. The positive trend will be strengthened by the development of new plant-based protein-rich products of high consumer acceptance, environmental sustainability, and income generation for farmers in European countries.

2. Methods

2.1. Literature Research

A systematic review (SR), across two bibliographic databases (ISI Web of Science™ and Scopus™) and one website search (Google Scholar), was used to identify studies exploring the agronomic practices on protein crops in European countries (we are including Turkey as European country, because it is a (founding) member of the Council of Europe, and a member of the EU Customs Union) and that were written in English, French, Spanish, and Italian (languages the authors felt competent to review) between 1985 and 2017 in peer-reviewed journals. Searches of academic databases were performed on 21 December 2017. In bibliographic databases the following search strings were used to search on 'topic words' combined with Boolean operators: (field OR cultivar* OR genotyp* OR ecotype* OR crop* OR farm* OR cultivar* OR accessions) AND (yield OR grain OR product* OR protein OR seed*OR ((water OR nitrogen) AND use AND efficiency)) AND (europ* OR Mediterranean NOT (Tunisia OR Algeria OR Morocco OR Egypt OR Lebanon OR Libya OR Syria OR Israel OR Palestine)) AND (legum* OR amaranth* OR (peas OR (Pisum AND sativum)) OR (bean OR (Vicia AND faba)) OR (lentil OR (Lens AND culinaris)) OR (lupin* OR (Lupinus AND (albus OR mutabilis))) OR (Chickpea OR (Cicer AND arietinum)) OR (quinoa* OR (Chenopodium and quinoa)). The wildcards * represent any number of characters. Because of the large number of treatments, genotypes, and investigation areas, we used abbreviations for each variable. The complete abbreviations list can be found in the Supplementary Materials (SM) 1.

2.2. Inclusion and Exclusion Criteria

We used a highly robust and rational systematic review methodology to synthesize the evidence from a wide range of sources. In this study, we constrained the SR by defining boundaries to include: (I) only studies that considered food crop production in specific locations in Europe; (II) studies conducted only under field conditions, but not under glasshouse conditions and pots; and (III) studies that focused on crop productivity, omitting forestry, fisheries, livestock, and other non-food crop agricultural sectors. Following SR convention, the search terms were based on the three PIO components (population, interventions, and outcome) (Table 1) and a list of references included in the SR meta-database is provided in SM2.

Table 1. Defining the PIO terms for the research 'question' used in this study

PIO	Description
Population	Agriculture—food crops under field conditions
	Crops included quinoa, amaranth, pea, faba bean, lupin, chickpea and lentil
	Europe: Study included all the countries in the continent
Intervention	Management included sowing date, sowing density, fertilizer, tillage, salinity, deficit irrigation, and weed control
Outcomes	Yield, yield gap, potential yield, farmer yield, and attainable yield

2.3. Screening

Following removal of duplicates, to extract yield information, data from accepted papers were entered into Endnote (online bibliographic management software) (version basic; Clarivate Analytics, https://access.clarivate.com/#/login?app=endnote), all the references retrieved and screened for relevance using the following inclusion criteria: every study identified was screened through three stages: title, abstract, and full text. At each level, records containing or likely to contain relevant information were retained and taken to the next stage.

2.4. Coding and Data Extraction

Meta-data (descriptive categorical information regarding citations, study setting, design, and methods) were extracted from included studies following full text assessment.

The treatments investigated (agronomic management) were recorded for each study as categorical variables where possible; in this case, a complete disjunctive coding of our variables (treatments investigated) was carried out. This means that variables are dichotomous, assuming value "1" if should the keyword be associated to the paper, and "0" if not. This coding was conducted according to methods described by Cuccurullo et al. [15].

Since meta-analysis considers each observation to be independent [16], data for different years or experimental conditions (i.e., cultivars or other experimental factors) within each publication were treated as independent observations. Data were obtained directly from tables and if data were provided in graphical form, means were extracted using WebPlotDigitizer [17].

2.5. Statistical Analysis

Multiple correspondence analysis (MCA) was performed on observations selected from literature sources analysis. The MCA is an exploratory multivariate technique used to simplify data visualization when individuals are described by categorical variables [18]. MCA allows the investigation of several qualitative parameters and permits a geometrical representation of all the information [19]. A MCA was performed to identify the overall correlation of the protein crops in the different agronomic management with environmental conditions (geographical and climatic region). To improve the quality (the total inertia) of the description provided by the MCA, an adjusted formula given by Benzécri [20] was used. This analysis was carried out using the software package FactoMineR [21] in R studio software [22].

3. Results

3.1. Screening Process

Schematic representation of the screening process is given in Figure 2. We ultimately identified and screened 2020 sources of literature (after removal of 1409 duplicates or non-journal papers), of which 42 were subsequently selected and analyzed, to provide 818 'observations'.

Our screening process reveals that accessible, published protein crop cropping systems research in European countries, of a high reporting standard suitable for this systematic review, is concentrated in the southern Europe (87% of observations) (Table 2 and Figure 3).

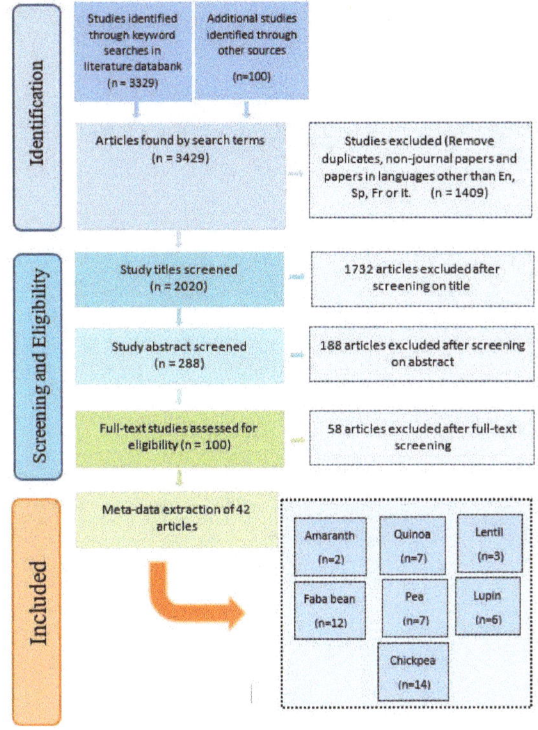

Figure 2. Selection of studies for inclusion in the systematic review (n represent the number of studies).

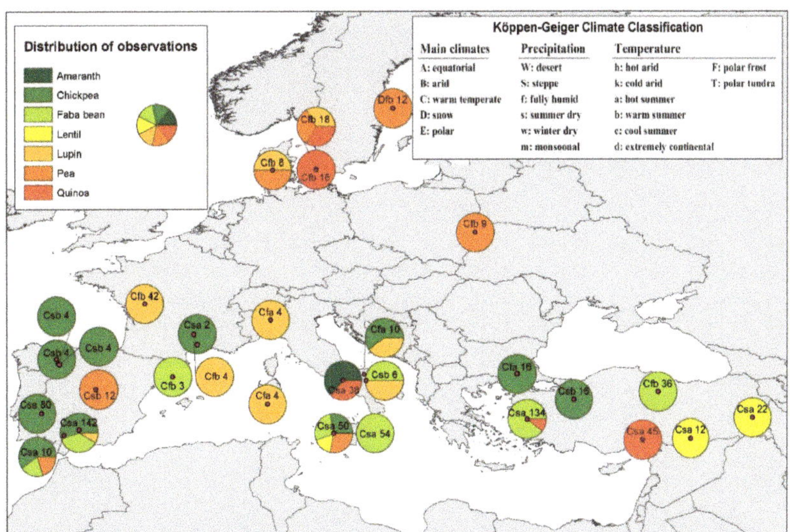

Figure 3. Distribution of doughnut chart of all observations (n = 818) by crops (amaranth, chickpea, faba bean, lentil, lupin, pea, and quinoa), by Köppen-Geiger climate classification zones (warm temperate climate zone (fully humid and summer dry, i.e., Cfa, Cfb, Csa, Csb, Csc); the snow climate zone (fully humid, i.e., Dfb)) and study sites on the map of Europe. The writing in each doughnut chart represents the climate classification and the total number of observations for each study area.

Table 2. Descriptive statistical parameters (minimum, maximum, mean, median, SD and CV) relative to yield (t ha^{-1}) for all crops and split into northern (NE), central (CE), and southern (SE) Europe.

European Region	No. of Countries	List of Countries	No. of Cases	No. of Observations	Crops	No. of Observations	Yield (t ha^{-1})				S.D.[1]	C.V.[2]
							Minimum	Maximum	Mean	Median		
Northern Europe (NE)	2	Sweden, Denmark	4	54	Lupin	10	0.98	3.95	2.83	2.96	1.03	36.18
					Peas	22	2.36	6.64	3.87	3.72	1.18	30.39
					Quinoa	22	1.54	2.27	1.82	1.76	0.21	11.42
Central Europe (CE)	2	France, Poland	4	57	Chickpea	2	2.86	3.18	3.02	3.02	0.23	7.49
					Lupin	46	2.15	4.67	3.51	3.44	0.58	16.47
					Peas	9	2.08	5.03	3.55	3.89	0.94	26.51
Southern Europe (SE)	5	Portugal, Spain, Italy, Greece, Turkey	34	707	Amaranth	24	1.3	2.65	1.94	1.93	0.41	21.39
					Chickpea	228	0.26	3.65	1.63	1.59	0.72	44.33
					Faba bean	284	0.01	8.29	3.36	2.98	1.77	52.63
					Lentil	42	0.01	2.18	0.78	0.73	0.60	77.32
					Lupin	28	0.64	3.99	2.20	1.78	1.18	53.45
					Peas	28	0.08	4.94	2.32	2.50	1.28	54.92
					Quinoa	73	0.87	3.31	2.11	1.95	0.56	26.54

[1] Standard deviation. [2] Coefficient of variation.

3.2. Geographical Distribution of Observations

Studies were reported from nine countries (Figure 3). In the following text, numbers in brackets indicate the number of individual observations in the categories described. Southern Europe (n = 706) was the most frequently studied region, with Italy (n = 166), Portugal (n = 80), Spain (n = 179), and Turkey (n = 265) being the most frequently studied countries.

The most commonly studied climate zones were Cfb (n = 136) with a marine mild climate with no dry season and warm summer and Csa (n = 589) with a warm temperate climate with dry and hot summer (Table 3).

Table 3. Number of observations included in the meta-dataset per Köppen-Geiger climate zone

Köppen-Geiger Climate Zone	Name of the Climate Zone	No. of Observations
Cfa	Cfa—Humid subtropical	34
Cfb	Cfb—Marin–mild winter	136
Csa	Csa—Interior mediterranean	589
Csb	Csb—Coastal mediterranean	46
Dfb	Dfb—Humid continental mild summer, wet all year	12

The most commonly studied crops were Faba bean (n = 284) and Chickpea (n = 230) (Table 2).

3.3. Management, Date, and Duration of Trials

Eight main groups of treatments were identified during the screening: deficit irrigation (Treatment A; n = 130), salinity (Treatment B; n = 6), tillage (Treatment C; n = 211), fertilization (Treatment D; n = 146), sowing density (Treatment E; n = 32), sowing date (Treatment F; n = 92), weed control (Treatment G; n = 71) and multiple interventions (Treatment AB/AD/CD/EF; n = 129). The number of studies reporting investigations of each group of treatments is shown in Figure 4.

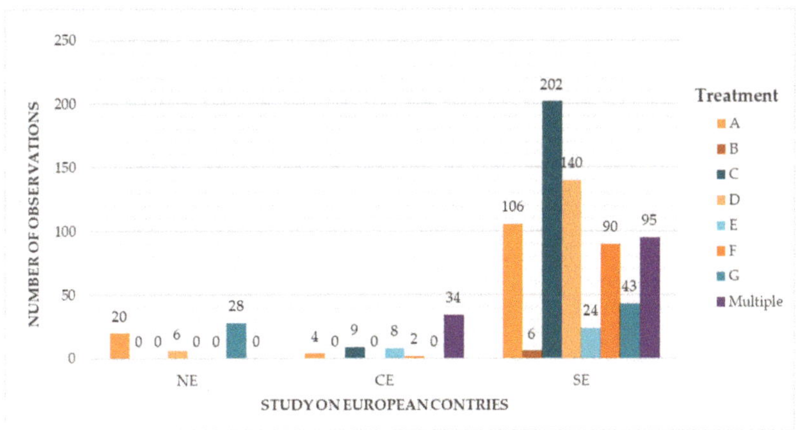

Figure 4. Number of studies undertaken across European countries. Numbers are separated by intervention group (treatment) investigated within each study. Studies may be present in more than one intervention category.

Treatment C (n = 202), Treatment D (n = 140), Treatment A (n = 106), and Treatment F (n = 90) were the most frequently studied treatments on southern Europe (SE). On northern Europe (NE) both Treatment A (n = 20) and G (n = 28) were the most frequently studied treatments and on central Europe (CE) we found only multiple interventions (e.g., Treatment EF) with 34 observations (data not shown).

A large number of studies was carried out over a 2-year period (n = 438), 136 for 1 year, 45 for 3 years, and 198 for more than 3 years (data not shown).

Forty-two papers were identified through cross-referencing dating from 1984 to 2014. An evident increase in the number of observations started after 1998. The peaks occurred in 2005 and 2010, while after 2010 there was a high drop in observations (data not shown). The longest-term studies occurred in southern Europe (18 years) in Italy [23] and in Spain (11–12 years; [24,25]).

Out of the 42 studies, 36 failed to report some critical information of measures of variability for yield mean values (e.g., standard deviation, standard error, and confidence intervals) and 18 observations failed to report the start year of treatment.

3.4. Distribution Frequencies of Yield

The descriptive statistical parameters, relative to yield (t ha^{-1}) for all crops and European regions, are shown in Table 2.

For northern Europe, most data were related to quinoa and pea (n = 22) with observations available from studies conducted in two countries in the region (Sweden, Denmark). In contrast, there was very limited published data on lupin (n = 10) and none for amaranth, chickpea, faba bean, and lentil. The average yields for pea and lupin were 3.87 and 2.83 t ha^{-1}, respectively.

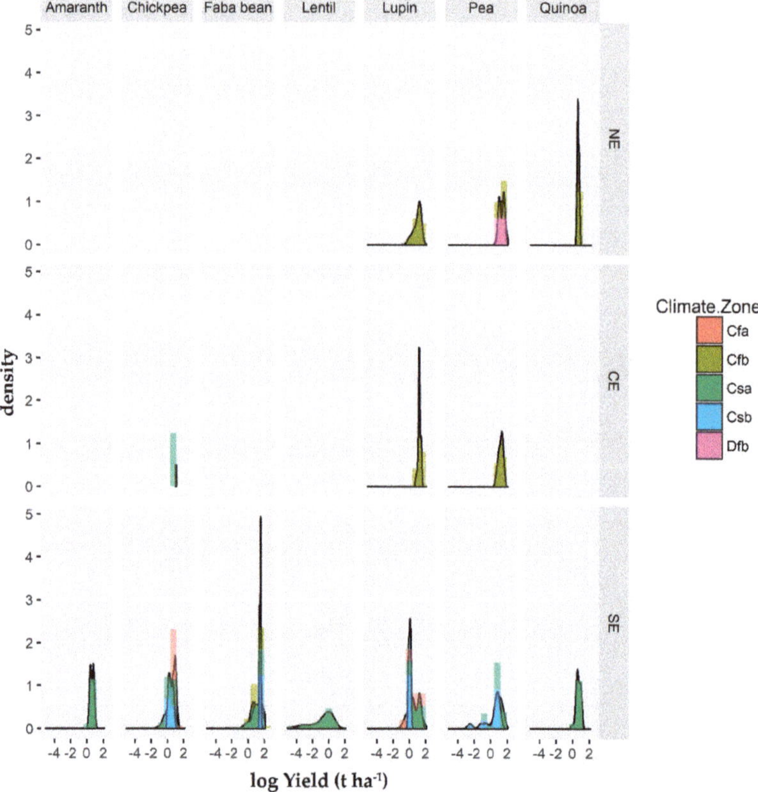

Figure 5. Distribution frequencies of log yield (t ha^{-1}) for all observations (n = 817) by crops (amaranth, chickpea, faba bean, lentil, lupin, pea, and quinoa) and European regions (northern (NE), central (CE), and southern (SE)) grouped by climate zone (Cfa: Humid subtropical; Cfb: Marin–mild winter; Csa: interior Mediterranean; Csb coastal Mediterranean; Dfb: humid continental mild summer, wet all year).

For central Europe, the average yield was relatively constant for different crops (chickpea, lupin, and pea). Lupin accounted for the largest number of observations (n = 46). Average yield in central

Europe for lupin and pea was similar to yields found in northern Europe, and higher than that observed in Southern Europe.

In Southern Europe, there were many different crops of our study (n = 706) with observations available from studies conducted in five countries of the region. Average yields varied from 0.78 t ha^{-1} for lentil to 3.36 t ha^{-1} for faba bean.

The Figure 5 shows that most of the yield observations was found in the south of Europe where all species were present. The plots convey the left skewness in the major part of the yield distributions for all crops, disaggregated into European regions.

3.5. Overall Yield across Factors of Variation

Figure 6 shows that the variation of yield due to environment, agronomic management, and crop factors was quite large.

Response yield varied across the five climatic zones classes with yield values for Dfb and Cfb ranging from 2.47 to 4.90 t ha^{-1} and 0.98 to 8.29 t ha^{-1}, respectively, with a higher percentage of values being below 4.9 and 3.9 t ha^{-1}, respectively (Figure 6a). In contrast, values of yield ranged from 0.64 to 3.99 t ha^{-1}, from 0.01 to 6.86 t ha^{-1} and from 0.08 to 4.87 t ha^{-1} for Cfa, Csa, and Csb, respectively. These results provide strong evidence to suggest that among the five climatic zones categories, Dfb and Cfb were the most productive climatic zones for all protein-crops.

Furthermore, this study found that the whisker values (the maximum and minimum values are displayed with vertical lines connecting the points to the center box) in the boxplot of the three-geographic zone categories can evaluate how yield variation is apportioned. Studies conducted in Northern Europe (NE) and Southern Europe (SE) showed the lowest yields which ranged from 0.98 to 6.64 t ha^{-1}, and 0.01 to 8.29 t ha^{-1}, respectively (Figure 6d). In contrast, values of yield ranged from 2.08 to 5.03 t ha^{-1} for central Europe (CE).

On the other hand, the variation of yield between protein-crops showed that faba bean, pea, and lupin were the most productive crops, with a higher percentage of values being below 4.9, 4.1 and 3.7 t ha^{-1}, respectively, and lentil the least one, with a higher percentage of values being below 1.2 t ha^{-1} (Figure 6b).

Figure 6c shows the yield variation between different agronomic management. Treatments AD, D, E, EF, and F were the agronomic interventions most affecting productive response and Treatment CD was the least one, with yield values ranging from 1.04 to 1.8 t ha^{-1} and with a higher percentage of values being below 1.2 t ha^{-1}.

3.6. Multiple Correspondence Analysis

To obtain a comprehensive view on the protein crop yield variation that occurred in the 44 cultivars, 8 ecotypes, 1 accession, and 14 genotypes as affected by climatic zone, geographic region, and agronomic management, the whole data set (categorical variables (column labels): protein crops, climatic zone, geographic region, and agronomic management); observations (row labels)) was subjected to multiple correspondence analysis (MCA). The first three dimensions components (dim.) explained 56.24% of the cumulative variance for our dataset, with Dim1 accounting for 27.91%, Dim2 for 16.30%, and Dim3 for 12.03% (Table 4). The loading plots (Figure 7) show the graphic display of the contribution and the cluster of variables on the plane defined by the two first axes of the analysis.

Table 4. Eigenvalues, relative and cumulative percentage of total variance for systematic review meta-database with respect to the three-dimension components (Dim1, Dim2, and Dim3)

Dimension Components	Dim1	Dim2	Dim3
Eigenvalue	0.741	0.566	0.486
Relative variance (%) *	27.91	16.30	12.03
Cumulative variance (%)	27.91	44.21	56.24

* Benzécri correction (1973).

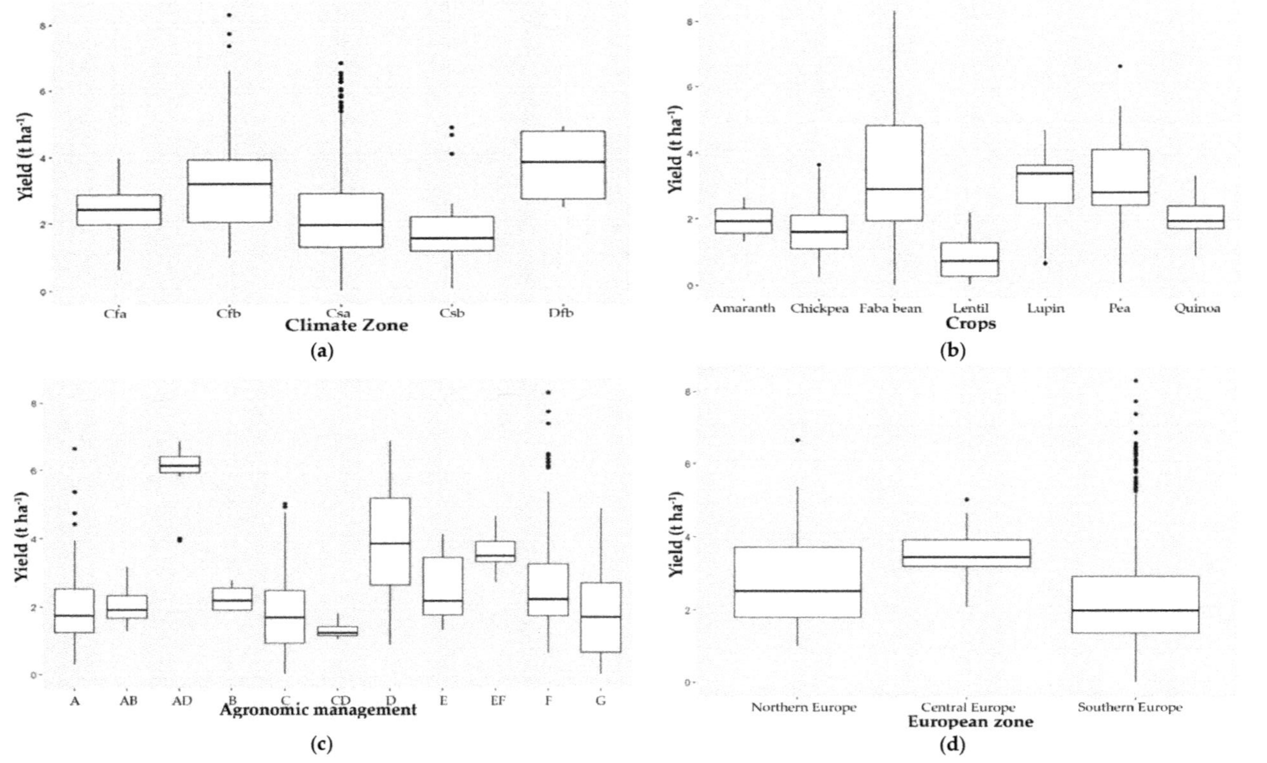

Figure 6. Box-plots of patterns of yield (t ha^{-1}) for all observations (n = 818) across: (**a**) different groups of climatic zones, Cfa: humid subtropical; Cfb: Marin–mild winter; Csa: interior Mediterranean; Csb: coastal Mediterranean.; Dfb: humid continental mild summer, wet all year (**b**) different group of Crops, (**c**) different groups of agronomic management A: deficit irrigation; AB: deficit irrigation and salinity; AD: deficit irrigation and fertilizer; B: salinity; C: tillage; CD: tillage and fertilizer; D: fertilizer; E: sowing density; EF: sowing density and sowing date; F: sowing date; G: weed control; and (**d**) different groups of European zone. Box edges represent the upper and lower quantile with median value shown in the middle of the box. The small circles on the boxplot relate to outliers.

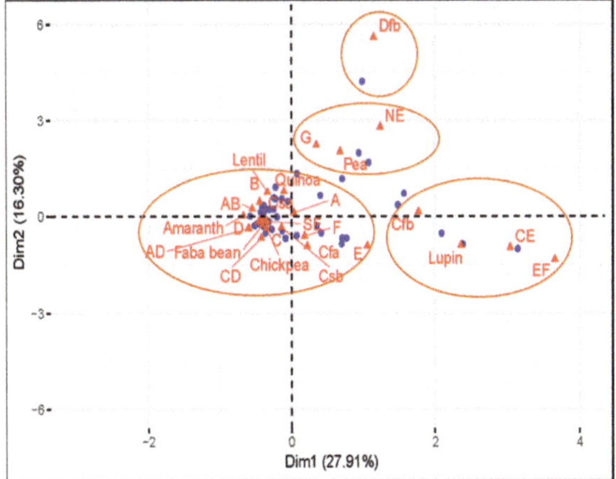

Figure 7. Multiple correspondence analysis plot of protein crops, European environments, and agronomic management factors induced by the first two dimensions components. Protein crops (quinoa, amaranth, pea, faba bean, lupin, chickpea, and lentil), European environments (geographic and climatic zones), and agronomic management (different agronomic practices) are plotted on the two first axes. Rows (observations) are represented by blue points and columns (categorical variables) by red triangles. Orange ellipses indicate different clusters analyzed.

Interpretation of correspondence analysis graphs is based on the distance between such proximate categorical variables (column labels: protein crops, climatic zone, geographic region, and agronomic management) or observations (row labels) to the origin of axes (the larger, the more significant). However, if we want to compare a row label to a column label, we need to:

1. Look at the length of the line connecting the row label to the origin. Longer lines indicate that the row label is highly associated with some of the column labels (i.e., it has at least one high residual).
2. Look at the length of the label connecting the column label to the origin. Longer lines again indicate a high association between the column label and one or more row labels.
3. Look at the angle formed between these two lines. Really small angles indicate association. 90° angles indicate no relationship. Angles near 180° indicate negative associations.

To interpret each axis (factor), it is necessary to consider different indicators, as absolute and relative contributions of each variable category. The former measures the extent to which one modality variable contributes to the determination of a specific factor. The latter is instead a quality indicator as it measures how much each factor contributes to the reproduction of the variable dispersion (Table 5). Particularly, we interpreted each factor considering categorical variables with a high absolute and relative contribution [15].

This path is mostly accounted for by the first (horizontal) axis and is strongly related to climatic zone (cfb) and positively correlated to agronomic management interaction (EF), geographic region (CE), and protein crops (Lupin). The second (vertical) axis was positively correlated to climatic zone (dfb), agronomic management (G), geographic region (NE), and protein crops (Pea) (Figure 7, Table 5).

The proximity between observations corresponds to shared-substance: observations are close to each other because a large proportion of articles treat the same search terms together; they are distant from each other when only a small fraction of articles discusses these search terms together. For example, the variables 'CE' and 'G' are far from each other, because no articles of central Europe discuss the G treatment.

Figure 7 shows that the most of observations has a large number of articles associated to agriculture management (A, AB, AD, B, C, CD, D, E, and F) and climatic zone (Csa, Csb, and Cfa) related to the southern Europe region. This information tells us that studies from southern Europe region represent the center of the research field and tend to resemble each other.

The partitioning in four clusters is represented on the map produced by the first two principal components (Figure 7). The graph shows that the four clusters are well-separated on the first two principal components.

Table 5. Contribution and value test of different sub categorical variables for systematic review meta-database with respect to the three-dimension components (Dim1 and Dim2 and Dim3).

Principal Components		Dim1		Dim2		Dim3	
Categorical variables	*Variables attributes*	Abs. Contr. [a]	V-Test [b]	Abs. Contr. [a]	V-Test	Abs. Contr. [a]	V-Test
	Amaranth	0.45	−3.35	0.001	0.15	**7.66**	−11.19
	Chickpea	1.15	−6.22	1.66	−6.54	**2.74**	16.78
	Faba bean	**1.95**	−8.51	0.98	−5.27	1.93	−6.86
Protein crops	Lentil	0.20	−2.24	**1.36**	5.15	0.01	0.41
	Lupin	**19.17**	22.72	3.49	−8.47	0.06	−1.03
	Pea	1.12	5.40	**13.22**	16.23	3.40	7.62
	Quinoa	0.05	−1.11	3.26	8.25	**9.30**	−12.93
European countries	Northern Europe	3.43	9.43	**22.96**	21.32	0.13	1.50
	Central Europe	**21.75**	23.78	2.79	−7.45	0.63	−3.28
	Southern Europe	**3.38**	−24.52	0.72	−9.92	0.02	1.36
	Cfa	0.06	1.27	1.53	-5.44	**2.72**	6.71
	Cfb	**17.60**	22.60	0.18	1.99	0.50	−3.07
Climatic zone	Csa	**4.57**	−19.90	0.16	−3.27	1.93	−10.48
	Csb	0.04	−0.99	0.28	−2.34	**17.23**	17.03
	Dfb	0.66	4.01	**20.34**	19.54	1.45	4.83
	A	0.003	0.31	0.12	1.59	**2.07**	6.26
	AB	0.84	−4.71	0.22	2.11	**15.55**	−16.40
	AD	**0.18**	−2.10	0.09	−1.27	0.24	−1.99
	B	0.05	−1.10	**0.07**	1.14	0.90	−3.79
	C	1.25	−6.40	0.19	−2.17	**0.63**	3.67
Agronomic management	CD	0.12	−1.71	0.38	−2.68	**12.98**	14.50
	D	1.31	−6.21	0.22	−2.23	**3.03**	−7.65
	E	**1.48**	6.11	1.41	−5.21	2.09	5.86
	EF	18.73	21.74	3.23	−7.90	1.41	−4.83
	F	0.12	1.80	1.78	−6.09	**1.30**	4.81
	G	0.34	3.00	**19.37**	19.80	0.11	1.38

Boldface contributions indicate the most relevant characters for each dimension component. Different climatic zones: Cfa: humid subtropical; Cfb: Marin–mild winter; Csa: interior Mediterranean; Csb coastal Mediterranean; Dfb: humid continental mild summer, wet all year, different group of agronomic management A: deficit irrigation; AB: deficit irrigation and salinity; AD: deficit irrigation and fertilizer; B: salinity; C: tillage; CD: tillage and fertilizer; D: fertilizer; E: sowing density; EF: sowing density and sowing date; F: sowing date; G: weed control; [a] Absolute contribution, [b] Value test: V-test statistic asymptotically follows a standard gaussian distribution, a value below −1.96 or above 1.96 indicates that the category has a coordinate significantly different to 0 and each category has positive or negative value for each dimension.

4. Discussion

The area of protein crops in Europe declined almost continuously over the last five decades, from 5.8 million ha in 1961 (4.7% of the arable area), when recording began, to 2.0 million ha in 2014 (1.6% of the arable area) [26]. A major underlying driver behind the reduction in the proportion of arable land used for protein crops is the increased comparative advantage in the production of starch-rich cereals in Europe over the production of protein-rich grain legumes [27].

Protein crops may produce high-quality protein for food and feed, increasing soil fertility and yields in subsequent crops, potentially reducing greenhouse gas emissions and supporting biodiversity [27]. However, there are several reasons why farmers may be hesitant to grow them: the quality and quantity of the yield, and thus the financial return, vary depending on the region and weather conditions [28].

4.1. European Environments

According to Reckling et al. [29], especially in the north of Europe, where all grain legumes are spring-sown, they have generally more unstable yield than those of autumn-sown crops, because they can be constrained by water deficits during crop establishment and subsequent growth stages. Winter crops are instead established in autumn and regrow quickly after winter without any delays due to soil tillage and seedbed preparation that can also reduce soil moisture. Another factor that can contribute to yield instability in grain legumes is the indeterminate growth habit that allows the crop to respond to good conditions such as high-water availability and adequate temperature or to stop growing and reproducing under poor conditions [30]. Differently from legumes, cereals can compensate in conditions of sufficient or insufficient water and nutrient supply through modifications in tillering and flower initiation. Finally, the symbiotic nitrogen fixation affects yield and can be reduced or fail in poor conditions resulting in greater yield instability.

According to De Visser [31], the large diversity of pedoclimatic conditions all over Europe, and also the different end user needs (food, feed/ruminants, feed/monogastrics), determine which protein crops are most easily adopted. Therefore, most of the published cropping systems studies in Europe, with a high reporting standard suitable for a SR, are concentrated in Southern Europe; perhaps because the manufacturers do not consider these protein crops as having enough of a market to justify investing in research and development in all European countries.

The most commonly studied climate zones for protein crop cultivation were geographical areas with a marine mild climate with no dry season and warm summer (Cfb)—most of them in Demark, France, and Poland—and areas with a warm temperate climate with dry and hot summer (Csa) especially in southern Europe. According to Malezieux et al. [32], in some parts of Europe where winters are not severe, autumn-sown grain legumes can be grown. In Mediterranean regions, grain legumes, such as pea, are grown as a cool-season crop harvested before summer drought, or before planting an irrigated, warm-season crop such as maize.

4.2. Protein Crops

The most commonly studied crops were faba bean and chickpea. Beans are preferred by arable farmers because this crop is easier to grow, and has more steady yield levels [7].

The most important parameters to be considered by growers are straw height, earliness of ripening, disease resistance, and yield. Earliness of ripening is important in particular, under North European conditions to anticipate harvest and avoid cooler weather when drying of the crop in the field is slow [33].

In our review, we have focused on yield, as it is the most important parameter, but there is still lack of knowledge concerning protein content in Europe.

In south of Europe, studies on lentil, lupin, and faba bean showed the most variable yields with a coefficient of variation (CV) value of 77%, 54%, and 53% respectively (Table 2), indicating that broad-leaved crops indeed have more unstable yields than cereals [27]. In the south, average yield was 0.78 t ha^{-1} for lentil and 3.36 t ha^{-1} for faba bean. Highest yields were identified for pea and lupin in northern Europe. For central Europe, the average yield was relatively constant for different crops (chickpea, lupin, and pea).

The variation of yield between protein crops showed that faba bean, pea, and lupin were the most productive crops, and lentil the lowest yielding.

At the European scale, the yield levels of field broad beans seem to be showing much potential in North West Europe [34] and the yield level of field peas is high, while lupins yield are considerably lower and consequently lupin crops are considered less attractive [7].

White lupin (*Lupinus albus* L.), yellow lupin (*L. luteus* L.), and narrow-leaved lupin (*L. angustifolius* L.), are native European legumes with a seed protein content high (up to 44%) [35]. Lupin is mainly cultivated in North and Central Europe [36], but the results from this study show that it could be cultivated also in south European countries. Lupin was cultivated in ancient times but is currently neglected [37].

Among the grain legume species, peas are the main grain legume produced in Europe: they can be grown almost anywhere [34]. Their high yield potential makes profitable use of fertile soils, most of the new varieties are easy to harvest, and peas can be used for several purposes [7].

There is a large variation in yield across Europe and across the different protein crops. Protein crops suffer from yield instability compared to cereals or rapeseeds, and yield fluctuations are one of the main reasons farmers give for not growing these crops [10]. The latter is a major obstacle in further expansion and a main target for improving protein crop production. According to Stoddard et al. [38], a lack of breeding resources (indeterminate growth habit, stress resistance, etc.) and knowledge gaps (low agronomic expertise, insufficient cooperation between farmers and other actors, etc.) are responsible for the fact that only 1.6% of EU arable land is currently used for legumes, despite their agronomic and environmental benefits.

4.3. Agronomic Managements

Recent studies outline a comparative lack of breeding investment in Europe to improve protein crops adaptation to local agroclimatic conditions and management techniques [12,39,40], such as crop protection or density and plant spacing, irrigation, tillage, fertilization, and harvesting techniques [41–43]. It is important to highlight that a general lack of specific agronomic references to manage protein crops may be a barrier for farmers cultivating these crops in Europe [44].

In our SR, tillage (n = 211), fertilizer (n = 146), deficit irrigation (n = 130), sowing density, and sowing date (both the last treatments n = 158) were the most productive agronomic management for protein crops in European countries.

According to Christopher and Lal [45], the cultivation and cropping may cause significant soil organic carbon (SOC) losses through decomposition of humus. The shift from pasture to cropping systems can lead to loss of soil C stocks between 25 and 43 percent [46]. Furthermore, the EU's agricultural policy has rewarded a wider range of options to increase soil carbon [47].

There is a general agreement on the influence of grain legumes on the properties of rhizosphere in terms of N supply, SOC, and P availability [48], the extent of the impact varied across legume species, soil properties, and climate conditions.

The soil type is the main determinant of plant growth, nutrient dynamics of the rhizosphere and microbial community structure. The depletion and accumulation of some macro- and micronutrients also differed between crop systems (i.e., monoculture, crop rotations) and soil management strategies (i.e., conventional tillage, conservation agriculture).

Soil tillage methods have complex effects on physical, chemical properties of soil, which alter in turn the biological properties. Protein crops, and in particular grain legumes, possess certain characteristics particularly suitable for sustainable cropping systems and conservation agriculture, and making them functional either as cash crop or as crop residue [49]. Conservation agriculture (CA) is based on minimal soil disturbance and permanent soil cover combined with rotations [50] and in Europe, the CA is applied in regions where soil erosion mitigation and protection against land degradation are important objectives.

Also, legumes' biological fixation of atmospheric N_2 can be affected by "starter-N" and tillage; Torabian et al. [51–53] showed that conservation tillage typically enhances nodulation and nitrogen fixation, through increased soil moisture retention and soil temperature, and increased soil microbial biomass. According Krishna [54], there are several legumes as pea, faba bean, soy bean, and forage legumes that need no fertilizer-N supply, perhaps except a starter-N in some locations to induce rapid rooting at seedling stage. Kitamura [55] showed that legumes preferentially use available soil nitrogen rather than fix atmospheric nitrogen. Thus, high levels of available soil nitrogen will greatly reduce the amount of nitrogen fixed by the legume. However, in low nitrogen soils, a low rate of starter-N placed away from the seed may boost seedling growth of the legume prior to the establishment of fully functioning nodules. On the other hand, legume-based systems improve various aspects of soil fertility, including the amount of nitrogen fixed into the soil and the high quality of the organic matter released to the soil in term of C/N ratio [49].

Many countries already depend on conservation agriculture. The SR research results show that the grain legumes like lentil, chickpea, pea, and faba bean play a major role in conservation agriculture in Spain, Italy, and Turkey (96% of evidences of tillage treatments in SE).

According to Stagnari et al. [49], the expansion of ecological approaches such as conservation agriculture opens up opportunities for the use of food legumes in sustainable cropping systems. In general, conservation agriculture is an environmentally sustainable production system that can increase the incorporation of grain legumes within large and small-scale farming.

Protein crops require an adequate supply of readily available nutrients for optimum growth and yield [56].

According to Da Silva et al. [57], considerable N is required in grain legumes at the beginning of pod fill as the translocation of N from vegetative parts to the pods is intensive. Studies on soybean have shown that this N drain may be high enough to decrease photosynthetic activity [58], induce premature leaf senescence, and reduce root activity [59,60]. Nutrients supplied through leaves may supplement rapidly those transferred from stems and roots, thus avoiding early leaf senescence [61].

By means of the inoculation of rhizobia or the application of N fertilizer with the use of ^{15}N labeled urea, during the late stages of growth, it was possible to enhance grain legume yields without necessarily inhibiting N_2 fixation [56]. According to the screening review results, these two methods were the most frequently studied treatment on SE especially in Italy and Spain.

Since indigenous rhizobia are not always in sufficient numbers, effective enough, or compatible with the specific legume crop to stimulate biological nitrogen fixation (BNF) and increase yields, inoculation of legumes with rhizobia is an important option for enhancing BNF in crop production systems [62]. The effectiveness of BNF is affected by agro-ecological factors. For instance, poor nodulation and poor plant vigor in beans grown in soil with low extractable P led to a poor BNF [63]. The idea of applying N through leaves to maximize bean yield is not recent [64,65]. The advantage of using urea is that it facilitates the accumulation of other nutrients such as Mn^{2+} and permits the transport of nutrients through a more permeable cell cuticle [66].

Seed yields of high-quality protein crops like quinoa strongly respond to N fertilization [67]; for this reason, the main field trials carried out in Europe during last years to evaluate adaptability of different quinoa varieties were also focused on the assessment of nitrogen requirements [68].

Sowing date is one of the most important management factors affecting protein crops production and quality [69]. In a given region, the optimum sowing date depends mainly upon the timing of rainfall [70]. In most cases, delaying sowing beyond the optimum period reduces crops yields [71,72]. As a consequence, delaying sowing date can cause significant differences of environmental conditions during grain filling, usually causing grains to grow with increasing temperatures and diminishing moisture conditions [73,74].

A number of studies, from our SR, has reported on the optimum sowing dates for legumes. Yield variation as a function of sowing date and sowing density ranged from 1.01 to 3.65 t ha^{-1} and from 1.45 to 2.29 t ha^{-1}, respectively. Furthermore, the early winter sowing date in Spain, and March sowing date in Greece, seemed to ensure the highest response for chickpea in the southern Europe. In general, early sowing resulted in seed yield increases and there was a frequency for seed yield to decrease with delay of sowing [56]. Quinoa grown under Mediterranean conditions produce higher seed yields if sown in April compared to May [75].

In addition, the sowing density is an important factor-affecting yield of grain legumes according to many studies, from inside and outside EU [76–78]. Therefore, yield response of seed legumes to seeding rates was discussed by several authors, and a significant effect of seeding rate on seed yield was found [79,80].

Water resources in the EU and especially in Mediterranean region become more and more scarce because of high demand for water due to population growth and increasing demand for food. Climate change has also aggravated the situation because of erratic rainfall and the succession of drought years [81].

This challenge is further compounded by the severe competition for land and water from industry and urban development [82]. Such competition pushes agriculture to marginal areas, where water-limiting conditions often constrain crop productivity. In these marginal areas (e.g., semi-arid environments), water limitation and year to year fluctuations of meteorological conditions tend to be large, and these variations significantly affect food security in rain-fed systems [83].

Droughts can negatively impact the yield of most protein crops, from C4 plant (e.g., amaranth) to C3 (e.g., quinoa and legumes) [84–86]. The yield of food legumes grown in arid to semi-arid environments or drylands such as the Mediterranean (e.g., faba beans, chickpea, and lentil) is usually variable or low due to terminal droughts that characterize these areas [87,88].

Currently, the economically viable approaches to support crop production under drought are still limited [89]. More importantly, it remains unclear how the impact of drought on legume production varies with legume species, regions, agroecosystems, soil texture, and drought timing.

The results of our systematic review and drought manipulation experiments across the EU region will allow to better characterize the factors that determine the magnitude of yield loss in legumes due to drought stress, which must be considered in agricultural planning to increase the resilience of legume production systems.

5. Conclusions

This systematic review details the setting for a large number of studies across a broad range of agronomic management, protein crops, and geographical locations.

The EU depends on imported high-protein plant products. The main reason for this is that protein crops in the EU are not competitive with the crops currently being produced. However, the competitiveness of crops can be expected to differ between regions within Europe because local conditions have a high influence on yield levels [3]. In addition to higher yield potential of cereals in Europe, farmers who grow protein crops face up to a range of agronomic challenges. The evidence from this SR confirms that protein crop yields are considered unstable, as pointed out in many studies [10,43,90].

Therefore, as in most of the cultivated areas, legume yield is unstable, legume production is limited. This constraint is explained by many other external factors, such as the historical dependence to the Common Agricultural Policy measures; low level of production of processed products; competition with soybean imports from the Americas or with protein-rich byproducts derived from non-legume crops. It is also possible that European farmers are not motivated to cultivate these species on soils of good quality in appropriate environments, and that they prefer growing more profitable major crops in these environments (e.g., wheat, maize, and rapeseed).

These agronomic challenges highlight a need for research to support crop development in order to increase and stabilize yields in relation to those of other crops. To do this we should answer this question: What are the main research needs (technical, social-economic) for protein crops to be competitive?

A useful approach to achieve this aim could be represented by exchange between several regions across Europe which will also contribute to increase the profitability of EU protein crops. Sharing knowledge on the use of varieties and best practices, even worst practices, is a key to success.

Access to practical knowledge and best practices of protein crop production is required. A useful approach to achieve this aim could be represented by a meta-regression analysis of mixed effects using the relative yield as effect size estimator. An effect size estimator is an index which allows us to compare the experimental treatment mean to the control treatment mean [91] and to quantify the magnitude of a treatment's effect. Because all of the studies we used did not report any measure of variance, an unweighted meta-regression could be performed. The meta-regression analysis would allow to investigate the interaction between different protein-crops, European region, and agronomic management and the effect of these factors on yield response.

Supplementary Materials: The following are available online at http://www.mdpi.com/2073-4395/9/6/292/s1.

Author Contributions: Conceptualization, M.H.S.; Methodology, M.H.S. and M.A.; Formal analysis, M.H.S. and M.A.; Investigation, M.H.S.; Resources, M.H.S.; Data curation, M.H.S.; Writing—original draft preparation, M.H.S.; Writing—review and editing, M.H.S., A.M.S., C.P., and A.L.; Visualization, M.H.S.; Supervision, A.L.

Funding: This research has been undertaken as part of the PROTEIN2FOOD project. This project has received funding from the European Union's Horizon 2020 research and innovation program under grant agreement no. 635727.

Acknowledgments: We are grateful to Maurizio Buonanno (CNR-ISAFOM) for plotting the map of selected studies to be included in the systematic review.

Conflicts of Interest: The authors declare no conflict of interest.

References

1. McCarthy, E.; Stephen, N.; O'Sullivan, S.N. How to feed the world in 2050? STOA Workshop 2013, European Parliament, Brussels. Available online: https://www.europarl.europa.eu/stoa/cms/home/events/workshops/feeding (accessed on 13 February 2019).

2. Van Gelder, J.W.; Herder, A. Soja Barometer 2012; Een Onderzoeksrapport voor de Nederlandse Sojacoalitie. Available online: https://www.bothends.org/nl/Actueel/Publicaties/Soy-Barometer-2012-English-version- (accessed on 13 February 2019).

3. Bues, A.; Preißel, S.; Reckling, M.; Zander, P.; Kuhlman, K.; Topp, K.; Watson, K.; Lindström, K.; Stoddard, FL.; Murphy-Bokern, D. Protein crops and their role in Europe. In *The Environmental Role of Protein Crops in the New Common Agricultural Policy*; Study 2013; European Union: Brussels, Belgium, 2013; Available online: http://www.europarl.europa.eu/studies (accessed on 24 June 2018).

4. Laaninen, T. *Imports of GM Food and Feed: Right of Member States to Opt Out 2015*; European Parliamentary Research Service: Brussels, Belgium, 2015; Available online: https://www.europarl.europa.eu/etudes/BRIE (accessed on 26 April 2019).

5. Halford, N.G. Legislation governing genetically modified and genome-edited crops in Europe: The need for change. *J. Sci. Food Agric.* **2019**, *99*, 8–12. [CrossRef] [PubMed]

6. Häusling, M. *The EU Protein Deficit: What Solution for a Long-Standing Problem?* 2010/2111 (INI); Committee on Agricultural and Rural Development, European Parliament: Brussels, Belgium, 2011.

7. De Visser, C. Starting paper Protein Crops. In *EIP-AGRI Focus Group Protein Crops: Final Report 2013*; European Commission: Brussels, Belgium, 2013; Available online: https://ec.europa.eu/eip/agriculture/en/publications/eip-agri-focus-group-protein-crops-final-report (accessed on 24 June 2018).

8. FAOstat. Statistics Database of the Food and Agriculture Organization of the United Nations 2017. Available online: http://www.fao.org/statistics/databases/en/ (accessed on 26 April 2019).

9. Bazile, D.; Jacobsen, S.E.; Verniau, A. The global expansion of quinoa: Trends and limits. *Front. Plant Sci. Sect. Crop Sci. Hortic.* **2016**, *7*, 622. [CrossRef] [PubMed]

10. Von Richthofen, J.-S.; Pahl, H.; Bouttet, D.; Casta, P.; Cartrysse, C.; Charles, R.; Lafarga, A. What do European farmers think about grain legumes? *Grain Legume* **2006**, *45*, 14–15. Available online: http://www.ias.csic.es/grainlegumesmagazine/ (accessed on 25 June 2018).

11. Voisin, A.S.; Guéguen, J.; Huyghe, C.; Jeuffroy, M.H.; Magrini, M.B.; Meynard, J.M.; Mougel, C.; Pellerin, S.; Pelzer, E. Legumes for feed, food, biomaterials and bioenergy in Europe: A review. *Agron. Sustain. Dev.* **2014**, *34*, 361–380. [CrossRef]

12. Stoddard, F.L. The Case Studies of Participant Expertise in Legume Futures; Legume Futures Report 1.2. 2013. Available online: http://www.legumefutures.de/ (accessed on 13 February 2019).

13. Sargeant, J.M.; Rajic, A.; Read, S.; Ohlsson, A. The process of systematic review and its application in agri-food public-health. *Prev. Vet. Med.* **2006**, *75*, 141–151. [CrossRef]

14. Sargeant, J.M.; O'Connor, AM. Introduction to systematic reviews in animal agriculture and veterinary medicine. *Zoonoses Public Health* **2014**, *61*, 3–9. [CrossRef] [PubMed]

15. Cuccurullo, C.; Aria, M.; And Sarto, F. Foundations and trends in performance management. A twenty-five years bibliometric analysis in business and public administration domains. *Scientometrics* **2016**, *108*, 595–611. [CrossRef]

16. Curtis, P.S.; Wang, X.Z. A meta-analysis of elevated CO_2 effects on woody plant mass, from and physiology. *Ecologia* **1998**, *113*, 299–313.

17. Rohatgi, A. WebPlotDigitizer 2011. Available online: http://arohatgi.info/WebPlotDigitizer/app3_12/ (accessed on 30 December 2017).

18. Lê, S.; Josse, J.; Husson, F. FactoMineR: An R package for multivariate analysis. *J. Stat. Softw.* **2008**, *25*, 1–18. [CrossRef]

19. Greenacre, M.; Hastie, T. The geometric interpretation of correspondence analysis. *J. Am. Stat. Assoc.* **1987**, *82*, 437–447. [CrossRef]

20. Benzécri, J.-P. Sur le calcul des taux d'inertie dans l'analyse d'un questionnaire: Addendum et Erratum à (Bin. Mult.). *Les Cahiers de l'Analyse des Données* **1979**, *4*, 377–378.

21. Kostov, B.; Bécue-Bertaut, M.; Husson, F. "Correspondence Analysis on Generalised Aggregated Lexical Tables (CA-GALT)" in the FactoMineR Package. *R. J.* **2015**, *7*, 109–117. [CrossRef]

22. R Core Team. *R: A Language and Environment for Statistical Computing*; R Foundation for Statistical Computing: Vienna, Austria, 2013.

23. Giambalvo, D.; Ruisi, P.; Saia, S.; Di Miceli, G.; Frenda, A.A.; Amato, G. Faba bean grain yield, N2 fixation, and weed infestation in a long-term tillage experiment under rainfed Mediterranean conditions. *Plant Soil* **2012**, *360*, 215–227. [CrossRef]

24. López-Bellido, R.J.; Lopez-Bellido, L.; Lopez-Bellido, F.J.; Castillo, J.E. Faba Bean (*Vicia faba* L.) Response to Tillage and Soil Residual Nitrogen in a Continuous Rotation with Wheat (*Triticum aestivum* L.) under Rainfed Mediterranean Conditions. *Agron. J.* **2003**, *95*, 1253–1261.

25. López -Bellido, L.; Lopez-Bellido, R.J.; Castillo, J.E.; Lopez-Bellido, F.J. Chickpea response to tillage and soil residual nitrogen in a continuous rotation with wheat. I. Biomass and seed yield. *Field Crop. Res.* **2004**, *88*, 191–200. [CrossRef]

26. FAOstat. Statistics Database of the Food and Agriculture Organization of the United Nations 2016. Available online: http://www.fao.org/statistics/databases/en/ (accessed on 30 October 2016).

27. Watson, C.A.; Reckling, M.; Preisse, S.; Bachinger, J.; Bergkvist, G.; Kuhlman, T.; Lindström, K.; Nemecek, T.; Topp, C.F.E.; Vanhatalo, A.; et al. *Chapter Four—Grain Legume Production and Use in European Agricultural Systems*. Advances in Agronomy, 1st ed.; Sparks, D.L., Ed.; Zoe Kruze: Cambridge, MA, USA, 2017; Volume 144, pp. 235–303.

28. Van Hoye, I. Growing Protein Crops Has a Profitable Future. Available online: https://ec.europa.eu/eip/agriculture/sites/agri-eip/files/2015-press20150611-ws_protein_crops_fin.pdf (accessed on 24 June 2018).

29. Reckling, M.; Döring, T.F.; Bergkvist, G.; Stoddard, F.L.; Watson, C.A.; Seddig, S.; Chmielewski, F.M.; Bachinger, J. Grain legume yields are as stable as other spring crops in long-term experiments across northern Europe. *Agron. Sustain. Dev.* **2018**, *38*, 63. [CrossRef] [PubMed]

30. Stoddard, F.L.; Balko, C.; Erskine, W.; Khan, H.R.; Link, W.; Sarker, A. Screening techniques and sources of resistance to abiotic stresses in cool-season food legumes. *Euphytica* **2006**, *147*, 167–186. [CrossRef]

31. De Visser, C. Protein Crops. Available online: https://ec.europa.eu/eip/agriculture/sites/agri-eip/files/fg2_protein_crops_starting_paper_2013_en.pdf (accessed on 24 June 2018).

32. Malezieux, E.; Crozat, Y.; Dupraz, C.; Laurans, M.; Makowski, D.; Ozier-Lafontaine, H.; Rapidel, B.; de Tourdonnet, S.; Valantin-Morison, M. Mixing plant species in cropping systems: Concepts, tools and models. A review. *Agron. Sustain. Dev.* **2009**, *29*, 43–62. [CrossRef]

33. Taylor, B.R.; Cormack, W.F. Choice of Cereal and Pulse Species and Varieties. In *Organic Cereals and Pulses*; Papers presented at conferences held at the Heriot-Watt University, Edinburgh, and at Cranfield University Silsoe Campus, Bedfordshire, 6 and 9 November 2001 (pp. 9–28); Chalcombe Publications: Southampton, UK, 2002.

34. EIP-AGRI. How to make protein crops profitable in the EU? 2015, Workshop. Budapest. Available online: https://ec.europa.eu/eip/agriculture/sites/agri-eip/files/field_event_attachments/report_ws_proteincrops_final_13022015.pdf (accessed on 25 June 2018).

35. Lucas, M.M.; Stoddard, F.L.; Annicchiarico, P.; Frías, J.; Martínez-Villaluenga, C.; Sussmann, D.; Pueyo, J.J. The future of lupin as a protein crop in Europe. *Front. Plant Sci.* **2015**, *6*, 705. [CrossRef]

36. Gresta, F.; Wink, M.; Prins, U.; Abberton, M.; Capraro, J.; Scarafoni, A.; Hill, G. Lupins. European cropping systems. In *Legumes in Cropping Systems*; CABI: Oxfordshire, UK, 2017; pp. 88–108.

37. Yeheyis, L.; Kijora, C.; Melaku, S.; Girma, A.; Peters, K.J. White lupin (*Lupinus albus* L.), the neglected multipurpose crop: Its production and utilization in the mixed crop-livestock farming system. *Livest. Res. Rural Dev.* **2010**, *22*, 74.

38. Stoddard, F.; Hovinen, S.; Kontturi, M.; Lindstrom, K.; Nykanen, A. Legumes in Finnish agriculture: History, present status and future prospects. *Agric. Food Sci.* **2009**, *18*, 191–205. [CrossRef]

39. Annicchiarico, P.; Iannucci, A. Breeding strategy for faba bean in southern Europe based on cultivar responses across climatically contrasting environments. *Crop Sci.* **2008**, *48*, 983–991. [CrossRef]

40. Lizarazo, C.I.; Lampi, A.M.; Liu, J.; Sontag-Strohm, T.; Piironen, V.; Stoddard, F.L. Nutritive quality and protein production from grain legumes in a boreal climate. *J. Sci. Food Agric.* **2015**, *95*, 2053–2064. [CrossRef] [PubMed]

41. Corre-Hellou, G.; Crozat, Y. N_2 fixation and N supply in organic pea (*Pisum sativum* L.) cropping systems as affected by weeds and peaweevil (*Sitona lineatus* L.). *Eur. J. Agron.* **2005**, *22*, 449–458. [CrossRef]

42. Jensen, E.S.; Peoples, M.B.; Hauggaard-Nielsen, H. Faba bean in cropping systems. *Field Crops Res.* **2010**, *115*, 203–216. [CrossRef]

43. Flores, F.; Nadal, S.; Solis, I.; Winkler, J.; Sass, O.; Stoddard, F.L.; Link, W.; Raffiot, B.; Muel, F.; Rubiales, D. Faba bean adaptation to autumn sowing under European climates. *Agron. Sustain. Dev.* **2012**, *32*, 727–734. [CrossRef]

44. Cernay, C.; Ben-Ari, T.; Pelzer, E.; Meynard, J.-M.; Makowski, D. Estimating variability in grain legume yields across Europe and the Americas. *Sci. Rep.* **2015**, *5*, 11171. [CrossRef]

45. Christopher, S.F.; Lal, R. Nitrogen management affects carbon sequestration in North American cropland soils. *Crit. Rev. Plant Sci.* **2007**, *26*, 45–64. [CrossRef]

46. Soussana, J.F.; Loiseau, P.; Vuichard, N.; Ceschia, E.; Balesdent, J.; Chevallier, T.; Arrouays, D. Carbon cycling and sequestration opportunities in temperate grasslands. *Soil Use Manag.* **2004**, *20*, 219–330. [CrossRef]

47. Freibauer, A.; Rounsevell, MD.; Smith, P.; Verhagen, J. Carbon sequestration in the agricultural soils of Europe. *Geoderma* **2004**, *122*, 1–23. [CrossRef]

48. Jensen, E.S.; Peoples, M.B.; Boddey, R.M.; Gresshoff, P.M.; Hauggaard-Nielsen, H.; Alves, B.J.; Morrison, M.J. Legumes for mitigation of climate change and the provision of feedstock for biofuels and biorefineries. A review. *Agron. Sustain. Dev.* **2012**, *32*, 329–364. [CrossRef]

49. Stagnari, F.; Maggio, A.; Galieni, A.; Pisante, M. Multiple benefits of legumes for agriculture sustainability: An overview. *Chem. Biol. Technol. Agric.* **2017**, *4*, 1–13. [CrossRef]

50. Hobbs, P.R.; Sayre, K.; Gupta, R. The role of conservation agriculture in sustainable agriculture. *Philos. Trans. R. Soc. Lond. Ser. B* **2008**, *363*, 543–555. [CrossRef] [PubMed]

51. Torabian, S.; Farhangi-Abriz, S.; Denton, M.D. Do tillage systems influence nitrogen fixation in legumes? A review. *Soil Tillage Res.* **2019**, *185*, 113–121. [CrossRef]

52. Okoth, J.O.; Mungai, N.W.; Ouma, J.P.; Baijukya, F.P. Effect of tillage on biological nitrogen fixation and yield of soybean ('*Glycine max* L. Merril') varieties. *Aust. J. Crop Sci.* **2014**, *8*, 1140.

53. Wiraguna, E. Enhancement of Fixation Nitrogen in Food Legumes. *J. Agric. Stud.* **2016**, *4*. [CrossRef]

54. Krishna, K.R. *Agricultural Prairies: Natural Resources and Crop Productivity*; CRC Press: Boca Raton, FL, USA, 2015; 514p.

55. Kitamura, Y. Use of starter nitrogen for establishing tropical legume-grass pasture. *Jarq-Japan Agric. Res. Q.* **1982**, *15*, 253–260.

56. Turk, M.A.; Tawaha, A.R.M. Impact of seeding rate, seeding date, rate and method of phosphorus application in faba bean (*Vicia faba* L. minor) in the absence of moisture stress. *Biotechnol. Agron. Soc. Environ.* **2002**, *6*, 171–178.

57. Da Silva, P.M.; Tsai, S.M.; Bonetti, R. Response to Inoculation and N Fertilization for Increased Yield and Biological Nitrogen Fixation of Common Bean (*Phaseolus vulgaris* L.). In *Enhancement of Biological Nitrogen Fixation of Common Bean in Latin America*; Springer: Dordrecht, The Netherlands, 1993; pp. 123–130.

58. Dunphy, E.J.; Hanway, J.J. Water-soluble carbohydrate accumulation in soybean plants. *Agron. J.* **1976**, *68*, 697–700. [CrossRef]

59. Harper, J.E. Soil and symbiotic nitrogen requirements for optimum soybean production. *Crop Sci.* **1974**, *14*, 255–260. [CrossRef]

60. Garcia, L.; Hanway, J.J. Foliar Fertilization of Soybeans During the Seed-filling Period 1. *Agron. J.* **1976**, *68*, 653–657. [CrossRef]

61. Frith, G.J.T.; Dalling, M.J. The Role of Peptide Hydrolases in Leaf Senescence. In *Senescence in Plants*; K V Thimann: Boca Raton, FL, USA, 1980; pp. 117–130.

62. Giller, K.E. *Nitrogen Fixation in Tropical Cropping Systems*, 2nd ed.; CABI Publishing: Wallingford, UK, 2001; pp. 271–297.

63. Amijee, F.; Giller, K.E. Environmental constraints to nodulation and nitrogen fixation of *Phaseolus vulgaris* L. in Tanzania, I. A survey of soil fertility, root nodulation and multilocational responses to rhizobium inoculation. *Afr. J. Crop Sci.* **1998**, *6*, 159–169. [CrossRef]

64. Miysaka, S.; Freire, E.S.; Mascarenhas, H.S. Modo e epoca de aplicacao de nitrogenio na cultura do feijoeiro. *Bragantia* **1963**, *22*, 511–519. [CrossRef]

65. Muraoka, T.; Neptune, A.M.L. Efeito da aplicacao foliar de polifosfato, superfosfato, ureia e yogen na producao de feijoeiro (*Phaseolus vulgaris* L.). *Rev. Bras. Cienc. Solo* **1978**, *2*, 51–55.

66. Malavolta, E. *Elementos de Nutricion de Plantas*; Ceres: Primavera do Leste, Brazil, 1981; pp. 80–94.

67. Kaul, H.P.; Kruse, M.; Aufhammer, W. Yield and nitrogen utilization efficiency of the pseudocereals amaranth, quinoa, and buckwheat under differing nitrogen fertilization. *Eur. J. Agron.* **2005**, *22*, 95–100.

68. Razzaghi, F.; Plauborg, F.; Jacobsen, S.E.; Jensen, C.R.; Andersen, M.N. Effect of nitrogen and water availability of three soil types on yield, radiation use efficiency and evapotranspiration in field-grown quinoa. *Agr. Water Manag.* **2018**, *109*, 20–29. [CrossRef]

69. McLeod, J.G.; Campbell, C.A.; Dyck, F.B.; Vera, C.L. Optimum seeding dates of winter wheat in southwestern Saskatchewan. *Agron. J.* **1992**, *84*, 86–90. [CrossRef]

70. Jackson, L.F.; Dubcovsky, J.; Gallagher, L.W.; Wennig, R.L.; Heaton, J.; Vogt, H.; Gibbs, L.K.; Kirby, D.; Canevari, M.; Carlson, H.; et al. Regional Barley and Common and Durum Wheat Performance Tests in California 2000; University of California, Agronomy Progress Report No. 272. Available online: http://agric.ucdavis.edu/crops/cereals/2000/oct2000.htm (accessed on 25 June 2018).

71. Anderson, W.K.; Smith, W.R. Yield advantage of two semi-dwarf compared with two tall wheats depends on sowing time. *Aust. J. Agric. Res.* **1990**, *41*, 811–826. [CrossRef]

72. Bassu, S.; Asseng, S.; Motzo, R.; Giunta, F. Optimising sowing date of durum wheat in a variable Mediterranean environment. *Field Crops Res.* **2009**, *111*, 109–118. [CrossRef]

73. Panozzo, J.F.; Eagles, H.A. Rate and duration of grain filling and grain nitrogen accumulation of wheat cultivars grown in different environments. *Aust. J. Agric. Res.* **1999**, *50*, 1007–1015. [CrossRef]

74. Subedi, K.D.; Ma, B.L.; Xue, A.G. Planting date and nitrogen effects on grain yield and protein content of spring wheat. *Crop Sci.* **2007**, *47*, 36–44. [CrossRef]

75. Pulvento, C.; Riccardi, M.; Lavini, A.; d'Andria, R.; Iafelice, G.; Marconi, E. Field trial evaluation of two chenopodium quinoa genotypes grown under rain-fed conditions in a typical Mediterranean environment in South Italy. *J. Agron. Crop Sci.* **2010**, *196*, 407–411. [CrossRef]

76. Tawaha, A.M.; El-Shatnawi, M.K.J. Response of lentil (Lens culinaris Medik) to plant density, sowing date, phosphorus fertilization and ethephon application in the absence of moisture stress. *J Agron. Crop Sci.* **2003**, *189*, 1–6.

77. Martin, I.; Tenoria, J.L.; Ayerbe, L. Yield growth and water use of conventional and semi leafless peas in semi-arid environments. *Crop Sci.* **1994**, *34*, 1576–1583. [CrossRef]

78. Noffsinger, L.S.; Santen, E. Yield and yield components of spring-sown white lupin in the southeastern USA. *Agron. J.* **1995**, *87*, 493–497. [CrossRef]

79. McEwen, J.; Yeoman, D.P.; Moffitt, R. Effect of seed rates, sowing dates and methods of sowing on autumn sown field beans (*Vicia faba* L.). *J. Agric. Sci. Camb.* **1988**, *110*, 345–352. [CrossRef]

80. Tawaha, A.M.; Turk, M.A. Crop-weed competition studies in faba bean (*Vicia faba* L.) under rainfed conditions. *Acta Agron. Hung.* **2001**, *49*, 299–303. [CrossRef]

81. Hirich, A. Effects of Deficit Irrigation using Treated Wastewater and Irrigation with Saline Water on Legumes, Corn and Quinoa Crops. Ph.D. Thesis, Institut Agronomique et vétérinaire Hassan II, Morocco, 2014.

82. Postel, S.L. Entering an Era of Water Scarcity: The Challenges Ahead. *Ecol. Appl.* **2000**, *10*, 941–948. [CrossRef]

83. Daryanto, S.; Wang, L.; Jacinthe, P.-A. Global Synthesis of Drought Effects on Food Legume Production. *PLoS ONE* **2015**, *10*, 1–16. [CrossRef]

84. Olesen, J.E.; Trnka, M.; Kersebaum, K.C.; Skjelvag, A.O.; Seguin, B.; Peltonen-Sainio, P.; Rossi, F.; Kozyra, J.; Micale, F. Impacts and adaptation of European crop production systems to climate change. *Eur. J. Agron.* **2011**, *34*, 96–112. [CrossRef]
85. Pandey, R.K.; Herrera, W.A.T.; Pendleton, J.W. Drought response of grain legumes under irrigation gradient: 1. Yield and yield components. *Agron. J.* **1984**, *76*, 549–553. [CrossRef]
86. Peterson, P.R.; Sheaffer, C.C.; Hall, M.H. Drought effects on perennial forage legume yield and quality. *Agron. J.* **1992**, *84*, 774–779. [CrossRef]
87. Karou, M.; Oweis, T. Water and land productivities of wheat and food legumes with deficit supplemental irrigation in a Mediterranean environment. *Agric. Water Manag.* **2012**, *107*, 94–103. [CrossRef]
88. Mafakheri, A.; Siosemardeh, A.; Bahramnejad, B.; Struik, P.C.; Sohrabi, E. Effect of drought stress on yield, proline and chlorophyll contents in three chickpea cultivars. *Aust. J. Crops Sci.* **2010**, *4*, 580–585.
89. Li, F.; Cook, S.; Geballe, G.T.; Burch, W.R., Jr. Rainwater Harvesting Agriculture: An Integrated System for Water Management on Rainfed Land in China's Semiarid Areas. *Ambio* **2000**, *29*, 477–483. [CrossRef]
90. Sass, O. Marktsituation und züchterische Aktivitäten bei Ackerbohnen und Körnererbsen in der EU (Market situation and breeding input in faba bean and field pea in the EU). *J. für Kulturpflanzen* **2009**, *61*, 306–308.
91. Osenberg, C.W.; Sarnelle, O.; Cooper, S.D.; Holt, R.D. Resolving ecological questions through meta-analysis: Goals, metrics, and models. *Ecology* **1999**, *80*, 1105–1117. [CrossRef]

Article

Effect of Soil Tillage and Crop Sequence on Grain Yield and Quality of Durum Wheat in Mediterranean Areas

Giancarlo Pagnani [1], Angelica Galieni [2], Sara D'Egidio [1], Giovanna Visioli [3], Fabio Stagnari [1,*] and Michele Pisante [1]

1   Faculty of Bioscience and Technologies for Food, Agriculture and Environment, University of Teramo, 64100 Teramo (TE), Italy
2   Council for Agricultural Research and Economics, Vegetable and Ornamental Crops Research Center, 63077 Monsampolo del Tronto (AP), Italy
3   Department of Chemistry, Life Sciences and Environmental Sustainability University of Parma, Parco Area delle Scienze 11/A, 43124 Parma, Italy
*   Correspondence: fstagnari@unite.it

Received: 8 August 2019; Accepted: 27 August 2019; Published: 28 August 2019

Abstract: Conservation agriculture (CA) can be very strategic in degradation prone soils of Mediterranean environments to recover soil fertility and consequently improve crop productivity as well as the quality traits of the most widespread crop, durum wheat, with reference to protein accumulation and composition. The results shown by two years of data in a medium long-term experiment (7-year experiment; split-plot design) that combined two tillage practices (conventional tillage (CT) and zero tillage (ZT)) with two crop sequences (wheat monocropping (WW) and wheat-faba bean (WF)) are presented. The combination ZT + WF (CA approach) induced the highest grain yields (617 and 370 g m^{-2} in 2016 and 2017, respectively), principally due to an increased number of ears m^{-2}; on the other hand, the lowest grain yield was recorded under CT + WW (550 and 280 g m^{-2} in 2016 and 2017, respectively). CA also demonstrated significant influences on grain quality because the inclusion of faba bean in the rotation favored higher N-remobilization to the grains (79.5% and 77.7% in 2017). Under ZT and WF, all gluten fractions (gliadins (Glia), high molecular-weight glutenins (GS), and low molecular-weight GS) as well as the GS/Glia ratio increased. In durum wheat-based farming systems in Mediterranean areas, the adoption of CA seems to be an optimal choice to combine high quality yields with improved soil fertility.

Keywords: no-tillage; conservation agriculture; durum wheat; gluten fractions; SDS-PAGE analysis

---

1. Introduction

Durum wheat is the main cereal crop grown within Mediterranean regions, although the erratic rainfall distribution and the extremely fluctuating temperatures during grain filling stages determine the instability of yields. In these areas, wheat is usually grown in monocropping or short rotations, in conventionally managed soils (ploughing and harrowing) characterized by severe water loss, erosion, organic matter depletion, and CO_2 release. Both the reduced soil fertility and effects of climatic change strongly affect durum wheat yields and quality traits [1]. To counteract such criticality, conservation agriculture (CA) practices-based on (i) minimum or no soil disturbance; (ii) permanent residue cover; and (iii) planned crop rotations or associations [2] should be applied; these aim to promote water conservation, biological processes, and organic matter accumulation due to reduced soil erosion and CO_2 emissions. Although the environmental sustainability of CA and its positive effects on crop yield in rainfed systems have been documented [2–6], data on CA impact on crop quality traits as well as on yield-quality relationships are missing.

In durum wheat, the protein content as well as the protein profiles play a key role in the technological quality. Gluten, which represents 80% of kernel proteins, can be classified into prolamins, namely gliadins (Glia) that are alcohol soluble proteins, and glutenins (GS) that are soluble in diluted acid or alkali or alcohol–water mixtures, while the rest 20% is composed by non-prolamins including albumins, globulins, and metabolic enzymes which are water and salt soluble [7]. Glia and GS exert the highest influence on the strength and elastic proprieties of the dough. Glia can be classified into S-rich prolamins (i.e., α-β-γ-gliadins, molecular weight 36–44 kDa), and S-poor prolamins (i.e., ω-gliadins, molecular weight 44–78 kDa). GS consist of different sub-units, high molecular weight GS (HMW-GS, 80–140 kDa), and low molecular weight GS (LMW-GS, 31–51 kDa), which form large aggregates joined by disulfide bonds. The GS/Glia ratio determines the functional proprieties of the gluten responsible for dough strength [8], and greater dough strength is observed when there are higher proportions of the HMW-GS than that of the LMW-GS polymer [9]. Several allelic variants of x and y genes coding for HMW-GS have been associated with flour technological quality [10] as well as the B-group LMW-GS, coded at Glu-3 loci at chromosome 1 (Glu-A3, Glu-B3, and Glu-B2) [11].

Aside from genetics, the main role in determining the amount and specificity of durum wheat proteins is caused environmentally and agronomically, especially in terms of the up or downregulation of specific Glia, HMW-GS, and LMW-GS sub-units that are related to flour quality [9,12–14]. In particular, temperatures, precipitation, and nitrogen (N) availability during grain filling are key factors for seed protein accumulation [15,16]. The synthesis and accumulation of gluten proteins is increased significantly by N availability, although the effect is principally driven by the efficient synchrony between soil N and plant N requirements [14,17,18]. Besides, fertilization strategies [18] as well as soil management, cover crops, and crop rotation are fundamental variables that affect physiological traits, and consequently, the quality (protein profile) and yields [19]. Crop rotation and residues influence carbon and N dynamics in soil [20], especially in low organic matter and N environments (i.e., Mediterranean areas); crop residues on soil surface are, indeed, not only mere soil protectors from erosion, but affect nutrient and water availability that in turn increase and stabilize yields [21–25].

However, the adoption of CA requires a transition phase (on average 7–8 years), often characterized by higher annual weed and disease pressures, slow rebuilding aggregates in soil, and lower and variable yields, so involve adaptations at the individual farm-level until the rehabilitation of soil-related functions [26].

Studies regarding the effect of CA practices on protein accumulation in durum wheat kernel, especially on storage proteins and their composition, are lacking. Durum wheat quality has been principally investigated in association with different tillage [27], while the effect of the interaction of the CA key practices in a holistic view on crop physiology, yield, and technological quality performances are still missing. To this end, we propose a comprehensive approach based on the combination of two tillage practices (conventional tillage and zero tillage) with two crop sequences (wheat monocropping and wheat-faba bean) including the CA approach in a medium-long term (7-years) experiment to investigate durum wheat responses. A short not-irrigated rotation, which is typical of the experimental area (Mediterranean area), was adopted (although it does not fully meet the requirements of CA). We raised the hypothesis that through the conservation approach (zero tillage and residue retention, crop rotation), already at the end of the transition phase, the soil quality amelioration reflected into improved crop performances in terms of:

(i)   growth, yield and yield components;
(ii) protein accumulation and profile total grain protein, Glia, LMW-GS, and HMW-GS, and the characterization of the single sub-units of gluten proteins; and
(iii) crop physiological status, efficiencies of biomass and nitrogen remobilization to kernels, reflectance-based vegetation indices.

2. Materials and Methods

2.1. Site Description

The results reported in this study were obtained during the 2015–2016 and 2016–2017 cropping seasons (referred below as 2016 and 2017, respectively) within a medium-long term experiment started in 2010–2011 in Teramo (Mosciano Sant'Angelo, Italy, 42°42′ N, 13°52′ E, 101 m above sea level). The principal soil characteristics, recorded at the beginning of the experiment, were as follows: 23% sand, 45% silt, and 32% clay, pH 8.0, 1.36% organic carbon, 19.0% total $CaCO_3$, 10.6% active $CaCO_3$, and 32.1 meq 100 g^{-1} cation-exchange capacity. The area has a typical Mediterranean climate with 732 mm of annual mean rainfall (58-year period) mainly concentrated between October and April; the mean of maximum temperatures ranges from 11 °C to 29 °C while the mean of minimum temperatures ranges from 2 °C to 17 °C. During crop cycles, meteorological data were recorded with a meteorological station situated at about 1 km from the experimental field; trends of average minimum and maximum monthly air temperatures and total rainfall are reported in Figure 1.

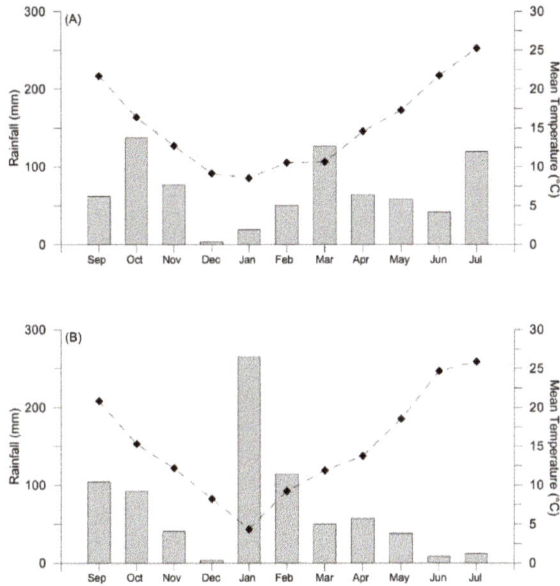

Figure 1. Monthly rainfall (bars) and monthly average temperatures (dots) as registered during the whole durum wheat crop cycle in the two different growing seasons, 2015–2016 (**A**) and 2016–2017 (**B**) at the experimental area.

2.2. Experimental Design and Agronomic Practices

The medium-long term experiment aimed to evaluate different management systems for durum wheat (scenarios) under a semi-arid environment by combining two soil tillage practices (ST, main treatment, main plots) with two crop sequences (secondary treatments) in a split-plot design with three replications. The soil tillage practices consisted of conventional tillage (CT) and zero tillage (ZT). CT and ZT were established in a 15,000 m^2 experimental field, properly spaced (i.e., field margins, edges, and traffic infrastructure) to avoid any overlays during tillage operations and were imposed starting from the beginning of the medium-long term experiment (2010–2011). It follows that the ZT treatment respected a six/seven-year period of conversion (2010–2016/2017) necessary to ensure reliable long-term production goals [28]. CT included complete soil inversion (moldboard ploughing) to a 35-cm depth during the summer, followed by secondary tillage in the autumn. ZT involved no soil disturbance, previous-crop residue retention, and a glyphosate (Roundup Power 2.0, Monsanto Agricoltura Italia S.p.A., Milano (MI), Italy) application at the rate of 2 L ha^{-1}, two weeks before sowing.

Within each soil tillage practice, two different crop sequences (CS, secondary treatment, sub-plots) were established: (i) durum wheat (*Triticum turgidum* L. subsp. *durum* (Desf.) Husn.) monocropping (WW); and (ii) a durum wheat-faba bean (*Vicia faba* L. var. *minor*) rotation (WF), rising to a total of 18 sub-plots. Each sub-plot consisted of an area of 1200 m^2 and was obtained by splitting the main plots into three areas of equal size that, according to the length of the crop cycles, were able to host both monocropping and rotation treatments (each crop present every year). Within the sub-plots, transects were set up in order to obtain two experimental units of 20 m^2 (2.5 m × 8.0 m) for sampling and data collection. In order to avoid edge-effects, the experimental units were delimited in the central area of each sub-plot. Edges between treatments were managed, leaving them free from crops and weeds using a brushcutter.

Treatments under comparison were obtained from the factorial combination of tillage and crop sequence, as follows: (i) conventional tillage and wheat monocropping (CT + WW); (ii) conventional tillage and wheat-faba bean rotation (CT + WF); (iii) zero tillage and wheat monocropping (ZT + WW); and (iv) zero tillage and wheat-faba bean rotation (ZT + WF). In the ZT + WF treatment, despite the short rotation, which was chosen to conveniently reflect the typical agronomic management of the area, all three principles of CA were respected.

The durum wheat and faba bean cultivars were "Saragolla" and "Protabath", respectively. Both crops were sown on 7 December and 23 November in 2016 and 2017, respectively, at a rate of 350 and 50 viable seeds m^{-2} for durum wheat and faba bean, respectively. A direct seeder (Gaspardo Direttissima, Gruppo Maschio Gaspardo S.p.A., Campodarsego, PD, Italy) was used in the ZT treatments while a mechanical seeder (S-SC MARIA 250, Gruppo Maschio Gaspardo S.p.A., Campodarsego, PD, Italy) was used in the CT treatments. In durum wheat, N was applied at the whole rate of 150 kg N ha^{-1}, split into half on 22 March 2016 and 28 March 2017 (beginning of stem elongation) as ammonium nitrate and half on 18 April 2016 and 20 April 2017 (emergence of head complete) as urea; no fertilization treatments were applied to the faba bean. In both years, the fungicide Sphere (Trifloxystrobin and Ciproconazolo, Bayer Crop Science Italia, MI, Italy) was applied at the dose of 1 L ha^{-1}, at the beginning of anthesis; herbicide treatments were not performed.

Immediately after sowing, the number of crop residues (thickness and weight) was measured for the ZT + WW and ZT + WF treatments on a sampling square of 0.25 m^2 for each experimental unit. Crop residues were also characterized for their N concentration (Kjeldahl procedure) and for their cellulose, hemicellulose, and lignin contents [29].

2.3. Plant Sampling and Growth Analysis

Phenological stages were constantly monitored on 10 randomly tagged plants per experimental unit and were scored following the Zadoks Decimal Code [30].

Starting from anthesis (29 April and 27 April for 2016 and 2017, respectively), 10 whole main wheat shoots within each experimental unit were randomly collected until the final harvest (23 June and 19 June for 2016 and 2017, respectively) at the following phenological stages: DC71, DC73, DC83, DC87, and DC91 in 2016 and DC71, DC77, DC85, DC87, and DC91 in 2017, which corresponded to 0, 165, 256, 404, 564, and 660 growing degree days (GDD, °C) in 2016 and to 0, 168, 310, 443, 523, and 630 GDD in 2017. Thermal time after anthesis (GDD) corresponded to the cumulative daily average air temperature exceeding 0 °C.

Sampled plants were separated into leaves, stems, and ears. Dry weight (DW) was measured after oven drying at 80 °C until reaching a constant weight.

According to Arduini et al. [31], the amount of vegetative DW remobilized into kernels during grain filling (DW remobilization efficiency, DWRemE, %) was calculated as the difference between the DW of the aerial plant part at heading and at physiological maturity.

At physiological maturity, one square meter of wheat plants was randomly harvested to measure the yield components (ears number m^{-2} and length (cm)) and yield (g m^{-2}) after hand threshing.

2.4. N Determination in Plant Tissues

In both 2016 and 2017, the N concentration in vegetative (leaves, stems) and reproductive (ears) organs was determined at the DC71 and DC91 durum wheat phenological phases, followed by the Kjeldahl method. N content was then calculated by multiplying the N concentration by DWs. At harvest, the grain protein content (GPC, %) was also calculated as the grain N concentration multiplied by 5.7 [32].

The N remobilization efficiency (NRemE, %) was calculated as DWRemE, according to Arduini et al. [31].

2.5. Analysis of Gluten Proteins

2.5.1. Extraction and Quantification

Grain sub-samples collected from each experimental unit were milled with Knifetec TM 1095 (Foss, Hillerød, Denmark) to obtain a fine powder, and 30 mg of wholemeal flour was used to evaluate the Glia, HMW-GS, and LMW-GS fractions by applying some modifications to the sequential procedure described in Singh et al. [33]. Flour was suspended in 1.5 mL of 50% (*v/v*) propan-2-ol for 20 min with continuous mixing at 65 °C; followed by centrifugation for 5 min at 10,000 rpm. The supernatant contained the extracted Glia component, which was collected and evaporated to dryness in a speed Vac concentrator (Svant Instruments, Farmingdale, NY, USA). A pellet containing the GS fraction, after washing twice with 50% (*v/v*) propan-2-ol, was suspended again within a solution of 55% (*v/v*) propan-2-ol, 0.08 M tris(hydroxymethyl) aminomethane hydrochloric acid pH 8.3, and 1% (*w/v*) dithiothreitol as a reducing agent and incubated for 30 min at 60 °C with continuous mixing. After centrifugation for 5 min at 14,000 rpm, the supernatant containing both the HMW-GS and LMW-GS fractions was transferred into a new tube. Acetone was then added to reach a final concentration of 40% (*v/v*) to precipitate HMW-GS. After centrifugation for 5 min at 13,000 rpm, the pellet was evaporated to dryness while the remaining supernatant containing the LMW-GS fraction was recovered into a new tube and precipitated by adding acetone to reach a final concentration of 80% (*v/v*). After centrifugation, the supernatant was discarded, and the pellet was air dried. Both protein fractions were quantified with the Bradford assay (Biorad Hercules, CA, USA) after dissolving pellets in (50:50 *v/v*) acetonitrile and H$_2$O with 0.1% (*v/v*) trifluoroacetic acid. Three technical replicates were performed for each sample.

2.5.2. Sub-Units Separation by SDS-PAGE and Densitometric Analyses

In order to deeply investigate the effects of the experimental treatments in terms of up or downregulation of single sub-units of the LMW-GS, HMW-GS, and Glia, Sodium Dodecyl Sulphate-PolyAcrylamide Gel Electrophoresis (SDS-PAGE)was performed on a Mini-PROTEAN Tetra Cell (Bio-Rad, CA, USA) on 8% and 12% acrylamide gel for the HMW-GS and LMW-GS and Glia, respectively, in 2017. Dried HMW-GS and LMW-GS and Glia (2.5 µg each) were suspended in 20 µL of loading buffer containing 2% (*w/v*) SDS, 0.02% (*w/v*) bromophenol blue, 0.1% β−mercaptoethanol, 0.05 M Tris-HCl pH 6.8, and 10% (*v/v*) glycerol, and boiled at 95 °C for 5 min before loading onto the gel. A ColorBurst™ Marker Electrophoresis High Range (Mw 30,000–220,000) was used to detect HMW-GS and a Molecular-Weight Marker® (Mw 14,000–66,000; Sigma Aldrich, St. Louis, MO, USA) to detect LMW-GS. After electrophoretic separation at 40 mA, the gels were stained with brilliant blue G-colloidal solution (Sigma Aldrich, St. Louis, MO, USA) fixed in 7% (*v/v*) acetic acid and 40% (*v/v*) methanol, and de-stained in 25% (*v/v*) methanol. Each protein sample (Glia, HMW-GS, and LMW-GS) was analyzed in three technical replicates. IMAGE lab 4.5.1 (Bio-Rad) software was used to identify protein molecular weights (MW) and for relative quantification of the LWM-GS and HMW-GS single protein sub-units on each gel.

2.6. Physiological Traits

At the DC85 developmental stage, canopy reflectance was measured using a HandHeld 2 Pro Portable FieldSpec Spectroradiometer (ADS Inc., Boulder, CO, USA). The normalized difference vegetation index (NDVI), green normalized difference vegetation index (GNDVI), water index (WI) and optimized soil adjusted vegetation index (OSAVI) were then calculated as follows [34–37]:

$$NDVI_{670} = (\rho800 - \rho670)/(\rho800 + \rho670) \tag{1}$$

$$GNDVI = (\rho750 - \rho550)/(\rho750 + \rho550) \tag{2}$$

$$WI = \rho900/\rho970 \tag{3}$$

$$OSAVI = (1 + 0.16)(\rho800 - \rho670)/(\rho800 - \rho670 + 0.16) \tag{4}$$

where p represents the reflectance measured at the specific wavelength.

The estimation of the chlorophyll content was obtained with the SPAD 502 plus portable chlorophyll meter (Konica Minolta, Inc., Tokyo, Japan). In order to avoid the natural yellowing of the older leaves, measurements were taken on the mid-section of the flag leaf, when present or otherwise, on the greenest leaf, for a total of 10 measurements for each experimental unit.

2.7. Statistical Analysis

Analysis of variance (ANOVA) was performed with the free Excel plugin DSAASTAT® VBA macro, version 1.1 (Pisa, Italy) [38], considering a multi-year split-plot design in order to test (*F*-test) the effects of growing season (year), ST, and CS as well as their interactions on the selected variables. Year was considered as a random factor. Densitometric data related to Glia, HMW-GS, and LMW-GS sub-units recorded in 2017 were analyzed using a two-way ANOVA considering a split-plot design.

When ANOVA detected significant differences, the means were separated by applying the Tukey's honestly significant difference (HSD) test (data in Tables). In the figures, the standard errors of the means are reported. Normality and homoschedasticity assumptions were tested with the Shapiro–Wilk and Bartlett tests, respectively (R software, Vienna, Austria) [39].

3. Results

3.1. Meteorological Data, Residue Retention, and Composition

Erratic amounts and distribution of rainfall characterized the experiments (Figure 1). In 2017, between anthesis and grain filling, the cumulative rainfall was 50% less than 2016 (290.2 mm vs. 153.8 mm in 2016 and 2017, respectively) while during the vegetative phase (sowing–anthesis period), values were higher and mainly concentrated in January (Figure 1). Generally, the monthly average temperatures were similar between the two years (15.3 and 15.0 °C for 2016 and 2017, respectively).

ZT favored the accumulation of residue mulching, which showed a remarkable thickness (0.93 cm on average) and weight (2.00 t ha^{-1} on average) (Table 1). Residues obtained under WW showed a higher thickness and weight than WF (1.40 vs. 0.46 cm and 3.06 vs. 0.95 t ha^{-1}) and were characterized by a greater percentage of hemicellulos, cellulose, and lignin, but with lower N content (Table 1).

Table 1. Thickness (cm), amount (t ha^{-1}), nitrogen (N, %), hemicellulose (%), cellulose (%), and lignin (%) contents of durum wheat and faba bean residues under zero tillage (ZT) treatments (durum wheat residues: ZT + wheat monocropping (ZT + WW); faba bean residues: ZT + wheat-faba bean rotation (ZT+WF)). Means ± standard errors (over growing seasons) are reported.

Crop Sequence	Thickness (cm)	Amount (t ha^{-1})	N (%)	Hemicellulose (%)	Cellulose (%)	Lignin (%)
ZT + WW	1.40 ± 0.12	3.06 ± 0.14	0.3 ± 0.02	24.4 ± 0.28	33.4 ± 0.11	15.2 ± 0.01
ZT + WF	0.46 ± 0.08	0.95 ± 0.07	1.68 ± 0.03	15.5 ± 0.34	23.4 ± 0.27	4.0 ± 0.03

3.2. Dry Mass Dynamics and Partitioning

The dynamics of the DW accumulation of the leaves, stems, and ears starting from 165 and 168 GDD after anthesis (2016 and 2017, respectively) are reported in Figure 2. Averaging over treatments, slightly higher dry aerial biomass values were obtained in 2016 (leaves: 0.53 vs. 0.38 g plant^{-1}; stem: 1.79 vs. 1.68 g plant^{-1}; ear: 2.20 vs. 2.17 g plant^{-1} in 2016 and 2017, respectively) with ZT often showing higher values than CT (at harvesting, ZT vs. CT: leaves +32%, stems +15%, ears +11%) (Figure 2).

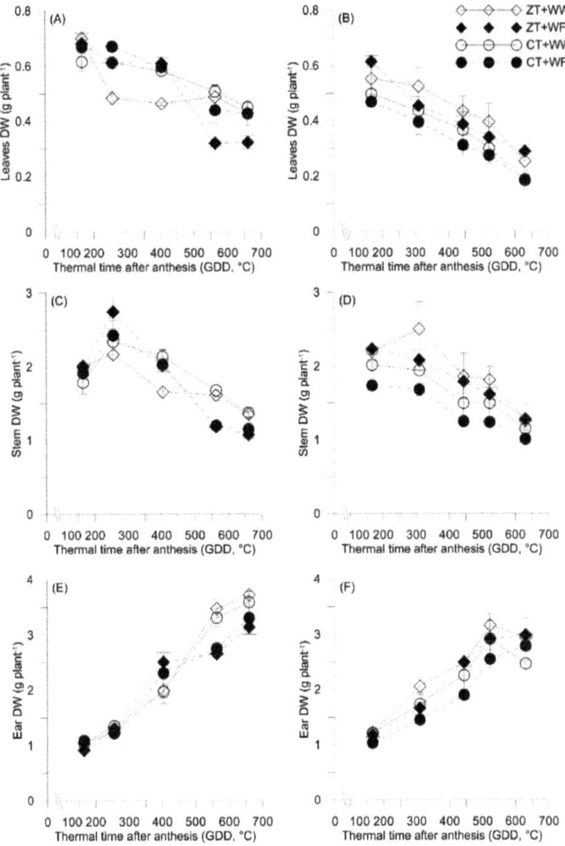

Figure 2. Dynamic of dry weight (DW, g plant^{-1}) of leaves, steam and ear against thermal time after anthesis (°C) of durum wheat in 2016 (A, C and E, respectively) and 2017 (B, D and F, respectively) affected by soil tillage practice and crop sequence (○ CT + WW; ● CT + WF; ◊ ZT + WW; ◆ ZT + WF). Treatments: CT, conventional tillage; ZT, zero tillage; WW, durum wheat monocropping; WF, durum wheat-faba bean rotation. Data are averages ± standard errors of n = 3 independent replicates.

During grain filling, 36.2% and 45.9% of the vegetative DW were remobilized in 2016 and 2017 (averaging over ST and CS), respectively (Table 2). Although differences between ST treatments were not significant (Table 3), ZT induced higher DWRemE than CT (43.7% vs 38.4%); furthermore, WF induced higher DWRemE than WW. Consequently, the highest value was registered for the ZT + WF combination (47.9%) while the lowest was obtained for CT + WW (23.3%) (Table 2).

Table 2. Dry weight (DW) remobilization efficiency (DWRemE, % of vegetative DW) and nitrogen (N) remobilization efficiency (NRemE, % of vegetative N) in the 2016 and 2017 cropping seasons. Means followed by different letters (upper case letters: main effects; lower case letters: effects of interaction) significantly differ according to Tukey's HSD test ($p < 0.05$). Treatments: CT, conventional tillage; ZT, zero tillage; WW, durum wheat monocropping; WF, durum wheat-faba bean rotation.

Year	Treatments	DWRemE (%)			NRemE (%)		
		ZT	CT	CS mean	ZT	CT	CS mean
2016	WW	34.4	23.3	28.8	74.2	73.6	73.9
	WF	47.9	39.3	43.6	82.4	76.6	79.5
	ST mean	41.2	31.3		78.3 A	75.1 B	
2017	WW	45.5	45.1	45.3	72.8	76.6	74.7
	WF	47.1	45.9	46.5	76.1	78.8	77.4
	ST mean	46.3	45.5		74.4 B	77.7 A	
2016–2017	WW	40.0	34.2	37.1	73.5	75.1	74.3
	WF	47.5	45.6	47.0	79.2	77.7	78.4
	ST mean	43.7	38.4		76.3	76.4	

Effects: ST, soil tillage; CS, crop sequence. Degrees of freedom: Year, 1; ST, 1; Year × ST, 1; CS, 1; Year × CS, 1; ST × CS, 1; Year × ST × CS, 1; Residual, 8.

Table 3. Summary of p-values from the analyses of variance for dry weight remobilization efficiency (DWRemE, % of vegetative DW), nitrogen remobilization efficiency (NRemE, % of vegetative N), yield (g m^{-2}), number of ears per square meter (ears, num m^{-2}), ear length (cm), N concentrations ([N]) in leaves (g kg^{-1}), [N] in stems (g kg^{-1}), and [N] in ears (g kg^{-1}), N content in leaves (g m^{-2}), N content in stems (g m^{-2}), N content in ears (g m^{-2}), grain protein content (GPC, %), gliadins (Glia, mg g^{-1}), high molecular weight glutenins (HMW-GS, mg g^{-1}), low molecular weight glutenins (LMW-GS, mg g^{-1}), normalized vegetation index (NDVI), green normalized difference vegetation index (GNDVI), soil-adjusted vegetation index optimized (OSAVI), water index (WI) and SPAD. Effects: ST, soil tillage; CS, crop sequence. Degrees of freedom: Year, 1; ST, 1; Year × ST, 1; CS, 1; Year × CS, 1; ST × CS, 1; Year × ST × CS, 1; Residual, 8.

	ST	Year × ST	CS	Year × CS	ST × CS	Year × ST × CS
DWRemE	0.446	0.133	0.449	0.057	0.669	0.800
NRemE	0.994	0.013	0.210	0.354	0.370	0.495
Yield (g m^{-2})	0.204	0.146	0.002	0.986	0.192	0.420
Ears (num m^{-2})	0.038	0.813	0.515	0.095	0.171	0.117
Ear length (cm)	0.734	0.001	0.129	0.125	0.706	0.005
[N] Leaves						
DC71[a]	0.880	0.033	0.449	0.012	0.428	0.165
DC91[a]	0.716	0.081	0.749	0.057	0.119	0.588
[N] Stems						
DC71	0.846	0.150	0.020	0.915	0.178	0.569
DC91	0.777	0.137	0.455	0.521	0.406	0.310
[N] Ears						
DC71	0.133	0.899	0.222	0.248	0.193	0.481
DC91	0.620	0.037	0.664	0.001	0.249	0.626
N content leaves						
DC71	0.230	0.141	0.628	0.040	0.528	0.423
DC91	0.651	0.022	0.006	0.989	0.295	0.367
N content stems						
DC71	0.267	0.164	0.230	0.325	0.538	0.512
DC91	0.411	0.023	0.195	0.741	0.179	0.629

[a] DC71 and DC91 phenological stages of durum wheat following the Zadoks Decimal Code. Further explanations are provided in the text.

Table 3. *Cont.*

	ST	Year × ST	CS	Year × CS	ST × CS	Year × ST × CS
N content ears						
DC71	0.249	0.088	0.208	0.234	0.976	0.024
DC91	0.007	0.928	0.347	0.500	0.646	0.727
GPC	0.004	0.892	0.035	0.583	0.347	0.055
Glia	0.396	0.005	0.016	0.887	0.670	0.004
HMW-GS	0.050	0.019	0.442	<0.001	0.479	<0.001
LMW-GS	0.083	0.004	0.590	<0.001	0.904	<0.001
GS/GLIA	0.019	0.264	0.743	<0.001	0.516	<0.001
HMW-G /LMW-GS	0.402	0.002	0.204	0.650	0.251	0.014
NDVI	0.870	<0.001	0.982	0.158	0.937	0.231
GNDVI	0.895	<0.001	0.838	0.007	0.666	0.012
OSAVI	0.969	<0.001	0.894	0.163	0.907	0.132
WI	0.338	<0.001	0.678	0.156	0.301	0.008
SPAD	0.801	0.004	0.168	0.056	0.720	0.001

[a] DC71 and DC91 phenological stages of durum wheat following the Zadoks Decimal Code. Further explanations are provided in the text.

3.3. Grain Yield and Yield Components

The higher grain yields were achieved during 2016 (570 vs. 320 g m^{-2} in 2016 and 2017, respectively) due to the most favorable climatic regime; regardless of the crop growing season, CS induced significant effects, with WF showing higher yields (Tables 3 and 4). In particular, the highest yield was achieved by the combination of ZT + WF (617 and 370 g m^{-2} in 2016 and 2017, respectively) and the worst by CT + WW (556 and 284 g m^{-2} in 2016 and 2017, respectively). ZT showed a significantly higher number of ears m^{-2} (295 vs. 253 and 304 vs. 267 ears m^{-2} for ZT and CT in 2016 and 2017, respectively) (Tables 3 and 4).

Table 4. Yield (g m^{-2}), number of ears per square meter (ears, num m^{-2}), and ear length (cm) as recorded at harvest in the 2016 and 2017 cropping seasons. Means followed by different letters (upper case letters: main effects; lower case letters: effects of interaction) significantly differ according to Tukey's HSD test ($p < 0.05$). Treatments: CT, conventional tillage; ZT, zero tillage; WW, durum wheat monocropping; WF, durum wheat-faba bean rotation.

Year	Treatments	Yield (g m^{-2})			Ears (num m^{-2})			Ear Length (cm)		
		ZT	CT	CS Mean	ZT	CT	CS Mean	ZT	CT	CS Mean
2016	WW	551	556	554	273	261	267	7.25 bc	7.50 b	7.37
	WF	617	561	589	316	245	281	7.28 bc	8.04 a	7.66
	ST mean	584	558		295	253		7.27	7.73	
2017	WW	319	284	302	296	276	286	6.62 d	5.38 f	6.00
	WF	370	303	337	312	259	286	7.13 c	5.73 e	6.43
	ST mean	344	294		304	267		6.88	5.56	
2016–2017	WW	435	420	428 B	285	269	277	6.93	6.44	6.69
	WF	494	432	463 A	314	252	283	7.20	6.88	7.04
	ST mean	464	426		299 A	261 B		7.06	6.66	

Effects: ST, soil tillage; CS, crop sequence. Degrees of freedom: Year, 1; ST, 1; Year × ST, 1; CS, 1; Year × CS, 1; ST × CS, 1; Year × ST × CS, 1; Residual, 8.

3.4. Nitrogen Status, Nitrogen Remobilization Efficiency, and Grain Protein Concentration

N concentration in all plant organs (i.e., leaves, stems, and ears) was higher in 2016. Considering the treatments, only the N concentrations of stems and leaves were significantly influenced by CS and by "Year × ST" and "Year × CS", respectively (Figure 3, Table 3). ZT induced a higher N concentration in stems in 2017, while the higher values for stems and leaves in 2016 was observed under WF.

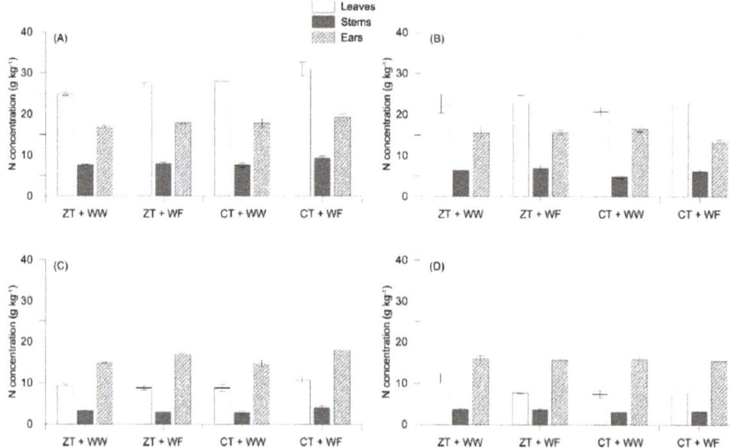

Figure 3. N concentration (g kg^{-1}) in leaves, stems and ears at DC71 and DC91 phenological stages of durum wheat in 2016 (A and C, respectively) and 2017 (B and D, respectively) cropping seasons. Treatments: CT, conventional tillage; ZT, zero tillage; WW, durum wheat monocropping; WF, durum wheat-faba bean rotation. Data are averages ± standard errors of n = 3 independent replicates.

At harvest, differences among the years and treatments reduced drastically. Averaging over treatments, the N concentration generally decreased in leaves (by 66% and 62% in 2016 and 2017, respectively) and stems (by 60% and 45% in 2016 and 2017, respectively), while remained substantially unchanged in the ears (17.9 and 16.1 g kg^{-1} in 2016; 15.3 and 15.7 g kg^{-1} in 2017) (Figure 3).

The N content was confirmed to be higher in 2016 and from post-anthesis to grain filling, it increased significantly in the ears (on average by 3-fold and 2.5-fold increase in 2016 and 2017, respectively) while it reduced in the stems (on average by 75% and 69% in 2016 and 2017, respectively) and leaves (on average by 80% and 84% in 2016 and 2017, respectively) (Figure 4). ZT always induced significantly higher N contents (Figure 4, Table 3).

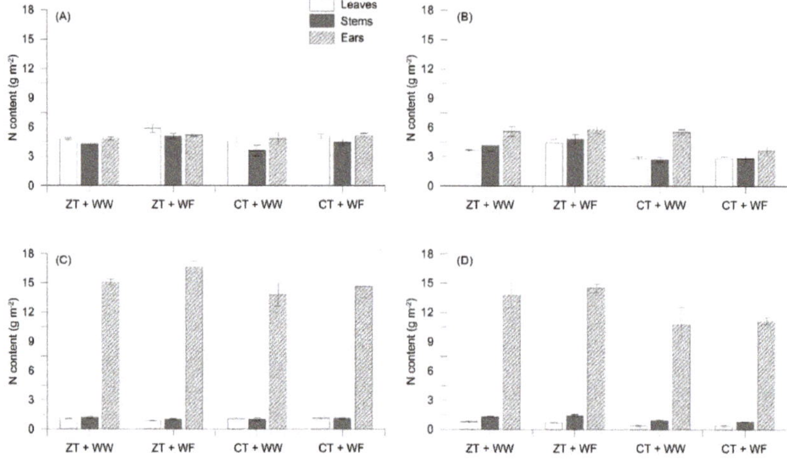

Figure 4. N content (g m^{-2}) in the leaves, stems, and ears at the DC71 and DC91 phenological stages of durum wheat in 2016 (A and C, respectively) and 2017 (B and D, respectively) cropping seasons. Treatments: CT, conventional tillage; ZT, zero tillage; WW, durum wheat monocropping; WF, durum wheat-faba bean rotation. Data are averages ± standard errors of n = 3 independent replicates.

The amount of N remobilized into kernels during grain filling were very similar between the years (77.0% and 76.0% in 2016 and 2017, respectively, averaged over ST and CS) (Table 2). The effect of ST was related to the crop growing season (see the significance of the interaction "Year × ST" in Table 3). Furthermore, although not significant, WF induced higher NRemE (79.5% vs. 74.6% in 2016 and 77.7% vs. 74.7% in 2017) (Table 2).

The GPC values were similar between the two years (Figure 5), while the effect of both ST and CS was significant ($p < 0.05$, Table 3). ZT induced higher values with respect to CT (12.07% vs. 10.32%) and WF than WW (11.52% vs. 10.87%); the highest values was obtained by the combination of ZT + WF (12.5%, averaging 2016 and 2017).

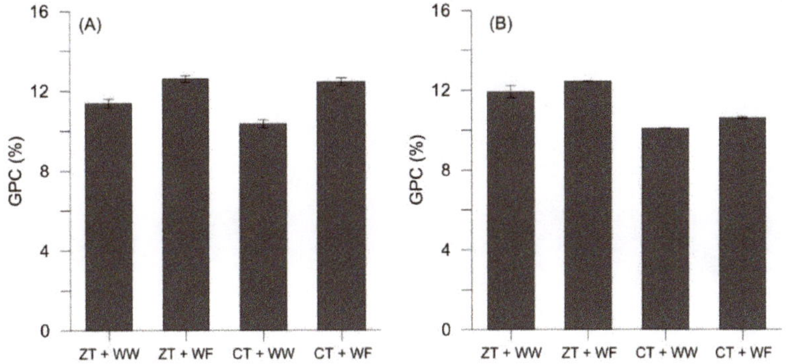

Figure 5. Grain protein content (GPC, %) as recorded for durum wheat at harvest in the 2016 (**A**) and 2017 (**B**) cropping seasons. Treatments: CT, conventional tillage; ZT, zero tillage; WW, durum wheat monocropping; WF, durum wheat-faba bean rotation. Data are averages ± standard errors of $n = 3$ independent replicates.

3.5. Gluten Proteins Content and Characterization

The effects of year, ST, and CS on gluten protein characteristics are reported in Table 3. All gluten fractions were significantly affected by the "Year × ST × CM" interaction; the highest Glia accumulation was obtained by CT + WF in 2017 and the lowest by ZT + WW in 2016 (Table 5). Conversely, the higher kernel accumulation of HMW-GS and LMW-GS were induced by ZT + WF in 2016, while the lowest were generally recorded under CT. A general trend showing higher total GS/Glia and HMW-GS/LMW-GS ratios under ZT with respect to CT was observed (Table 5).

In view of the obtained results, in 2017, the effects of ST and CS were also investigated in terms of the up or downregulation of single sub-units of the Glia, HMW-GS, and LMW-GS (Table 6, Figure 6). With respect to Glia, significant differences were observed in the molecular weight range of 28–41 KDa. Despite the interpretation of results being quite challenging, the inclusion of faba bean in the rotation led to 35 KDa and 28 KDa sub-units with a higher relative abundance (highest values reached by CT + WF treatment), while ZT significantly enhanced the 41 kDa subunit (Table 6, Figure 6). Results for the HWM-GS subunits were inconsistent. On the other hand, ZT induced the upregulation of the 42 kDa and 37 kDa (despite not significant) sub-units of LMW-GS (Table 6, Figure 6). Finally, WF significantly induced the downregulation of 42 KDa, 37 KDa, and 40 KDa sub-units of LMW-GS, while significantly enhancing the relative abundance of the 35 KDa sub-unit (+49% with respect to WW), with the highest values observed in the CT + WF treatment (Table 6, Figure 6).

Table 5. Gluten fractions (mg g^{-1} flour) (Glia: gliadins; HMW-GS: high molecular weight glutenins; LMW-GS: low molecular weight glutenins; Total GS-: HMW-GS + LMW-GS) and their ratios as recorded at harvest in the 2016 and 2017 cropping seasons. Means followed by different letters (upper case letters: main effects; lower case letters: effects of interaction) significantly differ according to Tukey's HSD test ($p < 0.05$). Treatments: CT, conventional tillage; ZT, zero tillage; WW, durum wheat monocropping; WF, durum wheat-faba bean rotation.

Year	Treatments	Glia			HMW-GS			LMW-GS			Total GS/GLIA			HMW-GS/LMW-GS		
		ZT	CT	CS Mean	ZT	CT	CS Mean	ZT	CT	CS Mean	ZT	CT	CS Mean	ZT	CT	CS Mean
2016	WW	9.0 c	9.4 bc	9.2	1.5 c	1.3 d	1.4	2.8 c	2.0 d	2.4	0.48 c	0.35 d	0.42	0.51 bc	0.62 a	0.57
	WF	9.8 bc	9.8 bc	9.8	2.3 a	1.1 e	1.7	3.6 a	2.2 d	2.9	0.60 a	0.33 de	0.47	0.65 a	0.50 c	0.58
	ST mean	9.4	9.6		1.9	1.5		3.2	2.1		0.54	0.34		0.58	0.56	
2017	WW	9.7 bc	10.3 b	10.0	1.9 b	1.1 e	1.5	3.2 b	2.1 d	2.6	0.52 b	0.31 ef	0.42	0.60 ab	0.51 bc	0.56
	WF	9.7 bc	11.5 a	10.6	1.9 b	1.1e	1.5	2.9 c	2.3 d	2.6	0.50 bc	0.29 f	0.40	0.67 a	0.48 c	0.57
	ST mean	9.7	10.9		1.9	1.1		3.0	2.2		0.51	0.30		0.64	0.49	
2016–2017	WW	9.4	9.8	9.6	1.7	1.2	1.45	3.0	2.1	2.5	0.50	0.33	0.42	0.55	0.56	0.56
	WF	9.8	10.7	10.2	2.1	1.1	1.6	3.2	2.3	2.7	0.55	0.31	0.43	0.66	0.49	0.58
	ST mean	9.6	10.3		1.9	1.1		3.1	2.2		0.53	0.32		0.61	0.53	

Effects: ST, soil tillage; CS, crop sequence. Degrees of freedom: Year, 1; ST, 1; Year × ST, 1; CS, 1; Year × CS, 1; ST × CS, 1; Year × ST × CS, 1; Residual, 8.

Table 6. Summary of *p*-values from two-way analysis of variance (ANOVA) on the relative abundances (%) of Glia, HMW-GS, and LMW-GS detected sub-units in 2017. Effects: ST, soil tillage; CS, crop sequence. Degrees of freedom: ST, 1; CS, 1; ST × CS, 1; Residual, 4.

	ST	CS	ST × CS
Glia sub-units			
46 KDa	0.118	0.297	0.513
44 KDa	0.893	0.235	0.194
41 KDa	0.009	0.879	0.015
39 KDa	0.297	0.023	0.213
35 KDa	0.038	0.011	0.591
28 KDa	0.043	0.019	0.008
HMW-GS sub-units			
83 KDa	0.033	0.748	0.288
74 KDa	0.041	0.753	0.276
LMW-GS sub-units			
42 KDa	0.023	0.027	0.067
41 KDa	0.421	0.759	0.860
40 KDa	0.125	0.036	0.089
37 KDa	0.050	0.002	0.142
35 KDa	0.058	<0.001	0.006

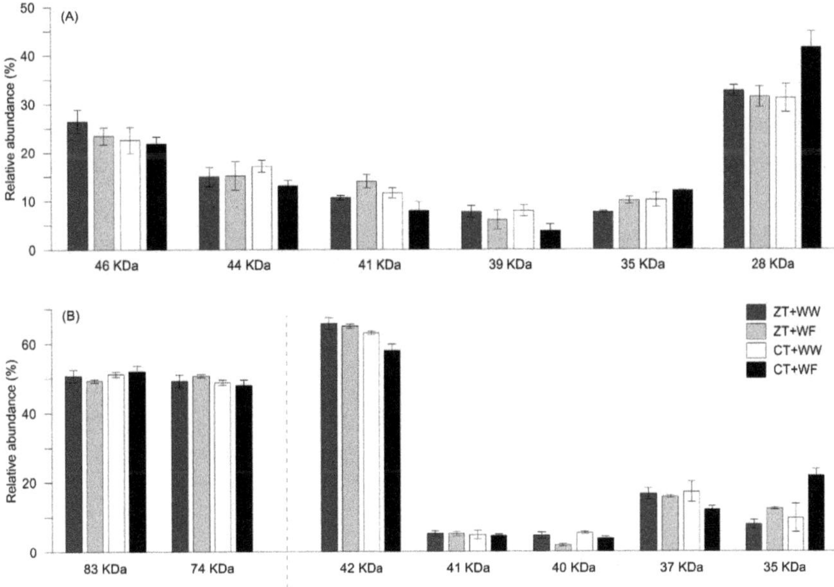

Figure 6. Relative abundances (%) obtained by the densitometric analysis of Glia (**A**), HMW-GS, and LMW-GS (**B**) sub-units of durum wheat in 2017. Treatments: CT, conventional tillage; ZT, zero tillage; WW, durum wheat monocropping; WF, durum wheat-faba bean rotation. Data are averages ± standard errors of *n* = 3 independent replicates.

3.6. Physiological Traits

The NDVI, GNDVI, and WI values were significantly higher in 2016. The combinations of ZT + WW and ZT + WF always registered the highest values and the interaction "Year × ST" was significant (*p* < 0.05) (Tables 3 and 7).

Table 7. Vegetation indices (NDVI: normalized vegetation index; GNDVI: green normalized difference vegetation index; OSAVI: soil-adjusted vegetation index optimized; WI: Water index) and SPAD values as recorded for durum wheat at the soft dough kernel stage (DC85) in 2016 and 2017 cropping seasons. Means followed by different letters (upper case letters: main effects; lower case letters: effects of interaction) significantly differ according to Tukey's HSD test ($p < 0.05$). Treatments: CT, conventional tillage; ZT, zero tillage; WW, durum wheat monocropping; WF, durum wheat-faba bean rotation.

Year		NDVI			GNDVI			OSAVI			WI			SPAD		
		ZT	CT	CS Mean	ZT	CT	CS Mean	ZT	CT	CS Mean	ZT	CT	CS Mean	ZT	CT	CS Mean
2016																
	WW	0.933	0.886	0.906	0.819a	0.763b	0.791	0.798	0.726	0.762	1.216a	1.032de	1.124	46.4d	50.1ab	48.3
	WF	0.935	0.900	0.918	0.829a	0.783b	0.806	0.798	0.747	0.773	1.224a	1.074c	1.149	48.1c	50.7a	49.4
	ST mean	0.934A	0.893B		0.824	0.773		0.798B	0.736C		1.220	1.053		47.3	50.4	
2017																
	WW	0.799	0.869	0.834	0.634d	0.723c	0.678	0.823	0.899	0.861	1.170b	1.051cd	1.105	48.4c	45.3d	46.85
	WF	0.799	0.854	0.826	0.642d	0.696c	0.669	0.824	0.883	0.854	1.157b	1.026e	1.091	48.9bc	48.9bc	48.9
	ST mean	0.799C	0.861B		0.638	0.710		0.823B	0.891A		1.164	1.039		48.7	48.7	
2016–2017																
	WW	0.866	0.877	0.871	0.726	0.743	0.734	0.810	0.813	0.812	1.124	1.111	1.175	47.4	47.7	47.55
	WF	0.867	0.877	0.872	0.736	0.740	0.738	0.811	0.815	0.813	1.149	1.091	1.120	48.5	49.8	49.2
	ST mean	0.866	0.877		0.736	0.741		0.811	0.814		1.137	1.101		48.0	48.7	

Effects: ST, soil tillage; CS, crop sequence. Degrees of freedom: Year, 1; ST, 1; Year × ST, 1; CS, 1; Year × CS, 1; ST × CS, 1; Year × ST × CS, 1; Residual, 8.

In the case of chlorophyll content in leaves, although the interaction "Year × ST × CS" was significant, it emerged that WF generally induced higher SPAD values (on average, 49.4 vs. 48.2 for WF and WW, respectively, in 2016 and 48.9 vs. 46.9 for WF and WW, respectively, in 2017).

4. Discussion

As it emerges from our work, CA promotes better crop physiological status, which in turns reflects positively on the accumulation of DWs and the quality traits of durum wheat. The determining factor was found in the build-up of a layer of crop residues, which enhance water retention, biological activity, and root development while reducing erosion; the latter acquire a pivotal role under the typical pedo-climatic conditions of Mediterranean environments. In such circumstances, Stagnari et al. [23] showed that 1.5 t ha^{-1} of straw was enough to produce significantly higher yields, although 2.5 t ha^{-1} also guaranteed effects on several crop physiological indicators. The ZT + WF approach appeared to give better performances, although faba bean accumulated residues were only 0.96 t ha^{-1}, much lower than the 3.02 t ha^{-1} obtained under ZT + WW. This can be ascribed to the chemical composition of the residues released. While straw exclusively (ZT + WW) stimulated N immobilization (high C/N ratio), reducing the immediate availability of N to the crop [40], the straw/faba bean mixture (ZT + WF) behaved differently, decomposing quicker and supplying a significant amount of N [41] ad humus. Indeed, the lower C/N ratio of the straw/faba bean mixture is known to result in higher humification rates [25,42–44] and easily releases macro and micronutrients [45]. The soil chemical analyses also revealed that, aside from the main effect of soil treatment, the combination of faba bean residues plus straw was more favorable to the build-up of soil organic carbon reserves (+18% on average under ZT; data not shown).

Although the observed differences were not significant, higher grain yields values were obtained under ZT, especially under drought conditions and in combination with WF (see higher gaps with CT in 2017) as observed in previous studies correlating yield performances with the crop water stress index [46]. The greater number of ears m^{-2} was obtained thanks to a higher seed emergence and better crop establishment probably favored by the reduction of water vapor loss from the first layers due to the residue accumulation (ZT) [23,47], although the presence of residues could sometimes promote the movement of phytotoxins toward seedling roots [48]. The CT approach (especially CT + WW) constrained crop growth before anthesis, thus reducing both the size and number of sinks (especially in 2016). Conversely, the higher wheat biomass reached at anthesis under ZT + WF resulted in both a greater accumulation of assimilates available for remobilization and assimilate demand during grain filling [49]. Such pre-anthesis biomass accumulation is critical for maintaining yields when adverse climatic conditions reduce post-anthesis photosynthesis and nutrient uptake [50].

The NDVI, GNDVI, and WI indexes were very successful in detecting plant physiological responses as well as in discriminating among treatments, thus confirming the higher efficiency of the combination of ZT + WF. The WI and NDVI have already been related to water status in plants [51,52] despite being strictly influenced by the relative water content and cell wall elasticity of leaf tissues [53]. Such physiological responses were also remarked by the SPAD readings, which are known to be a good indicator of chlorophyll concentration.

In general, there is a negative relationship between GPC and final grain yield, principally due to the energy constraints and N dilution effects. However, in accordance with other experimental trials [23,54], the higher protein content in kernels matched the higher yielding combination (ZT + WF), especially in the drought year (see 2017) [55]. The higher soil N availability, obtained with the introduction of a leguminous crop, and the ameliorated soil moisture conditions and nutrient release, due to no mechanical disturbance, most likely improved the wheat N and water uptake, favoring GPC accumulation [56] and protein quality [57] (i.e., promoting higher gluten fractions and its favorable characteristics [23,58]). During the transition phase to CA it is, indeed, exactly the management of N fertilization in wheat nutrition, which requires precautions [59,60]. We have previously demonstrated that about 150 kg N ha^{-1}, supplied as both calcium nitrate and urea, are required to improve N uptake and metabolic proteins for correct nutrient homeostasis, thus positively impacting on processing wheat quality [18]. In particular, the ratios of GS/Glia and HMW/LMW-GS are positively correlated to dough strength and the addition of a glutenin-rich fraction consisting of HMW-GS to the base semolina increases the mixograph dough strength [61]. The modulation of single protein sub-units of HMW-GS, LMW-GS, α-gliadins, γ-gliadins, and ω-gliadins in response to fertilization management are documented in the literature [62–64], while no information is available on the influence of soil practices and crop rotation on the accumulation of single gluten sub-units. Indeed, gluten protein accumulation is a complex process subjected to spatial and temporal regulation as well as to environmental signaling. Individual proteins within each gluten protein class accumulate to different levels and can be influenced by environmental changes, nutrient availability, and management practices to different extents, suggesting that the corresponding genes have different basal levels of expression and possibly different regulatory elements in the promoter sequences. Our results confirmed the complexity of these processes; however, preliminary conclusions can be drawn. The inclusion of a leguminous crop in the rotation allowed an upregulation of 28–35 KDa sub-units of the Glia fraction, which corresponded to those Glia mainly involved in gluten technological proprieties due to the presence of six (α/β-) and eight (γ-) cysteine residues in the C-terminal domain, with γ-gliadins forming intra- and inter-chain disulfide bonds interacting with the HMW and LMW polymers [65]. This could be related to the higher N remobilized to the developing grains. On the other hand, ZT seemed to improve the upregulation of 42 KDa and 37 KDa sub-units of LMWW-GS, which are the most abundant ones, as previously observed in response to the foliar application of N fertilizers [9]. However, further investigations should be aimed at assessing the effects of CA systems on the expression of gluten protein sub-units.

5. Conclusions

Although the transition phases to sustainable agricultural approaches is generally challenging, the application of CA techniques, in particular crop diversification, has allowed for higher durum wheat quality and yield to be achieved. Our work demonstrates that under Mediterranean climates, the application of zero tillage and the introduction of leguminous-based crop rotations exert positive externalities after a six/seven-year period of CA adoption.

The presence of a layer of crop residues and soil organic carbon accumulation is as important as no tillage to improve yield, especially during dry seasons (2017). In addition to a higher yield, ZT and WF practices also ensured higher GPC accumulation in the kernels, probably due to the improved water and nutrient availabilities in line with crop demand during its growth and development.

The ameliorated soil conditions favored the overall crop status, also detected by the reflectance-based indices, as shown with a greater DWRemE and NRemE to the developing grains. Interestingly, the gluten characteristics were also advantaged. Some gluten fractions (i.e., Glia, HMW-GS, and LMW-GS) increased under ZT + WF conditions; ZT seemed to favor the upregulation of the two sub-units of LMW-GS (i.e., 42 KDa, the most abundant sub-unit and a marker for wheat quality, and 37 KDa), while WF positively influenced the upregulation of Glia in the molecular weight range of 28–35 KDa (involved in gluten proprieties).

Under Mediterranean areas, the adoption of CA in durum wheat-based farming systems is a useful choice to combine interesting yield with quality traits (i.e., GPC and LMW-GS) as well as with soil, water, air, and biological benefits.

Author Contributions: Conceptualization, F.S., G.V., and A.G.; Methodology, F.S. and A.G.; Software, S.D. and G.P.; Validation, F.S., G.V., and A.G.; Formal Analysis, S.D. and G.P.; Investigation, G.P. and S.D.; Resources, F.S. and M.P.; Data Curation, F.S., G.V., and A.G.; Writing-Original Draft Preparation, A.G., S.D., F.S., and G.V.; Writing-Review & Editing, M.P.; Visualization, F.S., G.V., S.D., G.P., and A.G.; Supervision, M.P.

Funding: This research did not receive any specific grants from funding agencies in the public, commercial, or not-for-profit sectors.

Conflicts of Interest: The authors declare no conflicts of interest.

References

1. Anderson, W.K.; Impiglia, A. Management of dryland wheat. In *Bread Wheat Improvement and Production*; Curtis, B.C., Rajaram, S., Gomez Macpherson, H., Eds.; Plant Production and Protection Series, No. 30; FAO: Rome, Italy, 2002; pp. 407–432.
2. Pisante, M.; Stagnari, F.; Acutis, M.; Bindi, M.; Brilli, L.; Di Stefano, V.; Carozzi, M. Conservation agriculture and climate change. In *Conservation Agriculture*; Farooq, M., Siddique, K., Eds.; Springer International Publishing: Cham, Switzerland, 2015; pp. 579–620.
3. Hobbs, P.R. Conservation agriculture: What is it and why is it important for future sustainable food production? *J. Agric. Sci.* **2007**, *145*, 127. [CrossRef]
4. Pisante, M.; Stagnari, F.; Grant, C.A. Agricultural innovations for sustainable crop production intensification. *Ital. J. Agron.* **2012**, *7*, 40. [CrossRef]
5. Stagnari, F.; Jan, S.; Angelica, G.; Michele, P. Sustainable agricultural practices for water quality protection. In *Water Stress and Crop Plants: A Sustainable Approach*; Ahmad, P., Ed.; John Wiley & Sons: Chichester, UK, 2016; pp. 75–85.
6. Farooq, M.; Flower, K.; Jabran, K.; Wahid, A.; Siddique, K.H. Crop yield and weed management in rainfed conservation agriculture. *Soil Tillage Res.* **2011**, *117*, 172–183. [CrossRef]
7. Laino, P.; Shelton, D.; Finnie, C.; De Leonardis, A.M.; Mastrangelo, A.M.; Svensson, B.; Lafiandra, D.; Masci, D. Comparative proteome analysis of metabolic proteins from seeds of durum wheat (cv. Svevo) subjected to heat stress. *Proteomics* **2010**, *10*, 2359–2368. [CrossRef] [PubMed]
8. Varzakas, T.; Kozub, N.; Xynias, I.N. Quality determination of wheat: Genetic determination, biochemical markers, seed storage proteins—Bread and durum wheat germplasm. *J. Sci. Food Agric.* **2014**, *94*, 2819–2829. [CrossRef] [PubMed]
9. Visioli, G.; Bonas, U.; Dal Cortivo, C.; Pasini, G.; Marmiroli, N.; Mosca, G.; Vamerali, T. Variations in yield and gluten proteins in durum wheat varieties under late-season foliar versus soil application of nitrogen fertilizer in a northern Mediterranean environment. *J. Sci. Food Agric.* **2018**, *98*, 2360–2369. [CrossRef] [PubMed]
10. Sissons, J.M. Role of durum wheat composition on the quality of pasta and bread. *Food* **2008**, *2*, 75–90.
11. De Vita, P.; Nicosia, O.L.D.; Nigro, F.; Platani, C.; Riefolo, C.; Di Fonzo, N.; Cattivelli, L. Breeding progress in morpho-physiological, agronomical and qualitative traits of durum wheat cultivars released in Italy during the 20th century. *Eur. J. Agron.* **2007**, *26*, 39–53. [CrossRef]
12. Dupont, F.; Altenbach, S. Molecular and biochemical impacts of environmental factors on wheat grain development and protein synthesis. *J. Cereal Sci.* **2003**, *38*, 133–146. [CrossRef]
13. Graziano, S.; Marando, S.; Prandi, B.; Boukid, F.; Marmiroli, N.; Francia, E.; Pecchioni, N.; Sforza, S.; Visioli, G.; Gullì, M. Technological Quality and Nutritional Value of Two Durum Wheat Varieties Depend on Both Genetic and Environmental Factors. *J. Agric. Food Chem.* **2019**, *67*, 2384–2395. [CrossRef]
14. Visioli, G.; Galieni, A.; Stagnari, F.; Bonas, U.; Speca, S.; Faccini, A.; Pisante, M.; Marmiroli, N. Proteomics of Durum Wheat Grain during Transition to Conservation Agriculture. *PLoS ONE* **2016**, *11*, e0156007. [CrossRef] [PubMed]
15. López-Bellido, R.; López-Bellido, L. Efficiency of nitrogen in wheat under Mediterranean conditions: Effect of tillage, crop rotation and N fertilization. *Field Crops Res.* **2001**, *71*, 31–46. [CrossRef]
16. Dupont, F.M.; Hurkman, W.J.; Vensel, W.H.; Tanaka, C.; Kothari, K.M.; Chung, O.K.; Altenbach, S.B. Protein accumulation and composition in wheat grains: Effects of mineral nutrients and high temperature. *Eur. J. Agron.* **2006**, *25*, 96–107. [CrossRef]
17. Kindred, D.R.; Verhoeven, T.M.; Weightman, R.M.; Swanston, J.S.; Agu, R.C.; Brosnan, J.M.; Sylvester-Bradley, R. Effects of variety and fertiliser nitrogen on alcohol yield, grain yield, starch and protein content, and protein composition of winter wheat. *J. Cereal Sci.* **2008**, *48*, 46–57. [CrossRef]
18. Galieni, A.; Stagnari, F.; Visioli, G.; Marmiroli, N.; Speca, S.; Angelozzi, G.; D'Egidio, S.; Pisante, M. Nitrogen fertilisation of durum wheat: A case of study in Mediterranean area during transition to conservation agriculture. *Ital. J. Agron.* **2016**, *11*, 12–23. [CrossRef]
19. Ranaivoson, L.; Naudin, K.; Ripoche, A.; Affholder, F.; Rabeharisoa, L.; Corbeels, M. Agro-ecological functions of crop residues under conservation agriculture. A review. *Agron. Sustain. Dev.* **2017**, *37*, 26. [CrossRef]

20. Gao, Y.; Li, Y.; Zhang, J.; Liu, W.; Dang, Z.; Cao, W.; Qiang, Q. Effects of mulch, N fertilizer, and plant density on wheat yield, wheat nitrogen uptake, and residual soil nitrate in a dryland area of China. *Nutr. Cycl. Agroecosyst.* **2009**, *85*, 109–121. [CrossRef]

21. Baker, J.M.; Ochsner, T.E.; Venterea, R.T.; Griffis, T.J. Tillage and soil carbon sequestration—What do we really know? *Agric. Ecosyst. Environ.* **2007**, *118*, 1–5. [CrossRef]

22. Nishigaki, T.; Sugihara, S.; Kilasara, M.; Funakawa, S. Soil nitrogen dynamics under different quality and application methods of crop residues in maize croplands with contrasting soil textures in Tanzania. *Soil Sci. Plant Nutr.* **2017**, *63*, 288–299. [CrossRef]

23. Stagnari, F.; Galieni, A.; Speca, S.; Cafiero, G.; Pisante, M. Effects of straw mulch on growth and yield of durum wheat during transition to Conservation Agriculture in Mediterranean environment. *Field Crops Res.* **2014**, *167*, 51–63. [CrossRef]

24. Cherr, C.M.; Scholberg, J.M.S.; McSorley, R. Green Manure as Nitrogen Source for Sweet Corn in a Warm–Temperate Environment. *Agron. J.* **2006**, *98*, 1173–1180. [CrossRef]

25. Stagnari, F.; Pisante, M. Managing faba bean residues to enhance the fruit quality of the melon (*Cucumis melo* L.) crop. *Sci. Hortic.* **2010**, *126*, 317–323. [CrossRef]

26. Knowler, D.; Bradshaw, B. Farmers' adoption of conservation agriculture: A review and synthesis of recent research. *Food Policy* **2007**, *32*, 25–48. [CrossRef]

27. Campiglia, E.; Mancinelli, R.; De Stefanis, E.; Pucciarmati, S.; Radicetti, E. The long-term effects of conventional and organic cropping systems, tillage managements and weather conditions on yield and grain quality of durum wheat (*Triticum durum* Desf.) in the Mediterranean environment of Central Italy. *Field Crops Res.* **2015**, *176*, 34–44. [CrossRef]

28. Govaerts, B.; Sayre, K.D.; Deckers, J. Stable high yields with zero tillage and permanent bed planting? *Field Crops Res.* **2005**, *94*, 33–42. [CrossRef]

29. Ververis, C.; Georghiou, K.; Danielidis, D.; Hatzinikolaou, D.; Santas, P.; Santas, R.; Corleti, V. Cellulose, hemicelluloses, lignin and ash content of some organic materials and their suitability for use as paper pulp supplements. *Bioresour. Technol.* **2007**, *98*, 296–301. [CrossRef]

30. Zadoks, J.C.; Chang, T.T.; Konzak, C.F. A decimal code for the growth stages of cereals. *Weed Res.* **1974**, *14*, 415–421. [CrossRef]

31. Arduini, I.; Masoni, A.; Ercoli, L.; Mariotti, M. Grain yield, and dry matter and nitrogen accumulation and remobilization in durum wheat as affected by variety and seeding rate. *Eur. J. Agron.* **2006**, *25*, 309–318. [CrossRef]

32. Sosulski, F.W.; Imafidon, G.I. Amino acid composition and nitrogen-to-protein conversion factors for animal and plant foods. *J. Agric. Food Chem.* **1990**, *38*, 1351–1356. [CrossRef]

33. Singh, V.; Moreau, R.A.; Doner, L.W.; Eckhoff, S.R.; Hicks, K.B. Recovery of Fiber in the Corn Dry-Grind Ethanol Process: A Feedstock for Valuable Coproducts. *Cereal Chem. J.* **1999**, *76*, 868–872. [CrossRef]

34. Rouse, J.W.; Haas, R.H.; Schell, J.A.; Deering, D.W. Monitoring vegetation systems in the Great Plains with ERTS. In Proceedings of the Third ERTS Symposium, NASA SP-351, Washington, DC, USA, 10–14 December 1973; pp. 309–317.

35. Gitelson, A.A.; Merzlyak, M.N. Signature Analysis of Leaf Reflectance Spectra: Algorithm Development for Remote Sensing of Chlorophyll. *J. Plant Physiol.* **1996**, *148*, 494–500. [CrossRef]

36. Peñuelas, J.; Pinol, J.; Ogaya, R.; Filella, I. Estimation of plant water concentration by the reflectance Water Index WI (R900/R970). *Int. J. Remote Sens.* **1997**, *18*, 2869–2875. [CrossRef]

37. Rondeaux, G.; Steven, M.; Baret, F. Optimization of soil-adjusted vegetation indices. *Remote Sens. Environ.* **1996**, *55*, 95–107. [CrossRef]

38. Onofri, A. Routine statistical analyses of field experiments by using an Excel extension. In Proceedings of the 6th National Conference Italian Biometric Society: "La Statistica Nelle Scienze Della Vita e Dell'ambiente", Pisa, Italy, 20–22 June 2007; pp. 93–99.

39. R Core Team. *R: A Language and Environment for Statistical Computing*; R Foundation for Statistical Computing: Vienna, Austria, 2017.

40. Scott, B.J.; Eberbach, P.L.; Evans, J.; Wade, L.J. *Stubble Retention in Cropping Systems in Southern Australia: Benefits and Challenges*; EH Graham Centre Monograph No. 1; Industry and Investment NSW: Wagga, Australia, 2010.

41. Van Kessel, C.; Hartley, C. Agricultural management of grain legumes: Has it led to an increase in nitrogen fixation? *Field Crops Res.* **2000**, *65*, 165–181. [CrossRef]

42. Trinsoutrot, I.; Recous, S.; Bentz, B.; Lineres, M.; Cheres, M.; Cheneby, D.; Nicolardot, B. Biochemical Quality of Crop Residues and Carbon and Nitrogen Mineralization Kinetics under Nonlimiting Nitrogen Conditions. *Soil Sci. Soc. Am. J.* **2000**, *64*, 918–926. [CrossRef]

43. Hadas, A.; Kautsky, L.; Goek, M.; Kara, E.E. Rates of decomposition of plant residues and available nitrogen in soil, related to residue composition through simulation of carbon and nitrogen turnover. *Soil Biol. Biochem.* **2004**, *36*, 255–266. [CrossRef]

44. Galieni, A.; Stagnari, F.; Speca, S.; D'Egidio, S.; Pagnani, G.; Pisante, M. Management of crop residues to improve quality traits of tomato (*Solanum lycopersicum* L.) fruits. *Ital. J. Agron.* **2017**, *12*. [CrossRef]

45. Lupwayi, N.Z.; Clayton, G.W.; Lea, T.; Beaudoin, J.L. Soil microbial biomass, functional diversity and crop yields following application of cattle manure, hog manure and inorganic fertilizers. *Can. J. Soil Sci.* **2005**, *85*, 193–201. [CrossRef]

46. Amato, G.; Ruisi, P.; Frenda, A.S.; Di Miceli, G.; Saia, S.; Plaia, A.; Giambalvo, D. Long-Term Tillage and Crop Sequence Effects on Wheat Grain Yield and Quality. *Agron. J.* **2013**, *105*, 1317–1327. [CrossRef]

47. Wuest, S.B.; Albrecht, S.L.; Skirvin, K.W. Crop residue position and interference with wheat seedling development. *Soil Tillage Res.* **2000**, *55*, 175–182. [CrossRef]

48. Rebetzke, G.J.; Bruce, S.E.; Kirkegaard, J.A. Longer coleoptiles improve emergence through crop residues to increase seedling number and biomass in wheat (*Triticum aestivum* L.). *Plant Soil* **2005**, *272*, 87–100. [CrossRef]

49. Masoni, A.; Ercoli, L.; Mariotti, M.; Arduini, I. Post-anthesis accumulation and remobilization of dry matter, nitrogen and phosphorus in durum wheat as affected by soil type. *Eur. J. Agron.* **2007**, *26*, 179–186. [CrossRef]

50. Tahir, I.S.A.; Nakata, N. Remobilization of Nitrogen and Carbohydrate from Stems of Bread Wheat in Response to Heat Stress during Grain Filling. *J. Agron. Crop Sci.* **2005**, *191*, 106–115. [CrossRef]

51. Zhang, J.; Shi, L.; Shi, A.; Zhang, Q. Photosynthetic Responses of Four Hosta Cultivars to Shade Treatments. *Photosynthetica* **2004**, *42*, 213–218. [CrossRef]

52. Zhang, J.; Xu, Y.; Yao, F.; Wang, P.; Guo, W.; Li, L.; Yang, L. Advances in estimation methods of vegetation water content based on optical remote sensing techniques. *Sci. China Ser. E Technol. Sci.* **2010**, *53*, 1159–1167. [CrossRef]

53. Peñuelas, J.; Filella, I.; Biel, C.; Serrano, L.; Savé, R. The reflectance at the 950–970 nm region as an indicator of plant water status. *Int. J. Remote Sens.* **1993**, *14*, 1887–1905. [CrossRef]

54. Lovera, K.R.; Davies, W.P.; Cannon, N.D.; Conway, J.S. Influence of tillage systems and nitrogen management on grain yield, grain protein and nitrogen-use efficiency in UK spring wheat. *J. Agric. Sci.* **2016**, *154*, 1437–1452. [CrossRef]

55. Stagnari, F.; Galieni, A.; Pisante, M. Drought stress effects on crop quality. In *Water Stress and Crop Plants: A Sustainable Approach*; Parvaiz, A., Ed.; John Wiley & Sons, Ltd.: Amsterdam, The Netherlands, 2016; pp. 375–392.

56. Gao, X.; Lukow, O.M.; Grant, C.A. Grain concentrations of protein, iron and zinc and bread making quality in spring wheat as affected by seeding date and nitrogen fertilizer management. *J. Geochem. Explor.* **2012**, *121*, 36–44. [CrossRef]

57. Zhang, P.; Ma, G.; Wang, C.; Lu, H.; Li, S.; Xie, Y.; Zhu, Y.; Guo, T. Effect of irrigation and nitrogen application on grain amino acid composition and protein quality in winter wheat. *PLoS ONE* **2017**, *12*, e0178494. [CrossRef]

58. Mrabet, R.; Ibno-Namr, K.; Bessam, F.; Saber, N.; Ibno-Namr, K. Soil chemical quality changes and implications for fertilizer management after 11 years of no-tillage wheat production systems in semiarid Morocco. *Land Degrad. Dev.* **2001**, *12*, 505–517. [CrossRef]

59. Pellegrino, E.; Bosco, S.; Ciccolini, V.; Pistocchi, C.; Sabbatini, T.; Silvestri, N.; Bonari, E. Agricultural abandonment in Mediterranean reclaimed peaty soils: Long-term effects on soil chemical properties, arbuscular mycorrhizas and CO_2 flux. *Agric. Ecosyst. Environ.* **2015**, *199*, 164–175. [CrossRef]

60. Ciccolini, V.; Bonari, E.; Ercoli, L.; Pellegrino, E. Phylogenetic and multivariate analyses to determine the effect of agricultural land-use intensification and soil physico-chemical properties on N-cycling microbial communities in drained Mediterranean peaty soils. *Biol. Fertil. Soils* **2016**, *52*, 811–824. [CrossRef]

61. Edwards, N.M.; Mulvaney, S.J.; Scanlon, M.G.; Dexter, J.E. Role of gluten and its components in determining durum semolina dough viscoelastic proprieties. *Cereal Chem.* **2003**, *80*, 755–763. [CrossRef]

62. Hurkman, W.J.; Tanaka, C.K.; Vensel, W.H.; Thilmony, R.; Altenbach, S.B. Comparative proteomic analysis of the effect of temperature and fertiliser on gliadin and glutenin accumulation in the developing endosperm and flour from *Triticum aestivum* L. cv. Butte 86. *Proteome Sci.* **2013**, *11*, 8. [CrossRef] [PubMed]

63. Wan, Y.; Shewry, R.P.; Hawkesford, J. A novel family of γ-gliadin genes are highly regulated by nitrogen supply in developing wheat grain. *J. Exp. Bot.* **2013**, *64*, 161–168. [CrossRef] [PubMed]

64. Visioli, G.; Comastri, A.; Imperiale, D.; Paredi, G.; Faccini, A.; Marmiroli, N. Gel-based and gel-free analytical methods for the analysis of HMW-GS and LMW-GS in wheat flour. *Food Anal. Methods* **2016**, *9*, 469–474. [CrossRef]

65. Wieser, H. Chemistry of gluten proteins. *Food Microbiol.* **2007**, *24*, 115–119. [CrossRef] [PubMed]

Article

Long-Term Green Manure Rotations Improve Soil Biochemical Properties, Yield Sustainability and Nutrient Balances in Acidic Paddy Soil under a Rice-Based Cropping System

Muhammad Qaswar [1,†], Jing Huang [1,2,†], Waqas Ahmed [1], Dongchu Li [1,2], Shujun Liu [1,2], Sehrish Ali [1], Kailou Liu [1,3], Yongmei Xu [4], Lu Zhang [1,2], Lisheng Liu [1,2], Jusheng Gao [1,2,*] and Huimin Zhang [1,2,5,*]

[1] National Engineering Laboratory for Improving Quality of Arable Land, Institute of Agricultural Resources and Regional Planning, Chinese Academy of Agricultural Sciences, Beijing 100081, China; mqaswar2@gmail.com (M.Q.); huangjing@caas.cn (J.H.); waqasmalik184@hotmail.com (W.A.); lidongchu@caas.cn (D.L.); liushujun@caas.cn (S.L.); Sehrishbakhsh788@outlook.com (S.A.); liukailou@126.com (K.L.); zhanglu01@caas.cn (L.Z.); liulisheng@caas.cn (L.L.)

[2] National Observation Station of Qiyang Agri-Ecology System, Institute of Agricultural Resources and Regional Planning, Chinese Academy of Agricultural Sciences, Qiyang 426182, China

[3] Jiangxi Institute of Red Soil, National Engineering and Technology Research Center for Red Soil Improvement, Nanchang 330046, China

[4] Institute of Soil, Fertilizer and Agricultural Water Conservation, Xinjiang Academy of Agricultural Sciences, Urumqi 830091, China; xym1973@163.com

[5] College of Agriculture, Henan University of Science and Technology, Luoyang 471000, China

[*] Correspondence: gaojusheng@caas.cn (J.G.); zhanghuimin@caas.cn (H.Z.)

[†] The authors equally contributed to this work.

Received: 4 November 2019; Accepted: 18 November 2019; Published: 20 November 2019

Abstract: Cultivation of green manure (GM) crops in intensive cropping systems is important for enhancing crop productivity through soil quality improvement. We investigated yield sustainability, nutrient stocks, nutrient balances and enzyme activities affected by different long-term (1982–2016) green manure rotations in acidic paddy soil in a double-rice cropping system. We selected four treatments from a long-term experiment, including (1) rice-rice-winter fallow as a control treatment (R-R-F), (2) rice-rice-milkvetch (R-R-M), (3) rice-rice-rapeseed (R-R-R), and (4) rice-rice-ryegrass (R-R-G). The results showed that different GM rotations increased grain yield and the sustainable yield index compared with those of the R-R-F treatment. Compared with those of R-R-F, the average grain yield of early rice in R-R-M, R-R-R, and R-R-G increased by 45%, 29%, and 27%, respectively and that of late rice increased by 46%, 28%, and 26%, respectively. Over the years, grain yield increased in all treatments except R-R-F. Green manure also improved the soil chemical properties (SOM and total and available N and P), except soil pH, compared to those of the control treatment. During the 1983–1990 cultivation period, the soil pH of the R-R-M treatment was lower than that of the R-R-F treatment. The addition of green manure did not mitigate the soil acidification caused by the use of inorganic fertilizers. The soil organic matter (SOM), total nitrogen (TN) and total phosphorus (TP) contents and stocks of C, N and P increased over the years. Furthermore, GM significantly increased phosphatase and urease activities and decreased the apparent N and P balances compared with those in the winter fallow treatment. Variance partitioning analysis revealed that soil properties, cropping systems, and climatic factors significantly influenced annual grain yield. Aggregated boosted tree (ABT) analysis quantified the relative influences of the different soil properties on annual grain yield and showed that the relative influences of TN content, SOM, pH, and TP content on annual crop yield were 27.8%, 25.7%, 22.9%, and 20.7%, respectively. In conclusion, GM rotation is beneficial for sustaining high crop yields by improving soil biochemical properties and reducing N and P balances in acidic soil under double- rice cropping systems.

Agronomy **2019**, *9*, 780

Keywords: Acidic soil; crop rotation; enzyme activities; green manure; sustainable yield index; nutrient balance

1. Introduction

Long-term fertility experiments have been important resources for exploring nutrient cycling and performing overall assessments of soil fertility [1]. Long-term experiments provide a platform for investigating crop yield trends, productivity, soil quality, and other factors that contribute to agricultural sustainability [2]. Worldwide, irrigation and fertilization are the two most important factors for obtaining higher agricultural productivity and sustainability [3]. However, other factors, such as climate, pests, cultivars, soil types, and fertilization patterns, also contribute to changing yield trends and sustainability in long-term cropping systems [4].

During the past decades, due to the continuous and unwise use of fertilizers, rice (*Oryza sativa*) yield has increased significantly, but nutrient use efficiencies have decreased. This over fertilization has not only resulted in economic losses to farmers due to the misuse of expensive fertilizers but has also led to degradation of the soil and affected air, and water quality [5,6]. The negative effects of degraded soil quality, such as acidification and poor soil structure, on rice yield have been found in many studies [7,8]. Crop rotation with legumes such as green manure (GM) not only improves the soil quality and soil water content by increasing soil organic matter but also increases nutrient availability through atmospheric N fixation with rhizobia [9]. Studies have found that GM decreased the population, species, and density of unwanted weeds in rice fields [10]. Similarly, GM not only improves the soil biological (microbial activities) and physical properties [11,12] but also increases the carbon (C) and N cycling [13–15]. Furthermore, GM increases the quantity of water-stable macroaggregates in the topsoil and increases the capacity of microbes to mineralize organic N; consequently, it can enhance the N supply capacity and crop productivity [16,17]. Cultivation of GMs such as Chinese milkvetch (*Astragalus sinicus* L.) in paddy soils can exploit their natural qualities and improve rice productivity with minimal economic and environmental losses [18,19].

Previous studies of different long-term experiments at different locations have indicated decreasing yield trends after a few years of continuous cropping under rice-wheat (*Triticum aestivum*) rotations [20]. The cultivation of input-responsive high-yielding cultivars has been increased, compelling farmers to apply high doses of chemical fertilizers. Excessive use of inorganic fertilizers is a common practice to meet the nutrient need of high-yield varieties. Different long-term experiments have indicated that long-term continuous and excessive chemical fertilization degraded the soil quality and increased acidification, which resulted in poor crop sustainability [21,22].

The sustainable yield index (SYI) represents the actual yields over a long period. A higher sustainable yield index indicates that an area will produce an acceptable yield over the years through better cultivation practices [20–22]. SYI is a quantitative measure with which to assess the sustainability of an agricultural practice [23]. Long-term experiments provide one of the means of measuring the sustainability of an agricultural system [2]. Data records from long-term experiments can be used as early warning systems for the future [24]. In China, many long-term fertility experiments have indicated wide variability in crop yields that has been attributed to the continued reduction of soil fertility [6,25]. To monitor the long-term fertility and yield response of crops to fertilization, the first step in analyzing long-term experiments is to identify the sustainability of crop productivity. Previous studies have mainly focused on the effects of GM on rice production in short- term experiments or in a single cropping system [26,27]. However, little is known about the benefits of different GM rotations for sustainable rice productivity and nutrient balance, especially in paddy soils under long-term rice-based cropping systems [28–30]. The aims of this study were to investigate the effect of different green manure rotations on SYI, crop yield, apparent nutrient balance, and enzyme activities in an acidic paddy soil.

2. Materials and Methods

2.1. Site Description

A long-term (1982–2016) field experiment was conducted at the red soil experiment station of the Chinese Academy of Agricultural Sciences in Qiyang County (26°45′42″ N, 111°52′32″ E), Hunan Province, China. The climate in this region is a humid subtropical monsoon with a mean annual rainfall of 1290 mm. The main rainfall period is from April to the end of June, during the period of early-season rice cultivation. The drought period is from August to October during the late-season rice cultivation period, and the mean annual temperature is 17.8 °C. The mean annual temperature (MAT) and mean annual precipitation (MAP) for the experimental period are shown in Figure S1. The paddy soil of this region is categorized as a Ferralic Cambisol [31], which was made by Quaternary red clay. The initial soil properties were pH 6.1, soil organic matter (SOM) 20.4 g kg^{-1}, total N 0.94 g kg^{-1}, total P 0.65 g kg^{-1}, total K 10.6 g kg^{-1}, alkali-hydrolyzed N 148 mg kg^{-1}, Olsen-P 17.6 mg kg^{-1} and available K 175 mg kg^{-1} in the 0–20 cm soil layer.

2.2. Experimental Design

The four cropping treatments were rice-rice-winter fallow (R-R-F), rice-rice-milkvetch (*Astragalus sinicus* L.) (R-R-M), rice-rice-rapeseed (*Brassica napus* L.) (R-R-R) and rice-rice-ryegrass (*Lolium multiflorum*) (R-R-G). All treatments had three replicates arranged in a randomized complete block design. The area of each replicate was 37.5 m^2 (2.5 m × 15.0 m). Each replicate plot was separated from the adjacent replicates by a 60-cm cement barrier to prevent water and nutrient contamination from the nearby plot. The total fertilizer application rates for each rice growing season were 153 kg ha^{-1} N, 84 kg ha^{-1} P$_2$O$_5$, and 129 kg ha^{-1} K$_2$O. The compound fertilizer (600 kg ha^{-1}) containing 14% each of N, P$_2$O$_5$, and K$_2$O was applied as basal application 1 day before rice transplantation, while urea and potassium chloride for N (69 kg ha^{-1}) and K$_2$O (45 kg ha^{-1}), respectively, were topdressed 6 to 10 days after rice transplantation. Green manure crops were sown in the middle of October after the late rice was harvested and were plowed into the soil as a whole crop before the early rice transplantation (in the first week of April) every year. The seeding rates of milkvetch, rapeseed, and ryegrass were 37.5, 7.5, and 15.0 kg ha^{-1}, respectively. Early and late rice seedlings were transplanted with spacing of 20 cm × 20 cm and 20 cm × 25 cm, respectively, and each hill contained three seedlings. All rice straw (except the rice stubble) was removed from the plots after each seasonal rice harvest. No fertilizer was applied to the green manure crop in winter. The fresh biomass of the incorporated ryegrass, milkvetch, and rapeseed was 5700 kg ha^{-1}, 6000 kg ha^{-1}, and 6200 kg ha^{-1}, while the dry matter was 1126 kg ha^{-1}, 1276 kg ha^{-1}, and 1249 kg ha^{-1}, respectively. Milkvetch contained 43.7% C and 3.9% N contents, ryegrass contained 41.5% C and 1.8% N contents, rapeseed contained 42.3% C and 1.76% N contents in samples collected in 2017. Conventional routine management practices were used for pest management and irrigation. The rice varieties used locally were selected and were replaced every 3–5 years (Table S1). Flooding was maintained throughout the early rice season, and the field was drained at the late rice ripening stage [32,33].

2.3. Sampling and Analysis

Early and late rice grain and straw samples were manually harvested at six randomly selected points from each plot and weighed. Grain and straw samples were oven-dried at 105 °C for 30 min and then heated at 70 °C to a constant weight for dry matter and total N and P determination. Dried grain and straw samples were ground and digested with H$_2$SO$_4$-H$_2$O$_2$ at 260–270 °C. The total plant N and P were determined using the semimicro Kjeldahl digestion method and the vanadomolybdate yellow method, respectively (Jackson 1969; Nelson and Sommers 1980). The soil samples were collected every year from the starting date of the experiment (1983) to 2016. Soil samples were collected at a depth of 0–20 cm from five randomly selected points in each plot after late rice harvest, and plant materials and stones were removed from the soil. Soil samples were then mixed, air-dried, and sieved (0.2 mm) for the determination of soil chemical properties. Soil organic carbon (SOC) was determined with the potassium dichromate oxidation method [34], and total N (TN) was determined with the Kjeldahl method [35]. Soil total P (TP) was determined according to Murphy and Riley [36]. Alkali-hydrolyzable mineral N (AN) was determined according to Lu [37]. Available P (AP) was extracted using 0.5 mol L^{-1} $NaHCO_3$, and the measurement of AP was performed by the Olsen method [38].

For enzyme activity determination, soil samples were collected in 2017 after the late rice harvest. Enzyme activities were determined in field-moist soil samples. A portion of the subsamples was used to determine the soil moisture content before the analysis of enzyme activities by oven drying at 105 °C for 24 h. Acid phosphomonoesterase (AcP) and phosphodiesterase (DP) activities were determined following Tabatabai [39]. Briefly, 1 g of soil was incubated at 37 °C for 1 h after adding *p*- nitrophenyl phosphate as a substrate solution with a buffer solution of pH 6.5 for AcP. DP was assayed using a substrate solution of bis-p-nitrophenyl phosphate with a buffer solution of pH 8. Urease activity was determined following the colorimetric method [40]. Briefly, 5 g of moist soil was incubated at 37 °C for 2 h after adding urea solution as the substrate. A modified Berthelot reaction was followed to extract the released ammonium by potassium chloride solution. The activities of Acp and DP were expressed as $\mu g\ NP\ g^{-1}$ soil h^{-1}, and urease activity was expressed as $\mu g\ N\ g^{-1}$ soil $2h^{-1}$.

Stocks (S) of SOC, total N and total P were estimated as follows:

$$S = SOC \times BD \times H \times 10^{-1} \tag{1}$$

where, S is the stock (ton ha^{-1}) of soil organic carbon (SOC), total N or total P. The concentration of soil organic carbon (g kg^{-1}), total N (g kg^{-1}) or total P (g kg^{-1}) is used in the equation. BD is the soil bulk density (g cm^{-3}), and H is the depth of soil sampling (cm). BD is the soil bulk density (g cm^{-3}), and H is the depth of soil sampling (cm). In the present study, the soil bulk density was not measured directly from the field and was calculated using the following equation in ref. [41].

$$BD = -0.0048 \times SOC + 1.377 \tag{2}$$

The apparent nutrient balance (AB) (kg ha^{-1} $year^{-1}$) is the difference in nutrient inputs to the field through fertilization and nutrient outputs from the field in harvested biomass [42]. The apparent nutrient balance is based on the soil surface (lower plow) balance but does not incorporate the potential losses incurred from runoff or soil erosion. A positive value of the apparent nutrient balance indicates a nutrient surplus, and a negative value of the apparent nutrient balance (nutrient deficit) indicates declining soil fertility [42]. The environmental risk threshold values are 45 kg ha^{-1} $year^{-1}$ for the P balance and 180 kg ha^{-1} $year^{-1}$ for the N balance [43,44]. The apparent nutrient balance was estimated using the following equation [44]:

$$Apparent\ nutrient\ balance = F_{input} - N_{output} \tag{3}$$

where F_{input} is the nutrient input from fertilizers and N_{output} is nutrient offtake by the crop in the harvested biomass (nutrients assimilated into straw and grains of rice).

2.4. Sustainable Yield Index (SYI)

SYI is a quantitative measure with which to assess the sustainability of an agricultural practice [23]. The sustainable yield index (SYI) was estimated using the following equation [45]:

$$Sustainable\ Yield\ Index(SYI) = \frac{Ymean - \sigma}{Ymax} \tag{4}$$

where Y_{mean} is the mean yield of a treatment, σ is the treatment standard deviation, and Y_{max} is the maximum yield in the experiment over the years for each treatment.

2.5. Statistical Analysis

We evaluated the significant differences among treatments by one-way ANOVA, and interactive effects of treatments and cultivation years were evaluated by two-way ANOVA followed by least significant difference (LSD) test at the $p = 0.05$ level significance [46] using SPSS 16.0 software (SPSS, Chicago, IL, USA). Variance partitioning analysis (VPA) was performed to evaluate the contributions of soil factors (soil pH, SOM, total N, and total P), cropping systems, climatic factors (MAT and MAP), and their interactions in changing the annual rice yield over the cropping period (1984–2016) using CANOCO for Windows (version 5.0). Aggregated boosted tree (ABT) analysis was performed to investigate the relative influence of the important soil factors on annual crop yield using the 'gbmplus' package in R software (version 3.3.3) [47].

3. Results

3.1. Crop Yield and Sustainability Index

The results showed that long-term rotations of different green manure (GM) types in the double-rice cropping system significantly ($p \leq 0.05$) increased grain yield and sustainable yield index (SYI) compared to those in the winter fallow system (Table 1 and Figure 1). The highest increase in the average grain yield and SYI over the period of early and late season cropping was under the milkvetch rotation (R-R-M). Compared to R-R-F, the average increases in grain yield in R-R-M, R-R-R, and R-R-G were 45%, 29%, and 27%, respectively, in early rice and 46%, 28%, and 26%, respectively, in late rice. Compared to those in the winter fallow system, crop yields of both early and late rice were increased over the year, but the highest increase was under the R-R-M treatment (Figure 1). The increases in SYI in R-R-M, R-R-R. and R-R-G compared to R-R-F were 36%, 21%. and 23%, respectively, in early rice, 24%, 20.3%. and 10%, respectively, in late rice and 19%, 16%. and 9.5%, respectively, in double-rice cropping systems overall.

Table 1. Effect of long-term green manure rotations on the sustainability yield index (SYI) of double-rice cropping systems in acidic paddy soil.

Treatments	Average Grain Yield (kg ha^{-1})		Yield Sustainability Index		
	Early Rice Yield	**Late Rice Yield**	**Early Rice**	**Late Rice**	**Double Rice Cropping**
R-R-F	4532 ± 812 c	4022 ± 780 c	0.53 c	0.59 c	0.63 c
R-R-M	6559 ± 819 a	5859 ± 741 a	0.72 a	0.73 a	0.75 a
R-R-R	5853 ± 917 b	5135 ± 623 b	0.64 b	0.71 a	0.73 ab
R-R-G	5769 ± 930 b	5076 ± 728 b	0.65 b	0.65 b	0.69 b

Average grain yield is based on mean of grain yield of all cultivation years (1982–2016). Means followed by different letters are significantly ($p \leq 0.05$) different from each other according to Tukey's LSD test.

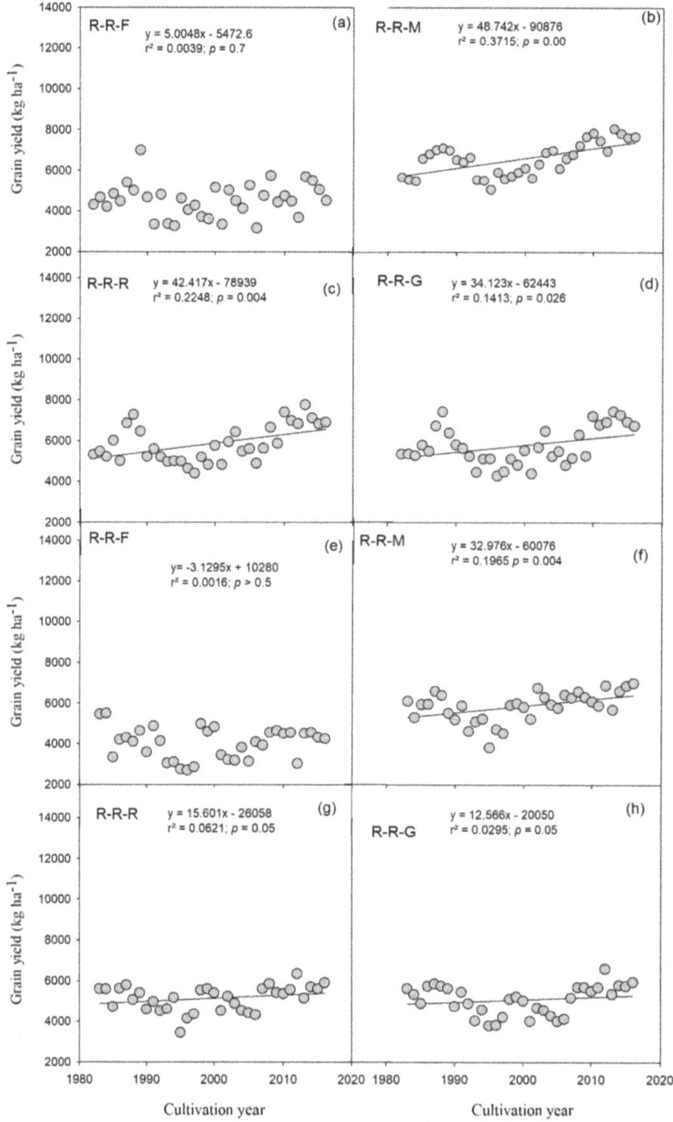

Figure 1. Linear regression showing trends of rice grain yields over the years (1982–2016) under green manure rotations in the rice-based cropping system.

3.2. Soil Nutrient Content, Nutrient Stocks, and Apparent Nutrient Balance

The treatment × year interaction significantly affected soil properties (Table 2). During the fertilization period of 1983–1990, soil pH was not significantly different between the R-R-G and R-R-R treatments. Among all treatments, during 1983–1990, the soil pH was lowest under the R-R-M treatment. The soil pH under the R-R-F and R-R-M treatments during 1983–1990 was decreased by 0.98% and 2% compared to the initial soil pH. During the period of 1991–2000, the soil pH among all treatments was decreased compared to the soil pH during 1983–1990, and it was highest under the R-R-R treatment. In all treatments, over the years, the SOM content was higher than the initial SOM content. Compared to the initial SOM content, the increases in SOM content under the R-R-F, R-R-M, R-R-G, and R-R-R treatments were 7.3%, 23%, 12.2%, and 12.7%, respectively, during 1983–1990 and 11.3%, 20%, 15.6%, and 16.6%, respectively, during 1991–2000. The respective increases in SOM content during 2001–2016 under the R-R-F, R-R-M, R-R-G, and R-R-R treatments were 33.8%, 40.6%, 36.7%, and 41.2%, respectively, compared to the initial SOM content. The soil TN content in all treatments was higher in all fertilization periods compared to the initial soil TN content. Compared to that in the R-R-F treatment, the increase in the soil TN content under the R-R-M, R-R-G, and R-R-R treatments was 15.9%, 14%, and 7.5%, respectively, during the period 1983–1990, by 23%, 7.0%, and 16%, respectively, during 1991–2000, and by 9.8%, 4.6%, and 3.9%, respectively, during 2001–2016. Over the years, the soil TP content increased in all treatments compared with the initial soil TP content. The soil TP content in the R-R-F, R-R-M, R-R-G, and R-R-R treatments increased by 6%, 7.7%, 12%, and 7.7%, respectively, during 1983–1990. During 1991–2000, the TP content was highest under the R-R-R treatment and did not show a significant difference among the R-R-F, R-R-M, and R-R-G treatments. During 2001–2016, the soil TP content also did not show a significant difference between the R-R-F and R-R-M treatments or between the R-R-G and R- R-R treatments. After 34 years of cropping, the increases in AN under R-R-M, R-R-R and R-R-G compared to R-R-F were 22%, 6.4%, and 16.5%, respectively, in early rice and 11%, 6%, and 5.6%, respectively, in late rice (Figure 2). Similarly, compared to R-R-F, the increases in Olsen-P under R-R-M, R-R-R, and R-R-G were 10.5%, 25.7%, and 22%, respectively, in early rice and 11.7%, 19.3%, and 25.5%, respectively, in late rice. Green manure increased stocks of C, N, and P compared to those in the winter fallow treatment (Figure 3). On average, across the years, the soil C and N stocks were not different between the R-R-R and R-R-G treatments. Over the years, soil C, N, and P stocks increased in all treatments, and the lowest increase was under the R-R-F treatment.

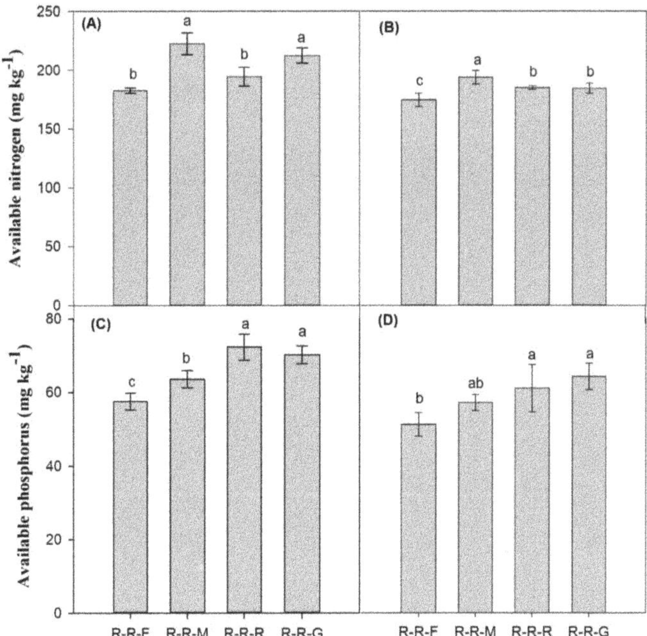

Figure 2. Effect of different long-term green manure rotations on soil available N after early rice harvest (**A**) and after late rice harvest (**B**), and available P after early rice harvest (**C**) and after late rice harvest (**D**). Error bars represent ± standard deviations; different letters over the bars indicate significant ($p \leq$ 0.05) differences according to Tukey's LSD test.

Table 2. Effect of long-term green manure rotations on soil pH, soil organic matter (SOM), total nitrogen, and total phosphorus content in acidic paddy soil.

Year	Treatment	pH	SOM (g kg^{-1})	TN (g kg^{-1})	TP (g kg^{-1})
Initial values		6.1	20.4	0.94	0.65
1983–1990	R-R-F	6.04 ± 0.06 bc	21.9 ± 0.34 c	1.07 ± 0.03 c	0.69 ± 0.01 b
	R-R-M	5.98 ± 0.07 c	25.1 ± 0.29 a	1.24 ± 0.03 a	0.70 ± 0.02 b
	R-R-G	6.07 ± 0.07 ab	22.9 ± 0.84 b	1.22 ± 0.02 a	0.74 ± 0.02 a
	R-R-R	6.13 ± 0.01 a	23.0 ± 0.57 b	1.15 ± 0.01 b	0.70 ± 0.02 b
1991–2000	R-R-F	5.45 ± 0.03 c	22.7 ± 0.45 c	1.13 ± 1.01 d	0.90 ± 0.02 b
	R-R-M	5.70 ± 0.13 a	24.5 ± 0.44 a	1.39 ± 0.02 a	0.91 ± 0.02 b
	R-R-G	5.57 ± 0.04 b	23.6 ± 0.51 b	1.21 ± 0.02 c	0.89 ± 0.04 b
	R-R-R	5.80 ± 0.04 a	23.8 ± 0.35 b	1.31 ± 0.02 b	0.96 ± 0.04 a
2001–2016	R-R-F	5.78 ± 0.05 ns	27.3 ± 0.52 c	1.67 ± 0.03 c	1.11 ± 0.03 b
	R-R-M	5.83 ± 0.08	28.7 ± 0.34 a	1.84 ± 0.09 a	1.13 ± 0.08 b
	R-R-G	5.77 ± 0.04	27.9 ± 0.50 a	1.75 ± 0.03 b	1.26 ± 0.08 a
	R-R-R	5.87 ± 0.08	28.8 ± 0.43 b	1.74 ± 0.04 b	1.22 ± 0.07 a
Two-way ANOVA	Treatment (T)	***	***	***	***
	Year (Y)	***	***	***	***
	T × Y	***	***	***	*

Values are means ± standard deviations. Means followed by different letters are significantly ($p \leq 0.05$) different from each other according to Tukey's LSD test. * Significant ($p \leq 0.05$). *** Very highly significant ($p \leq 0.001$). ns: nonsignificant ($p > 0.05$). Note: We divided the long-term (1983–2016) data into three groups by taking the average of years 1983–1990, 1991–2000, and 2001–2016 to better describe the long-term changes in the results.

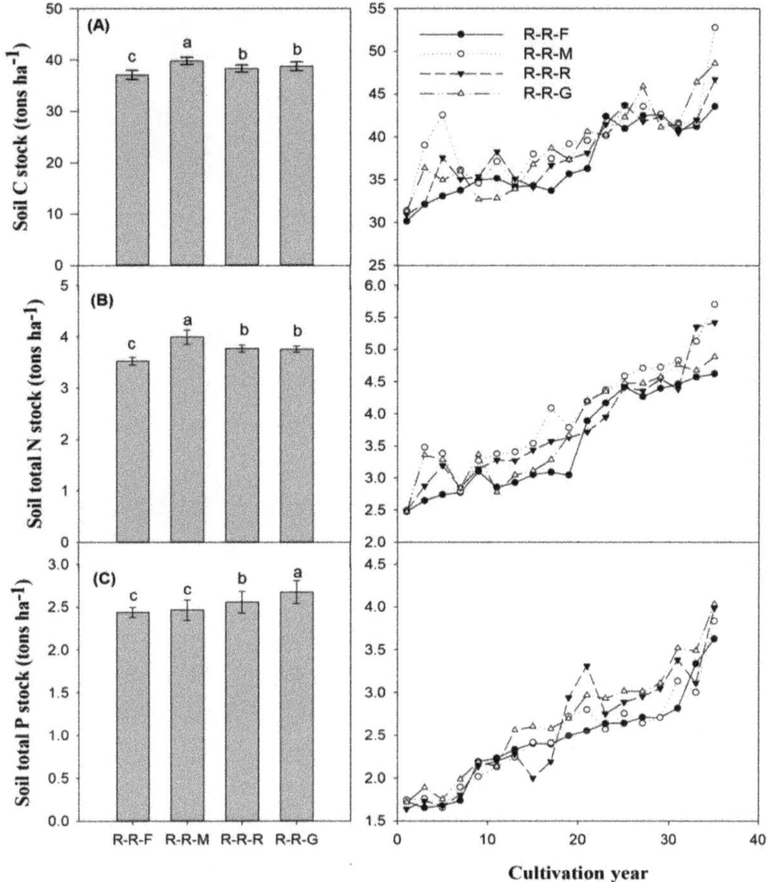

Figure 3. Effect of different long-term green manure rotations on stocks of soil organic carbon (**A**), total N (**B**) and total P (**C**) in acidic paddy soil. Error bars represent ± standard deviations; different letters over the bars indicate significant ($p \leq 0.05$) differences according to Tukey's LSD test.

GM rotations significantly ($p \leq 0.05$) influenced the annual apparent N and P balance in the long-term experiment on the double-rice cropping system (Figure 4). In the control treatment (R- R- F), the apparent N and P balances were highest compared with those in all other GM treatments. The annual N balance in the R-R-M, R-R-R and R-R-R treatments was decreased by 38%, 21%, and 24%, respectively, and the annual P balance was decreased by 36%, 27%, and 34%, respectively, compared with those of R-R-F in the double-rice cropping system.

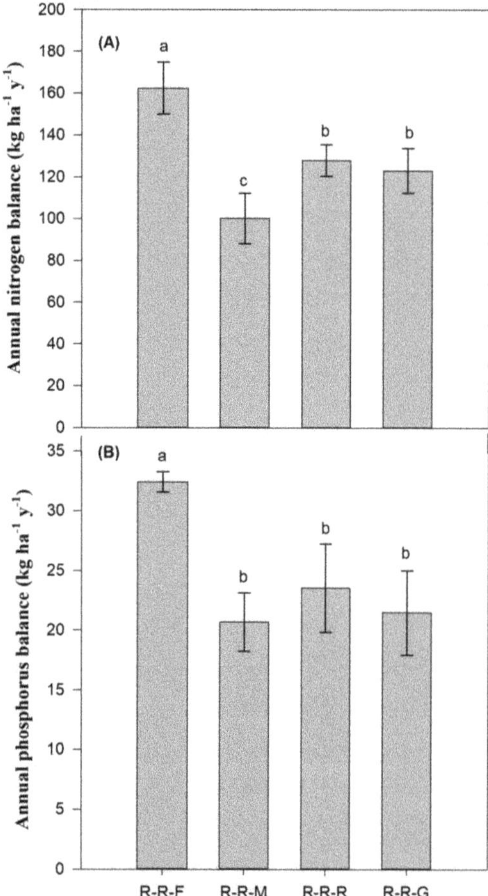

Figure 4. Effect of different long-term green manure rotations on annual apparent (**A**) N and (**B**) P balances in acidic paddy soil. Error bars represent ± standard deviations; different letters over the bars indicate significant ($p \leq 0.05$) differences according to Tukey's LSD test.

3.3. Enzyme Activities

Different long-term GM rotations significantly influenced phosphatase and urease activities (Figure 5). Acid phosphomonoesterase (AcP), phosphodiesterase (DP), and urease activities were highest in the R-R-M treatment and lowest in the R-R-F treatment. Compared with those in the control, the R-R-M, R-R-R, and R-R-G treatments increased AcP activity by 68%, 52%, and 50%, DP activity by 83%, 70%, and 65% and urease activity by 72%, 20%, and 25%, respectively. Available N and P significantly positively affected AcP and DP activities (Figure S2). Available N also showed significant positive linear relationship with urease activities.

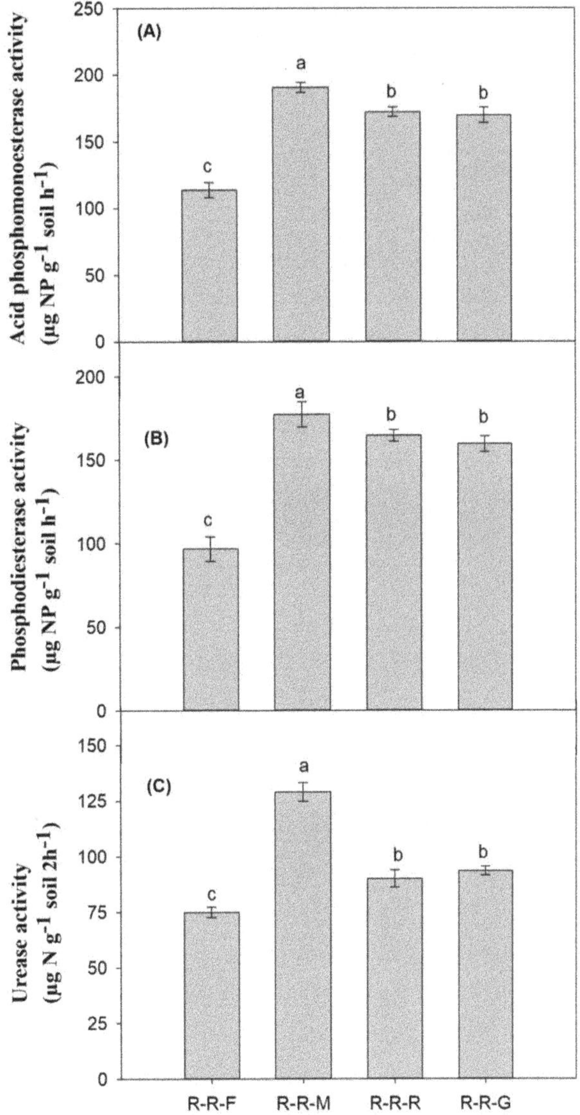

Figure 5. Effect of different long-term green manure rotations on enzyme activities in acidic paddy soil. Error bars represent ± standard deviations; different letters over the bars indicate significant ($p \leq 0.05$) differences according to Tukey's LSD test.

3.4. Factors Influencing Crop Yield

We investigated the contributions of different factors influencing crop yield by variance partitioning analysis and aggregated boosted tree (ABT) analysis. The changes in early and late rice yield over the long period of cropping were due to variations in different soil and climatic factors in the long-term double-cropping system. Variance partitioning analysis showed that soil properties (pH, SOM, total N, and total P) accounted for 11.3% of the variation, the cropping system accounted for 8.8%, and climatic factors (MAT and MAP) accounted for 4.6% of the variation in annual grain yield. The total variance explained by VPA was 36%, and the unexplained proportion was 64% for annual grain yield (Figure 6). Aggregated boosted tree (ABT) analysis indicated that soil TN, SOM, soil pH, and soil TP contents were the most influential factors on crop yield (Figure 7).

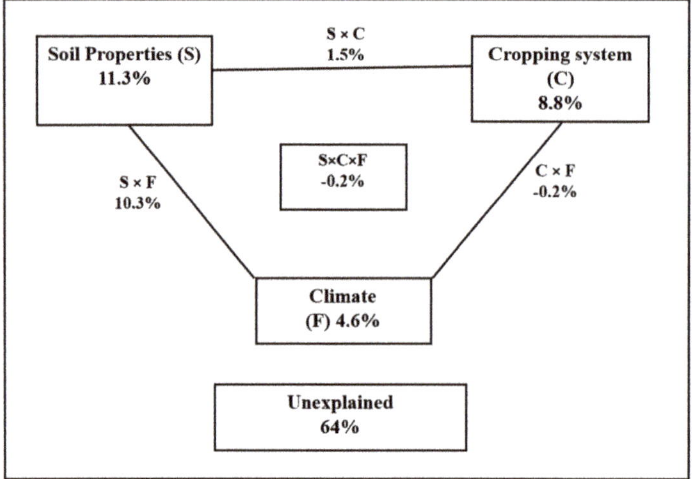

Figure 6. Variance partitioning analysis (VPA) indicating the effects of soil properties, cropping systems, and climatic factors on annual crop yield under long-term green manure rotations in rice-based cropping systems.

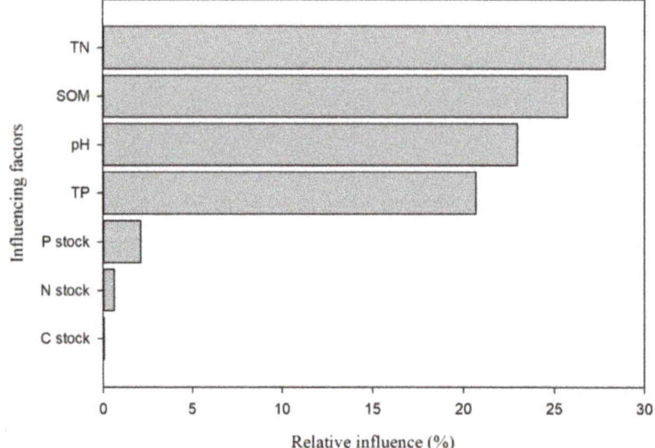

Figure 7. Relative influence (%) of different predictors on annual crop yield by aggregated boosted tree (ABT) analysis. Long-term data (1983–2016) were used for ABT analysis.

4. Discussion

The sustainability yield index (SYI) is considered a major indicator of agricultural sustainability for crop production and soil fertility management [48,49]. A high SYI indicates a more sustainable cropping system [29]. In this study, different long-term (34-year) GM rotations significantly ($p \leq 0.05$) increased grain yield and sustainability yield index (SYI) values compared to those of a winter fallow system in acidic soil under a double-rice cropping system (Table 1 and Figure 1). The highest increases in grain yield and SYI for both early and late rice were under the milkvetch rotation because milkvetch is a leguminous species that provides a large quantity of biologically fixed N for better plant growth and production [33]. GM rotation improves the soil properties and nutrient supply capacity [28]. Yang et al. [32] found that GM increased both the quality and quantity of SOM and had a positive impact on soil nutrient availability, which helped to increase the sustainability of rice yield. However, in previous studies, long-term experiments showed declining trends in grain yield under rice-rice and rice-wheat cropping systems [20,50]. In our results, the R-R-F treatment decreased the crop yield over the years (Figure 1). It has been shown that in double-cropping systems, long-term inorganic fertilization may decrease the crop yield over the years due to decreasing soil quality, including through increasing soil acidification [29,51,52]. In long-term experiments, different factors such as climate and soil properties influence productivity over the long cultivation period. In our study, soil properties, cropping systems, and climate significantly influenced crop yield. These results are consistent with previous studies, indicating the influence of soil properties and climate (mean annual temperature and mean annual precipitation) on rice yield [53,54]. However, in our VPA analysis results, the unexplained proportion was 64%, which was higher than the explained proportion. This might be due to the same rates of inorganic fertilization being applied in all cropping treatments because fertilizer is one of the main influencing factors on crop yield in long-term fertility experiments [55,56].

Soil fertility is one of the main indicators for measuring the sustainability of cropping systems [28,57]. Our results showed that GM increased nutrient contents and stocks compared to those under the winter fallow treatment, and the highest increase in available N content was under the R-R-M treatment (Figure 2). Non-leguminous cover crops decrease nutrient leaching and enhance nutrient availability [58]. Leguminous cover crops enhance N availability and therefore increase crop yield by improving N use efficiency [27]. The ability of legumes to fix atmospheric molecular dinitrogen into mineral N for plant uptake makes legumes the most efficient GM [59]. The decreasing trend in soil pH could have been due to long-term synthetic fertilization [28]. Lin et al. [51] also observed a declining trend in soil pH in paddy soil under long-term fertilization. Many studies observed soil acidification caused by long- term inorganic fertilization [51,60]. Moreover, plants release net H^+ ions; on the other hand, when the uptake of anions exceeds the cation uptake, they release a net excess of OH^- or HCO_3^- [61]. It was also observed in previous studies that N fertilization shifted the soil to the Al^{3+} buffering stage. Al is released in soil solution from the surface of clay minerals during the hydrolysis process of Al hydroxides at a relatively low soil pH, which may also decrease the amount of base cations and increase the soil acidity [62]. However, the alkaline nature of GM neutralizes the protons in acidic soil and helps to improve soil pH [63]. GM can also help to improve soil physical and chemical properties by improving soil porosity [17,64] and structure, reducing leaching [65] and increasing water holding capacity [32,66], thereby providing suitable conditions for proper plant growth to sustain long-term crop yield.

Nutrient release in soil from decaying SOM under long-term GM application increases the soil enzyme activities [67]. Leguminous cover crops change the microbial community in rhizosphere soil and increase the enzyme activities [33]. Moreover, available N and P showed significant positive relationship with enzyme activities (Figure S2), which indicate that higher enzyme activities enhance the nutrient availability and their uptake by crop, that may reduce the nutrient balance in the present study. The decreases in the annual N and P balances were consistent with the results of Balík et al., [68] but different from the results found by Ladha et al., [69] probably because the crop yield in the study by Ladha et al. [69] was not significantly different after GM rotation compared with the yields obtained by chemical fertilization. In this study, the lowest annual N balance was under the Chinese milkvetch rotation, possibly because the Chinese milkvetch increased crop yield and N uptake more than the ryegrass and rapeseed rotations [33,70,71]. In addition, the magnitude of the annual P balance reduction was less than that of the annual N balance (Figure 4). This finding was supported by a previous study conducted in Tianjin [72]. However, the annual P balances among all three GM treatments were not significantly different. The annual P balance is mainly controlled by soil abiotic factors (soil properties and climate) or biological factors (microorganisms) and is less affected by the GM species [73], which was consistent with our findings.

5. Conclusions

We have concluded that green manure rotation in a rice-based cropping system significantly increased crop yield sustainability by increasing soil nutrient availability and enzyme activities in acidic paddy soil. The highest increase in crop yield was under the rice-rice-milkvetch rotation system. However, green manure crops did not mitigate soil acidity due to the long-term addition of inorganic fertilizer during the early and late rice seasons. Green manure also decreased the apparent N and P balances by increasing the N and P uptake compared to those in winter fallow in rice-based cropping systems. The soil properties, cropping system, and climatic conditions significantly contributed to changing the annual crop yield over the experimental period. Among the soil properties, TN, SOM, pH, and TP were the most influential factors on crop yield. Therefore, the cultivation of green manure in rice-based cropping systems could be helpful for enhancing crop yield sustainability but might not be very effective for mitigating soil acidity under long-term inorganic fertilization in acidic paddy soil. Moreover, GM crops could be helpful for minimizing the environmental losses of N and P by decreasing the apparent N and P balances.

Supplementary Materials: The following are available online at http://www.mdpi.com/2073-4395/9/12/780/s1, Figure S1: Long-term (1983–2016) mean annual temperature and precipitation during the period of experiment of GM rotation in double rice cropping system, Figure S2: Linear regression analysis indicating relationships of enzyme activities with available nitrogen and available Phosphorus under long-term green manure rotation in rice based cropping system, Table S1: The name of cultivars of rice sown in this experiment from 2009 to 2018.

Author Contributions: Conceptualization, M.Q. and J.H.; methodology, D.L.; software, S.L.; validation, S.A. and K.L.; formal analysis, Y.X.; investigation, L.L.; resources, J.G.; data curation, M.Q.; writing—original draft preparation, M.Q.; writing—review and editing, W.A. and L.Z.; visualization, S.L.; supervision, H.Z.; project administration, H.Z.; funding acquisition, H.Z.

Funding: This research was supported by the National Key Research and Development Program of China (2016YFD0300901, 2016YFD0300902 and 2017YFD0800101), the National Natural Science Foundation of China (Nos. 41671301, 41371293 and 41561070), and the Fundamental Research Funds for Central Non-profit Scientific Institution (No.161032019035, No.161032019020).

Acknowledgments: We acknowledge the staff at the Qiyang long-term experiment sites for the technical support in collecting soil samples.

Conflicts of Interest: The authors declare no conflict of interest.

References

1. Shahid, M.; Shukla, A.K.; Bhattacharyya, P.; Tripathi, R.; Mohanty, S.; Kumar, A.; Lal, B.; Gautam, P.; Raja, R.; Panda, B.B.; et al. Micronutrients (Fe, Mn, Zn and Cu) balance under long-term application of fertilizer and manure in a tropical rice-rice system. *J. Soils Sediments* **2016**, *16*, 737–747. [CrossRef]

2. Rasmussen, P.E. Long-term agroecosystem experiments: Assessing agricultural sustainability and global change. *Science.* **1998**, *282*, 893–896. [CrossRef] [PubMed]

3. Liang, H.; Hu, K.; Batchelor, W.D.; Qi, Z.; Li, B. An integrated soil-crop system model for water and nitrogen management in North China. *Sci. Rep.* **2016**, *6*, 25755. [CrossRef] [PubMed]

4. Singh, R.J.; Ghosh, B.N.; Sharma, N.K.; Patra, S.; Dadhwal, K.S.; Meena, V.S.; Deshwal, J.S.; Mishra, P.K. Effect of seven years of nutrient supplementation through organic and inorganic sources on productivity, soil and water conservation, and soil fertility changes of maize-wheat rotation in north-western Indian Himalayas. *Agric. Ecosyst. Environ.* **2017**, *249*, 177–186. [CrossRef]

5. Melero, S.; Porras, J.C.R.; Herencia, J.F.; Madejon, E. Chemical and biochemical properties in a silty loam soil under conventional and organic management. *Soil Tillage Res.* **2006**, *90*, 162–170. [CrossRef]

6. Liu, C.-A.; Li, F.-R.; Zhou, L.-M.; Zhang, R.-H.; Yu, J.; Lin, S.-L.; Wang, L.-J.; Siddique, K.H.M.; Li, F.-M. Effect of organic manure and fertilizer on soil water and crop yields in newly-built terraces with loess soils in a semi-arid environment. *Agric. Water Manag.* **2013**, *117*, 123–132. [CrossRef]

7. Guo, J.H.; Liu, X.J.; Zhang, Y.; Shen, J.L.; Han, W.X.; Zhang, W.F.; Christie, P.; Goulding, K.W.T.; Vitousek, P.M.; Zhang, F.S. Significant acidification in major Chinese croplands. *Science.* **2010**, *327*, 1008–1010. [CrossRef]

8. Liu, X.; Zhang, Y.; Han, W.; Tang, A.; Shen, J.; Cui, Z.; Vitousek, P.; Erisman, J.W.; Goulding, K.; Christie, P.; et al. Enhanced nitrogen deposition over China. *Nature* **2013**, *494*, 459–462. [CrossRef]

9. Nyfeler, D.; Huguenin-Elie, O.; Suter, M.; Frossard, E.; Lüscher, A. Grass–legume mixtures can yield more nitrogen than legume pure stands due to mutual stimulation of nitrogen uptake from symbiotic and non-symbiotic sources. *Agric. Ecosyst. Environ.* **2011**, *140*, 155–163. [CrossRef]

10. Norsworthy, J.K.; Brandenberger, L.; Burgos, N.R.; Riley, M. Weed suppression in Vigna unguiculata with a spring-seeded brassicaceae green manure. *Crop Prot.* **2005**, *24*, 441–447. [CrossRef]

11. Zhang, Q.-C.; Shamsi, I.H.; Xu, D.-T.; Wang, G.-H.; Lin, X.-Y.; Jilani, G.; Hussain, N.; Chaudhry, A.N. Chemical fertilizer and organic manure inputs in soil exhibit a vice versa pattern of microbial community structure. *Appl. Soil Ecol.* **2012**, *57*, 1–8. [CrossRef]

12. Bedini, S.; Avio, L.; Sbrana, C.; Turrini, A.; Migliorini, P.; Vazzana, C.; Giovannetti, M. Mycorrhizal activity and diversity in a long-term organic Mediterranean agroecosystem. *Biol. Fertil. Soils* **2013**, *49*, 781–790. [CrossRef]

13. Tejada, M.; Gonzalez, J.L.; García-Martínez, A.M.; Parrado, J. Effects of different green manures on soil biological properties and maize yield. *Bioresour. Technol.* **2008**, *99*, 1758–1767. [CrossRef] [PubMed]

14. Piotrowska, A.; Wilczewski, E. Effects of catch crops cultivated for green manure and mineral nitrogen fertilization on soil enzyme activities and chemical properties. *Geoderma* **2012**, *189–190*, 72–80. [CrossRef]

15. Bedada, W.; Karltun, E.; Lemenih, M.; Tolera, M. Long-term addition of compost and NP fertilizer increases crop yield and improves soil quality in experiments on smallholder farms. *Agric. Ecosyst. Environ.* **2014**, *195*, 193–201. [CrossRef]

16. Liu, Y.-R.; Li, X.; Shen, Q.-R.; Xu, Y.-C. Enzyme activity in water-stable soil aggregates as affected by long-term application of organic manure and chemical fertiliser. *Pedosphere* **2013**, *23*, 111–119. [CrossRef]

17. Kumar, S.; Patra, A.K.; Singh, D.; Purakayastha, T.J. Long-term chemical fertilization along with farmyard manure enhances resistance and resilience of soil microbial activity against heat stress. *J. Agron. Crop Sci.* **2014**, *200*, 156–162. [CrossRef]

18. Crews, T.; Peoples, M. Legume versus fertilizer sources of nitrogen: Ecological tradeoffs and human needs. *Agric. Ecosyst. Environ.* **2004**, *102*, 279–297. [CrossRef]

19. Voisin, A.-S.; Guéguen, J.; Huyghe, C.; Jeuffroy, M.-H.; Magrini, M.-B.; Meynard, J.-M.; Mougel, C.; Pellerin, S.; Pelzer, E. Legumes for feed, food, biomaterials and bioenergy in Europe: A review. *Agron. Sustain. Dev.* **2014**, *34*, 361–380. [CrossRef]

20. Ram, S.; Singh, V.; Sirari, P. Effects of 41 years of application of inorganic fertilizers and farm yard manure on crop yields, soil quality, and sustainable yield index under a rice-wheat cropping system on Mollisols of North India. *Commun. Soil Sci. Plant Anal.* **2016**, *47*, 179–193. [CrossRef]

21. Choudhary, M.; Panday, S.C.; Meena, V.S.; Singh, S.; Yadav, R.P.; Mahanta, D.; Mondal, T.; Mishra, P.K.; Bisht, J.K.; Pattanayak, A. Long-term effects of organic manure and inorganic fertilization on sustainability and chemical soil quality indicators of soybean-wheat cropping system in the Indian mid-Himalayas. *Agric. Ecosyst. Environ.* **2018**, *257*, 38–46. [CrossRef]

22. Das, A.; Sharma, R.P.; Chattopadhyaya, N.; Rakshit, R. Yield trends and nutrient budgeting under a long-term (28 years) nutrient management in rice-wheat cropping system under subtropical climatic condition. *Plant Soil Environ.* **2014**, *60*, 351–357. [CrossRef]

23. Wanjari, R.H.; Singh, M.V.; Ghosh, P.K. Sustainable yield index: An approach to evaluate the sustainability of long-term intensive cropping systems in India. *J. Sustain. Agric.* **2004**, *24*, 39–56. [CrossRef]

24. Dawe, D.; Dobermann, A.; Moya, P.; Abdulrachman, S.; Singh, B.; Lal, P.; Li, S.Y.; Lin, B.; Panaullah, G.; Sariam, O. How widespread are yield declines in long-term rice experiments in Asia? *Field Crop. Res.* **2000**, *66*, 175–193. [CrossRef]

25. Xin, X.; Qin, S.; Zhang, J.; Zhu, A.; Yang, W.; Zhang, X. Yield, phosphorus use efficiency and balance response to substituting long-term chemical fertilizer use with organic manure in a wheat-maize system. *F. Crop. Res.* **2017**, *208*, 27–33. [CrossRef]

26. Ashraf, M.; Mahmood, T.; Azam, F.; Qureshi, R.M. Comparative effects of applying leguminous and non-leguminous green manures and inorganic N on biomass yield and nitrogen uptake in flooded rice (Oryza sativa L.). *Biol. Fertil. Soils* **2004**, *40*, 147–152. [CrossRef]

27. Zhao, X.; Wang, S.; Xing, G. Maintaining rice yield and reducing N pollution by substituting winter legume for wheat in a heavily-fertilized rice-based cropping system of southeast China. *Agric. Ecosyst. Environ.* **2015**, *202*, 79–89. [CrossRef]

28. Xie, Z.; Tu, S.; Shah, F.; Xu, C.; Chen, J.; Han, D.; Liu, G.; Li, H.; Muhammad, I.; Cao, W. Substitution of fertilizer-N by green manure improves the sustainability of yield in double-rice cropping system in south China. *Field Crop. Res.* **2016**, *188*, 142–149. [CrossRef]

29. Yadav, R.; Dwivedi, B.; Pandey, P. Rice-wheat cropping system: Assessment of sustainability under green manuring and chemical fertilizer inputs. *Field Crop. Res.* **2000**, *65*, 15–30. [CrossRef]

30. Hong, X.; Ma, C.; Gao, J.; Su, S.; Li, T.; Luo, Z.; Duan, R.; Wang, Y.; Bai, L.; Zeng, X. Effects of different green manure treatments on soil apparent N and P balance under a 34-year double-rice cropping system. *J. Soils Sediments* **2019**, *19*, 73–80. [CrossRef]

31. Baxter, S. *World Reference Base for Soil Resources*; World Soil Resources Report 103; Food and Agriculture Organization of the United Nations: Rome, Italy, 2006; p. 132. ISBN 92-5-10511-4.

32. Yang, Z.; Xu, M.; Zheng, S.; Nie, J.; Gao, J.; Liao, Y.; Xie, J. Effects of long-term winter planted green manure on physical properties of reddish paddy soil under a double-rice cropping system. *J. Integr. Agric.* **2012**, *11*, 655–664. [CrossRef]

33. Zhang, X.; Zhang, R.; Gao, J.; Wang, X.; Fan, F.; Ma, X.; Yin, H.; Zhang, C.; Feng, K.; Deng, Y. Thirty-one years of rice-rice-green manure rotations shape the rhizosphere microbial community and enrich beneficial bacteria. *Soil Biol. Biochem.* **2017**, *104*, 208–217. [CrossRef]

34. Pages, A.L.; Miller, R.H.; Dennis, R.K. *Methods of Soil Analysis. Part 2 Chemical Methods*; Soil Science Society of America Inc.: Madison, WI, USA, 1982.

35. Black, C.A. *Methods of Soil Analysis Part II. Chemical and Microbiological Properties*; American Society of Agriculture: Madison, WI, USA, 1965.

36. Murphy, J.; Riley, J.P. A modified single solution method for the determination of phosphate in natural waters. *Anal. Chim. Acta* **1964**, *27*, 31–36. [CrossRef]

37. Lu, R.K. *Analytical Methods of Soil Agricultural Chemistry*; China Agricultural Science and Technology Press: Beijing, China, 2000.

38. Olsen, S.R. *Estimation of Available Phosphorus in Soils by Extraction with Sodium Bicarbonate*; US Dept. of Agriculture: Washington, DC, USA, 1954.

39. Tabatabai, M.A. *Methods of Soil Analysis. Part 2: Microbiological and Biochemical Properties*; Mickelson, S.H.J.M.B., Ed.; Soil Science Society of America: Madison, WI, USA, 1994; pp. 775–833.

40. Kandeler, E.; Gerber, H. Short-term assay of soil urease activity using colorimetric determination of ammonium. *Biol. Fertil. Soils* **1988**, *6*, 68–72. [CrossRef]

41. Song, G.; Li, L.; Pan, G.; Zhang, Q. Topsoil organic carbon storage of China and its loss by cultivation. *Biogeochemistry* **2005**, *74*, 47–62. [CrossRef]

42. OECD. *OECD Compendium of Agri-Environmental Indicators*; OECD: Paris, France, 2013; ISBN 9789264181151.

43. Bai, Z.; Li, H.; Yang, X.; Zhou, B.; Shi, X.; Wang, B.; Li, D.; Shen, J.; Chen, Q.; Qin, W.; et al. The critical soil P levels for crop yield, soil fertility and environmental safety in different soil types. *Plant Soil* **2013**, *372*, 27–37. [CrossRef]

44. Ouyang, W.; Li, Z.; Liu, J.; Guo, J.; Fang, F.; Xiao, Y.; Lu, L. Inventory of apparent nitrogen and phosphorus balance and risk of potential pollution in typical sloping cropland of purple soil in China—A case study in the Three Gorges Reservoir region. *Ecol. Eng.* **2017**, *106*, 620–628. [CrossRef]

45. Singh, P.R.; Rao, S.K.; Das Bhaskarrao, U.M.; Ready, M.N. *Sustainability Index under Different Management: Annual Report*; CRIDA: Hydarabad, India, 1990.

46. Steel, R.G.D.; Torrie, J.H.; Dickey, D.A. *Principles and Procedures of Statistics: A Biological Approach*; McGraw-Hill Book Compay: New York, NY, USA, 1997.

47. De'Ath, G. Boosted trees for ecological modeling and prediction. *Ecology* **2007**, *88*, 243–251. [CrossRef]

48. Manna, M.C.; Swarup, A.; Wanjari, R.H.; Ravankar, H.N.; Mishra, B.; Saha, M.N.; Singh, Y.V.; Sahi, D.K.; Sarap, P.A. Long-term effect of fertilizer and manure application on soil organic carbon storage, soil quality and yield sustainability under sub-humid and semi-arid tropical India. *Field Crop. Res.* **2005**, *93*, 264–280. [CrossRef]

49. Tirol-Padre, A.; Ladha, J.K. Integrating rice and wheat productivity trends using the SAS mixed-procedure and meta-analysis. *Field Crop. Res.* **2006**, *95*, 75–88. [CrossRef]

50. Shahid, M.; Nayak, A.K.; Shukla, A.K.; Tripathi, R.; Kumar, A.; Mohanty, S.; Bhattacharyya, P.; Raja, R.; Panda, B.B. Long-term effects of fertilizer and manure applications on soil quality and yields in a sub-humid tropical rice-rice system. *Soil Use Manag.* **2013**, *29*, 322–332. [CrossRef]

51. Lin, H.; Jing, C.M.; Wang, J.H. The influence of long-term fertilization on soil acidification. *Adv. Mater. Res.* **2014**, *955–959*, 3552–3555. [CrossRef]

52. Cai, Z.; Wang, B.; Xu, M.; Zhang, H.; He, X.; Zhang, L.; Gao, S. Intensified soil acidification from chemical N fertilization and prevention by manure in an 18-year field experiment in the red soil of southern China. *J. Soils Sediments* **2015**, *15*, 260–270. [CrossRef]

53. Prabnakorn, S.; Maskey, S.; Suryadi, F.X.; Fraiture, C. De Rice yield in response to climate trends and drought index in the Mun River Basin, Thailand. *Sci. Total Environ.* **2018**, *621*, 108–119. [CrossRef] [PubMed]

54. Oladele, S.O.; Adeyemo, A.J.; Awodun, M.A. Influence of rice husk biochar and inorganic fertilizer on soil nutrients availability and rain-fed rice yield in two contrasting soils. *Geoderma* **2019**, *336*, 1–11. [CrossRef]

55. Gai, X.; Liu, H.; Liu, J.; Zhai, L.; Yang, B.; Wu, S.; Ren, T.; Lei, Q.; Wang, H. Long-term benefits of combining chemical fertilizer and manure applications on crop yields and soil carbon and nitrogen stocks in North China Plain. *Agric. Water Manag.* **2018**, *208*, 384–392. [CrossRef]

56. Meena, B.P.; Biswas, A.K.; Singh, M.; Chaudhary, R.S.; Singh, A.B.; Das, H.; Patra, A.K. Long-term sustaining crop productivity and soil health in maize–chickpea system through integrated nutrient management practices in Vertisols of central India. *Field Crop. Res.* **2019**, *232*, 62–76. [CrossRef]

57. Tirol-Padre, A.; Ladha, J.K.; Regmi, A.P.; Bhandari, A.L.; Inubushi, K. Organic Amendments Affect Soil Parameters in Two Long-Term Rice-Wheat Experiments. *Soil Sci. Soc. Am. J.* **2007**, *71*, 442. [CrossRef]

58. Doltra, J.; Gallejones, P.; Olesen, J.E.; Hansen, S.; Frøseth, R.B.; Krauss, M.; Stalenga, J.; Jończyk, K.; Martínez-Fernández, A.; Pacini, G.C. Simulating soil fertility management effects on crop yield and soil nitrogen dynamics in field trials under organic farming in Europe. *Field Crop. Res.* **2019**, *233*, 1–11. [CrossRef]

59. Thorup-Kristensen, K.; Magid, J.; Jensen, L.S. Catch crops and green manures as biological tools in nitrogen management in temperate zones. In *Advances in Agronomy*; Academic Press: Cambridge, MA, USA, 2003; Volume 79, pp. 227–302. ISBN 0120007975.

60. Zhu, Q.; Liu, X.; Hao, T.; Zeng, M.; Shen, J.; Zhang, F.; Vries, W. De Modeling soil acidi fi cation in typical Chinese cropping systems. *Sci. Total Environ.* **2018**, *613–614*, 1339–1348. [CrossRef]

61. Tang, C.; Conyers, M.K.; Nuruzzaman, M.; Poile, G.J.; Liu, D.L. Biological amelioration of subsoil acidity through managing nitrate uptake by wheat crops. *Plant Soil* **2011**, *338*, 383–397. [CrossRef]

62. Stevens, C.J.; Dise, N.B.; Gowing, D.J. Regional trends in soil acidification and exchangeable metal concentrations in relation to acid deposition rates. *Environ. Pollut.* **2009**, *157*, 313–319. [CrossRef] [PubMed]

63. Rukshana, F.; Butterly, C.R.; Xu, J.-M.; Baldock, J.A.; Tang, C. Organic anion-to-acid ratio influences pH change of soils differing in initial pH. *J. Soils Sediments* **2013**, *14*, 407–444. [CrossRef]

64. Srinivasarao, C.; Venkateswarlu, B.; Lal, R.; Singh, A.K.; Kundu, S.; Vittal, K.P.R.; Patel, J.J.; Patel, M.M. Long-term manuring and fertilizer effects on depletion of soil organic carbon stocks under pearl millet-cluster beans-castor rotation in Western India. *Land Degrad. Dev.* **2014**, *25*, 173–183. [CrossRef]

65. Thorup-Kristensen, K.; Dresbøll, D.B.; Kristensen, H.L. Crop yield, root growth, and nutrient dynamics in a conventional and three organic cropping systems with different levels of external inputs and N re-cycling through fertility building crops. *Eur. J. Agron.* **2012**, *37*, 66–82. [CrossRef]

66. Fließbach, A.; Oberholzer, H.-R.; Gunst, L.; Mäder, P. Soil organic matter and biological soil quality indicators after 21 years of organic and conventional farming. *Agric. Ecosyst. Environ.* **2007**, *118*, 273–284. [CrossRef]

67. Piotrowska-Długosz, A.; Wilczewski, E. Changes in enzyme activities as affected by green-manure catch crops and mineral nitrogen fertilization. *Zemdirbyste Agric.* **2014**, *101*, 139–146. [CrossRef]

68. Balík, J.; Černý, J.; Tlustoš, P.; Zitková, M. Nitrogen balance and mineral nitrogen content in the soil in a long experiment with maize under different systems of N fertilization. *Plant Soil Environ.* **2011**, *49*, 554–559. [CrossRef]

69. Ladha, J.K.; Dawe, D.; Ventura, T.S.; Singh, U.; Ventura, W.; Watanabe, I. Long-term effects of urea and green manure on rice yields and nitrogen balance. *Soil Sci. Soc. Am. J.* **2000**, *64*, 1993. [CrossRef]

70. Zhang, X.-X.; Gao, J.-S.; Cao, Y.-H.; Ma, X.-T.; He, J.-Z. Long-term rice and green manure rotation alters the endophytic bacterial communities of the rice root. *Microb. Ecol.* **2013**, *66*, 917–926. [CrossRef]

71. Gabriel, J.L.; Muñoz-Carpena, R.; Quemada, M. The role of cover crops in irrigated systems: Water balance, nitrate leaching and soil mineral nitrogen accumulation. *Agric. Ecosyst. Environ.* **2012**, *155*, 50–61. [CrossRef]

72. Yang, J.; Gao, W.; Ren, S.R. Response of soil phosphorus to P balance under long-term fertilization in fluvo-aquic soil. *Sci. Agric. Sin.* **2015**, *48*, 4738–4747.

73. Blake, L.; Mercik, S.; Koerschens, M.; Moskal, S.; Poulton, P.R.; Goulding, K.W.T.; Weigel, A.; Powlson, D.S. Phosphorus content in soil, uptake by plants and balance in three European long-term field experiments. *Nutr. Cycl. Agroecosyst.* **2000**, *56*, 263–275. [CrossRef]

Review

Mechanisms of Nitrogen Use in Maize

Aziiba Emmanuel Asibi [1,2], Qiang Chai [1,2,* and Jeffrey A. Coulter [3,*

[1]   College of Agronomy, Gansu Agricultural University, Lanzhou 730010, China; aziibason4u@yahoo.com
[2]   Gansu Provincial Key Laboratory of Aridland Crop Science, Lanzhou 730070, China
[3]   Department of Agronomy and Plant Genetics, University of Minnesota, St. Paul, MN 55108, USA
*   Correspondence: chaiq@gsau.edu.cn (Q.C.); jeffcoulter@umn.edu (J.A.C.)

Received: 22 October 2019; Accepted: 18 November 2019; Published: 20 November 2019

Abstract: Nitrogen (N) fertilizers are needed to enhance maize (*Zea mays* L.) production. Maize plays a major role in the livestock industry, biofuels, and human nutrition. Globally, less than one-half of applied N is recovered by maize. Although the application of N fertilizer can improve maize yield, excess N application due to low knowledge of the mechanisms of nitrogen use efficiency (NUE) poses serious threats to environmental sustainability. Increased environmental consciousness and an ever-increasing human population necessitate improved N utilization strategies in maize production. Enhanced understanding of the relationship between maize growth and productivity and the dynamics of maize N recovery are of major significance. A better understanding of the metabolic and genetic control of N acquisition and remobilization during vegetative and reproductive phases are important to improve maize productivity and to avoid excessive use of N fertilizers. Synchronizing the N supply with maize N demand throughout the growing season is key to improving NUE and reducing N loss to the environment. This review examines the mechanisms of N use in maize to provide a basis for driving innovations to improve NUE and reduce risks of negative environmental impacts.

Keywords: maize production; nitrogen use efficiency; nitrogen nutrition

---

1. Nitrogen Application in Crop Production

Crop production is the principal cause of human modification of the global nitrogen (N) cycle [1,2]. Nitrogen is a crucial input in agriculture to sustain agricultural livelihoods and support the increasing human population [3]. From 1930 to 1960, N fertilizer use escalated from 1.3 to 10.2 million metric tons (MMt) [4]. During this time of increased N fertilizer use, there was a concomitant increase in N losses to the environment, leaching of nitrate into groundwater, aquatic eutrophication, emission of ammonia and nitrous oxide, and soil acidification [5–7]. The global demand for N fertilizers was 112 MMt in 2014 [8] and is predicted to reach 240 MMt by 2050 [9].

Since the 1960s, the global use of N fertilizers has grown nearly seven-fold [10,11], but the overall yield increase was only 2.4-fold [12,13]. Worldwide, more than 50% to 75% of applied N fertilizer is not taken up by crops [14–16], and recovery of applied N by maize (*Zea mays* L.) rarely exceeds 50% [17,18]. Nitrogen use efficiency (NUE) can be defined as the increase in grain yield per unit of applied N and is the product of absorption efficiency (increase in absorbed N per unit of applied N) and utilization efficiency (increase in grain yield per unit of absorbed N) [19,20].

Synthetic fertilizers are now the primary source of N applied to cropland, although organic N from livestock manure remains important [20]. The environmental cost of excess N application in Europe is estimated at 78 to 357 billion US dollars per year [21]. Nitrogen availability to plants is the difference between N supply and losses due to pathways such as leaching, runoff, ammonia volatilization, additional gaseous N losses, and immobilization, highlighting the multifaceted interaction among the soil, plant, and atmosphere in regulating N availability and uptake by plants [22]. Nitrogen is an essential nutrient for maize and a key determinant of grain yield, particularly through its role in photosynthesis and other biological processes such as absorption of water and minerals, vacuole storage, and xylem transport. However, N application is a great concern in maize production due to the negative effects of excess application on groundwater quality [23]. Mineral N fertilizers manufactured by the Haber–Bosh process can represent up to 50% of the operational cost in crop production, depending on the crop [24]. Comparatively, the overall cost of reduced N fertilizer requirements for grain crops due to crop rotation with legumes is just 4% to 71% of the cost of an equivalent amount of synthetic N fertilizer [25,26]. Legumes can also increase the yield of subsequent grain crops and have a delayed N release compared to synthetic fertilizers [27,28]. Therefore, it is environmentally beneficial to include legumes in cropping systems [20,26].

2. Nitrogen Use Efficiency in Maize

Cereal grains such as maize, rice (*Oryza sativa* L.), and wheat (*Triticum aestivum* L.) provide 60% to 94% of the world's nourishment [20,29]. Among these, maize ranks second to wheat in global production, with 500 MMt produced annually on 130 million hectares [30]. Maize is used for livestock feed, biofuel, and human consumption [31,32], and global demand for maize is anticipated to increase by 16% by 2027, due primarily to a > 50% increase in consumption by livestock [33].

Nitrogen accumulation in aboveground maize biomass typically increases as the N level in the soil increases [34], and aboveground biomass can be reduced if the soil N supply is insufficient for crop demand [35]. Aboveground biomass production also influences maize N uptake potential [36]. Nitrogen accumulated in maize biomass is partitioned into grain and Stover, with luxury N uptake occurring when N supply exceeds the minimum requirements for maximum grain yield [37,38]. Several factors can affect N uptake by maize, of which soil moisture, temperature, structure, and bulk density are most important [39,40]. Thus, improvement in aboveground biomass production requires adequate N supply and uptake by maize.

Increases in maize grain yield are associated with increases in aboveground biomass [41–43], and grain yield improvements are associated with increased harvest index [41,44]. Harvest index is the ratio of grain to total aboveground biomass (grain, cob, and Stover) per unit area at physiological maturity [45,46]. Harvest index of maize produced in favorable growing environments is about 50% [46]. High levels of harvest index in maize typically occur with intermediate values of total aboveground biomass relative to grain and decreases when individual plants are small (as with high plant density) or large (as with low plant density) [47]. Harvest index can increase with heterosis if the increase in kernel number is proportionately greater than that in total aboveground dry matter accumulation during the grain-filling period [48,49]. Kernel number is related to dry matter accumulation [50] and partitioning during the periods of anthesis and kernel establishment [51,52].

Nitrogen use efficiency has two components: (i) the efficiency of crop uptake of applied N and (ii) the transformation efficiency of total aboveground crop N uptake to grain yield [53]. Nitrogen use efficiency in maize is multifaceted and rests on N availability in the soil and rhizosphere, and the utilization ability of the crop [22,54]. This is complicated by the many soil and environmental factors that control the rates and products of N mineralization [55]. The use of soil N by maize involves numerous steps, including uptake, translocation, and remobilization (Figure 1). This occurs during two phases: (i) the N assimilation phase and (ii) the remobilization phase [22,38,56]. During the N assimilation and remobilization phases, young roots and leaves are sink organs that efficiently absorb and assimilate inorganic N for amino acids and protein synthesis.

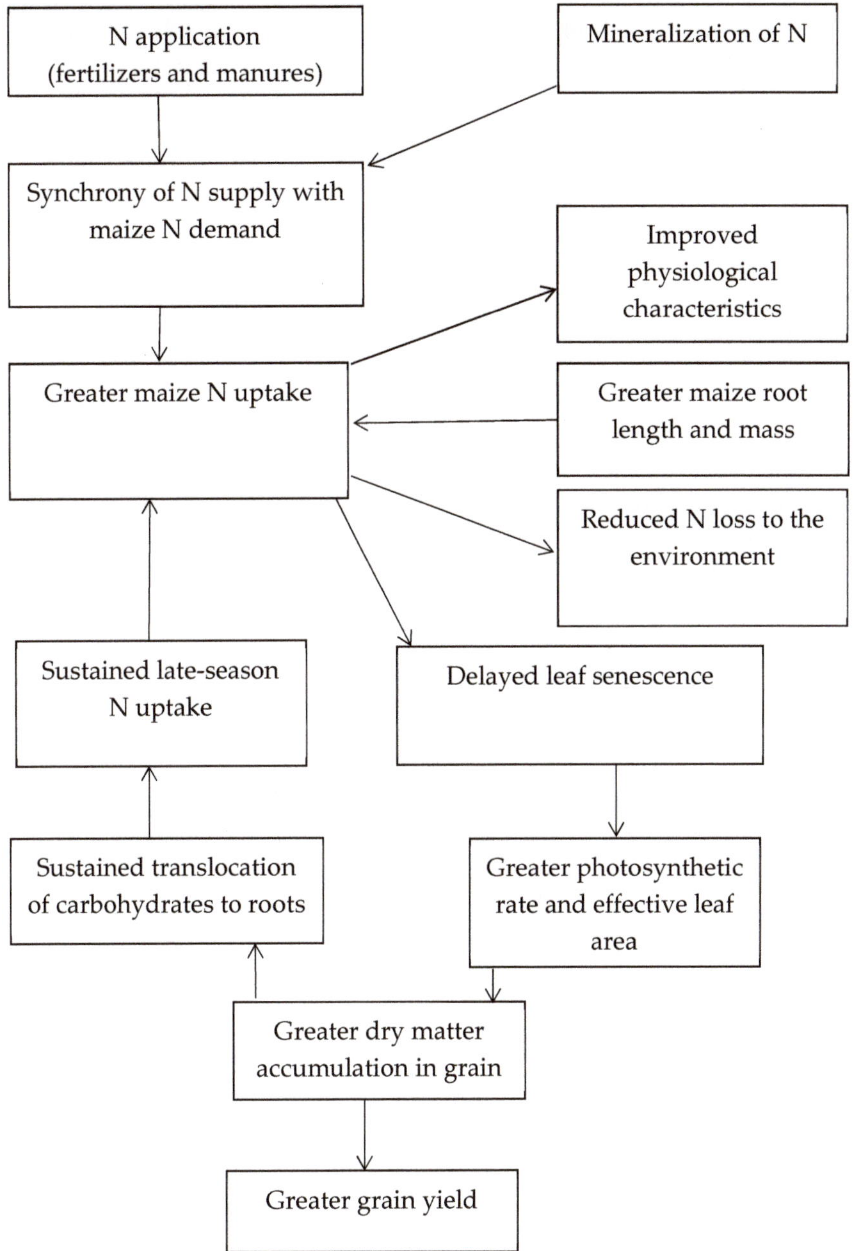

Figure 1. Diagram of nitrogen (N) use in maize.

The components of NUE include recovery efficiency of applied nitrogen (RE), physiological efficiency of applied nitrogen (PE), and agronomic efficiency of applied nitrogen (AE) [38]. The recovery efficiency of applied N reflects the efficiency of aboveground N uptake per unit of N applied and is closely associated with assimilation capacity and decreases at greater N supply [57]. Physiological efficiency is the efficiency by which N in aboveground plant tissues is converted to grain, and AE is grain yield per unit of N applied [58,59]. High NUE in maize is a combination of high N uptake and high N utilization efficiency [60]. With a high level of N application, genetic differences in maize grain yield are mainly due to variation in N uptake [53]. Since RE in maize is commonly < 50% [17,18], there is a great opportunity to enhance maize RE and yield and reduce environmental impact [61,62].

3. Nitrogen Assimilation and Utilization by Maize

Nitrate and ammonium are the forms of inorganic N most greatly absorbed by the roots of maize [63]. In warm and moist conditions, ammonium is oxidized to nitrate in soils almost as rapidly as it is formed [64]. Thus, nitrate is usually the dominant form of plant-available N in soils [65]. Assimilated nitrate is stored in cell vacuoles, where it serves as a reservoir for N remobilization when N uptake declines [22]. This pool of nitrate is the major source of N that contributes to grain filling. The absorption of nitrate is a two-step process that involves a collective set of enzymes, namely nitrate reductase and nitrite reductase, which catalyze and activate the reduction of nitrate to ammonium. The reduction of nitrate to ammonium is the most limiting phase in N assimilation [66]. Nitrate reductase increases leaf nitrate quantity at initial growth stages, and therefore can be used as a selection criterion for maize genotypes with superior N absorption and grain yield [38,67].

Within plants, ammonium is incorporated into amino acids through glutamine synthetase and glutamate synthase [68]. These are the main enzymes that convert inorganic N to organic molecules for other metabolic activities [69]. Glutamine synthetase concentration is a key regulator of plant growth that is controlled by N accumulation in plants and can be used as an indicator of grain yield and to differentiate sink leaves versus source leaves [70,71]. The transformation of glutamate to glutamine in leaves by glutamine synthetase is active in new vegetative tissues and is positively correlated with post-anthesis N uptake and negatively correlated with the percentage of N in grain from remobilization [72]. All assimilated N is directed through the processes catalyzed by glutamine synthetase, and nitrate accumulation catalyzed by nitrate reductase and glutamine synthetase is a key factor regulating NUE in maize [56,67].

Maize growth is dependent on the quantity of photosynthetically captured active radiation and the efficiency by which it is converted to aboveground biomass [73]. Enhanced photosynthetic capacity can contribute to greater biomass and grain yield [74]. The amount of intercepted photosynthetically active radiation is influenced by the size and architecture of the crop canopy [75]. The distribution of N within the canopy, with its significance for leaf and canopy photosynthesis, is a foremost factor regulating radiation use efficiency in maize [76]. Following anthesis, the rate of maize N uptake declines, and hybrids with greater N uptake have delayed leaf senescence (Figure 1) and prolonged effective leaf area for photosynthesis [77].

Maize hybrids with delayed senescence have a greater ability to take up N during the grain-filling period since continued leaf activity stimulates uptake of N [78]. During the grain-filling period, a decline in N supply decreases dry matter partitioning to grain [79]. Maize genotypes with longer stay-green maintain leaf chlorophyll for a greater duration, which can increase grain yield by 10% to 12% [80]. Figure 1 is a simple diagram illustrating the mechanisms of N use in maize.

4. Nitrogen Remobilization in Maize

Efficient N remobilization and leaf longevity are key features contributing to an improved yield of contemporary maize hybrids [81,82]. The magnitude and duration of active leaf tissue affects N uptake and the quantity of N accessible for remobilization [77]. Remobilization of N is greater when post-anthesis N accumulation is less and is negatively correlated with leaf senescence [83]. Extended duration of leaf metabolic activity enhances the ratio of source leaves to sink leaves throughout the grain-filling period [81,82]. When maize N uptake is sustained throughout the grain-filling period, less N is mobilized from vegetative organs, thereby increasing leaf area duration, delaying senescence, and enhancing dry matter accumulation (Figure 1) [13]. High maize yield is attributed to the maintenance of photosynthate source and N uptake throughout the grain-filling period [84]. During maize reproductive growth, increasing and sustaining the capacity for N uptake in soils with low N supply is an avenue for improving NUE, and is dependent on the ability of the plant to supply the root system with the needed assimilates for root development and N uptake for sustained photosynthesis [85]. At a given level of N supply, increases in grain yield among maize genotypes are not always due to improvements in N assimilation and usage, but rather enhanced NUE as a result of a more effective N remobilization [86].

Remobilization of N by maize early in the growing season contributes to vegetative development and formation of the reproductive sink capacity, which affects grain yield [87]. Improved synchrony between N supply and crop N uptake is key to enhancing NUE [88]. Shortage of N uptake due to physiological restraints can arise even when there is a large sink capacity for grain protein synthesis, resulting in N loss, enhanced remobilization of N from vegetative tissues to grain, and low NUE [89]. Relationships between N accessibility, late-season N uptake by maize, and remobilization of N are influenced by soil N supply [84,90]. Low N supply following anthesis can lead to early leaf senescence since the development of grain requires more N than the maintenance of vegetative tissues [91].

Maize hybrids with rapid early-season N uptake, efficient N remobilization, and complete leaf senescence can perform well in environments with low N supply late in the growing season, while hybrids that are proficient in late-season N uptake require conditions with greater soil N availability late in the growing season to optimize grain yield [80]. Maize hybrids with enhanced NUE have the capacity to achieve greater yield at a given level of N supply, particularly at low levels, through enhanced N uptake and more efficient translocation of N to grain [19,91,92]. There is an inverse relationship between N remobilization and N assimilation in maize, which results from a correlation between photosynthesis and N uptake [83]. Growing environments that are favorable for post-anthesis photosynthesis promote N uptake and are therefore associated with reduced N remobilization. When growing environments are unfavorable for photosynthesis, post-anthesis N uptake in maize is reduced; therefore, the quantity of N remobilization is positively associated with aboveground N content at anthesis, and grain N content is associated with the level of post-anthesis N uptake and assimilation [83]. Disparities in post-anthesis N uptake in grain are mainly due to differences in total aboveground maize N uptake following anthesis [93].

During senescence, photosynthesis is reduced and translocation of carbohydrates to roots declines, contributing to a reduction in N uptake and an upsurge in N remobilization [94]. Genotypic differences in the photosynthetic rate at the cellular level of maize are associated with differences in N remobilization and can be used in breeding programs to improve NUE [95]. A substantial fraction of the N in maize leaf tissue is involved in photosynthetic processes, and the photosynthetic rate is closely associated with leaf N content [96]. The level of N remobilization from the stalk of maize is related to the level of N remobilization from leaves and N accumulation at flowering [97]. Following anthesis, leaf N content quickly declines due to remobilization to grain, especially in conditions with low N supply [98]. With greater soil N supply, there is generally no relationship between leaf N content and N remobilization [98]. Nitrogen supply from floral initiation to anthesis regulates kernel number and subsequently kernel weight [99]. Nitrogen uptake and remobilization are independently inherited traits, so favorable alleles could be combined when breeding for NUE [100].

5. Nitrogen Nutrition Index

Nitrogen nutrition index defines the N status of a crop and is the ratio between aboveground N uptake and the critical N uptake required for maximum aboveground biomass. Aboveground N accumulation in maize occurs most rapidly from mid-vegetative growth until anthesis and then decreases until senescence [101]. During maize vegetative growth, the amount of structural tissues increases more rapidly compared to metabolic tissues, such that N content is greatest in young leaves [102,103]. Nitrogen content in leaves declines as leaf area increases, demonstrating that N content depends on plant biomass [76,104]. Optimal N fertilizer application in maize requires knowledge of the adequate N content for a given amount of maize biomass, along with a rapid and accurate method to determine N content and biomass [105]. Nitrogen nutrition index provides the critical N content for a given amount of biomass [106].

Leaf and canopy N content in maize are positively correlated with their chlorophyll content and are indicators of photosynthetic ability [107,108]. Canopy chlorophyll can be used to estimate canopy-level N content in maize [107]. However, at the canopy level, the estimation of chlorophyll and N content is much more challenging [109,110]. There is a close relationship between leaf N content and maize grain yield [111]. Critical values or ranges of leaf chlorophyll meter readings in maize have been used to determine N deficiency to improve N management [112–114]. However, knowledge about the relationship between the N nutrition index and spectral canopy reflectance is limited [115].

6. Grain Nitrogen Harvest in Maize

Nitrogen harvest index is the proportion of N in grain relative to total aboveground biomass and is an indicator of N translocation efficiency [116]. Nitrogen harvest index is not directly linked to grain yield [45]. However, grain yield and grain N content are important for assessing NUE and serve as selection criteria for breeding programs and N management practices [117]. Nitrogen harvest index is greater in maize hybrids that sustain a high rate of photosynthesis during the grain-filling period, leading to greater NUE [118].

7. Nitrogen Supply

Nitrogen availability is one of the most significant constraints to crop growth [119], and the application of N through mineral and organic fertilizers plays a vital role in sustaining crop production [12]. Nitrogen application can play a significant role in improving soil fertility [120]. Fertilization of N can increase grain yield and biomass in maize [121,122]. The leading source of applied N worldwide is synthetic fertilizer, followed closely by livestock manure [123–125]. Application of livestock manure with a high carbon-to-N ratio can lead to N immobilization in the soil, thereby restricting maize N uptake and potentially yield in the short-term [126]. Soil organic matter can serve as a temporary sink for available soil N not utilized by maize to reduce N losses [127]. Organic fertilizers can lead to greater soil total N than mineral fertilizers after several cropping seasons [128,129], increase organic matter in soil, enhance soil porosity, and improve soil water holding capacity [130], thereby enhancing maize growth and nutrient uptake [131]. Application of composted manure for maize production has been shown to produce less greenhouse gas emissions compared to mineral N fertilizer in maize cropland due to slower release of available N [126]. Additionally, the application of organic N (manure) has been shown to increase uptake of N, P, K, Ca, Mg, Zn, Fe, and Cu of maize grown on Alfisol soils compared to NPK [131].

Mineral fertilizers can enhance maize grain yield and N uptake but also result in greater residual NO_3^--N levels in the soil compared to organic fertilizers [132]. The application of mineral fertilizers can directly or indirectly cause changes to the chemical, physical, and biological properties of the soil [133]. The usage of mineral N, P, and K fertilizers can increase the availability of N, P, K, Ca, Mg, Fe, Cu, Mn, and Zn to maize [134,135]. The application of Zn has been shown to facilitate the tasseling/silking of maize in combination with greater N doses [136,137]. Magnesium stimulates a large number of enzymes in maize, can increase the rate of mineral N transformation into proteins [138], and plays a key role in chlorophyll and the development of kernels in maize [139]. Potassium deficiency strongly affects N metabolism and photosynthesis in maize [140]. The application of inorganic N fertilizers has been associated with loss of soil organic carbon through enhanced respiration and greenhouse gas emissions [141,142]. This loss reduces soil productivity and agronomic efficiency of applied N and changes the balance of N and carbon in soil in favor of greenhouse gas emissions [143]. However, there can be a positive collaboration between organic manures and urea as a N source [144], and integrated organic and inorganic N application can lead to greater maize yield compared to sole application of organic or inorganic N sources [145].

Due to changes in the form of N in the soil, maize NUE is low and rarely exceeds 60% [146,147]. Nitrogen takes on nine different forms in the soil, conforming to different oxidative conditions (nitrate, nitrogen dioxide, nitrite, nitric oxide, nitrous oxide, dinitrogen, ammonia, ammonium, and organic N) [148]. Nitrogen mineralization is the conversion of organic N to inorganic forms, immobilization is the uptake or assimilation of inorganic forms of N by microbes and other soil heterotrophs, nitrification is the conversion of ammonium to nitrite and then nitrate, and denitrification is the conversion of nitrate to nitrous oxide and then dinitrogen gas [148]. Most of the N in the soil originates as dinitrogen gas, and this inert N cannot be used by plants until it is transformed to ammonium or nitrate [149]. Investigations using ion-selective microelectrode methods demonstrated that maize absorption is suppressed from the root apex to 60 mm behind the root apex when ammonium and nitrate are both supplied at the same time [150]

Mineralization of N is influenced by several factors, including the source and carbon-to-N ratio of organic material, soil texture, temperature, water content, and pH, the microbial community, and agronomic practices [151]. Mineralization of organic amendments intensifies available N and, depending on the timing of mineralization, crop requirements for N, and sources and amounts of applied N can be a key component of an integrated N management approach to achieve high NUE while minimizing N loss [152]. However, accurate estimation of N mineralization from soil and organic sources of applied N is challenging due to the large number of site, environmental, and agronomic factors influencing it [153]. Integrated application of organic and mineral fertilizers at appropriate rates can be an effective strategy for improving maize N uptake and yield [154–156].

8. Maize Adaptation to Low Nitrogen Supply

Nitrogen supply affects root mass and morphology, which are important for acquiring spatially heterogeneous N in soil, particularly for newly mineralized N [157]. There is a close relationship between root length and N accumulation in maize seedlings [158,159]. Root size is vital for N uptake over the entire growth period of maize, but the initial establishment of a large root system is indispensable [159]. In response to low N supply, maize develops a greater root-to-shoot ratio and undergoes a slower rate of phenological development, with a greater proportion of root biomass enhancing the absorption capacity of N [160].

At the same quantity of accumulated N, maize genotypes with enhanced NUE typically allocate a greater proportion of N to the root system and develop a greater root system compared to less efficient genotypes [161–163]. This may be related to their capacity to sustain root development after anthesis, especially under conditions of low N supply [164]. Grain yield of maize is influenced by N supply, level of N uptake, and capacity to effectively partition N to grain during the grain-filling period [165]. High-yielding maize hybrids typically have greater yield response to the N application rate than low-yielding hybrids [19]. Sustainable post-anthesis uptake of N and maintenance of green leaf area until maturity are key traits of maize hybrids with high yield potential [166]. These traits are especially important for root longevity and N uptake in conditions of low N supply [166,167].

The efficiency of maize to utilize soil N depends on its ability to acquire, utilize, and translocate N, which is influenced by root morphology and the biochemical and physiological processes involved in nitrate assimilation [168]. Nitrogen uptake efficiency (total aboveground N accumulation at a given level of N supply) is separated into constituent traits, which describe root characteristics associated with genetic variation in N uptake [169]. Root tissue that is active in N acquisition, and the absorption rate per unit of root mass, influence the rate of N uptake, and both component traits differ genetically [169]. Morphological development of the root system affects the processes involved in nitrate acquisition [169]. Differential lateral root development between genotypes affects the translocation of solutes from the root system [170], as there is greater activity of nitrate reduction in root apices compared to basal parts [171].

9. Strategies to Improve Nitrogen Use Efficiency in Maize

Improvement in maize grain yield since the 1930s has been one of the greatest achievements in agriculture [46]. High NUE is critical for sustainable maize production [151], but globally, maize RE is only about 35% to 55% [61,62]. There is an opportunity to increase the NUE of maize through enhanced understanding of the mechanisms of N use to guide appropriate N application. Nitrogen use efficiency in maize is affected by soil N supply; hence, understanding the factors influencing N absorption, assimilation, and mobilization are key to increasing NUE to circumvent excessive N application and the subsequent negative environmental impacts (Figure 1) [172]. Increasing N supply results in reductions in all indices of NUE (i.e., RE, PE, and AE) [173]. One strategy to reduce N losses is to increase NUE through management practices to enhance the synchrony between N supply and crop demand throughout the growing season [174–176]. It is also important to recognize the critical stages of maize phenological development that are associated with NUE [125]. Further research should be conducted to assess the effects of the different fertilizer types on the mechanisms of N use in maize.

Since the maize harvest index is already high, increasing N uptake, aboveground biomass, and grain yield are paths to increasing NUE [177]. Developing hybrids capable of taking up more N or utilizing assimilated N more effectively can improve NUE, while knowledge on the physiology and genetics of N uptake and utilization is critical for the development of N-efficient hybrids [19,38,54]. Nitrogen use efficiency is controlled by a complex set of interactions among genotype, growing environment, and agronomic management [20,178]. Therefore, more extensive screening of a wide range of genotypes covering genetic diversity is needed to identify specific elements controlling NUE and productivity [38,178].

Sustainability is founded on the standard that present needs are met without compromising the needs of the future. Integrated collaboration between public and private sectors will improve the understanding and management of the biological and agronomic factors controlling NUE [20]. This will require a multidisciplinary approach, integrating expertise from crop developmental biology, physiology, genomics, genetics, breeding, simulation modeling, and agronomy [179]. Therefore, rigorous collaboration among stakeholders (i.e., environmental agencies, policy-makers, researchers, farm advisors, and farmers) and skillful implementation of indicators and guidelines are needed to improve NUE in maize production [180].

Author Contributions: Conceptualization, A.E.A., Q.C., and J.A.C.; methodology, A.E.A.; software formatting/writing style, J.A.C.; validation, Q.C. and J.A.C.; formal analysis, Q.C. and J.A.C.; investigation, A.E.A.; resources, Q.C.; data collection/information, A.E.A.; writing, A.E.A.; editing, J.A.C.; visualization, A.E.A. and J.A.C.; supervision, Q.C. and J.A.C.; project administration, Q.C. and J.A.C.

Funding: This research received no external funding.

Conflicts of Interest: The authors declare no conflict of interest.

Abbreviations

agronomic efficiency of applied nitrogen	AE
nitrogen	N
nitrogen use efficiency	NUE
physiological efficiency of applied nitrogen	PE
recovery efficiency of applied nitrogen	RE

References

1. Smil, V. Nitrogen in crop production: An account of global flows. *Glob. Biogeochem. Cycles* **1999**, *13*, 647–662. [CrossRef]

2. UNEP. *Reactive Nitrogen in the Environment, Too Much or Too Little of a Good Thing*; The Woods Hole Research Center: Falmouth, MA, USA; UNEP DTIE Sustainable Consumption and Production Branch: Paris, France, 2007; Volume 51, ISBN 978 92 807 2783 8.

3. Spiertz, J.H.J. Nitrogen, sustainable agriculture and food security: A review. In *Sustainable Agriculture*; Lighthouse, E., Navarrete, M., Debaeke, P., Véronique, S., Alberola, C., Eds.; Springer: Dordrecht, The Netherlands, 2009; Volume 30, pp. 635–651.

4. Frink, C.R.; Waggoner, P.E.; Ausubel, J.H. Nitrogen fertilizer: Retrospect and prospect. *Proc. Natl. Acad. Sci. USA* **1999**, *96*, 1175–1180. [CrossRef] [PubMed]

5. Guo, J.H.; Liu, X.J.; Zhang, Y.; Shen, J.L.; Han, W.X.; Zhang, W.F. Significant acidification in major Chinese croplands. *Science* **2010**, *327*, 1008–1010. [CrossRef] [PubMed]

6. Hoang, V.N.; Alauddin, M. Assessing the eco-environmental performance of agricultural production in OECD countries: The use of nitrogen flows and balance. *Nutr. Cycl. Agroecosyst.* **2010**, *87*, 353–368. [CrossRef]

7. Ju, X.; Lu, X.; Gao, Z.; Chen, X.; Su, F.; Kogge, M. Processes and factors controlling N_2O production in an intensively managed low carbon calcareous soil under sub-humid monsoon conditions. *Environ. Pollut.* **2011**, *159*, 1007–1016. [CrossRef]

8. FAO. *World Fertilizer Trends and Outlook to 2018*; Food and Agriculture Organization of the United Nations-Rome: Rome, Italy, 2015; Available online: http://www.fao.org/3/a-i4324e.pdf (accessed on 20 July 2019).

9. Tilman, D. Global environmental impacts of agricultural expansion: The need for sustainable and efficient practices. *Proc. Natl. Acad. Sci. USA* **1999**, *96*, 5995–6000. [CrossRef]

10. Goulding, K.; Jarvis, S.; Whitmore, A. Optimizing nutrient management for farm systems. *Philos. Trans. R. Soc. B* **2008**, *363*, 667–680. [CrossRef]

11. Zahoor, A.; Otten, A.; Wendisch, V.F. Metabolic engineering of Corynebacterium glutamicum for glycolate production. *J. Biotechnol.* **2014**, *192*, 366–375. [CrossRef]

12. Tilman, D.; Cassman, K.G.; Matson, P.A.; Naylor, R.; Polasky, S. Agricultural sustainability and intensive production practices. *Nature* **2002**, *418*, 671–677. [CrossRef]

13. Glass, A.D.M. Nitrogen use efficiency of crop plants: Physiological constraints upon nitrogen absorption. *Crit. Rev. Plant Sci.* **2003**, *22*, 453–470. [CrossRef]

14. Hodge, A.; Robinson, D.; Fitter, A. Are microorganisms more effective than plants at competing for nitrogen? *Trends Plant Sci.* **2000**, *5*, 304–308. [CrossRef]

15. Asghari, H.R.; Cavagnaro, T.R. Arbuscular mycorrhizas enhance plant interception of leached nutrients. *Funct. Plant Biol.* **2011**, *38*, 219–226. [CrossRef]

16. Modolo, L.V.; Da-Silva, C.J.; Brandão, D.S.; Chaves, I.S. A mini review on what we have learned about urease inhibitors of agricultural interest since mid-2000s. *J. Adv. Res.* **2018**, *13*, 29–37. [CrossRef] [PubMed]

17. Abbasi, M.K.; Tahir, M.M.; Rahim, N. Effect of N fertilizer source and timing on yield and N use efficiency of rainfed maize (*Zea mays* L.) in Kashmir-Pakistan. *Geoderma* **2013**, *195*, 87–93. [CrossRef]

18. Conant, R.T.; Berdanier, A.B.; Grace, P.R. Patterns and trends in nitrogen use and nitrogen recovery efficiency in world agriculture. *Glob. Biogeochem. Cycles* **2013**, *27*, 558–566. [CrossRef]

19. Mi, G.; Chen, F.; Zhang, F. Physiological and genetic mechanisms for nitrogen-use efficiency in maize. *J. Crop Sci. Biotechnol.* **2007**, *10*, 57–63.

20. Hirel, B.; Tétu, T.; Lea, P.J.; Dubois, F. Improving nitrogen use efficiency in crops for sustainable agriculture. *Sustainability* **2011**, *3*, 1452–1485. [CrossRef]

21. Sutton, M.A.; Oenema, O.; Erisman, J.W.; Leip, A.; Van Grinsven, H.; Winiwarter, W. Too much of a good thing. *Nature* **2011**, *472*, 159–161. [CrossRef]

22. Kaur, A.; Bedi, S. Nitrogen use Efficiency and Source-sink Relations in Maize. *J. Plant Sci. Res.* **2012**, *28*, 219.

23. Schröder, J.J.; Neeteson, J.J.; Withagen, J.C.M.; Noij, I.G.A.M. Effects of N application on agronomic and environmental parameters in silage maize production on sandy soils. *Field Crop. Res.* **1998**, *58*, 55–67. [CrossRef]

24. Reganold, J.P.; Papendick, R.I.; Parr, F.F. Sustainable agriculture. *Sci. Am.* **1990**, *262*, 112–120. [CrossRef]

25. Wichern, F.; Mayer, J.; Joergensen, R.G.; Müller, T. Release of C and N from roots of peas and oats and their availability to soil microorganisms. *Soil Biol. Biochem.* **2007**, *39*, 2829–2839. [CrossRef]

26. Fustec, J.; Lesuffleur, F.; Mahieu, S.; Cliquet, J.B. Nitrogen rhizodeposition of legumes. A review. *Agron. Sustain. Dev.* **2010**, *30*, 57–66. [CrossRef]

27. Bundy, L.G.; Kelling, K.A.; Good, L.W. *Using Legumes as a Nitrogen Source*; University of Wisconsin Extension, Cooperative Extension, Mifflin St.: Madison, WI, USA, 1992; Volume 3517, pp. 1–3.

28. Coulter, J.A.; Sheaffer, C.C.; Wyse, D.L.; Haar, M.J.; Porter, P.M.; Quiring, S.R.; Klossner, L.D. Agronomic performance of cropping systems with contrasting crop rotations and external inputs. *Agron. J.* **2011**, *103*, 182–192. [CrossRef]

29. Ranum, P.; Peña-Rosas, J.P.; Garcia-Casal, M.N. Global maize production, utilization, and consumption. *Ann. N. Y. Acad. Sci.* **2014**, *1312*, 105–112. [CrossRef]

30. Dowswell, C. *Maize in the Third World*; CRC Press-Taylor and Francis Group: New York, NY, USA; Routledge: Abingdon, UK; New York, NY, USA, 2019; pp. 1–2.

31. Chen, X.P.; Cui, Z.L.; Vitousek, P.M.; Cassman, K.G.; Matson, P.A.; Bai, J.S. Integrated soil–crop system management for food security. *Proc. Natl. Acad. Sci. USA* **2011**, *108*, 6399–6404. [CrossRef]

32. Grassini, P.; Cassman, K.G. High-yield maize with large net energy yield and small global warming intensity. *Proc. Natl. Acad. Sci. USA* **2012**, *109*, 1074–1079. [CrossRef]

33. *OECD-FAO Agricultural Outlook 2018–2027*; OECD Publishing: Paris, France; Food and Agriculture Organization of the United Nations: Rome, Italy, 2018. [CrossRef]

34. Worku, M.; Bänziger, M.; Friesen, D.; Horst, W.J. Nitrogen uptake and utilization in contrasting nitrogen efficient tropical maize hybrids. *Crop Sci.* **2007**, *47*, 519–528. [CrossRef]

35. Kiniry, J.R.; McCauley, G.; Xie, Y.; Arnold, J.G. Rice parameters describing crop performance of four USA cultivars. *Agron. J.* **2001**, *93*, 1354–1361. [CrossRef]

36. Peng, Y.; Niu, J.; Peng, Z.; Zhang, F.; Li, C. Shoot growth potential drives N uptake in maize plants and correlates with root growth in the soil. *Field Crop. Res.* **2010**, *115*, 85–93. [CrossRef]

37. Byers, D.L. Evolution in heterogeneous environments and the potential of maintenance of genetic variation in traits of adaptive significance. *Genetica* **2005**, *123*, 107. [CrossRef] [PubMed]

38. Hirel, B.; Le Gouis, J.; Ney, B.; Gallais, A. The challenge of improving nitrogen use efficiency in crop plants: Towards a more central role for genetic variability and quantitative genetics within integrated approaches. *J. Exp. Bot.* **2007**, *58*, 2369–2387. [CrossRef] [PubMed]

39. Masclaux-Daubresse, C.; Daniel-Vedele, F.; Dechorgnat, J.; Chardon, F.; Gaufichon, L.; Suzuki, A. Nitrogen uptake, assimilation and remobilization in plants: Challenges for sustainable and productive agriculture. *Ann. Bot.* **2010**, *105*, 1141–1157. [CrossRef] [PubMed]

40. Hammad, H.M.; Farhad, W.; Abbas, F.; Fahad, S.; Saeed, S.; Nasim, W.; Bakhat, H.F. Maize plant nitrogen uptake dynamics at limited irrigation water and nitrogen. *Environ. Sci. Pollut. Res.* **2017**, *24*, 2549–2557. [CrossRef] [PubMed]

41. Duvick, D.N. The contribution of breeding to yield advances in maize (*Zea mays* L.). *Adv. Agron.* **2005**, *86*, 83–145.

42. Ciampitti, I.A.; Vyn, T.J. A comprehensive study of plant density consequences on nitrogen uptake dynamics of maize plants from vegetative to reproductive stages. *Field Crop. Res.* **2011**, *121*, 2–18. [CrossRef]

43. Ciampitti, I.A.; Vyn, T.J. Physiological perspectives of changes over time in maize yield dependency on nitrogen uptake and associated nitrogen efficiencies: A review. *Field Crop. Res.* **2012**, *133*, 48–67. [CrossRef]

44. Echarte, L.; Andrade, F.H. Harvest index stability of Argentinean maize hybrids released between 1965 and 1993. *Field Crop. Res.* **2003**, *82*, 1–12. [CrossRef]

45. Sinclair, T.R. Historical changes in harvest index crop nitrogen accumulation. *Crop. Sci.* **1998**, *38*, 638–643. [CrossRef]

46. Tollenaar, M.; Lee, E.A. Dissection of physiological processes underlying grain yield in maize by examining genetic improvement and heterosis. *Maydica* **2006**, *51*, 399.

47. Vega, C.R.C.; Sadras, V.O.; Andrade, F.H.; Uhart, S.A. Reproductive allometry in soybean, maize and sunflower. *Ann. Bot.* **2000**, *85*, 461–468. [CrossRef]

48. Echarte, L.; Luque, S.; Andrade, F.H.; Sadras, V.O.; Cirilo, A.; ME Otegui, M.E.; Vega, C.R.C. Response of maize kernel number to plant density in Argentinean hybrids released between 1965 and 1995. *Field Crop. Res.* **2000**, *68*, 1–8. [CrossRef]

49. Tollenaar, M.; Dwyer, L.M.; Stewart, D.W.; Ma, B.L. Physiological parameters associated with differences in kernel set among maize hybrids. In *Physiology and Modeling Kernel Set in Maize*; Westgate, M.A., Boote, K.J., Eds.; CSSA Spec. CSSA/ASA/SSSA: Madison, WI, USA, 2000; Volume 51, pp. 115–130.

50. Andrade, F.H.; Vega, C.; Uhart, S.; Cirilo, A.; Cantarero, M.; Valentinuz, O. Kernel number determination in maize. *Crop Sci.* **1999**, *39*, 453–459. [CrossRef]

51. Vega, C.R.C.; Andrade, F.H.; Sadras, V.O.; Uhart, S.A.; Valentinuz, O.R. Seed number as a function of growth. A comparative study in soybean, sunflower and maize. *Crop Sci.* **2001**, *41*, 748–754. [CrossRef]

52. Echarte, L.; Andrade, F.H.; Vega, C.R.C.; Tollenaar, M. Kernel number determination in Argentinean maize hybrids released between 1965 and 1993. *Crop Sci.* **2004**, *44*, 1654–1661. [CrossRef]

53. Moll, R.H.; Kamprath, E.J.; Jackson, W.A. Analysis and interpretation of factors which contribute to efficiency of nitrogen utilization. *Agron. J.* **1982**, *74*, 562–564. [CrossRef]

54. Good, A.G.; Shrawat, A.K.; Muench, D.G. Can less yield more? Is reducing nutrient input into the environment compatible with maintaining crop production? *Trends Plant Sci.* **2004**, *9*, 597–605. [CrossRef]

55. Seyfried, M.S.; Rao, P.S.C. Kinetics of nitrogen mineralization in Costa Rican soils: Model evaluation and pretreatment effects. *Plant Soil* **1988**, *106*, 159–169. [CrossRef]

56. Hirel, B.; Lea, P.J. Ammonium assimilation. In *Plant Nitrogen*; Springer: Belin/Heidelberg, Germany, 2001; pp. 79–99.

57. Huggins, D.R.; Pan, W.L. Key indicators for assessing nitrogen use efficiency in cereal-based agro-ecosystems. *J. Crop Prod.* **2003**, *8*, 157–185. [CrossRef]

58. Dobermann, A. Nitrogen use efficiency: State of the art. In *IFA International Workshop on Enhanced-Efficiency Fertilizers, Frankfurt*; International Fertilizer Industry Association: Paris, France, 2005; pp. 28–30.

59. Yan, L.; Zhang, Z.D.; Zhang, J.J.; Gao, Q.; Feng, G.Z.; Abelrahman, A.M.; Chen, Y. Effects of improving nitrogen management on nitrogen utilization, nitrogen balance, and reactive nitrogen losses in a Mollisol with maize monoculture in Northeast China. *Environ. Sci. Pollut. Res.* **2016**, *23*, 4576–4584. [CrossRef]

60. Wiesler, F.; Behrens, T.; Horst, W.J. The Role of N Efficient Cultivars in Sustainable Agriculture. *Sci. World J.* **2001**, *1*, 61–69. [CrossRef] [PubMed]

61. Kumar, M.; Rajput, T.; Kumar, R.; Patel, N. Water and nitrate dynamics in baby corn (*Zea mays* L.). under different fertigation frequencies and operating pressures in semi-arid region of India. *Agric. Water Manag.* **2016**, *163*, 263–274. [CrossRef]

62. Couto-Va´zquez, A.; Gonza´lez-Prieto, S.J. Fate of ^{15}N-fertilizers in the soil-plant system of a forage rotation under conservation and plough tillage. *Soil Tillage Res.* **2016**, *161*, 10–18. [CrossRef]

63. Trenkel, M.E. *Slow and Controlled-Release and Stabilized Fertilizers: An Option for Enhancing Nutrient Use Efficiency in Agriculture*; International Fertilizer Association: Paris, France, 2010.

64. Schmidt, E.L. Nitrification in soil. In *Nitrogen in Agricultural Soils*; Stevenson, F.J., Ed.; ASA, CSSA, and SSSA: Madison, WI, USA, 1982; Volume 22, pp. 253–288.

65. Kaboneka, S.; Sabbe, W.E.; Mauromoustakos, A. Carbon decomposition kinetics and nitrogen mineralization from corn, soybean, and wheat. *Commun. Soil Sci. Plant Anal.* **1997**, *28*, 1359–1373. [CrossRef]

66. Kelly, J.T.; Bacon, R.K.; Wells, B.R. Genetic variability in nitrogen utilization at four growth stages in soft red winter wheat. *J. Plant Nutr.* **1995**, *18*, 969–982. [CrossRef]

67. Hirel, B.; Bertin, P.; Quillere, I.; Bourdoncle, W.; Attagnant, C.; Dellay, C.; Gouy, S.; Retailliau, C.; Falque, M.; Gallais, A. Towards a better understanding of the genetic and physiological basis for nitrogen use efficiency in maize. *Plant Physiol.* **2001**, *125*, 1258–1270. [CrossRef]

68. Lea, P.J.; Miflin, B.J. Glutamate synthase and the synthesis of glutamate in plants. *Plant Physiol. Biochem.* **2003**, *41*, 555–564. [CrossRef]

69. Fuentes, S.I.; Allen, D.J.; Ortiz-Lopez, A.; Hernández, G. Over-expression of cytosolic glutamine synthetase increases photosynthesis and growth at low nitrogen concentrations. *J. Exp. Bot.* **2001**, *52*, 1071–1081. [CrossRef]

70. Kichey, T.; Heumez, E.; Pocholle, P.; Pageau, K.; Vanacker, H.; Dubois, F.; Le Gouis, J.; Hirel, B. Combined agronomic and physiological aspects of nitrogen management in wheat (*Triticum aestivum* L). Dynamic and integrated views highlighting the central role for the enzyme glutamine synthetase. *New Phytol.* **2006**, *169*, 265–278. [CrossRef]

71. Forde, B.G.; Lea, P.J. Glutamate in plants: Metabolism, regulation and signaling. *J. Exp. Biol.* **2007**, *58*, 2339–2358. [CrossRef]

72. Cren, M.; Hirel, B. Glutamine synthetase in higher plants: Regulation of gene and protein expression from the organ to the cell. *Plant Cell Physiol.* **1999**, *40*, 1187–1193. [CrossRef]

73. Cirilo, A.G.; Dardanelli, J.; Balzarini, M.; Andrade, F.H.; Cantarero, M.; Luque, S.; Pedrol, H.M. Morpho-physiological traits associated with maize crop adaptations to environments differing in nitrogen availability. *Field Crop. Res.* **2009**, *113*, 116–124. [CrossRef]

74. Borrell, A.; Hammer, G. Nitrogen dynamics and the physiological basis of stay-green in sorghum. *Crop Sci.* **2000**, *40*, 1295–1307. [CrossRef]

75. Maddonni, G.A.; Otegui, M.E.; Cirilo, A.G. Plant population density, row spacing and hybrid effects on maize canopy architecture and light attenuation. *Field Crop. Res.* **2001**, *71*, 183–193. [CrossRef]

76. Gastal, F.; Lemaire, G. Nitrogen uptake and distribution in crops: An agronomical and ecophysiological perspective. *J. Exp. Bot.* **2002**, *53*, 789–799. [CrossRef]

77. Ma, B.L.; Dwyer, L.M. Nitrogen uptake and use in two contrasting maize hybrids differing in leaf senescence. *Plant Soil* **1998**, *199*, 283–291. [CrossRef]

78. Hikosaka, K. Interspecific difference in the photosynthesis–nitrogen relationship: Patterns, physiological causes, and ecological importance. *J. Plant Res.* **2004**, *117*, 481–494. [CrossRef]

79. Andrade, F.H.; Echarte, L.; Rizzalli, R.; Della, M.A.; Casanovas, M. Kernel number prediction in maize under nitrogen or water stress. *Crop Sci.* **2002**, *42*, 1173–1179. [CrossRef]

80. Spano, G.; Di Fonzo, N.; Perrota, C.; Platani, C.; Ronga, G.; Lawlor, Q.W.; Napier, J.A.; Shewry, P.R. Physiological characterization of 'stay green' mutants in durum wheat. *J. Exp. Bot.* **2003**, *54*, 1415–1420. [CrossRef]

81. Racjan, I.; Tollenaar, M. Source-sink ratio and leaf senescence in maize I. Dry matter accumulation and partitioning during grain filling. *Field Crop. Res.* **1999**, *60*, 245–253.

82. Racjan, I.; Tollenaar, M. Source-sink ratio and leaf senescence in maize II. Nitrogen metabolism during grain filling. *Field Crop. Res.* **1999**, *60*, 255–265.

83. Coque, M.; Gallais, A. Genetic variation for N-remobilization and post-silking N-uptake in a set of maize recombinant inbred lines I: Evaluation by ^{15}N labeling, heritabilities and correlations among traits for test cross performance. *Crop Sci.* **2007**, *47*, 1787–1796. [CrossRef]

84. Borras, L.; Slafer, G.A.; Otegui, M.E. Seed dry weight response to source-sink manipulations in wheat, maize, and soybean: A quantitative reappraisal. *Field Crop. Res.* **2004**, *86*, 131–146. [CrossRef]

85. Pang, X.P.; Letey, J. Organic farming challenge of timing nitrogen availability to crop nitrogen requirements. *Soil Sci. Soc. Am. J.* **2000**, *64*, 247–253. [CrossRef]

86. Purcino, A.A.C.; Arellano, C.; Athwal, G.S.; Huber, S.C. Nitrate effect on carbon and nitrogen assimilating enzymes of maize hybrids representing seven eras of maize breeding. *Maydica* **1998**, *43*, 83–94.

87. Halvorson, A.D.; Nielsen, D.C.; Reule, C.A. Nitrogen fertilization and rotation effects on no-till dry land wheat production. *Agron. J.* **2004**, *96*, 1196–1201. [CrossRef]

88. Peoples, M.B.; Freney, J.R.; Mosier, A.R. Minimizing gaseous losses of nitrogen. In *Nitrogen Fertilization in the Environment*; Bacon, P.E., Ed.; Marcel Dekker Inc.: New York, NY, USA, 1995; pp. 565–602.

89. Rizzi, E.; Balconi, C.; Bosio, D.; Nembrini, L.; Morselli, A.; Motto, M. Accumulation and partitioning of nitrogen among plant parts in the high and low protein strains of maize. *Maydica* **1996**, *41*, 325–332.

90. Subedi, K.D.; Ma, B.L. Nitrogen uptake and partitioning in stay-green and leafy maize hybrids. *Crop Sci.* **2005**, *45*, 740–747. [CrossRef]

91. Sattelmacher, B.; Horst, W.J.; Becker, H.C. Factors that contribute to genetic variation for nutrient efficiency of crop plants. *Zeitschrift für Pflanzenernährung und Bodenkunde* **1994**, *157*, 215–224. [CrossRef]

92. Akintoye, H.A.; Kling, J.; Gand, L.E.O. N-use efficiency of single, double and synthetic maize lines grown at four N levels in three ecological zones of West Africa. *Field Crop. Res.* **1999**, *60*, 189–199. [CrossRef]

93. Gallais, A.; Coque, M.; Quilléré, I.; Le Gouis, J.; Prioul, J.L.; Hirel, B. Estimating proportions of N-remobilization and of post-silking N-uptake allocated to maize kernels by ^{15}N labeling. *Crop Sci.* **2007**, *47*, 685–691. [CrossRef]

94. Tolley-Henry, L.; Raper, C.D., Jr. Soluble carbohydrate allocation to roots, photosynthetic rate of leaves, and nitrate assimilation as affected by nitrogen stress and irradiance. *Bot. Gaz.* **1991**, *152*, 23–33. [CrossRef] [PubMed]

95. Pons, T.L.; van der Werf, A.; Lambers, H. Photosynthetic nitrogen use efficiency of inherently low- and fast-growing species: Possible explanations for observed differences. In *A Whole Plant Perspective on Carbon-Nitrogen Interactions*; Roy, J., Garnier, E., Eds.; SPB Academic Publishing: The Hague, The Netherlands, 1994; pp. 61–77.

96. Evans, J.R. Photosynthesis and nitrogen relationships in leaves of C$_3$ plants. *Oecologia* **1989**, *78*, 9–19. [CrossRef] [PubMed]

97. Gallais, A.; Coque, M.; Quilléré, I.; Prioul, J.L.; Hirel, B. Modelling post-silking N-fluxes in maize using ^{15}N-labeling-field experiments. *New Phytol.* **2006**, *172*, 696–707. [CrossRef] [PubMed]

98. Paponov, I.A.; Engels, C. Effect of nitrogen supply on leaf traits related to photosynthesis during grain filling in two maize genotypes with different N efficiency. *J. Plant Nutr. Soil Sci.* **2003**, *166*, 756–763. [CrossRef]

99. Pearson, C.J.; Jacobs, B.C. Yield components and nitrogen partitioning of maize in response to nitrogen before and after anthesis. *Aust. J. Agric. Res.* **1987**, *38*, 1001–1009. [CrossRef]

100. McKendry, A.L.; McVetty, P.B.E.; Evans, L.E. Selection criteria for combining high grain yield and high grain protein concentration in bread wheat. *Crop Sci.* **1995**, *35*, 1597–1602. [CrossRef]

101. Mistele, B.; Schmidhalter, U. Estimating the nitrogen nutrition index using spectral canopy reflectance measurements. *Eur. J. Agron.* **2008**, *29*, 184–190. [CrossRef]

102. Grindlay, D.J.C.; Sylvester-Bradley, R.; Scott, R.K. The relationship between canopy green area and nitrogen in the shoot. In *Diagnostic Procedures for Crop N Management*; Lemaire, G., Burns, I.G., Eds.; INRA Editions: Paris, France, 1995; pp. 53–60.

103. Lemaire, G.; Gastal, F.; Plenet, D. Dynamics of N uptake and N distribution in plant canopies. Use of crop N status index in crop modeling. In *Diagnostic Procedures for Crop N Management*; Lemaire, G., Bruns, I.G., Eds.; INRA Editions: Paris, France, 1995; pp. 15–29.

104. Eichelmann, H.; Oja, V.; Rasulov, B.; Padu, E.; Bichele, I.; Pettai, P.; Mänd, P.; Kull, O.; Laisk, A. Adjustment of leaf photosynthesis to shade in a natural canopy: Reallocation of nitrogen. *Plant Cell Environ.* **2005**, *28*, 389–401. [CrossRef]

105. Schmidhalter, U. Development of a quick on-farm test to determine nitrate levels in soil. *J. Plant Nutr. Soil Sci.* **2005**, *168*, 432–438. [CrossRef]

106. Houlès, V.; Guérif, M.; Mary, B. Elaboration of a nitrogen nutrition indicator for winter wheat based on leaf area index and chlorophyll content for making nitrogen recommendations. *Eur. J. Agron.* **2007**, *27*, 1–11. [CrossRef]

107. Baret, F.; Houlès, V.; Guérif, M. Quantification of plant stress using remote sensing observations and crop models: The case of nitrogen management. *J. Exp. Bot.* **2007**, *58*, 869–880. [CrossRef] [PubMed]

108. Schlemmer, M.; Gitelson, A.; Schepers, J.; Ferguson, R.; Peng, Y.; Shanahan, J.; Rundquist, D. Remote estimation of nitrogen and chlorophyll contents in maize at leaf and canopy levels. *Int. J. Appl. Earth Obs. Geoinf.* **2013**, *25*, 47–54. [CrossRef]

109. Sripada, R.P.; Schmidt, J.P.; Dellinger, A.E.; Beegle, D.B. Evaluating multiple indices from a canopy reflectance sensor to estimate corn N requirements. *Agron. J.* **2008**, *100*, 1553–1561. [CrossRef]

110. Gitelson, A.A. Remote Sensing estimation of crop biophysical characteristics at various scales. In *Chapter 15 in Hyperspectral Remote Sensing of Vegetation*; Thenkabail, P.S., Lyon, J.G., Huete, A., Eds.; CRC Press-Taylor and Francis Group: Boca Raton, FL, USA; London, UK; New York, NY, USA, 2011; pp. 329–358.

111. Walters, D.T. Diagnosis of nitrogen deficiency in maize and the influence of hybrid and plant density. In Proceedings of the North Central Extension-Industry Soil Fertility Conference, Des Moines, IA, USA, 19–20 November 2003; Volume 19.

112. Zebarth, B.J.; Younie, M.; Paul, J.W.; Bittman, S. Evaluation of leaf chlorophyll index for making fertilizer nitrogen recommendations for silage corn in a high fertility environment. *Commun. Soil Sci. Plant Anal.* **2002**, *33*, 665–684. [CrossRef]

113. Argenta, G.; Ferreira da Silva, P.R.; Sangoi, L. Leaf relative chlorophyll content as an indicator parameter to predict nitrogen fertilization in maize. *Ciênc Rural* **2004**, *34*, 1379–1387. [CrossRef]

114. Rashid, M.T.; Voroney, P.; Parkin, G. Predicting nitrogen fertilizer requirements for corn by chlorophyll meter under different N availability conditions. *Can. J. Soil Sci.* **2005**, *85*, 149–159. [CrossRef]

115. Link, A.; Jasper, J. Site-specific N fertilization based on remote sensing: Is it necessary to take yield variability into account? In Proceedings of the 4th European Conference on Precision Agriculture, Berlin, Germany, 15–19 June 2003; Stafford, J., Werner, A., Eds.; Academic Publishers: Wageningen, The Netherlands, 2003; pp. 353–359.

116. Soon, Y.K.; Clayton, G.W. Eight years of crop rotation and tillage effects on crop production and N fertilizer use. *Can. J. Soil Sci.* **2002**, *82*, 165–172. [CrossRef]

117. Silla, F.; Escudero, A. Nitrogen-use efficiency: Trade-offs between N productivity and mean residence time at organ, plant and population levels. *Funct. Ecol.* **2004**, *18*, 511–521. [CrossRef]

118. Echarte, L.; Rothstein, S.; Tollenaar, M. The response of leaf photosynthesis and dry matter accumulation to nitrogen supply in an older and a newer maize hybrid. *Crop Sci.* **2008**, *48*, 656–665. [CrossRef]

119. Masclaux, C.; Quilleré, I.; Gallais, A.; Hirel, B. The challenge of remobilization in plant nitrogen economy. A survey of physio-agronomic and molecular approaches. *Ann. Appl. Biol.* **2001**, *138*, 69–81. [CrossRef]

120. Habtegebrial, K.; Singh, B.R.; Haile, M. Impact of tillage and nitrogen fertilization on yield, nitrogen use efficiency of tef (*Eragrostis tef* (Zucc.) Trotter) and soil properties. *Soil Tillage Res.* **2007**, *94*, 55–63. [CrossRef]

121. Ogola, J.B.O.; Wheeler, T.R.; Harris, P.M. Effects of nitrogen and irrigation on water use of maize crops. *Field Crop. Res.* **2002**, *78*, 105–117. [CrossRef]

122. Anatoliy, G.K.; Thelen, K.D. Effect of winter wheat crop residue on no-till corn growth and development. *Agron. J.* **2007**, *99*, 549–555.

123. Hooda, P.S.; Edwards, A.C.; Anderson, H.A.; Miller, A. A review of water quality concerns in livestock farming areas. *Sci. Total Environ.* **2000**, *250*, 143–167. [CrossRef]

124. Robertson, G.P.; Vitousek, P.M. Nitrogen in agriculture: Balancing the cost of an essential resource. *Ann. Rev. Environ. Resour.* **2009**, *34*, 97–125. [CrossRef]

125. Lucas, F.T.; Borges, B.M.M.N.; Coutinho, E.L.M. Nitrogen fertilizer management for maize production under tropical climate. *Agron. J.* **2019**, *111*, 2031–2037. [CrossRef]

126. Nyamangara, J.; Piha, M.I.; Giller, K.E. Effects of combined cattle manure and mineral nitrogen on maize N uptake and grain yield. *Afr. Crop Sci. J.* **2003**, *11*, 289–300. [CrossRef]

127. Rimski-Korsakov, H.; Rubio, G.; Lavado, R.S. Fate of the nitrogen from fertilizers in field-grown maize. *Nutr. Cycl. Agroecosyst.* **2012**, *93*, 253–263. [CrossRef]

128. Huang, B.; Sun, W.; Zhao, Y.; Zhu, J.; Yang, R.; Zou, Z.; Su, J. Temporal and spatial variability of soil organic matter and total nitrogen in an agricultural ecosystem as affected by farming practices. *Geoderma* **2007**, *139*, 336–345. [CrossRef]

129. Khan, Z.; Shah, P.; Arif, M. Management of organic farming: Effectiveness of farm yard manure (FYM) and nitrogen for maize productivity. *Sarhad J. Agric.* **2000**, *16*, 461–465.

130. Gangwar, K.S.; Singh, K.K.; Sharma, S.K.; Tomar, O.K. Alternative tillage and crop residue management in wheat after rice in sandy loam soils of Indo-Gengetic plains. *Soil Tillage Res.* **2006**, *88*, 242–252. [CrossRef]

131. Ayeni, L.S.; Adetunji, M.T.; Ojeniyi, S.O.; Ewulo, B.S.; Adeyemo, A.J. Comparative and cumulative effect of cocoa pod husk ash and poultry manure on soil and maize nutrient contents and yield. *Am.–Euras. J. Sustain. Agric.* **2008**, *2*, 92–97.

132. Biau, A.; Santiveri, F.; Mijangos, I.; Lloveras, J. The impact of organic and mineral fertilizers on soil quality parameters and the productivity of irrigated maize crops in semiarid regions. *Eur. J. Soil Biol.* **2012**, *53*, 56–61. [CrossRef]

133. Belay, A.; Claassens, A.; Wehner, F.C. Effect of direct nitrogen and potassium and residual phosphorus fertilizers on soil chemical properties, microbial components and maize yield under long-term crop rotation. *Biol. Fertil. Soils* **2002**, *35*, 420–427.

134. Ayeni, L.S.; Adetunji, M.T. Integrated application of poultry manure and mineral fertilizer on soil chemical properties, nutrient uptake, yield and growth components of maize. *Nat. Sci.* **2010**, *8*, 60–67.

135. Ali, K.; Munsif, F.; Zubair, M.; Hussain, Z.; Shahid, M.; Din, I.U.; Khan, N. Management of organic and inorganic nitrogen for different maize varieties. *Sarhad J. Agric.* **2011**, *27*, 525–529.

136. Rafiq, M.A.; Ali, A.; Malik, M.A.; Hussain, M. Effect of fertilizer levels and plant densities on yield and protein contents of autumn planted maize. *Pak. J. Agric. Sci.* **2010**, *47*, 201–208.

137. Asif, M.; Saleem, M.F.; Anjum, S.A.; Wahid, M.A.; Bilal, M.F. Effect of nitrogen and zinc sulphate on growth and yield of maize (*Zea mays* L.). *J. Agric. Res.* **2013**, *51*, 03681157.

138. Pessarakli, M. (Ed.) *Handbook of Plant and Crop Physiology*; Eastern Hemisphere Distribution, Marcel Dekker Inc.: New York, NY, USA; Basel, Switzerland, 2002; Volume 940.

139. Potarzycki, J. Effect of magnesium or zinc supplementation at the background of nitrogen rate on nitrogen management by maize canopy cultivated in monoculture. *Plant Soil Environ.* **2011**, *57*, 19–25. [CrossRef]

140. Qu, C.; Liu, C.; Ze, Y.; Gong, X.; Hong, M.; Wang, L.; Hong, F. Inhibition of nitrogen and photosynthetic carbon assimilation of maize seedlings by exposure to a combination of salt stress and potassium-deficient stress. *Biol. Trace Elem. Res.* **2011**, *144*, 1159–1174. [CrossRef] [PubMed]

141. Mulvaney, R.L.; Khan, S.A.; Ellsworth, T.R. Synthetic nitrogen fertilizers deplete soil nitrogen: A global dilemma for sustainable cereal production. *J. Environ. Qual.* **2009**, *38*, 2295–2314. [CrossRef] [PubMed]

142. Khan, S.A.; Mulvaney, R.L.; Ellsworth, T.R.; Boast, C.W. The myth of nitrogen fertilization for soil carbon sequestration. *J. Environ. Qual.* **2007**, *36*, 1821–1832. [CrossRef] [PubMed]

143. Mapanda, F.; Wuta, M.; Nyamangara, J.; Rees, R.M. Effects of organic and mineral fertilizer nitrogen on greenhouse gas emissions and plant-captured carbon under maize cropping in Zimbabwe. *Plant Soil* **2011**, *343*, 67–81. [CrossRef]

144. Yang, J.Y.; Huffman, E.C.; Jong, R.D.; Kirkwood, V.; MacDonald, K.B.; Drury, C.F. Residual soil nitrogen in soil landscapes of Canada as affected by land use practices and agricultural policy scenarios. *Land Use Policy* **2007**, *24*, 89–99. [CrossRef]

145. Khan, H.Z.; Malik, M.A.; Saleem, M.F. Effect of rate and source of organic material on the production potential of spring maize (*Zea mays* L.). *Pak. J. Agric. Sci.* **2008**, *45*, 40–43.

146. Grant, C. Policy aspects related to the use of enhanced-efficiency fertilizers: View point of the scientific community. In *IFA International Workshop on Enhanced-Efficiency Fertilizers, Frankfurt*; International Fertilizer Industry Association: Paris, France, 2005; pp. 28–30.

147. Stevens, W.B.; Hoeft, R.G.; Mulvaney, R.L. Fate of Nitrogen15 in a long-term nitrogen rate study: II. Nitrogen uptake efficiency. *Agron. J.* **2005**, *97*, 1046–1053. [CrossRef]

148. Robertson, G.P.; Groffman, P.M. Nitrogen transformations. In *Soil Microbiology, Ecology and Biochemistry*; Academic Press: New York, NY, USA, 2007; pp. 341–364.

149. Provin, T.; Hossner, L.R. What Happens to Nitrogen in Soils? In *Texas A&M AgriLife Extension*; The Texas A&M University Systems: College Station, TX, USA, 2001.

150. Jackson, L.E.; Burger, M.; Cavagnaro, T.R. Roots, nitrogen transformations, and ecosystem services. *Annu. Rev. Plant Biol.* **2008**, *59*, 341–363. [CrossRef]

151. Griffin, T.S. Nitrogen availability. In *Nitrogen in Agricultural Soils*; Schepers, J.S., Raun, W.R., Eds.; ASA, CSSA, and SSSA: Madison, WI, USA, 2008; Volume 49, pp. 616–646.

152. Ma, B.L.; Dwyer, L.M.; Gregorich, E.G. Soil nitrogen amendment effects on seasonal nitrogen mineralization and nitrogen cycling in maize production. *Agron. J.* **1999**, *91*, 1003–1009. [CrossRef]

153. Schröder, J.J.; Neeteson, J.J.; Oenema, O.; Struik, P.C. Does the crop or the soil indicate how to save nitrogen in maize production? Reviewing the state of the art. *Field Crop. Res.* **2000**, *66*, 151–164. [CrossRef]

154. Rusinamhodzi, L.; Corbeels, M.; Zingore, S.; Nyamangara, J.; Giller, K.E. Pushing the envelope? Maize production intensification and the role of cattle manure in recovery of degraded soils in smallholder farming areas of Zimbabwe. *Field Crop. Res.* **2013**, *147*, 40–53. [CrossRef]

155. Dunjana, N.; Nyamugafata, P.; Shumba, A.; Nyamangara, J.; Zingore, S. Effects of cattle manure on selected soil physical properties of smallholder farms on two soils of Murewa, Zimbabwe. *Soil Use Manag.* **2012**, *28*, 221–228. [CrossRef]

156. Negassa, W.; Gebrekidan, H.; Friesen, D.K. Integrated use of farmyard manure and NP fertilizers for maize on farmers' fields. *J. Agric. Rural Dev. Trop. Subtrop.* **2005**, *106*, 131–141.

157. Bengough, A.G.; Bransby, M.F.; Hans, J.; McKenna, S.J.; Roberts, T.J.; Valentine, T.A. Root response to soil physical conditions: Growth dynamics from field to cell. *J. Exp. Bot.* **2006**, *57*, 437–447. [CrossRef]

158. Li, Y.; Mi, G.H.; Chen, F.; Lao, X.; Zhang, F. Genotypic difference of nitrogen recycling between root and shoot of maize seedlings. *Acta Phytophysiol. Sin.* **2001**, *27*, 226–230.

159. Wang, Y.; Mi, G.H.; Chen, F.J.; Zhang, J.H.; Zhang, F.S. Response of root morphology to nitrate supply and its contribution to nitrogen accumulation in maize. *J. Plant Nutr.* **2004**, *27*, 2189–2202. [CrossRef]

160. Marschner, H.; Kirkby, E.A.; Cakmak, I. Effect of mineral nutritional status on shoot-root partitioning of photoassimilates and cycling of mineral nutrients. *J. Exp. Bot.* **1996**, *47*, 1255–1263. [CrossRef]

161. Chun, L.; Chen, F.; Zhang, F.; Mi, G.H. Root growth, nitrogen uptake and yield formation of hybrid maize with different N efficiency. *Plant Nutr. Fert. Sci.* **2005**, *11*, 615–619.

162. Tian, Q.Y.; Chen, F.J.; Zhang, F.S.; Mi, G.H. Genotypic difference in nitrogen acquisition ability in maize plants is related to the coordination of leaf and root growth. *J. Plant Nutr.* **2006**, *29*, 317–330. [CrossRef]

163. Niu, J.; Chen, F.J.; Mi, G.H.; Li, C.J.; Zhang, F.S. Transpiration, and nitrogen uptake and flow in two maize (*Zea mays* L.). inbred lines as affected by nitrogen supply. *Ann. Bot.* **2007**, *99*, 153–160. [CrossRef]

164. Mackay, A.D.; Barber, S.A. Effects of nitrogen on root growth of two corn genotypes in the field. *Agron. J.* **1986**, *78*, 699–703. [CrossRef]

165. Mkhabela, M.S.; Mkhabela, M.S.; Pali-Shikhulu, J. Response of maize (Zea mays L.) cultivars to different levels of nitrogen application in Swaziland. In Proceedings of the Seventh Eastern and Southern Africa Regional Maize Conference, Nairobi, Kenya, 5–11 February 2001; Volume 11, pp. 377–381.

166. Mi, G.H.; Liu, J.A.; Chen, F.J.; Zhang, F.S.; Cui, Z.L.; Liu, X.S. Nitrogen uptake and remobilization in maize hybrids differing in leaf senescence. *J. Plant Nutr.* **2003**, *26*, 237–247. [CrossRef]

167. Horst, W.J.; Behrens, T.; Heuberger, H.; Kamh, M.; Reidenbach, G.; Wiesler, F. Genotypic differences in nitrogen use efficiency in crop plants. In *Innovative Soil-Plant Systems for Sustainable Agricultural Production*; Lynch, J.M., Schepers, J.S., Unver, I., Eds.; Organisation for Economic Co-operation and Development: Paris, France, 2003; pp. 75–92.

168. Miflin, B.J.; Habash, D.Z. The role of glutamine synthetase and glutamate dehydrogenase in nitrogen assimilation and possibilities for improvement in the nitrogen utilization of crops. *J. Exp. Bot.* **2002**, *53*, 979–987. [CrossRef] [PubMed]

169. Pan, W.L.; Jackson, W.A.; Moll, R.H. Nitrate uptake and partitioning by corn (Zea mays L.). root systems and associated morphological differences among genotypes and stages of root development. *J. Exp. Bot.* **1985**, *36*, 1341–1351. [CrossRef]

170. Peterson, C.A.; Emanuel, M.E.; Humphreys, G.B. Pathway of movement of apoplastic fluorescent dye tracers through the endodermis at the site of secondary root formation in corn (Zea mays L.) and broad bean (Vicia faba). *Can. J. Bot.* **1981**, *59*, 618–625. [CrossRef]

171. Oaks, A.; Aslam, M.; Boesel, I. Ammonium and amino acids as regulators of nitrate reductase in corn roots. *Plant Physiol.* **1977**, *59*, 391–394. [CrossRef]

172. Lewis, C.E.; Noctor, G.; Causton, D.; Foyer, C. Regulation of assimilate partitioning in leaves. *Aust. J. Plant Physiol.* **2000**, *27*, 507–519. [CrossRef]

173. Ortiz-Monasterio, J.I.; Sayre, K.D.; Rajaram, S.; McMahon, M. Genetic progress in wheat yield and nitrogen use efficiency under four nitrogen regimes. *Crop Sci.* **1997**, *37*, 898–904. [CrossRef]

174. Wivstad, M.; Dahlin, A.S.; Grant, C. Perspectives on nutrient management in arable farming systems. *Soil Use Manag.* **2005**, *21*, 113–121. [CrossRef]

175. Djaman, K.; Irmak, S.; Martin, D.L.; Ferguson, R.B.; Bernards, M.L. Plant nutrient uptake and soil nutrient dynamics under full and limited irrigation and rainfed maize production. *Agron. J.* **2013**, *105*, 527–538. [CrossRef]

176. Qiu, S.J.; He, P.; Zhao, S.C.; Li, W.J.; Xie, J.G.; Hou, Y.P.; Jin, J.Y. Impact of nitrogen rate on maize yield and nitrogen use efficiencies in northeast China. *Agron. J.* **2015**, *107*, 305–313. [CrossRef]

177. Sinclair, T.R.; Vadez, V. Physiological traits for crop yield improvement in low N and P environments. *Plant Soil* **2002**, *245*, 1–15. [CrossRef]

178. Cañas, R.A.; Amiour, N.; Quilleré, I.; Hirel, B. An integrated statistical analysis of the genetic variability of nitrogen metabolism in the ear of three maize inbred lines (Zea mays L.). *J. Exp. Bot.* **2010**, *62*, 2309–2318. [CrossRef]

179. Stark, C.H.; Richards, K.G. The continuing challenge of agricultural nitrogen loss to the environment in the context of global change and advancing research. *Dyn. Soil Dyn. Plant* **2008**, *2*, 1–12.

180. Delgado, J.A.; Shaffer, M.; Hu, C.; Lavado, R.; Cueto-Wong, J.; Joosse, P.; Sotomayer, D.; Colon, W.; Follett, R.; DelGrosso, S.; et al. An index approach to assess nitrogen losses to the environment. *Ecol. Eng.* **2008**, *32*, 108–120. [CrossRef]

 *agronomy*

Article

Compost as a Substitute for Mineral N Fertilization? Effects on Crops, Soil and N Leaching

Carmelo Maucieri *, Alberto Barco and Maurizio Borin

Department of Agronomy, Food, Natural Resources, Animals and Environment (DAFNAE),
Viale dell'Università 16, 35020 Legnaro (PD), Italy; albertobarco92@gmail.com (A.B.);
maurizio.borin@unipd.it (M.B.)
* Correspondence: carmelo.maucieri@unipd.it; Tel.: +39-049-827-2859

Received: 6 March 2019; Accepted: 12 April 2019; Published: 15 April 2019

Abstract: A three-year study was conducted to test the fertilization properties of different types of compost as the total or partial mineral nitrogen fertilization substitute in an herbaceous crop succession (*Zea mays* L., *Triticum aestivum* L. and *Helianthus annus* L.). Four types of compost (i. green cuttings and depuration sludge, ii. green cuttings, organic fraction of municipal wastes and other organic materials, iii. green cuttings, iv. green cuttings and organic fraction of municipal wastes) and eight fertilization treatments (combining: unfertilized control, 100% mineral fertilization, 100% compost, and 50% compost +50% mineral fertilization) were evaluated in terms of: (i) crop yields and nitrogen uptake, (ii) soil organic carbon and nitrate nitrogen soil contents variation, and (iii) residual nitrate nitrogen leached at the end of the experiment. Maize grain yield ranged from 5.2 ± 1.0 Mg ha^{-1} to 7.4 ± 0.7 Mg ha^{-1} with the highest value in the mineral fertilization treatment and the lowest values in the 100% compost fertilization. Wheat and sunflower grain yields were not significantly different among control, mineral, compost, or mineral/compost fertilization treatments with average values of 5.1 ± 0.7 Mg ha^{-1} and 2.3 ± 0.3 Mg ha^{-1}, respectively. Cumulative crop yield at the end of the three years was not affected by the compost type, but was affected by fertilization treatment (highest values with mineral and 50% compost +50% mineral fertilization). The compost application did not highlight a relevant effect on soil organic carbon. Under 100% of compost fertilization, the crops did not take up a large amount of the N supplied, but it did not generate an increase of NO_3-N leaching in the percolation water. Obtained results show the good fertilization properties of compost whereas the amendment property was not relevant, probably due to the low rates applied and the short experimental period.

Keywords: *Zea mais* L.; *Triticum aestivum* L.; *Helianthus annuus* L.; organic fertilization; mineral N fertilization

1. Introduction

In the last decade, abundant soil mineral fertilizations have led to several issues (costs, nitrate pollution, and loss of soil carbon) [1]. To maintain or increase optimum soil fertility, one possible practice is maintenance of the humic acid content by recycling plant and algal matter, or by adding outside sources of decomposed plant or algal matter such as composts, mulch, peat, or lignite coals [2]. In this scenario, fertilization with organic matter represents an alternative for the sustainability of agro-ecosystems [3–5] even though its environmental impact should also be carefully evaluated [6]. Organic matter composting is one of the best-known and well-established processes that allow the stabilization and sanitation of organic wastes through accelerated aerobic decomposition under controlled conditions. This results in a product called compost [7]. Due to its main characteristic, which is high content of stable organic nitrogen (N) [8] and stabilized organic matter, the use of compost in agriculture is a way to increase the organic matter content in the soil for long periods. Therefore, this allows soil fertility maintenance and/or recovery [9] and better crop yields [10].

Compost characteristics are influenced by the organic matter used. Therefore, different types of organic materials can produce compost with different composition, especially in terms of nutrients. Considering the effect on crops, Montemurro et al. [11], using municipal solid waste compost alone or in association with mineral fertilizer on sunflower, found similar oil and protein yield performance as that with mineral fertilization. Higher tomato yield was reported by Islam et al. [12] when vermi-compost was used in association with mineral fertilizers with respect to vermi-compost alone. Positive effects on wheat and sunflower production and quality characteristics were reported by Fecondo et al. [13] using bio-waste compost. Green compost, alone or in partial substitution for mineral fertilization, was tested on cabbage by Nicoletto et al. [14] who reported comparable yield but higher content of antioxidants, phenolic acids, and ascorbic acid with respect to mineral fertilization. Instead, neutral or antagonistic interactive effects have been reported on many plant growth traits combining compost with biochar for their fertilization [15]. Three different composts (food processing industry residues, municipal waste, green cuttings residues, the organic fraction of municipal waste solid residues, and municipal sludge and green cuttings residues) since N source were investigated on potatoes, lolium, and rye succession by Passoni and Borin [16]. These experts reported a specific response of potato yield to compost type, with the best results for the compost derived from food processing industry residues and municipal waste. Gobbi et al. [17], using different types of spent mushroom substrate, alone or combined with mineral fertilization, observed positive effects on lettuce and leek performance with comparable yields to mineral fertilization without differences among compost types. Furthermore, the short-cycle crops preferred mixed fertilization because organic substrates usually require a long time for organic matter mineralization whereas organic fertilization is more effective for long-cycle crops such as leeks [17]. In addition, the dose of compost should be taken into account as highlighted by Ponchia et al. [18] who, using different doses of green waste compost as medium to grow rooted cuttings of rose and *Abelia × grandiflora*, concluded that the use of compost has to be carefully evaluated since there is a clearly different species tolerance, particularly in younger plants. This overview, thus, indicates great potential for compost as a nutrient source, but contrasting results in relation to crop and compost origin and characteristics. The question of whether compost can be used as a total or partial substitute of mineral N is still open.

Although several studies have been conducted to evaluate the effect of compost application on herbaceous crops yield, only a few papers focused on the comparison of more than one type of compost with variable results. For this reason, the main aim of this study was to compare different compost types as total or partial mineral N fertilization substitute in an herbaceous crop succession (maize, wheat, and sunflower) in North-East Italy, considering crop response, soil organic carbon content, and concentration of N in the percolation water and soil.

2. Materials and Methods

2.1. Site Description and Experiment Setup

The experiment was conducted at the University of Padua "L. Toniolo" Experimental Farm, Veneto Region, North-East Italy (45°11′ N, 11°21′ E, 6 m a.s.l.) during a three-year research program from 2006 to 2008. The site consisted of 48 growth boxes, each with a 4 m^2 surface area, arranged in two symmetric lines of 24 boxes installed at 1.3 m above ground level to avoid water table interaction, especially during the winter, and with the bottom open, to allow water percolation. A semi-automatic tension-controlled ceramic suction plate system was installed in 16 boxes (2 replicates per treatment) at 0.90 m depth to collect the percolation water. The system was controlled by an electric vacuum pump that was manually activated to regulate the ceramic plate suction at 0.02 Mbar. At the beginning of the experimental period, all growth boxes were filled with Fluvi-Calcaric Cambisol [19], whose physical-chemical features related to the arable layer (0–30 cm), carried out following the Italian official methods [20,21], are reported in Table 1. The soil has a clay-loam texture, with a sub-basic reaction (7.7), soil organic matter content of 2.1%, and C:N ratio of 8.7.

Table 1. Main features of soil used during the experiment.

Parameters	Average Value
Sand (%)	34.7
Silt (%)	55.3
Clay (%)	10.0
pH	7.7
Organic carbon (%)	1.2
Organic matter (%)	2.1
Total $CaCO_3$ (%)	26.4
Active $CaCO_3$ (%)	2.7
Total nitrogen (g kg^{-1})	1.4
C/N ratio	8.7
Available phosphorus (mg P_2O_5 kg^{-1})	81.1
Available Potassium (mg K_2O kg^{-1})	2.6
Bulk density (Mg m^{-3})	1.2
Volumetric field capacity % (10 kPa)	32.0
Volumetric wilting point % (1500 kPa)	8.0

Three crops were cultivated during the experimental period: maize (*Zea mays* L.) (Pioneer "Costanza" FAO 600), wheat (*Triticum aestivum* L.) (cv. Blasco), and sunflower (*Helianthus annus* L.) (PR64H41 variety). The agronomic operations performed in each crop cycle are reported in Table 2. Maize was sown on 16 May 2006 at 7.5 plants m^{-2} sowing density and was harvested at complete grain maturation stage on 3 October 2006. After maize harvesting, soil in each growth box was tilled manually using spade. Crop residues (stalks) were incorporated directly into the soil in the 0 to 20 cm layer. After an adequate soil preparation, wheat was sown at a rate of 250 kg seeds ha^{-1} on 3 November 2006 and was harvested on 18 June 2007. In autumn 2007, the soil was tilled manually using spade and wheat residues (straw), were incorporated in the 0–20 cm layer. Sunflower was sown on 28 April 2008 at 9 plants m^{-2} and harvested on 2 October 2008. The crop residues buried in the soil at the end of maize and wheat growing seasons were 90% of produced residues. The 10% of maize stalks and wheat straw was instead used for analysis.

To partially or completely substitute the crop N mineral requirements, four different types of compost were used as organic fertilizer: (i) compost derived from green cuttings and depuration sludge (G+S), (ii) from green cuttings, organic fraction of municipal wastes, and other organic materials (G+F+O), (iii) from green cuttings (G), and (iv) from green cuttings and organic fraction of municipal wastes (G+F).

Eight different fertilization treatments, replicated six times, were compared in each crop cycle, adopting a completely randomized blocks experimental design. The fertilization scheme included: (i) 50% of N supplied through G compost and 50% through mineral fertilization (G50), (ii) 50% of N supplied through G + F + O compost and 50% through mineral fertilization (GFO50), (iii) 50% of N supplied through G + S compost and 50% through mineral fertilization (GS50), (iv) 50% of N supplied through G + F compost and 50% through mineral fertilization (GF50), (v) 100% of N supplied through G + F + O fertilization (GFO100), (vi) 100% of N supplied through G compost (G100), vii) 100% of N supplied through mineral fertilization (Min100), and (viii) unfertilized treatment for comparison (C).

Table 2. Main agronomic operations carried out during the three-year crop succession.

Month	2006	2007	2008
January	-	Mineral fertilization (27th)	-
Febuary	-	-	-
March	-	-	Soil till (20th) Organic and mineral fertilization (31st)
April	Soil till (3rd)	-	Sunflower sowing, herbicide distribution (Global 0.4 L ha^{-1}, Erbifos 1 L ha^{-1}) (28th)
May	Organic and mineral fertilization (15th) Maize sowing (16th)	-	Fungicide distribution (King 250 mL hl^{-1}) (29th)
June	Irrigation (40 mm) (16th) Mineral fertilization (29th)	Wheat harvesting (18th)	Mineral fertilization (5th) Fungicide distribution (King 250 mL hl^{-1}) (9th)
July	Herbicide distribution (Silver, 1 kg ha^{-1}) (4th) Irrigation (40 mm) (21st) Irrigation (40 mm) (28th)	-	-
August	-	-	-
September	-	-	Sunflower harvesting (2nd)
October	Maize harvesting (3rd) Soil till (10th)	-	-
November	Organic and mineral fertilization (2nd) Wheat sowing (3rd)	-	-
December	-	-	-

The dry matter and N concentration of each type of compost were determined before starting the experiment (Table 3) to calculate the amount of compost to distribute in each fertilization treatment (Table 4).

Crop P and K requirements were satisfied by adding mineral fertilizers to the content of composts to avoid these elements that may be limiting and to evaluate the effect of compost only in terms of N fertilization. The amount of P and K supplied was chosen according to the standard recommendation for crops typical of the Po Valley area. In particular, maize received 250 kg N ha^{-1}, 100 kg P ha^{-1}, and 100 kg K ha^{-1} and wheat received 150 kg N ha^{-1}, 100 kg P ha^{-1}, and 100 kg K ha^{-1}, which is the typical fertilization of the experimental area. Sunflower received 250 kg N ha^{-1}, 100 kg P ha^{-1}, and 100 kg K ha^{-1}. The sunflower N fertilization quantity was established to maximize yield by taking into account the maximum N uptake available in literature [22,23]. Mineral N was supplied as ammonium nitrate, P as triple superphosphate, and K as potassium sulphate. Maize was the only irrigated crop, with 120 mm of water supplied through artificial irrigation, divided equally on three different dates in June and July to replenish crop evapotranspiration.

Table 3. Main chemical-physical characteristics of the compost types used in the trial.

Compost	TN (g kg^{-1})	Dry Matter %	Salinity (dS m^{-1}) *	Organic Carbon (g kg^{-1}) *	C/N
GS	17 ± 3	70.9 ± 2.9	1.76	265	15.5
GFO	27 ± 2	65.8 ± 1.6	2.47	250	11.0
G	19 ± 2	60.2 ± 2.9	1.08	275	18.0
GF	20 ± 1	75.6 ± 12.2	3.69	250	11.0

TN and Organic Carbon values are expressed on a dry matter basis. * Value determination in an average sample of compost used in the three experimental years. GS: Compost derived from green cutting and depuration sludge. GFO: Compost derived from green cutting, organic fraction of municipal wastes, and other organic materials. G: Compost derived from green cutting. GF: Compost derived from green cutting and organic fraction of municipal wastes.

Table 4. Fertilization schemes adopted for all studied crops.

Treatment	Crop	Fresh Compost (Mg ha^{-1})	Compost N (kg ha^{-1})	Mineral N (kg ha^{-1})
G50	*Z. mays*	11.5	125	125
GFO50	*Z. mays*	7.2	125	125
GS50	*Z. mays*	9.4	125	125
GF50	*Z. mays*	7.0	125	125
GFO100	*Z. mays*	14.4	250	-
G100	*Z. mays*	18.7	250	-
Min100	*Z. mays*	-	-	250
C	*Z. mays*	-	-	-
G50	*T. aestivum*	5.0	75	75
GFO50	*T. aestivum*	4.4	75	75
GS50	*T. aestivum*	6.4	75	75
GF50	*T. aestivum*	5.1	75	75
GFO100	*T. aestivum*	8.8	150	-
G100	*T. aestivum*	12.8	150	-
Min100	*T. aestivum*	-	-	150
C	*T. aestivum*	-	-	-
G50	*H. annus*	11.5	125	125
GFO50	*H. annus*	6.5	125	125
GS50	*H. annus*	12.9	125	125
GF50	*H. annus*	9.6	125	125
GFO100	*H. annus*	13.0	250	-
G100	*H. annus*	25.8	250	-
Min100	*H. annus*	-	-	250
C	*H. annus*	-	-	-

G50: 50% of N from G compost and 50% of N from mineral fertilization (MF). GFO50: 50% of N from G + F + O compost and 50% of N from MF. GS50: 50% of N from G + S compost and 50% of N from MF. GF50: 50% of N from G + F compost and 50% of N from MF. GFO100: 100% of N from G + F + O fertilization. G100: 100% of N from G compost. Min100: 100% of N from MF. C: Un-fertilized treatment.

2.2. Meteorological Data

The experiment has been carried out under humid subtropical climate (Cfa) conditions, according to the Köppen climate classification [24]. The following data were collected from 1 January 2006 to 31 December 2008 at the ARPAV weather station located near the experimental site: rain (mm), average air temperature (°C), and global radiation (MJ m^{-2}).

During the whole experimental period, global radiation showed the classic seasonal bell-shaped trend with a progressive increase from the beginning of the year until July when the highest values were recorded, and then a sloping decrease until December (Figure 1a). The average global radiation values measured in 2006 and 2008 of 11.4 MJ m^{-2} and 12.3 MJ m^{-2}, respectively, were lower than the 1995 to 2005 decade average (13.9 MJ m^{-2}), which was in line with the average value obtained in the second experimental year (13.1 MJ m^{-2}).

Average air temperature followed the same monthly trend as global radiation, with a maximum value in July and minimum in January (Figure 1b). In general, for all experimental years, average monthly air temperature values were in line with the corresponding monthly values calculated for the long-term period (1995–2005).

During the entire experimental period, total rainfall was 2279.2 mm, with a different distribution and amount over the three experimental years (Figure 1c). During the first and second years, the annual rainfall of 699.2 mm and 640.0 mm was lower than the long-term average (860.2 mm), whereas values recorded in the third year (940.0 mm) exceeded the long-term average. Focusing attention on the different cropping seasons, maize (16 May–3 October 2006) received 410.8 mm of rainfall, wheat (3 November 2006–18 June 2007) received 441.6 mm of rainfall, and sunflowers (28 April 2008–2 October 2008) received 358.4 mm of rainfall.

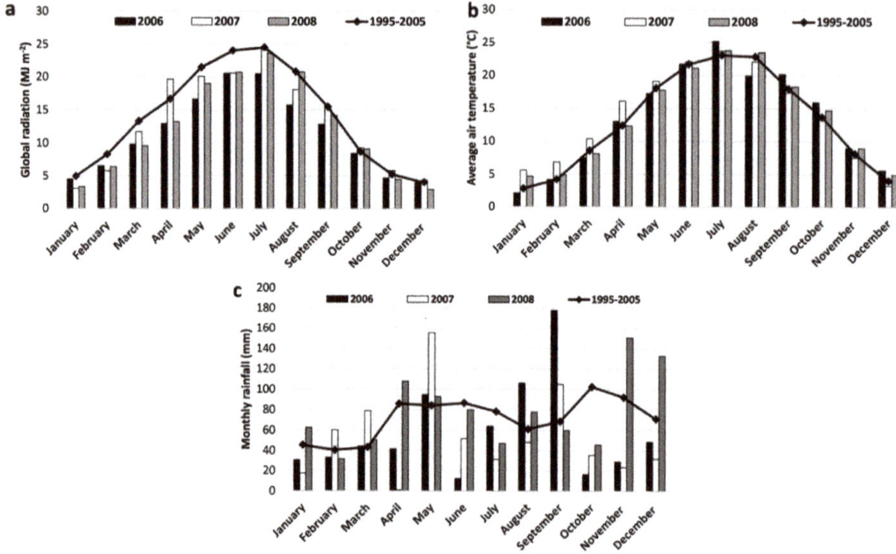

Figure 1. Average global radiation (MJ m^{-2}) (**a**), air temperature (°C), (**b**) and cumulated monthly rainfall (mm) (**c**) recorded during the experimental period of plants harvesting and measurements.

At the end of each crop cycle, the above-ground biomass produced by each crop was determined by cutting all plants in the growth boxes at about 10 cm above the soil level. The collected above-ground biomass was weighed onsite to determine the total fresh biomass production. It was then manually separated into marketable fresh biomass (grain) and crop residues. The dry biomass production was assessed by drying a biomass sample from each growth box and each biomass fraction in a forced-air oven at 65 °C until constant weight was reached. Dry biomass was milled to 2 mm and an average sample for each studied treatment was analyzed to determine N concentration through the Kjeldhal method. The total N uptake in the above-ground biomass was determined as the product of dry biomass production and nutrient concentration. This allowed the following indexes to be calculated:

- Apparent N Balance (ANB) using the equation:

$$\text{ANB} = \text{N input} - \text{N uptake} \tag{1}$$

- N Use Efficiency (NUE) was evaluated with the approach suggested by Fageria et al. [25] calculating: Agronomic efficiency (AE), Physiological efficiency (PE), Agrophysiological efficiency (APE), Apparent recovery efficiency (ARE), and Utilization efficiency (UE) using the following equations:

$$\text{AE (mg } \textit{grain yield} \text{ mg}^{-1} \textit{ N applied}) = (Y_f - Y_c)/N_a \tag{2}$$

$$\text{PE (mg } \textit{total biomass} \text{ mg}^{-1} \textit{ N uptake}) = (BY_f - BY_c)/(N_f - N_c) \tag{3}$$

$$\text{APE (mg } \textit{grain yield} \text{ mg}^{-1} \textit{ N uptake}) = (Y_f - Y_c)/(N_f - N_c) \tag{4}$$

$$\text{ARE (\%)} = [(N_f - N_c)/N_a] \times 100 \tag{5}$$

$$\text{UE (mg mg}^{-1}) = \text{PE} \times \text{ARE} \tag{6}$$

where Y_f is the grain yield harvested in the fertilized treatments (mg), Y_c is the grain yield of the unfertilized control treatment (mg), and N_a is the quantity of N applied (mg), BY_f is the biological yield (total biomass) of the fertilized treatments (mg), BY_c is the biological yield of the unfertilized control treatment (mg), N_f is the N uptake (total biomass) of the fertilized treatments (mg), and N_c is the N uptake (total biomass) of the unfertilized control treatment (mg). The indexes were calculated for the different treatments and applied for single crops during the entire succession.

2.3. Soil and Percolation Water Analysis

To monitor the evolution of organic carbon (OC) and nitrate N (NO_3-N) over time, soil samples were taken before the beginning (March 2006) and after the end (November 2008) of the experiment. Soil sampling was performed in each growth box, at 0–20 cm and 20–50 cm depths. After collection, samples were air-dried for about a week and then manually sieved at 2 mm. The soil OC and NO_3-N were determined through the Walkley-Black method and spectrophotometric analysis [26], respectively.

To assess the NO_3-N concentration in percolation water among the studied treatments as a potential residual environmental effect of the different fertilization management, 96 samples were collected six times during the sunflower growing season (from April 2008 to October 2008). The NO_3-N concentration was determined following the Cataldo method [26].

2.4. Statistical Analysis

The normality of data was checked with the Kolmogorov-Smirnov test. As the aboveground biomass production, soil OC and NO_3-N data showed a normal distribution. They were analyzed statistically with one-way analysis of variance (ANOVA) and the differences between average values were found by the Tukey's honestly significant difference (HSD) test, $p < 0.05$. Biomass production was evaluated for each species comparing fertilization treatments whereas soil OC and NO_3-N data were evaluated in time for each fertilization treatment using experimental years as treatment. On the contrary, the percolation water NO_3-N concentration data did not follow a normal distribution, so they were processed by the non-parametric Kruskal-Wallis statistical test, $p < 0.05$.

3. Results

3.1. Biomass Production

The maize stalk biomass and grain productions ranged from 20.1 ± 1.6 Mg ha^{-1} to 25.5 ± 1.6 Mg ha^{-1} and from 5.2 ± 1.0 Mg ha^{-1} to 7.4 ± 0.7 Mg ha^{-1}, respectively. The significantly highest (ANOVA, $p < 0.05$) total biomass (stalk + grain) productions were in the mineral fertilization and 50% compost+50% mineral fertilization treatments. In addition, the significantly lowest (ANOVA, $p < 0.05$) biomass productions were harvested under compost fertilization (GFO100 and G100 treatments) and unfertilized treatment (Figure 2).

Both wheat total and straw biomasses showed the same statistical trend among fertilization treatments (Figure 3). The significantly highest values were recorded in the treatments where mineral fertilizers were used at either 50% or 100%, without any significant differences among them, whereas the significantly lowest ones (ANOVA, $p < 0.05$) were obtained in the unfertilized treatment (Figure 3) and grain yield did not give significant differences among treatments with an average value of 5.1 ± 0.7 Mg ha^{-1} (Figure 3).

Despite sunflower total biomass (on average 7.0 ± 0.8 Mg ha^{-1}), stalk (on average 4.7 ± 0.6 Mg ha^{-1}) and grain (on average 2.3 ± 0.3 Mg ha^{-1}) yields showed big differences among treatments. The ANOVA statistical test did not indicate any significant differences among them, which is probably attributed to the high variability within each treatment.

At the end of the three-year crop succession, the cumulated biomass production of all the treatments with 50% N from compost and 50% from mineral fertilizer (on average 58.3 Mg ha^{-1}) did not differ with respect to that obtained with 100% mineral N (58.7 Mg ha^{-1}). The application of N by compost alone gave lower results (on average 50.7 Mg ha^{-1}), which is similar to those obtained in the unfertilized control (Figure 4).

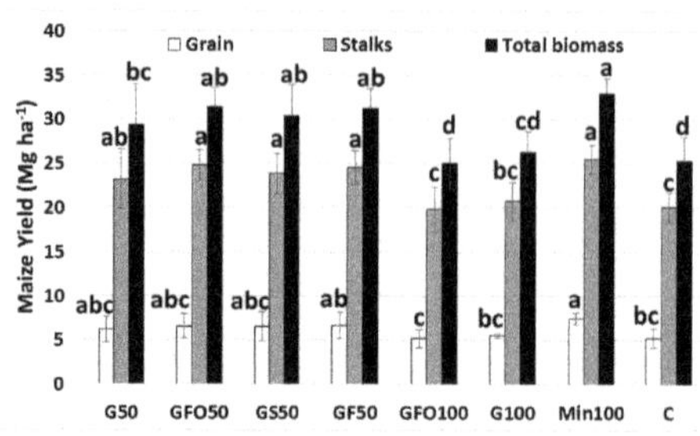

Figure 2. Maize dry biomass production: total biomass, stalk biomass, and grain yield. Histograms indicate average values while bars indicate standard deviation. Different letters indicate significant differences according to Tukey's HSD test at $p < 0.05$ between treatments of the same biomass component. G50: 50% of N from G compost and 50% of N from mineral fertilization (MF). GFO50: 50% of N from G + F + O compost and 50% of N from MF. GS50: 50% of N from G + S compost and 50% of N from MF. GF50: 50% of N from G + F compost and 50% of N from MF. GFO100: 100% of N from G + F + O fertilization. G100: 100% of N from G compost. Min100: 100% of N from MF. C: Un-fertilized treatment.

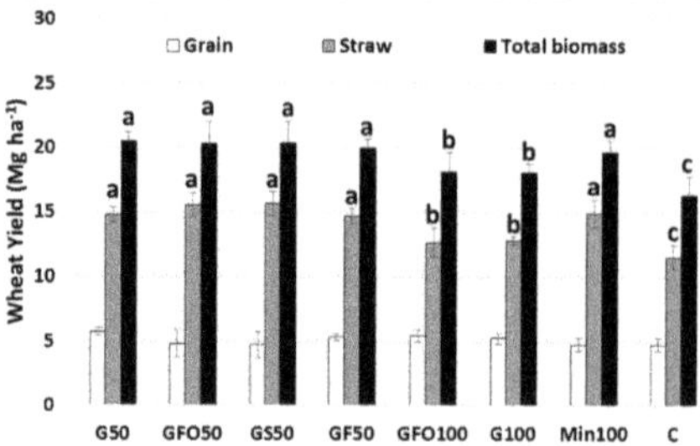

Figure 3. Wheat dry biomass production: grain, straw, and total biomass. Histograms indicate average values while bars indicate standard deviation. Different letters indicate significant differences, according to Tukey's HSD test at $p < 0.05$ between treatments of the same biomass component. G50: 50% of N from G compost and 50% of N from mineral fertilization (MF). GFO50: 50% of N from G + F + O compost and 50% of N from MF. GS50: 50% of N from G + S compost and 50% of N from MF. GF50: 50% of N from G + F compost and 50% of N from MF. GFO100: 100% of N from G + F + O fertilization. G100: 100% of N from G compost. Min100: 100% of N from MF. C: Un-fertilized treatment.

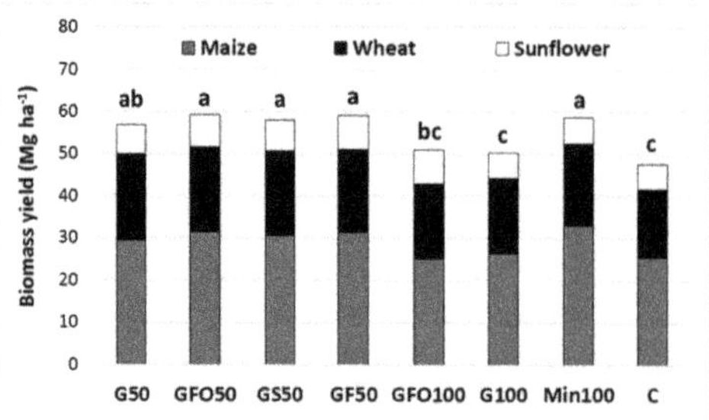

Figure 4. Cumulated biomass production (Mg ha^{-1}) over the entire crop succession. Different letters indicate significant differences, according to Tukey's HSD test at $p < 0.05$. G50: 50% of N from G compost and 50% of N from mineral fertilization (MF). GFO50: 50% of N from G + F + O compost and 50% of N from MF. GS50: 50% of N from G + S compost and 50% of N from MF. GF50: 50% of N from G + F compost and 50% of N from MF. GFO100: 100% of N from G + F + O fertilization. G100: 100% of N from G compost. Min100: 100% of N from MF. C: Un-fertilized treatment.

3.2. N Percentage Concentration, Uptake, Apparent Nitrogen Balance, and Nitrogen Use Efficiency

The N concentration of both grain and residues of all crops showed the same trend among the studied treatments. The highest values were recorded in the 50% compost +50% mineral fertilization and 100% mineral fertilization treatments, whereas the lowest ones were recorded in the 100% compost fertilization and unfertilized treatments (Table 5). As for the N concentration, N uptake varied greatly between the compared treatments. All crops showed the highest N uptake under the 50% compost+50% mineral fertilization, which was comparable to that obtained with 100% mineral fertilization. Instead, the N uptake obtained under 100% compost N fertilization was lower than those recorded in the presence of mineral fertilizer and strictly comparable to the unfertilized treatment (Table 5).

The significantly highest (ANOVA, $p < 0.05$) three years cumulative N uptake was obtained under 100% mineral fertilization treatment (598.3 kg ha^{-1}) followed by 50% compost +50% mineral fertilization (on average 536.3 kg ha^{-1}). Instead, the significantly lowest (ANOVA, $p < 0.05$) cumulative N uptakes were recorded for organic fertilization and unfertilized treatments without any significant difference among them (on average 337.4 kg ha^{-1}) (Figure 5). The cumulative N supplied during the three-year study was higher than crop uptake for all treatments, leaving in the soil 52 kg N ha^{-1}, 114 kg N ha^{-1} and 294 kg N ha^{-1} in 100% mineral, 50% compost +50% mineral, and 100% compost fertilization treatments, respectively.

Table 5. Nitrogen (N) concentration (%) in the grain and residues of the studied crops.

Crop	Treatment	N Concentration (%)		N Uptake (kg ha^{-1})	
		Grain	Residues	Grain	Residues
	G50	1.5	0.8	90.9 ± 21.1	176.6 ± 25.5
	GFO50	1.5	0.5	96.9 ± 20.6	133.0 ± 9.4
	GS50	1.5	0.5	98.8 ± 25.2	128.8 ± 12.3
	GF50	1.4	0.6	96.0 ± 21.4	155.9 ± 12.2
Z. mays	GFO100	1.0	0.3	53.7 ± 10.7	64.9 ± 8.3
	G100	1.1	0.4	61.3 ± 2.6	84.1 ± 8.6
	Min100	1.7	0.8	126.6 ± 11.6	193.6 ± 12.5
	C	1.1	0.3	59.3 ± 12.1	65.0 ± 5.3
	Average	1.4	0.5	85.5 ± 28.6	125.2 ± 48.5
	G50	1.9	0.6	95.5 ± 17.0	88.4 ± 7.8
	GFO50	2.0	0.5	106.2 ± 14.7	74.2 ± 10.5
	GS50	2.0	0.6	101.7 ± 16.4	88.2 ± 13.4
	GF50	2.0	0.6	106.3 ± 14.7	88.3 ± 12.4
T. aestivum	GFO100	1.7	0.4	85.1 ± 10.5	51.3 ± 4.5
	G100	1.6	0.3	82.6 ± 9.4	40.2 ± 5.8
	Min100	2.1	0.7	98.4 ± 23.9	105.5 ± 7.5
	C	1.7	0.4	85.6 ± 6.7	54.4 ± 5.1
	Average	1.9	0.5	95.2 ± 9.6	73.8 ± 22.8
	G50	2.1	1.2	48.3 ± 12.2	54.9 ± 11.9
	GFO50	1.7	1.1	41.6 ± 9.1	56.5 ± 9.9
	GS50	2.0	1.1	46.5 ± 9.6	55.4 ± 11.1
	GF50	2.0	1.1	50.1 ± 11.3	62.6 ± 10.8
H. annus	GFO100	1.1	1.0	27.9 ± 6.8	53.5 ± 11.6
	G100	1.5	1.3	29.7 ± 7.2	51.1 ± 9.4
	Min100	2.5	1.5	47.9 ± 14.1	62.8 ± 9.6
	C	0.9	0.8	16.8 ± 5.0	30.9 ± 10.6
	Average	1.7	1.1	38.6 ± 12.3	53.5 ± 10.0

G50: 50% of N from G compost and 50% of N from mineral fertilization (MF). GFO50: 50% of N from G + F + O compost and 50% of N from MF. GS50: 50% of N from G + S compost and 50% of N from MF. GF50: 50% of N from G + F compost and 50% of N from MF. GFO100: 100% of N from G + F + O fertilization. G100: 100% of N from G compost. Min100: 100% of N from MF. C: Un-fertilized treatment.

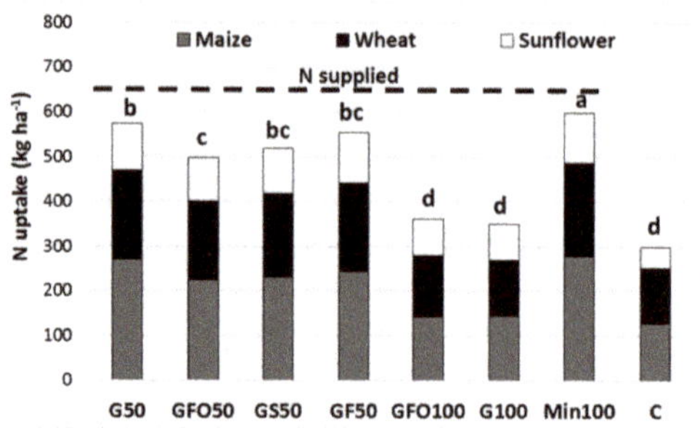

Figure 5. Cumulated nitrogen uptake (kg ha^{-1}) over the entire crop succession. Different letters indicate significant differences according to Tukey's HSD test at $p < 0.05$. G50: 50% of N from G compost and 50% of N from mineral fertilization (MF). GFO50: 50% of N from G + F + O compost and 50% of N from MF. GS50: 50% of N from G + S compost and 50% of N from MF. GF50: 50% of N from G + F compost and 50% of N from MF. GFO100: 100% of N from G + F + O fertilization. G100: 100% of N from G compost. Min100: 100% of N from MF. C: Un-fertilized treatment.

Table 6 reports the calculation of the most common indexes to define the apparent N balance and N use efficiency. Particularly, in the maize cropping season, the ANB ranged from −128.0 kg ha^{-1} to 108.3 kg ha^{-1}, with negative values obtained for unfertilized, 100% mineral fertilization, G compost+mineral fertilization, and G+F compost+mineral fertilization schemes. During the wheat cropping season, the N uptake exceeded N supplied through fertilization where the fertilization scheme included mineral fertilizer application (both 100% mineral fertilizer and 50% compost+50% mineral fertilizer), which provided negative ANB values. On the contrary, strictly positive ANB values were obtained under 100% compost fertilization. The ANB values of sunflower were always higher than 135.0 kg ha^{-1} except for the unfertilized treatment that gave a negative value. The AE varied following a specific treatment trend (Table 6) and, on the average of the studied treatments, maize showed the highest AE (on average 4.0 mg mg^{-1}), which was followed by the wheat (on average 2.6 mg mg^{-1}) and sunflower (on average 1.3 mg mg^{-1}). The highest ARE and UE values of cereals were obtained under 100% mineral fertilization and 50% compost+50% mineral fertilization treatments while the lowest values were obtained with the application of 100% compost organic matter. ARE of sunflower followed the same trend as maize and wheat, while the UE showed the highest values in the 50% organic +50% mineral fertilization treatments, and values became negative under 100% organic or 100% mineral fertilizations.

Table 6. Apparent nitrogen balance (ANB), agronomic efficiency (AE), physiological efficiency (PE), agro-physiological efficiency (APE), apparent recovery efficiency (ARE), and utilization efficiency (UE) indexes were calculated for the different crops and for the entire crop succession.

Crop	Treatment	ANB (kg ha^{-1})	AE (mg mg^{-1})	PE (mg mg^{-1})	APE (mg mg^{-1})	ARE (%)	UE (mg mg^{-1})
	G50	−17.6	3.5	27.9	6.1	57.3	16.0
	GFO50	20.1	5.1	56.5	12.0	42.2	23.8
	GS50	22.4	5.1	48.3	12.2	41.3	20.0
Z. mays	GF50	−1.9	5.5	45.2	10.7	51.0	23.0
	GFO100	131.4	−0.5	69.7	20.5	−2.3	−1.6
	G100	104.7	1.0	42.8	12.4	8.4	3.6
	Min100	−70.3	8.5	38.4	10.9	78.4	30.1
	C	−124.3	-	-	-	-	-
	G50	−48.3	6.5	57.1	13.2	49.2	28.1
	GFO50	−25.6	0.3	79.5	1.0	34.0	27.1
	GS50	−37.2	−0.3	65.2	−0.6	41.7	27.2
T. aestivum	GF50	−48.6	3.6	49.9	7.3	49.4	24.7
	GFO100	11.6	4.6	129.6	49.8	9.2	12.0
	G100	24.2	0.4	1499.8	426.2	0.8	11.9
	Min100	−58.8	0.1	40.2	0.2	56.1	22.6
	C	−124.6	-	-	-	-	-

Table 6. *Cont.*

Crop	Treatment	ANB (kg ha^{-1})	AE (mg mg^{-1})	PE (mg mg^{-1})	APE (mg mg^{-1})	ARE (%)	UE (mg mg^{-1})
	G50	146.7	1.2	17.4	5.6	22.2	3.9
	GFO50	151.9	1.8	29.2	8.9	20.1	5.9
	GS50	148.1	1.6	21.2	7.2	21.7	4.6
H. annus	GF50	137.3	2.4	29.1	9.1	26.0	7.6
	GFO100	168.6	2.5	54.8	18.7	13.5	7.4
	G100	169.2	0.04	−2.2	0.3	13.2	−0.3
	Min100	139.2	−0.3	−0.3	−1.1	25.2	−0.1
	C	−47.7	-	-	-	-	-
	G50	77.1	4.2	28.2	9.9	41.9	11.8
	GFO50	151.8	3.5	41.8	11.6	30.4	12.7
	GS50	130.2	3.5	38.0	10.4	33.8	12.8
Crop succession	GF50	95.7	4.4	31.8	11.4	39.1	12.4
	GFO100	288.4	3.8	102.9	40.8	9.4	9.7
	G100	299.6	1.3	16.8	16.9	7.7	1.3
	Min100	51.7	0.3	21.5	0.6	45.9	9.9
	C	−300.3	-	-	-	-	-

G50: 50% of N from G compost and 50% of N from mineral fertilization (MF). GFO50: 50% of N from G + F + O compost and 50% of N from MF. GS50: 50% of N from G + S compost and 50% of N from MF. GF50: 50% of N from G + F compost and 50% of N from MF. GFO100: 100% of N from G + F + O fertilization. G100: 100% of N from G compost. Min100: 100% of N from MF. C: Un-fertilized treatment.

3.3. Soil and Percolation Water Analysis

At the beginning of the trial, soil OC was, on average, 1.07 ± 0.07% and 0.88 ± 0.10% in the 0–20 cm and 20–50 cm soil profiles, respectively. All fertilization treatments showed a significant increase (ANOVA, $p < 0.05$) of soil OC at the end of the trial with an average increase of 60.7% and 53.4% in the 0 to 20 cm and 20 to 50 cm soil profiles, respectively.

Soil NO_3-N content was not significantly different between the beginning and end of the experimental period in compost fertilized treatments at both 50% and 100%. Instead, a clear trend was observed in 100% mineral fertilization and unfertilized control treatments with an increase and decrease of soil NO_3-N concentrations, respectively (Table 7).

The median NO_3-N concentrations in percolation water ranged between 10.1 mg L^{-1} and 22.2 mg L^{-1}, with significant differences among the studied treatments. Although the significantly highest NO_3-N concentration (Kruskal-Wallis, $p < 0.05$) was determined in the treatment of G50 (50% compost + 50% mineral fertilization), this trend was not confirmed in other similar treatments (GFO50, GS50, and GF50), where lower values were measured (Figure 6).

Table 7. Soil nitrate nitrogen (0–50 cm soil profile) at the beginning and end of the trial (mean ± SD).

Treatment	NO_3-N (mg kg^{-1}) Start	NO_3-N (mg kg^{-1}) End	ANOVA
G50	2.49 ± 0.37	2.43 ± 0.51	ns
GFO50	3.73 ± 0.78	2.54 ± 0.37	ns
GS50	3.27 ± 0.54	2.68 ± 0.29	ns
GF50	2.52 ± 0.44	2.96 ± 0.67	ns
GFO100	2.19 ± 0.52	1.75 ± 0.17	ns
G100	2.56 ± 0.57	2.29 ± 0.36	ns
Min100	2.54 ± 0.36	4.45 ± 0.91	*
C	3.24 ± 0.52	1.84 ± 0.39	*
Average	3.09	2.62	-

* Significant difference among treatment for each soil layer at $p < 0.05$ Tukey (HSD) test. ns = not significant. G50: 50% of N from G compost and 50% of N from mineral fertilization (MF). GFO50: 50% of N from G + F + O compost and 50% of N from MF. GS50: 50% of N from G + S compost and 50% of N from MF. GF50: 50% of N from G + F compost and 50% of N from MF. GFO100: 100% of N from G + F + O fertilization. G100: 100% of N from G compost. Min100: 100% of N from MF. C: Un-fertilized treatment.

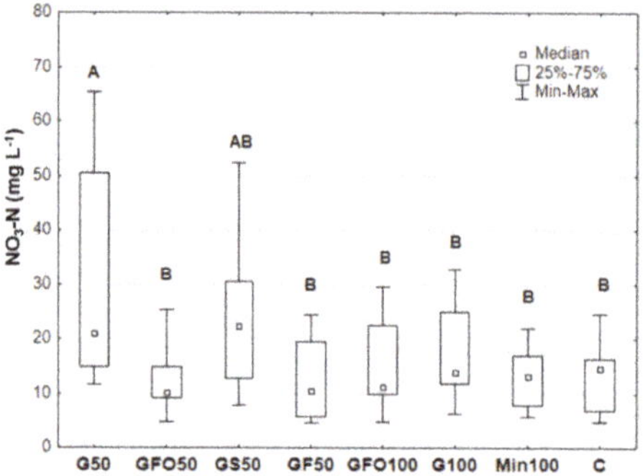

Figure 6. Nitrate nitrogen concentration in percolation water during the sunflower growing season. Different letters among treatments indicate significant differences according to the Kruskal-Wallis test, $p < 0.05$. G50: 50% of N from G compost and 50% of N from mineral fertilization (MF). GFO50: 50% of N from G + F + O compost and 50% of N from MF. GS50: 50% of N from G + S compost and 50% of N from MF. GF50: 50% of N from G + F compost and 50% of N from MF. GFO100: 100% of N from G + F + O fertilization. G100: 100% of N from G compost. Min100: 100% of N from MF. C: Un-fertilized treatment.

4. Discussion

Maize and wheat grain average yields obtained in the current study were similar to those obtained under open field conditions in the same area with conventional agricultural practices [27]. Despite all compost types used in the experiment containing more than 1% of N (dry weight basis), according to the typical range for agricultural compost used as fertilizer (1–3% of N) [28], the results suggested the necessity of associating mineral fertilization with the organic one to maximize cereal yields at least during the short-term period (first and second cropping seasons of this study). In addition, the C/N of compost used to fertilize crops should be taken into account. Mature compost (with low C/N ratio) gives a small increase in plant N availability, whereas immature compost (with high C/N ratio) results in net N immobilization [29,30]. However, initial immobilization is more likely with less mature compost and with a high C/N ratio since, in mature compost, N–immobilisation had already taken place during the composting process [31]. In this context, cereals, particularly maize and wheat, have high N requirements that are mainly satisfied by promptly available N sources such as mineral fertilizers rather than the mineralized organic N typical of organic fertilizers such as compost. The organic matter turnover requires a longer time. For this purpose, Eghball et al. [32,33] reported average N percentage mineralization rates of 11% and 18% for composted manure following the first year after distribution. Similarly, Preusch et al. [34] recorded average N mineralization rates of 7% to 9% when the composted poultry litter was distributed on silt loam soils, whereas average values were 1% to 5% when the same organic material was applied to sandy loam soils.

The unsatisfactory cereal yields obtained with 100% compost fertilization in this study are in agreement with Martínez-Blanco et al. [7], who reported a crop yield reduction of 138% after a single compost fertilization in the short-term period. On the contrary, in the third cropping season, the comparable sunflower grain yield obtained among the treatments could be justified considering not only the relatively low N requirements typical of the crop but also the achievement of the compost residual effect.

Wheat and maize, both of the *Poaceae* family, were differently affected by the fertilization treatments. Maize was influenced in both stalk and grain yield, while wheat was influenced only in straw production. The different behavior of the two crops could be attributed to the completely different environmental conditions that affected the crop cycles (autumn-winter-spring seasons for wheat, spring-summer seasons for maize) and, consequently, the different pathway through which N is made available for crops. In particular, wheat fully exploited N supplied by the mineral fertilization treatment is done at the end of January (100% mineral fertilization and 50% organic+50% mineral fertilization treatments)

during the vegetative growth, which, in the study region, occurs during the cold months (from January until mid-April), for increasing the vegetative biomass (straw). On the contrary, in the 100% compost fertilization and unfertilized treatments, N was not immediately available for crop absorption due to the reduced organic matter mineralization rates, which penalizes the production of straw biomass. Wheat grain yield and protein content strictly depends on N availability at flowering. Since no mineral fertilizations were done during this phenological phase, in all studied treatments, N was made available for crops by the mineralization of previous crop residues or compost organic matter. Despite microbial activity being greater in this phase than during the vegetative growth period, it probably remained insufficient to fully satisfy crop N requirements, which justifies the absence of significant differences in wheat grain yield among the fertilization treatments.

Maize N requirements during the phenological phase of vegetative growth (from May until the end of June) were mainly satisfied by promptly available N supplied through mineral fertilization (50% organic+50% mineral fertilization and 100% mineral fertilization treatments), as proven by the greater stalks' biomass produced by maize under 100% mineral fertilization than 100% organic fertilization and unfertilized treatments. Instead, during the reproductive phase, at ear formation, maize exploited both (i) residual available N supplied with mineral fertilization (part is leached through water percolation in the previous period) and (ii) N derived from mineralization of the organic matter (crop residues and compost). This explanation justifies the restricted differences on the crop yield between 50% organic+50% mineral fertilization and 100% organic fertilization treatments.

Considering the agronomic indexes calculated, ANB, ARE, and UE followed a specific crop and treatment trend. On the average of studied treatments, the relatively low ANB values of maize and wheat (8.1 and −38.4, respectively) and the high average value calculated for sunflower (126.7) indicated the excellent aptitude of cereals to uptake available N compounds from soil and their translocation into the biomass with respect to sunflowers.

In the maize and wheat cropping seasons, the negative ANB values obtained under mineral fertilization treatment indicated that N uptake through crop biomass exceeded the N applied through fertilization. Therefore, this subtracted soil N reserves. Instead, the strictly positive values calculated for 100% compost fertilization treatments indicated, as previously reported, that N released by compost organic matter mineralization was not immediate and, therefore, was not enough to satisfy cereals' N requirements. Despite Passoni and Borin [16] reported a peak of NO_3-N release in the soil through organic matter mineralization just after compost distribution, our results suggested that part of N supplied through compost fertilization remained in the soil in the organic form, probably due to the slow mineralization rate of soil organic matter [33]. Similarly, the high ARE and UE values calculated for maize and wheat fertilized with 100% mineral fertilization or 50% compost +50% mineral fertilization suggested a great N use efficiency likely due to the relatively immediate availability of N supplied with mineral fertilizers. Instead, the low ARE and UE values calculated under 100% organic fertilization treatments suggested that a single application of compost at the beginning of the cropping season was not enough to promptly satisfy the crop N requirements, which reduces the crop N use efficiency. Sunflower exhibited a completely different behavior with respect to maize and wheat with the comparable ANB and ARE obtained for 100% compost and 50% compost +50% mineral fertilization schemes. The comparable results between compost and miner fertilization suggested that, after three years of organic fertilizer application, the turnover of supplied organic matter with compost reached a virtual equilibrium between the stable organic matter and the mineralized one. Therefore, this allows a suitable availability of N for cultivated crops.

The significant soil organic carbon increase measured between the beginning and end of the experiment on the average of treatments suggested the interesting amendment effect of crop residues (straw biomass) when correctly incorporated in the soil whereas it indicated that compost has no evident effects on soil properties in the short term when applied at low doses. Our results may appear to disagree with the literature data that highlight the amendment potential of compost [35]. However, it has to be taken into account that the approach of this paper considers compost as the N fertilization source and, as a consequence, it was applied in a quantity many times lower than in studies where compost is studied as an amendment.

The relatively low cumulated N uptake in crop biomass obtained under 100% compost fertilization suggested relatively high N residues in the soil after harvesting. However, this peculiar situation could allow a probable contamination of groundwater through NO_3-N leaching, especially during an abundant rainy period. The results obtained disclaimed this hypothesis. The comparable soil NO_3-N concentration measured between 100% compost and unfertilized treatments at both the beginning and end of this study indicated that N supplied through composted material remained in the soil mainly in the organic form [16], which was less mobile than the nitric form. Therefore, this reduces groundwater contamination risk. The potential contamination of groundwater depends mainly on the organic N mineralization rate of the amendment used. The not significant different N leaching between soil amended with compost, at the end of three experimental years, was reported by Mamo et al. [36]. Instead, Santos et al. reported the possible compost use as organic fertilizers in the partial substitution of mineral fertilizers, without a significant nitrate leaching risk [37].

5. Conclusions

The results obtained indicated that the crop response to compost fertilization is not unique. Indeed, organic fertilization with compost represents a valid substitute for mineral fertilization in wheat and sunflower whereas it should be complemented by mineral fertilization to maximize maize yield. The lower performance of maize than the other species was probably due to its position as the first species in the crop succession when compost organic matter turnover had not reached an equilibrium between the stable fraction and the mineralized one yet. Under 100% of compost fertilization, the crops did not take up a large amount of N, but it did not generate an increase of NO_3-N leaching in the percolation water. Long-term studies to better evaluate the compost as a substitute for mineral N fertilization of herbaceous crops are desirable.

Author Contributions: Conceptualization, M.B. Formal analysis, C.M. and A.B. Resources, M.B.; Writing—original draft preparation, C.M. and A.B. Writing—review and editing, C.M., A.B. and M.B. Visualization, C.M. and A.B. Supervision, M.B.

Funding: This research was supported by Veneto Agricoltura, Project "Valorization and environmental impact of the use of compost in the fertilization of field crops".

Conflicts of Interest: The authors declare no conflict of interest.

References

1. Khan, S.A.; Mulvaney, R.L.; Ellsworth, T.R.; Boast, C.W. The myth of nitrogen fertilization for soil carbon sequestration. *J. Environ. Qual.* **2007**, *36*, 1821–1832. [CrossRef] [PubMed]
2. Susic, M. Replenishing humic acids in agricultural soils. *Agronomy* **2016**, *6*, 45. [CrossRef]
3. Camacho Barboza, J.; Zanin, G.; Ponchia, G.; Sambo, P. Use of anaerobic digested residues in open field horticulture in the Veneto region, Italy. *Acta Hortic.* **2014**, *1018*, 133–138. [CrossRef]
4. Bassan, A.; Sambo, P.; Zanin, G.; Evans, M.R. Rice hull-based substrates amended with anaerobic digested residues for tomato transplant production. *Acta Hortic.* **2014**, *1018*, 573–581. [CrossRef]
5. Zanin, G.; Gobbi, V.; Coletto, L.; Passoni, M.; Nicoletto, C.; Ponchia, G.; Sambo, P. Use of organic fertilizers in nursery production of ornamental woody species. *Acta Hortic.* **2016**, *1112*, 379–386. [CrossRef]
6. Takakai, F.; Nakagawa, S.; Sato, K.; Kon, K.; Sato, T.; Kaneta, Y. Net greenhouse gas budget and soil carbon storage in a field with paddy–upland rotation with different history of manure application. *Agriculture* **2017**, *7*, 49. [CrossRef]
7. Martínez-Blanco, J.; Lazcano, C.; Christensen, T.H.; Muñoz, P.; Rieradevall, J.; Møller, J.; Antón, A.; Boldrin, A. Compost benefits for agriculture evaluated by life cycle assessment. A review. *Agron. Sustain. Dev.* **2013**, *33*, 721–732. [CrossRef]
8. Horrocks, A.; Curtin, D.; Tregurtha, C.; Meenken, E. Municipal compost as a nutrient source for organic crop production in New Zealand. *Agronomy* **2016**, *6*, 35. [CrossRef]

9. Fecondo, G.; Guastadisegni, G.; D'Ercole, M.; Del Bianco, M.; Buda, P.A. Utilizzo di compost di qualità su colture arboree per il miglioramento della fertilità del terreno. *Ital. J. Agron.* **2008**, *1*, 31–36. [CrossRef]

10. Takakai, F.; Kikuchi, T.; Sato, T.; Takeda, M.; Sato, K.; Nakagawa, S.; Kon, K.; Sato, T.; Kaneta, Y. Changes in the nitrogen budget and soil nitrogen in a field with paddy–upland rotation with different histories of manure application. *Agriculture* **2017**, *7*, 39. [CrossRef]

11. Montemurro, F.; Maiorana, M.; Convertini, G.; Fornaro, F. Improvement of soil properties and nitrogen utilisation of sunflower by amending municipal solid waste compost. *Agron. Sustain. Dev.* **2005**, *25*, 369–375. [CrossRef]

12. Islam, M.A.; Islam, S.; Akter, A.; Rahman, M.H.; Nandwani, D. Effect of organic and inorganic fertilizers on soil properties and the growth, yield and quality of tomato in Mymensingh, Bangladesh. *Agriculture* **2017**, *7*, 18. [CrossRef]

13. Fecondo, G.; Bucciarelli, S.; Di Paolo, E.; Ghianni, G. Biowaste compost effects on productive and qualitative characteristics of some field crops and on soil fertility. *Ital. J. Agron.* **2015**, *10*, 85–89. [CrossRef]

14. Nicoletto, C.; Santagata, S.; Sambo, P. Effect of compost application on qualitative traits in cabbage. *Acta Hortic.* **2013**, *1005*, 389–395. [CrossRef]

15. Seehausen, M.L.; Gale, N.V.; Dranga, S.; Hudson, V.; Liu, N.; Michener, J.; Thurston, E.; Williams, C.; Smith, S.M.; Thomas, S.C. Is there a positive synergistic effect of biochar and compost soil amendments on plant growth and physiological performance? *Agronomy* **2017**, *7*, 13. [CrossRef]

16. Passoni, M.; Borin, M. Effects of different composts on soil nitrogen balance and dynamics in a biennial crop succession. *Compost Sci. Util.* **2009**, *17*, 108–116. [CrossRef]

17. Gobbi, V.; Bonato, S.; Nicoletto, C.; Zanin, G. Spent mushroom substrate as organic fertilizer: Vegetable organic trials. *Acta Hortic.* **2016**, *1146*, 49–56. [CrossRef]

18. Ponchia, G.; Passoni, M.; Bonato, S.; Nicoletto, C.; Sambo, P.; Zanin, G. Evaluation of compost and anaerobic digestion residues as a component of growing media for ornamental shrub production. *Acta Hortic.* **2017**, *1168*, 71–78. [CrossRef]

19. IUSS Working Group WRB. *World Reference Base for Soil Resources 2014, Update 2015 International Soil Classification System for Naming Soils and Creating Legends for Soil Maps*; World Soil Resources Reports No. 106; FAO: Rome, Italy, 2015.

20. Ministero delle Politiche Agricole. Metodi Ufficiali di Analisi Fisica del Suolo. 1997. Available online: https://www.gazzettaufficiale.it/eli/id/1997/09/02/097A6592/sg (accessed on 28 October 2005).

21. Ministero delle Politiche Agricole. Metodi Ufficiali di Analisi Chimica del Suolo. 1999. Available online: https://www.gazzettaufficiale.it/eli/id/1999/10/21/099A8497/sg (accessed on 28 October 2005).

22. Loubser, H.L.; Human, J.J. The effect of nitrogen and phosphorus fertilization on the nitrogen absorption by sunflowers. *J. Agron. Crop Sci.* **1993**, *170*, 39–48. [CrossRef]

23. Ruffo, M.L.; García, F.O.; Bollero, G.A.; Fabrizzi, K.; Ruiz, R.A. Nitrogen balance approach to sunflower fertilization. *Commun. Soil Sci. Plant Anal.* **2003**, *34*, 2645–2657. [CrossRef]

24. Köppen, W. Das geographische System der Klimate (Handbuch der Klimatologie, Bd. 1, Teil C). 1936. Available online: http://koeppen-geiger.vu-wien.ac.at/pdf/Koppen_1936.pdf (accessed on 27 March 2019).

25. Fageria, N.K.; De Morais, O.P.; Dos Santos, A.B. Nitrogen use efficiency in upland rice genotypes. *J. Plant Nutr.* **2010**, *33*, 1696–1711. [CrossRef]

26. Cataldo, D.A.; Maroon, M.; Schrader, L.E.; Youngs, V.L. Rapid colorimetric determination of nitrate in plant tissue by nitration of salicylic acid. *Commun. Soil Sci. Plant Anal.* **1975**, *6*, 71–80. [CrossRef]

27. Dal Ferro, N.; Zanin, G.; Borin, M. Crop yield and energy use in organic and conventional farming: A case study in north-east Italy. *Eur. J. Agron.* **2017**, *86*, 37–47. [CrossRef]

28. Lakhdara, A.; Rabhia, M.; Ghnayaa, T.; Montemurro, F.; Jedidi, N.; Abdellya, C. Effectiveness of compost use in salt-affected soils. *J. Hazard. Mater.* **2009**, *171*, 29–37. [CrossRef]

29. Teutscherova, N.; Vazquez, E.; Santana, D.; Navas, M.; Masaguer, A.; Benito, M. Influence of pruning waste compost maturity and biochar on carbon dynamics in acid soil: Incubation study. *Eur. J. Soil Biol.* **2017**, *78*, 66–74. [CrossRef]

30. Wilson, C.; Zebarth, B.J.; Burton, D.L.; Goyer, C.; Moreau, G.; Dixon, T. Effect of Diverse Compost Products on Potato Yield and Nutrient Availability. *Am. J. Potato Res.* **2019**. [CrossRef]

31. Amlinger, F.; Götz, B.; Dreher, P.; Geszti, J.; Weissteiner, C. Nitrogen in biowaste and yard waste compost: Dynamics of mobilisation and availability—A review. *Eur. J. Soil Biol.* **2003**, *39*, 107–116. [CrossRef]

32. Eghball, B. Nitrogen mineralization from field applied beef cattle feedlot manure or compost. *Commun. Soil Sci. Plant Anal.* **2000**, *17*, 989–998. [CrossRef]

33. Eghball, B.; Wienhold, B.J.; Gilley, J.E.; Eigenberg, R.A. Mineralization of manure nutrients. *J. Soil Water Conserv.* **2002**, *57*, 470–473.

34. Preusch, P.L.; Adler, P.R.; Sikora, L.J.; Tworkosoky, T.J. Nitrogen and phosphorus availability in composted and uncomposted poultry litter. *J. Environ. Qual.* **2002**, *31*, 2051–2057. [CrossRef]

35. Hergreaves, J.C.; Adl, M.S.; Warman, P.R. A review of the use of composted municipal solid waste in agriculture. *Agric. Ecosyst. Environ.* **2008**, *123*, 1–14. [CrossRef]

36. Mamo, M.; Rosen, C.J.; Halbach, T.R. Nitrogen availability and leaching from soil amended with municipal solid waste compost. *J. Environ. Qual.* **1999**, *28*, 1074–1082. [CrossRef]

37. Santos, A.; Fangueiro, D.; Moral, R.; Bernal, M.P. Composts Produced from Pig Slurry Solids: Nutrient Efficiency and N-Leaching Risks in Amended Soils. *Front. Sustain. Food Syst.* **2018**, *2*, 8. [CrossRef]

 *agronomy*

Article

Maize Straw Returning Approaches Affected Straw Decomposition and Soil Carbon and Nitrogen Storage in Northeast China

Ping Tian, Pengxiang Sui, Hongli Lian, Zhengyu Wang, Guangxin Meng, Yue Sun, Yingyan Wang, Yehan Su, Ziqi Ma, Hua Qi * and Ying Jiang *

College of Agronomy, Shenyang Agricultural University, Shenyang 110866, China;
tianping23@stu.syau.edu.cn (P.T.); suipengxiang@stu.syau.edu.cn (P.S.); 2019200057@stu.syau.edu.cn (H.L.);
wangzhengyu27@stu.syau.edu.cn (Z.W.); 20172202241@stu.syau.edu.cn (G.M.);
2017220244@stu.syau.edu.cn (Y.S.); wangyingyan@stu.syau.edu.cn (Y.W.); 2017220228@stu.syau.edu.cn (Y.S.);
2018240174@stu.syau.edu.cn (Z.M.)
* Correspondence: qihua10@syau.edu.cn (H.Q.); jiangying@syau.edu.cn (Y.J.); Tel.: +86-024-88487135 (Y.J.)

Received: 3 November 2019; Accepted: 25 November 2019; Published: 28 November 2019

Abstract: The characterization of straw decomposition and the resulting carbon (C) and nitrogen (N) release is crucial for understanding the effects of different straw returning methods on the immobilization and cycling of soil organic carbon (SOC) and soil total nitrogen (STN). In 2017–2018, a field micro-plot experiment was carried out in northeastern China to investigate the effects of different straw returning approaches on straw decomposition, C release, N release, SOC, STN, and the soil C–N ratio. Six straw returning treatments were applied: straw mixed with soil (SM) and straw buried in the soil (SB) at soil depths of 10 (O), 30 (T), and 50 cm (F). The results indicate that the straw decomposition proportion (SD), C release, and N release in SM + O were higher than that in SM + T and SM + F. Moreover, SOC and STN concentrations and the soil C–N ratio were significantly enhanced by SM/B + O in the 0–20 cm soil layers, SM/B + T in the 20–30 cm soil layer, and SM/B + F in the 40–60 cm soil layers. In the 0–50 cm soil profile, the highest SOC stocks were obtained using SB + T. The STN stocks were also significantly affected by the straw returning depth, but the effect was inconsistent between the two years. SD had a positive relationship with SOC and STN in the 0–20 cm soil layers; conversely, they were negatively related in the 30–60 cm soil layers. The results of this study suggest that straw buried in the soil to a depth not exceeding 30 cm might be an optimal straw returning approach in northeastern China.

Keywords: crop residue incorporation; straw decomposition; residue C and N release; SOC and STN stocks

1. Introduction

Feeding the increasing population has required intensified conventional tillage and high nitrogen input for food production, resulting in the extensive loss of productive soil [1]. For instance, the massive loss of soil organic carbon (SOC) due to conventional tillage management has caused an increase in soil erosion and the destruction of soil structure [2], which, in turn, are related to the decreases in cropland productivity that are clearly observed in parts of developing countries [3]. The ongoing land degradation in northeast China—a key base of commodity grain, accounting for approximately 30% of maize of China [4]—has threatened sustainable crop production and even national food security [5]. Thus, for several years, maize producers have been encouraged to incorporate the resulting crop residue, which is a highly accessible form of organic matter, into agricultural soils to maintain the soil organic matter (SOM) content [6,7] and thus contribute to soil quality improvement [8] and identify the ill effects of burning maize residues in the field [9].

To date, maize straw incorporation by rotary or plow tillage remains the predominant approach for residue return because the surface retention with no-till systems is associated with low decay rates due to the low average temperature and low precipitation conditions in northeast China [10]. Different management practices result in the varying placement of crop residue in agricultural field soils; for example, residues can be incorporated into soil or surface mulch. Furthermore, in an agrosystem, the complex feedback between the climate, the soil type, and management factors make it difficult to predict the effects of crop residue placement on its decomposition rate and the ultimate fate of the derived nutrients [11]. In particular, the soil microenvironments for biological and chemical processes located in the soil surface differ from those in shallow or deep soils; thus, the depth also influences the decomposition and nutrient release of incorporated crop residues [12]. Recently, studies have shown that the decay rate and nutrient release of crop residues in the soil are site-specific because the decomposer community [13] and nutrient sources vary between the topsoil and subsoil [11,14,15]. Additionally, straw decomposition is accompanied by the release of straw-derived carbon and nitrogen, which, in turn, interact with the residue decomposition process [16,17]. Therefore, in the maize production of northeast China, it is critical to clarify the effects of different straw incorporation approaches on the decay proportion and straw-derived C and N release levels, as well as the interactions between the release levels and the soil content of C and N.

Straw returning methods not only affect the straw decomposition rate but also influence the cycle and immobilization of SOC and soil total nitrogen (STN) in field ecosystems [18]. The responses of SOC and STN dynamics to crop residue input can be influenced by many factors, such as the soil's moisture, temperature, and biochemical properties, which all vary depending on the straw returning method and the depth of soil incorporation [19]. Leaving crop residue on the soil surface tends to concentrate C on the soil surface, while incorporating straw residue into the soil by tillage is usually considered to contribute to C losses due to increased SOM decomposition rates [20] as tillage disturbs the soil's structural stability, redistributes organic matter, and influences the microbial activity throughout the soil profile [8]. However, several studies have also observed increases in SOC content in deeper layers from the incorporation of residues in the soil [21–23]; thus, the total SOC stored in the soil by surface retention may be the same as that resulting from incorporation approaches when the whole soil profile is studied [24]. In addition to the inconsistent effects on SOC content, the variable effect of straw incorporation management strategies on STN has also been emphasized [25,26]. Crop residues incorporated into the soil at different depths influence STN through highly complex mechanisms involving N mineralization and immobilization [27], as well as N leaching and denitrification losses in soil [18]. In addition, the concentration of STN has been statistically correlated with SOC content in general soil types [26], and the soil C–N ratio may reflect the interaction between SOC and STN under tillage practices for straw incorporation. Thus, it is necessary to investigate residues incorporated at various returning depths by sampling the entire plow depth to accurately assess the influence of residue incorporation practices on SOC, STN, and the soil C–N ratio [28].

Given this high variability, the effects of straw incorporation practices on SOC and STN stocks, as reported by the studies mentioned above, have been conflicting [25,26]. For instance, Xue et al. [29] investigated a paddy cropping system in southern China, and their results indicated that the application of residue with no-till farming enhanced the SOC and STN stocks in the 0–10 cm layer, whereas plow tillage increased the SOC and STN stocks in the 0–50 cm profile [29]. However, Dikgwatlhe et al. [26] observed no significant differences among straw returning treatments for SOC stocks in the 0–50 cm soil profile, but they noted that STN stocks increased [26]. Additionally, an earlier study reported that SOC and STN stocks in the 0–50 cm soil profile did not significantly differ between soils subjected to straw returning by rotary tillage and that by moldboard plow tillage [25]. The different findings for SOC and STN stocks among the previous studies are likely the result of variations in factors such as climate, soil type, and tillage intensity. To date, although the benefits of crop residue retention have been well documented in the previous literature, few studies have assessed the influence of straw returning practices on SOC and STN stocks. It is critical to obtain comprehensive knowledge of the effects of different returning methods and depths on SOC and STN dynamics and their correlation with the straw decay proportion in different soil layers. Therefore, the objectives of the present study were to (1) determine the effects of different straw returning methods and depths on the decomposition and C and N release proportions of maize straw residue, (2) evaluate the effects of straw returning approaches on SOC and STN stocks in the 0–50 cm soil profile, (3) investigate the response of SOC and STN concentrations in different soil layers on straw incorporation and their interactions with the straw decomposition ratio during the decay process, and (4) identify a suitable straw returning approach for the sustainability of maize production in northeastern China.

2. Materials and Methods

2.1. Site Description

The field experiment was conducted in 2017 and 2018 at the Experimental Station of Shenyang, Agricultural University, Shenyang, Liaoning province, China (41°82′ N, 123°56′ E; 43 m above sea level). This area is located in a flat region that is characterized by a sub-humid warm-temperate continental climate. The average annual temperature was 9.17 °C, with a frost-free period of 155–180 days. According to measurements taken before this study, the concentration of SOC, STN, available phosphorus, available potassium, and soil bulk density in the soil layer (0–20 cm) were 10.81g kg^{-1}, 0.92 g kg^{-1}, 51.17 mg kg^{-1}, 128.49 mg kg^{-1}, and 1.43 g cm^{-3}, respectively. The main crop in this region is spring maize (*Zea mays* L.), and the crop-planting pattern is one harvest per year. All the water required for crop growth was provided by natural precipitation in this study. During the experimental period, precipitation and temperature were measured using an automatic weather station (5TM, Decagon, Washington, USA) around the experimental site (Figure 1). The mean daily air temperature during the experiment was 9.26 and 9.09 °C; the highest temperatures were 26.36 and 27.21 °C, and the lowest temperatures were −8.77 and −12.20 °C in July and January 2017 and 2018, respectively. The total precipitation levels in 2017 and 2018 were 456.4 and 505.2 mm, 58.81% and 73.12% of which occurred from June to August, while 9.03% and 11.05% of the total precipitation occurred in May, which is a critical period for straw decomposition. The annual precipitation was lower than the mean annual precipitation of 714 mm [30], indicating that the experimental periods were seasons characterized by poor rainfall.

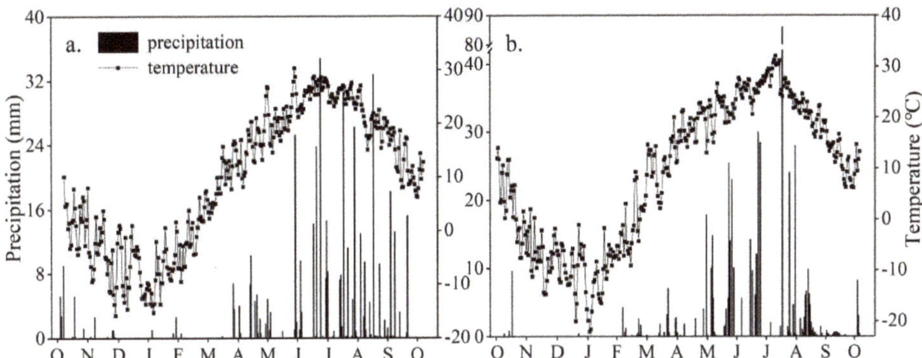

Figure 1. Mean daily precipitation and temperature during (**a**) October 2016–October 2017 and (**b**) October 2017–October 2018 at the experimental site.

2.2. Experimental Design

This study employed a two-factor (straw returning method, M; straw returning depth, D) design in randomized complete blocks with three replicates. Micro-plots [31] were used for six straw returning treatments: straw mixed (SM) with soil at a depth of 10, 30, and 50 cm (O, T, and F) and straw buried

(SB) in soil at a depth of 10, 30, and 50 cm (O, T, and F). The SM and SB treatments as straw returning methods (M) were intended to simulate straw returning by rotary tillage and plow tillage in the field, respectively. All micro-plots were made by using stainless-steel plates (length, width, and height of $1.5 \times 1.2 \times 0.7$ m) without bottoms; undisturbed soil was employed in all the micro-plots. Before being returned into the soil, the maize straw was chopped into 3–5 cm pieces. The maize straw was then manually incorporated into the soil in the plots on 24 October 2016 and 24 October 2017. The C and N contents of the straw for incorporation were 9.51 g N kg^{-1} and 415.75 g C kg^{-1}(C–N = 44:1) in 2017 and 9.70 g N kg^{-1} and 443.01 g C kg^{-1}(C/N, 46:1) in 2018.

The C and N release dynamics of maize straw from different treatments were investigated by using the litterbag decomposition technique described by Varela et al. [32] and Xu et al. [18] with a few modifications. Briefly, air-dried straw (53 g per treatment in nylon bags, which is equivalent to the amount of straw returned to the field) was chopped and placed into nylon mesh bags (20×30 cm, 0.1 mm mesh). For the SM + O treatment, one nylon bag of straw was placed into the soil at a depth of 0–10 cm and a slope of 30°. For the SM + T and SM + F treatments, three and five nylon bags, respectively, with the same amount of straw (a total of 53 g of straw was equally separated into three or into five nylon bags) were set up with the same method per 10 cm of the soil profile. For the SB + O, SB + T, and SB + F treatments, one nylon bag loaded with 53 g of maize straw was placed in the 10, 30, and 50 cm soil layer, respectively, with no slope. All the litterbags had three replications. After crop harvest, the straw was collected from the litterbag and then shaken gently over a 1 mm sieve and spray-rinsed to remove the adhering soil. The straw samples were oven-dried in envelopes at 60 °C until the weight was constant. The samples were then weighed and ground to pass through a 0.15 mm sieve for further chemical analysis.

Basal fertilizer with 75 kg ha^{-1} N, 90 kg ha^{-1} P, and 90 kg ha^{-1} K was used when maize was sowed, and 150 kg ha^{-1} N was applied as topdressing around the middle of June. The variety of maize used was Zhengdan 958 (Jinboshi, Zhengzhou, China), and it was sowed at a rate of 67,500 plants ha^{-1} at the end of April and harvested at the end of September. The management practices for controlling pests, disease, and weeds were according to local practices for high-yield production.

2.3. Sampling and Analysis Methods

The weight loss of the straw in the nylon bag was assumed to be the amount of straw that decomposed during the experimental period and was calculated using Equation (1). The amount of N or C lost from the straw was calculated as the difference between the initial N or C contained in the input straw and the N or C recovered from the treated straw; it was determined using Equation (2).

$$SD = \frac{W_0 - W_t}{W_0} \times 100\% \tag{1}$$

$$NR = \frac{W_0 \times C_0 - W_t \times C_t}{W_0 \times C_0} \times 100\% \tag{2}$$

where SD is the straw decomposition proportion (%); NR is the nutrient release proportion (%); W_0 and W_t are the initial and remaining straw weights (g), respectively; and C_0 and C_t are the nutrient concentrations (g kg^{-1}) in the initial and remaining straw, respectively.

Soil sampling and analysis were performed immediately after harvest in 2017 and 2018. In each plot, six soil samples were collected using a coring tube (5 cm in diameter) from depths between 0 and 60 cm at 10 cm intervals. The soil samples that were collected from two points in each plot with replication were mixed to produce a composite sample. The samples were air-dried with plant stubbles and pebbles discarded and then ground through a 0.15 mm sieve for SOC and STN determination.

SOC and STN concentrations (g kg^{-1}) were determined using the $K_2Cr_2O_7$-H_2SO_4 digestion method [26] and the Kjeldahl method [33], respectively. The soil C–N ratio was calculated according to the values from the SOC and STN measurements. SOC and STN stocks were calculated by the equivalent soil mass (ESM) method to eliminate the uncertainties associated with the fixed-depth method and Equation (3) according to the description by Ellert et al. and Xue et al. [29,34].

$$M_{element} = \sum_{i=1}^{n} \left[M_{soil,\,i} \times conc_i + (M_{o,i} - M_{soil,i}) \times conc_{i+1} \right] \times 0.001 \tag{3}$$

where i is the soil layer (i = 1, 2, 3, 4, and 5 represent the 0–10, 10–20, 20–30, 30–40, and 40–50 cm soil layers, respectively); $M_{element}$ is the SOC or STN stocks (Mg ha^{-1}); $M_{soil,i}$ is soil mass per unit area in the ith layer (Mg ha^{-1}), which is calculated by Equation (4); $M_{o,i}$ is the equivalent soil mass of each layer; and $conc_i$ and $conc_{i+1}$ are the concentrations of SOC or STN in the ith and i+1th layers, respectively (g kg^{-1}).

$$M_{soil,\,i} = \rho_{b,\,I} \times T_i \times 10000 \tag{4}$$

where $M_{soil,i}$ is soil mass per unit area in the ith layer (Mg ha^{-1}), $\rho_{b,i}$ is soil bulk density in the ith layer (g cm^{-3}), and T_i is the thickness of the ith layer (m).

2.4. Statistical Analysis

The effects of the different treatments on all the data were analyzed by ANOVA using the SPSS 23.0 (SPSS Inc., Chicago, Illinois, USA) software, and effects of years were analysed separately. The SPSS procedure was used to analyze the variance and determine the statistical significance of the treatment. Duncan's multiple range test was used to compare the treatment means at a 95% confidence level.

3. Results

3.1. Straw Decomposition Proportion

The ANOVA results demonstrate that the straw returning depth had a consistently significant effect on the maize straw decomposition proportion in both study years, but the straw returning method and its interaction effects with the returning depth only exhibited a significant difference in 2017, with no significant differences observed in 2018 (Figure 2). Under the SM treatments, the straw decomposition proportion decreased with the increasing depth of maize residue incorporation (O, T, and F treatments), and the tendency was similar in 2017 and 2018. Compared with the SM + T and SM + F treatments, the breakdown of maize residue in the SM + O treatment was 5.81% and 9.14% higher in 2017, and 22.09% and 56.28% higher in 2018. In 2017, a slight variation in the residue decomposition proportion was found between the SB + O, SB + T, and SB + F treatments, whereas a significant reduction was observed between the same treatments in 2018, with a significant difference between the SB + T and SB + F treatments in 2018, but not 2017 ($P < 0.05$). Overall, the shallow residue incorporation (O treatment) had the highest decomposition proportion under the SM and SB management approaches in both seasons. Moreover, the maize straw decomposition proportion in 2017 was generally higher, ranging from 63.25% to 69.09% between the treatments, while the results in 2018 ranged from 39.82% to 62.23%.

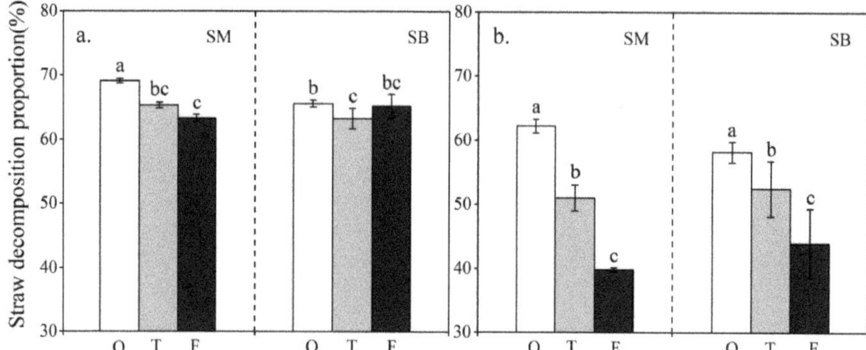

Figure 2. The effect of different straw returning management approaches on the straw decomposition proportion in (**a**) 2017 and (**b**) 2018. ANOVA results: straw returning depth (D) ***, straw returning method (M) *, D × M ** in 2017; D ***, M ns, D × M ns in 2018 (* $P < 0.05$; ** $P < 0.01$; *** $P < 0.001$; ns, not significant). SM and SB indicate straw mixed with soil and buried in soil, respectively, and O, T, and F indicate straw incorporated into the soil (mixed or buried) at depths of 10, 30, and 50 cm, respectively. The bars with different letters indicate statistically significant differences between treatments at $P < 0.05$.

3.2. Straw-Derived C and N Release

In both study years, the straw-derived C release proportion during the decay process was significantly influenced by the straw incorporation depth but not by the returning method. Their interaction effects differed significantly in 2017, but no difference between them was observed in 2018 (Figure 3). Generally, the proportion of C released by straw decay had a similar tendency to the straw decomposition indicator between treatments and between years. Additionally, in 2017, the C release proportion under the SM + O treatment was higher than that under the SM + T and SM + F treatments, but it only slightly fluctuated and lacked statistical significance under the SB treatments. In 2018, under SM and SB returning practices, the C release proportion significantly decreased (by 45.08–66.07% and 46.32–62.88%, respectively) as the depth of straw incorporation increased.

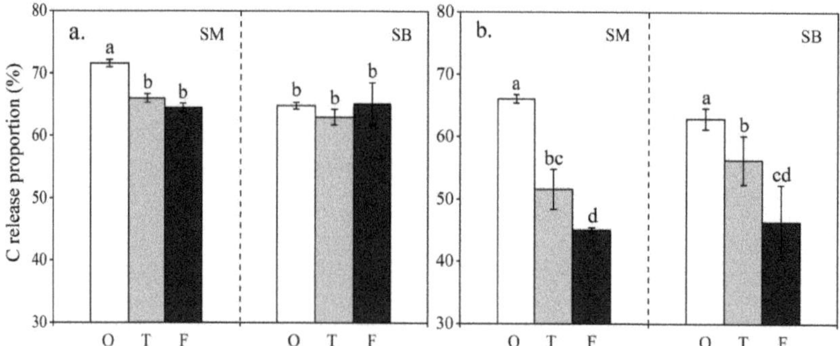

Figure 3. The effect of different straw returning management approaches on the straw carbon release proportion in (**a**) 2017 and (**b**) 2018. ANOVA results: straw returning depth (D) ***, straw returning method (M) ns, D × M ** in 2017; D ***, M ns, D × M ns in 2018 (** $P < 0.01$; *** $P < 0.001$; ns, not significant). SM and SB indicate that the straw was mixed with soil and buried in soil, respectively, and O, T, and F indicate that straw was incorporated into the soil (mixed or buried) at depths of 10, 30, and 50 cm, respectively. The bars with different letters indicate statistically significant differences between treatments in each year at $P < 0.05$.

In 2017, the proportion of straw-derived N release exhibited a decreasing tendency with incorporation depth under the SM treatments; the SM + F treatment was found to cause a significantly lower N release proportion than the SM + O treatment, but there was no difference between the returning depth treatments under SB management approaches. In 2018, the N release proportion under SM + O was 0.66-fold and 2.30-fold higher than the results under the SM + T and SM + F treatments, respectively. Under the SB treatment, the N release proportion under SB + O was significantly higher than that under the SB + T treatment, and there was no difference between SB + T and SB + F. Thus, according to the ANOVA results, the straw returning depth resulted in significant differences in the N release proportion in 2018 and no difference in 2017. The returning methods had no significant influence on straw-derived N release, but the interactive effects between the straw incorporation depth and the methods consistently showed significant differences in both experimental periods (Figure 4).

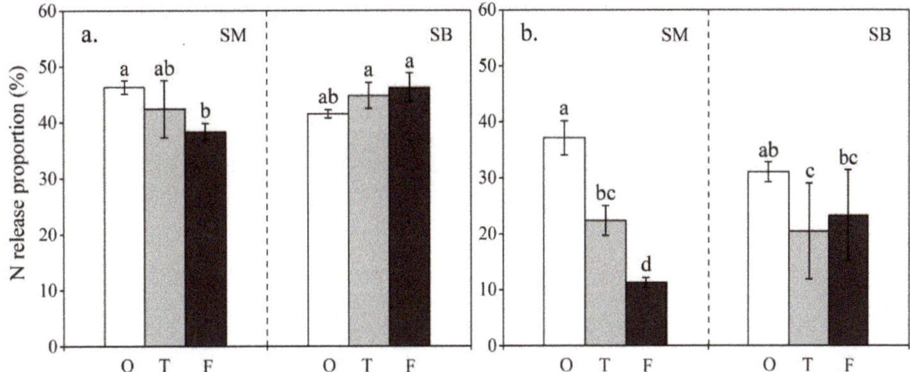

Figure 4. The effects of different straw returning management approaches on the straw nitrogen release proportion in (**a**) 2017 and (**b**) 2018. ANOVA results: straw returning depth (D) ns, straw returning method (M) ns, D × M ** in 2017; D ***, M ns, D × M * in 2018 (* $P < 0.05$; ** $P < 0.01$; *** $P < 0.001$; ns, not significant). SM and SB indicate that the straw was mixed with soil and buried in soil, respectively, and O, T, and F indicate that the straw was incorporated into the soil (mixed or buried) at depths of 10, 30, and 50 cm, respectively. The bars with different letters indicate statistically significant differences between the treatments in each study year at $P < 0.05$.

3.3. Soil Organic Carbon (SOC) and Soil Total Nitrogen (STN) Concentration

Regardless of whether the maize residue was mixed or buried in the soil, the SOC concentration throughout the soil profile (0–60 cm) was significantly affected by the straw returning approaches over the experimental period, but there was no difference between SM$_{mean}$ and SB$_{mean}$ in either study year (Figure 5). Generally, the changes in SOC concentration in the 0–60 cm soil profile in the SM and SB treatments followed a similar trend. In both study years, in the soil samples taken from a depth greater than 30 cm (the SM/B + O and SM/B + T treatments), the SOC concentration sharply decreased (Figure 5a,b,d,e), which caused a corresponding marked reduction in the SM$_{mean}$ and SB$_{mean}$ of the SOC content (Figure 5c,f); however, the same trend did not occur under the SM/B + F treatment. In the 0–20 cm soil layer, the SOC content under the SM/B + O treatment was significantly higher than that under the SM/B + T and SM/B + F treatments in both 2017 and 2018 ($P < 0.05$, Figure 5a,b,d,e). In both study years, the SOC content in the 20–30 cm soil layer was highest under the SM/B + T treatment, followed by that under the SM/B + O and SM/B + F treatments. In contrast, at depths below 30 cm, the SOC content was clearly higher under the SM/B + F treatment compared with that under the SM/B + O and SM/B + T treatments.

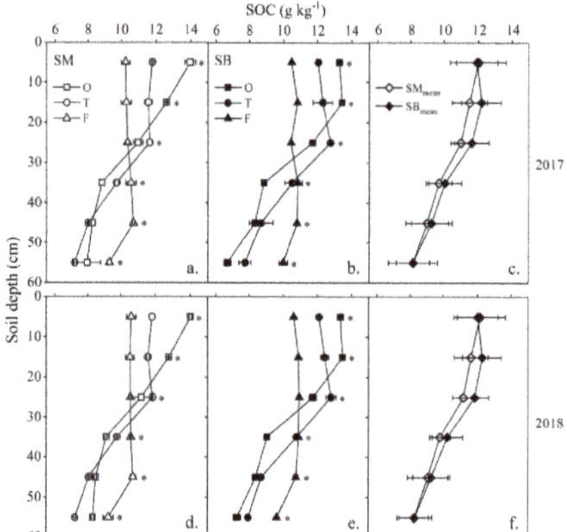

Figure 5. Soil organic carbon (SOC) content in the 0–60 cm soil profile under different straw returning strategies in (**a–c**) 2017 and (**d–f**) 2018. SM and SB indicate straw mixed with soil and buried in soil; O, T, and F indicate straw incorporated (mixed or buried) into the soil at depths of 10, 30, and 50 cm, respectively. SM_{mean} is the mean of SM + O, SM + T, and SM +F; SB_{mean} is the mean of SB + O, SB + T, and SB + F; * indicates significant differences ($P < 0.05$) between the straw incorporation depth treatments for the same soil layer.

The trends and significant differences in the STN concentration were generally similar to those of the SOC content between different straw returning treatments and soil layers in the 0–60 cm soil profile. In both years of the experiment, the average STN values for SM and SB were very close to each other and linearly decreased as the depth of the soil layers increased (Figure 6). In the 0–20 cm soil layer, the STN content under the SM/B + O treatment was significantly higher than that under the SM/B + T and SM/B + F treatments (Figure 6a,b,d,e). Under the SM/B + F treatment, in both years, the STN content tended to slightly fluctuate while gradually declining with increasing depth in the 0–60 cm soil profile; this pattern was similar to the changes in SOC content.

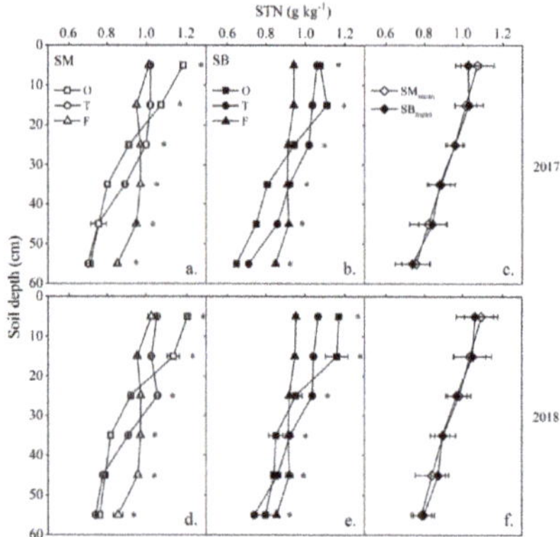

Figure 6. Soil total nitrogen (STN) content in the 0–60 cm soil profile under different straw returning strategies in (**a–c**) 2017 and (**d–f**) 2018. SM and SB indicate straw mixed with soil and buried in soil; O, T, and F indicate straw incorporated (mixed or buried) into the soil at depths of 10, 30, and 50 cm, respectively; SM_{mean} is the mean of SM + O, SM + T, and SM + F; SB_{mean} is the mean of SB + O, SB + T, and SB + F; * indicates significant difference ($P < 0.05$) between the straw incorporation depth treatments for the same soil layer.

3.4. Soil Organic Carbon (SOC) and Soil Total Nitrogen (STN) Stocks

The mean SOC stocks under the SB treatments were significantly higher than that under the SM treatments in the 10–20, 20–30, and 30–40 cm soil layers as well as the 0–50 cm soil profile in both study years and in the 40–50 cm soil layer in 2018 (Table 1). In the 0–50 cm soil profile, the highest SOC stocks among all treatments were obtained with SB + T, with values of 77.38 and 77.14 Mg ha^{-1} in 2017 and 2018, respectively. Moreover, the SOC stocks in different soil layers and the overall 0–50 cm soil profile were significantly influenced by the straw returning depth treatments. Of all the soil layers and straw returning depths, the highest SOC stocks were in the 0–20 and 30–50 cm soil layers under the SM/B + O and SM/B + F treatments, respectively.

The mean STN stock in the overall 0–50 cm soil profile did not significantly differ between the SM and SB treatments, but it significantly differed in the 0–10 cm soil layer in both years and in the 10–20 and 40–50 cm layers in 2017 and 2018, respectively (Table 1). In the 0–50 cm soil profile, among all the treatments, the STN stocks were the highest under SB + T (6.74 Mg ha^{-1}) and SB+O (6.76 Mg ha^{-1}) in 2017 and 2018, respectively. In both study seasons, similar to the SOC stocks, the STN stocks in the 0–50 cm soil profile, were strongly influenced by the straw incorporation depths. In the 0–20 cm soil layers, the STN stocks were greater under SM/B + O than those treated with SM/B + F; conversely, in the 30–50 cm soil layers, the STN stocks were higher under the SM/B + F treatments than those under the SM/B + O treatments. In both 2017 and 2018, of the three returning depth treatments in the SM incorporation system, the overall (i.e., in the 0–50 cm profile) STN stocks were the highest for SM + F. However, when the straw was treated using the SB approach, the overall STN stocks were higher under the SB + O/T treatments than the SB + F treatment.

Table 1. Soil Organic Carbon (SOC) and Soil Total Nitrogen (STN) stocks in the 0–50 cm profile under different straw returning approaches in 2017 and 2018.

Year	Treatment		SOC (Mg ha^{-1}), Depth (cm)						STN (Mg ha^{-1}), Depth (cm)					
			0–10	10–20	20–30	30–40	40–50	0–50	0–10	10–20	20–30	30–40	40–50	0–50
2017	SM	O	18.57 a	16.65 b	15.88 d	12.62 d	11.38 bc	75.10 b	1.57 a	1.41 b	1.32 d	1.15 d	1.04 d	6.49 b
		T	15.79 c	15.39 c	16.48 c	13.86 c	11.04 c	72.56 c	1.37 c	1.36 d	1.43 b	1.27 c	1.04 d	6.47 b
		F	13.73 d	13.67 e	15.04 e	15.11 ab	14.64 a	72.19 c	1.35 c	1.26 e	1.41 b	1.38 a	1.30 a	6.70 a
		Mean	16.03 A	15.24 B	15.80 B	13.86 B	12.35 A	73.28 B	1.43 A	1.34 B	1.39 A	1.27 A	1.13 A	6.56 A
	SB	O	17.79 b	17.83 a	16.92 b	12.64 d	11.37 bc	76.55 a	1.44 b	1.47 a	1.37 c	1.15 d	1.03 d	6.46 b
		T	16.13 c	16.41 b	18.15 a	14.80 b	11.89 b	77.38 a	1.42 b	1.38 c	1.46 a	1.31 b	1.17 c	6.74 a
		F	14.06 d	14.36 d	15.18 e	15.50 a	14.86 a	73.96 b	1.26 d	1.25 e	1.32 d	1.30 bc	1.26 b	6.39 c
		Mean	15.99 A	16.20 A	16.75 A	14.31 A	12.71 A	75.96 A	1.37 B	1.37 A	1.38 A	1.25 A	1.15 A	6.53 A
2018	SM	O	17.96 a	16.84 b	15.95 cd	12.93 d	11.54 c	75.22 c	1.55 a	1.49 a	1.32 c	1.16 d	1.07 d	6.59 bc
		T	15.23 d	15.42 c	16.50 bc	13.82 c	10.94 d	71.91 e	1.36 c	1.37 b	1.48 a	1.29 b	1.06 d	6.56 c
		F	13.66 e	13.99 e	15.09 e	15.03 b	14.43 a	72.20 e	1.32 d	1.27 c	1.39 b	1.38 a	1.29 a	6.65 abc
		Mean	15.62 A	15.42 B	15.85 B	13.93 B	12.30 B	73.11 B	1.41 A	1.38 A	1.40 A	1.28 A	1.14 B	6.61 A
	SB	O	17.21 b	17.88 a	16.75 b	12.82 d	11.35 c	76.01 b	1.51 b	1.54 a	1.36 b	1.21 c	1.14 c	6.76 a
		T	15.62 c	16.56 b	18.03 a	15.10 b	11.83 b	77.14 a	1.37 c	1.39 b	1.47 a	1.30 b	1.16 c	6.69 ab
		F	13.70 e	14.50 d	15.65 de	15.48 a	14.56 a	73.89 d	1.23 e	1.26 c	1.31 c	1.30 b	1.25 b	6.35 d
		Mean	15.51 A	16.31 A	16.81 A	14.47 A	12.58 A	75.68 A	1.37 B	1.40 A	1.38 A	1.27 A	1.18 A	6.60 A

ANOVA for SOC stocks (0–50 cm): straw returning depth (D) ***, straw returning method (M) ***, D × M ** in 2017, D ***, M ***, D × M *** in 2018. ANOVA for STN stocks (0–50 cm): D ***, M ns, D × M *** in 2017; D **, M ns, D × M *** in 2018 (** $P < 0.01$; *** $P < 0.001$; ns, not significant). SM and SB indicate that the straw was mixed with soil and buried in soil, and O, T, and F indicate that straw was incorporated (mixed or buried) into the soil at depths of 10, 30, and 50 cm, respectively. The different lowercase within column for each year indicate significant differences statistically ($P < 0.05$) between straw returning depths (O, T, and F) and different capital letters within column for each year indicate significant differences statistically ($P < 0.05$) between soil returning approaches (SM and SB).

3.5. Soil C–N Ratio

In both years, the straw returning depths had significant effects on the soil C–N ratio in the different soil layers of the 0–60 cm soil profile. The mean soil C–N ratio did not significantly differ between the SB and SM incorporation methods, except for that in the 10–20 cm soil layer in 2018 (Table 2). At the 0–10 and 10–20 cm soil depths, the highest soil C–N ratios were obtained using SB + O in 2017, SM + O in 2018, and SB + T in 2018. Of all treatments in this study, the soil C–N ratios were highest under the SB + T treatment in the 20–30 cm soil layer and the SB + F treatment in the 30–60 cm soil layers with a 10 cm interval. In both experimental seasons, the mean soil C–N ratio at different sampling depths was lower in the SM treatments than the SB treatments, except for in the 40–50 cm soil layer in 2018.

Table 2. Soil C–N ratio in the 0–60 cm profile under different straw returning approaches in 2017 and 2018.

Year	Treatment		Soil Depth (cm)					
			0–10	10–20	20–30	30–40	40–50	50–60
2017	SM	O	11.81 b	11.76 ab	12.03 b	11.02 b	10.92 abc	11.13 ab
		T	11.57 bc	11.35 bc	11.66 c	10.90 b	10.63 bc	10.27 b
		F	10.11 e	10.82 c	10.67 d	10.89 b	11.27 ab	10.89 ab
		Mean	11.16 A	11.31 A	11.46 A	10.93 A	10.94 A	10.76 A
	SB	O	12.33 a	12.10 a	12.38 a	10.96 b	11.07 ab	10.30 b
		T	11.35 cd	11.86 ab	12.50 a	11.41 ab	10.11 c	10.86 ab
		F	11.11 d	11.48 b	11.43 c	11.92 a	11.78 a	11.76 a
		Mean	11.60 A	11.81 A	12.10 A	11.43 A	10.99 A	10.97 A
2018	SM	O	11.65 a	11.26 bc	12.14 a	11.15 b	10.74 c	10.90 a
		T	11.22 b	11.29 bc	11.20 bc	10.75 c	10.36 b	9.76 b
		F	10.33 c	11.03 c	10.85 c	10.89 bc	11.18 b	10.81 a
		Mean	11.06 A	11.19 B	11.40 A	10.93 A	10.76 A	10.49 A
	SB	O	11.41 ab	11.62 ab	12.33 a	10.66 c	9.97 c	9.60 b
		T	11.34 b	11.91 a	12.35 a	11.69 a	10.14 bc	10.72 a
		F	11.14 b	11.50 abc	11.91 ab	11.93 a	11.70 a	11.27 a
		Mean	11.30 A	11.68 A	12.20 A	11.43 A	10.60 A	10.53 A

SM and SB indicate that straw was mixed with soil and buried in soil, and O, T, and F indicate that straw was incorporated into the soil (mixed or buried) at depths of 10, 30, and 50 cm, respectively. The different lowercase within column for each year indicate significant differences statistically ($P < 0.05$) between straw returning depths (O, T, and F) and different capital letters within column for each year indicate significant differences statistically ($P < 0.05$) between soil returning approaches (SM and SB).

3.6. Relationship Between the Straw Decomposition Proportion and the Soil Organic Carbon (SOC) and Soil Total Nitrogen (STN) Concentrations

Figures 7 and 8 show the fitted correlations between the straw decomposition proportion (SD) and the SOC and STN concentrations, respectively. The results indicate that the SD–SOC and SD–STN relationships had parallel tendencies throughout the 0–60 cm soil profile. In the 0–10 and 10–20 cm soil layers, SD had a significantly positive linear correlation with both SOC and STN, except for the result between SD and STN in the 0–10 cm soil layer. SD had significant parabolic relationships with both SOC and STN in the 20–30 cm soil layer. However, SD had significantly negative linear correlations between SOC and STN in the 30–40, 40–50, and 50–60 cm soil layers, except for the result between SD and SOC in the 50–60 cm soil layer.

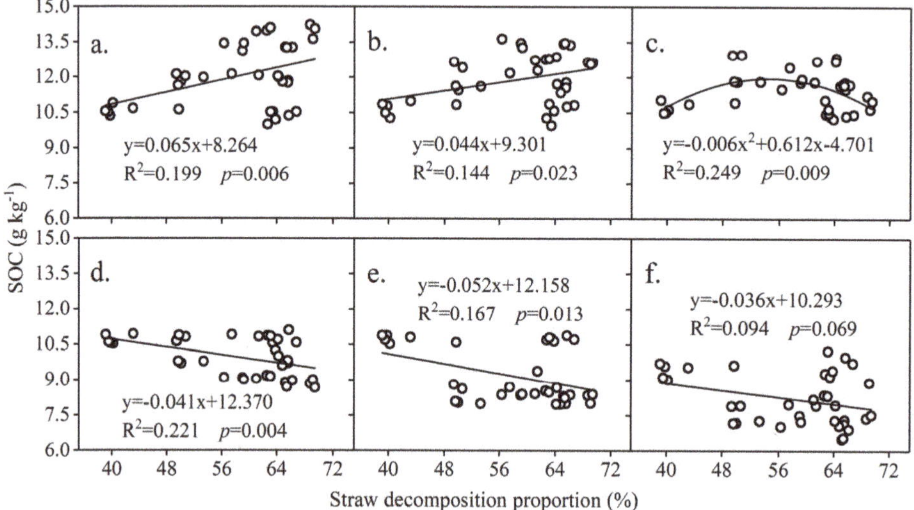

Figure 7. Relationships between the straw decomposition proportion and the SOC concentration at the soil depth of (**a**) 0–10 cm, (**b**)10–20 cm, (**c**) 20–30 cm, (**d**) 30–40 cm, (**e**) 40–50 cm, and (**f**) 50–60 cm ($n = 36$).

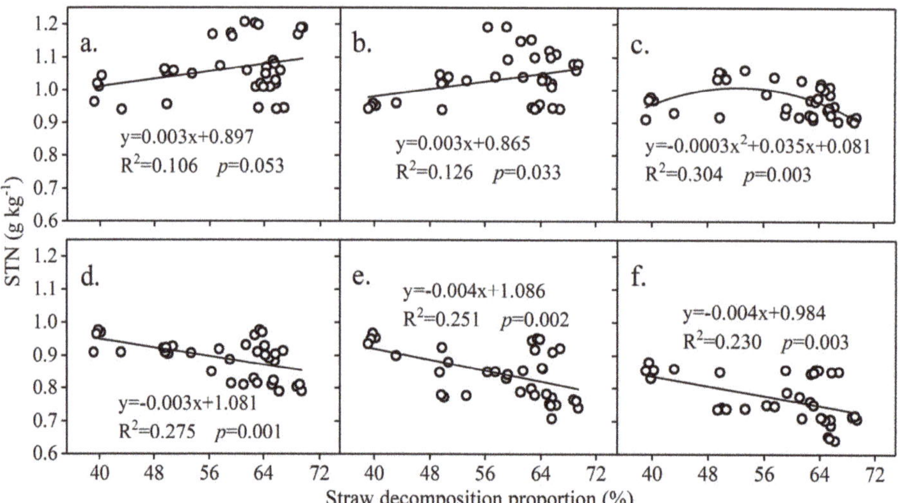

Figure 8. Relationships between the straw decomposition proportion and the STN concentration at the soil depth of (**a**) 0–10 cm, (**b**) 10–20 cm, (**c**) 20–30 cm, (**d**) 30–40 cm, (**e**) 40–50 cm, and (**f**) 50–60 cm ($n = 36$).

4. Discussion

4.1. Straw Decomposition Proportion and C/N Release

The straw decomposition process depends strongly on climatic environmental conditions; among all possible factors, temperature and precipitation are viewed as the primary factors that affect the process [19]. In the present study, the overall annual mass loss of the straw shows that the decomposition proportions were higher in 2017 relative to those in 2018 (Figure 2), likely because the mean air temperature in 2017 (9.26 °C) was higher than that in 2018 (9.09 °C). A previous study showed that a higher air temperature with appropriate precipitation contributed to crop residue decomposition [35]. The straw decomposition proportion was influenced by the returning methods, but the results were inconsistent between the two years, which indicates that annual conditions affected the straw decay process, as well. In addition, straw decomposition generally decreased as the returning depths increased, which is in agreement with previous litterbag studies [18,35,36]. These findings may be explained by the variations in the soil temperature and moisture [37], as well as differences in the community of straw decomposers [13], at the different straw incorporation depths used in this study.

The changing trend of C release among all treatments was extremely similar to the overall response observed in straw decomposition. On average, the C release proportions found in 2017 were higher than the measurements in 2018, with respective ranges of 62.96–71.62% and 45.08–66.07%. Similar to the C release derived from the decay of maize straw, the N release proportions from maize residue under the different treatments in 2017 were generally higher than the results in 2018, with ranges of 38.32–46.33% and 11.25–37.11%, respectively (Figure 4). This discrepancy implies that straw-derived N release during the decay process was dependent on annual conditions at various straw returning depths, whereas C release from straw incorporation was not as affected. Straw decomposition and the C and N release proportions were higher at shallower returning depths, which is primarily because deep tillage creates conditions that are more anaerobic compared with the conditions in shallow tillage, and anaerobes decompose less straw-derived C than aerobes [37].

4.2. Soil Organic Carbon (SOC), Soil Total Nitrogen (STN), and C–N Ratio

Straw distribution in the plow layer can change with different incorporation strategies; accordingly, straw distribution affects the spatial distributions of SOC and STN [38]. In this study, the SOC and STN contents were significantly high in soil layers close to the location of straw incorporation, regardless of the straw returning method (Figures 5 and 6). These findings are in agreement with an earlier study [25] that reported that farmland SOC and STN concentrations throughout the soil profile depended on the straw placement from the straw incorporation practice. In addition, Turmel et al. [8] pointed out that the SOC and STN levels can be controlled by organic matter inputs, native SOM decomposition rates, or both. We found that the C and N release levels were higher in treatments with straw incorporation at the shallowest depth (SM/B + O), and conversely, the C and N release levels were lower when straw incorporation was deeper (SM/B + F) (Figures 3 and 4). On the other hand, deep straw incorporation practices cause greater soil disturbance, resulting in increased native SOM mineralization relative to shallow straw treatments [39]. Thus, the C and N released from straw and native soil organic matter mineralization likely contributed to the distinctive SOC and STN behaviors between straw returning depth treatments in different soil layers, especially in the 0–20 cm and 40–60 cm soil layers.

The soil C–N ratio affects C and N cycling in an ecosystem, C and N interactions, and the stability of SOM in the soil profile [40,41]. Similar to SOC and STN, the soil C–N ratio in the different soil layers was significantly affected by straw returning depths; furthermore, the ratio was generally higher in soil layers close to the location of straw incorporation. These findings for the soil C–N ratio are in accord with the results reported in a previous study [38], which suggested that crop residues are conducive to an improvement in the soil C–N ratio. Such improvements may be explained by the higher straw-derived C release compared with N release in the present study (Figures 3 and 4). Additionally, the increasing C sequestration and net N mineralization that generally occur in a soil layer with a high C–N ratio due to straw incorporation may account for the difference in the soil C–N ratio between the straw returning treatments in the different soil layers investigated [27,42].

4.3. Soil Organic Carbon (SOC) and Soil Total Nitrogen (STN) Stocks

Similar to the SOC and STN concentrations reported above, SOC and STN stocks were significantly increased in soil layers close to the location of straw incorporation in both years of the study period. Furthermore, with the increasing depth in the overall 0–50 cm soil profile, both SOC and STN stocks under the SM/B + O treatments gradually diminished compared with those under SM/B + T and SM/B + F treatments. This result is supported by findings in previous studies [29,43,44] and mainly attributed to the higher straw-derived C and N release rates caused by the increased SOC and STN concentration associated with the SM/B + O treatments: thus, stocks in the upper soil improve, regardless of the changes in the soil bulk density (Figure S1). In addition, SOC and STN stocks tend to be uniformly distributed throughout the whole soil profile when straw incorporation is deeper, which is in accord with results from a previous study [26]. However, for the 0–50 cm soil profile, SOC stocks associated with SM/B + F were generally lower than the two treatments with a shallower straw return. This suggests that straw incorporation at depths greater than 30 cm may not favor C sequestration. Moreover, in both years, markedly higher SOC stocks were obtained using the SB returning method, indicating that the residue burying practice is a possible alternative method for straw incorporation in the present study area. These findings are likely related to another observation in this study: Straw incorporated into the deep subsoil layer was associated with accelerated SOM mineralization due to enhanced soil profile disturbance [39] and straw decomposition under poorly available nutrient conditions [45,46].

4.4. Relationships Between Straw Decomposition and Soil Organic Carbon (SOC) and Soil Total Nitrogen (STN) Concentrations

To provide further insight into the reason for the relatively low SOC stocks resulting from straw incorporation at a soil depth greater than 30 cm, we analyzed the relationship between the straw decay proportion (SD) and the SOC and STN concentrations in different soil layers using the overall data for both years. Similar changing trends of SOC and STN concentrations were correlated with the straw decay rate in the different soil layers (Figures 7 and 8). The correlation was positive and linear in the 0–20 cm soil layers but negative and linear in the 30–60 cm layers, while a parabolic relationship was revealed for the 20–30 cm soil layer. This strongly suggests the process of fresh straw input and decomposition generated priming effects (PEs), which define the changes in the SOM decomposition resulting from the addition of organic or mineral substances to the soil [47]. In a previous short-term study, in the topsoil (0–20 cm), negative PEs with C and N immobilization were usually found because microbes were provided with sufficient and available nutrients for decomposition [48]; therefore, the SOC and STN content increased as the SD increased. However, in the subsoil (30–60 cm), positive PEs might have resulted from the straw decaying process, especially in the late stage, during which, more recalcitrant C-derived from straw is decomposed in anaerobic conditions [37], and fewer nutrients are available [14], resulting in a trend that is opposite to that in the topsoil. These results suggest that the nutrient level in the subsoil needs to be considered when practicing deep straw incorporation in farmland.

5. Conclusions

When straw was mixed with soil, the straw decomposition proportion and C and N release markedly declined as the returning depth increased; however, they were variable between the two years when the straw was buried in the soil at different depths. Moreover, maize straw incorporated into the soil at 10, 30, and 50 cm depths tended to increase SOC and STN concentrations and the soil C–N ratio in the 0–20, 20–30, and 40–60 cm soil layers, regardless of the straw returning method employed. Thus, SOC and STN stocks in the 0–20 cm soil layers increased in the shallow straw returning approaches (SM/B + O), but straw buried at a depth of 50 cm strongly enriched them in the deep soil layers (30–50 cm). In the 0–50 cm soil profile, the highest SOC stocks were 77.38 Mg ha^{-1} in 2017 and 77.14 Mg ha^{-1} in 2018, which were both obtained using the SB + T treatment. Burying straw in the soil significantly increased the SOC stocks compared with mixing the straw with the soil. The STN stock was also significantly affected by the straw returning depth, but the effect was inconsistent between the two years. Interesting results for the interactions between the straw decomposition proportion and SOC and STN were found: They were positively correlated in the 0–20 cm soil layer but negatively correlated in the 30–60 cm soil layers. Taken together, our results indicate that straw could be incorporated into the soil, in practice, through plow tillage in northeastern China, and the incorporation depths not exceeding 30 cm may be beneficial for sustainable maize production systems considering soil quality conservation.

Supplementary Materials: The following are available online at http://www.mdpi.com/2073-4395/9/12/818/s1, Figure S1: Changes in soil bulk density in the 0–60 cm profile under different straw returning approaches in 2017 and 2018. SM and SB indicate straw mixed with soil and buried in the soil; O, T, and F indicate straw incorporated into the soil (mixed or buried) at depths of 10, 30, and 50 cm, respectively. SM_{mean} is the mean of SM + O, SM + T, and SM + F; SB_{mean} is the mean of SB + O, SB + T, and SB + F; * indicates significant differences ($P < 0.05$) between straw incorporation depths (treatments) in the same soil layer.

Author Contributions: Designed and performed most of the experiments, analyzed the data and wrote the manuscript, P.T., H.Q., and Y.J.; helped with the data collection and analysis, editing of the manuscript as well, H.L., Z.W., G.M., Y.S. (Yue Sun), Y.W., Y.S. (Yehan Su), Z.M. and P.S. All authors read and approved the manuscript.

Funding: This research was funded by the Special Fund for Agro-scientific Research in the Public Interest (201503116); The National Key Research and Development Program of China (2016YFD0300801, 2016YFD0300103).

Conflicts of Interest: The authors declare no conflict of interest. The funders had no role in the design of the study; in the collection, analyses, or interpretation of data; in the writing of the manuscript, or in the decision to publish the results.

References

1. FAO Soil Health. *Technologies that Save and Grow*; FAO: Rome, Italy, 2011.
2. Melero, S.; López-Garrido, R.; Murillo, J.M.; Moreno, F. Conservation tillage: Short- and long-term effects on soil carbon fractions and enzymatic activities under Mediterranean conditions. *Soil Tillage Res.* **2009**, *104*, 292–298. [CrossRef]
3. Kaisuer, J. Wounding Earth's fragile skin. *Science* **2004**, *304*, 1616–1618. [CrossRef] [PubMed]
4. Yin, X.G.; Olesen, J.E.; Wang, M.; Kersebaum, K.C.; Chen, H.; Baby, S.; Öztürk, I.; Chen, F. Adapting maize production to drought in the Northeast Farming Region of China. *Eur. J. Agron.* **2016**, *77*, 47–58. [CrossRef]
5. Liu, X.B.; Zhang, X.Y.; Herbert, S.J. Feeding China's growing needs for grain. *Nature* **2010**, *465*, 420. [CrossRef] [PubMed]
6. Freibauer, A.; Rounsevell, M.D.A.; Smith, P.; Verhagen, J. Carbon sequestration in the agricultural soils of Europe. *Geoderma* **2004**, *122*, 1–23. [CrossRef]
7. Lu, F.; Wang, X.K.; Han, B.; Ou Yang, Z.Y.; Duan, X.N.; Zheng, H.; Miao, H. Soil carbon sequestrations by nitrogen fertilizer application, straw return and no-tillage in China's cropland. *Glob. Change Biol.* **2009**, *15*, 281–305. [CrossRef]
8. Turmel, M.S.; Speratti, A.; Baudron, F.; Verhulst, N.; Govaerts, B. Crop residue management and soil health: A systems analysis. *Agric. Syst.* **2015**, *134*, 6–16. [CrossRef]
9. Zhang, X.X.; Ma, F. Emergy Evaluation of Different Straw Reuse Technologies in Northeast China. *Sustainability* **2015**, *7*, 11360–11377. [CrossRef]
10. Tian, P.; Jiang, Y.; Sun, Y.; Ma, Z.Q.; Sui, P.X.; Mei, N.; Qi, H. Effect of straw return methods on maize straw decomposition and soil nutrients contents. *Chin. J. Eco-Agric.* **2019**, *27*, 100–108, (in Chinese with English abstract).
11. Helgason, B.L.; Gregorich, E.G.; Janzen, H.H.; Ellert, B.H.; Lorenz, N.; Dick, R.P. Long-term microbial retention of residue C is site-specific and depends on residue placement. *Soil Biol. Biochem.* **2014**, *68*, 231–240. [CrossRef]
12. Cookson, W.R.; Beare, M.H.; Wilson, P.E. Effects of prior crop residue management on microbial properties and crop residue decomposition. *Appl. Soil Ecol.* **1998**, *7*, 179–188. [CrossRef]
13. Osono, T. Role of phyllosphere fungi of forest trees in the development of decomposer fungal communities and decomposition processes of leaf litter. *Can. J. Microbiol.* **2006**, *52*, 701–716. [CrossRef] [PubMed]
14. Fontaine, S.; Barot, S.; Barré, P.; Bdioui, N.; Mary, B.; Rumpel, C. Stability of organic carbon in deep soil layers controlled by fresh carbon supply. *Nature* **2007**, *450*, 277–280. [CrossRef] [PubMed]
15. Yadvinder-Singh; Gupta, R.K.; Jagmohan-Singh; Gurpreet-Singh; Gobinder-Singh; Ladha, J.K. Placement effects on rice residue decomposition and nutrient dynamics on two soil types during wheat cropping in rice–wheat system in northwestern India. *Nutr. Cycl. Agroecosyst.* **2010**, *88*, 471–480. [CrossRef]

16. Curtin, D.; Selles, F.; Wang, H.; Campbell, C.A.; Biederbeck, V.O. Carbon dioxide emissions and transformation of soil carbon and nitrogen during wheat straw decomposition. *Soil Sci. Soc. Am. J.* **1998**, *62*, 1035–1041. [CrossRef]

17. Gómez-Muñoz, B.; Hatch, D.J.; Bol, R.; García-Ruiz, R. Nutrient dynamics during decomposition of the residues from a sown legume or ruderal plant cover in an olive oil orchard. *Agric. Ecosyst. Environ.* **2014**, *184*, 115–123. [CrossRef]

18. Xu, Y.H.; Chen, Z.M.; Fontaine, S.; Wang, W.J.; Luo, J.F.; Fan, J.L.; Ding, W.X. Dominant effects of organic carbon chemistry on decomposition dynamics of crop residues in a Mollisol. *Soil Biol. Biochem.* **2017**, *115*, 221–232. [CrossRef]

19. Bradford, M.A.; Berg, B.; Maynard, D.S.; Wieder, W.R.; Wood, S.A. Understanding the dominant controls on litter decomposition. *J. Ecol.* **2016**, *104*, 229–238. [CrossRef]

20. Reicosky, D.C. *Tillage-induced CO2 emissions and carbon sequestration: Effect of secondary tillage and compaction*; Kluwer Academic Publishers: Dordrecht, The Netherlands, 2003; pp. 291–300.

21. Yang, X.M.; Wander, M.M. Tillage effects on soil organic carbon distribution and storage in a silt loam soil in Illinois. *Soil Tillage Res.* **1999**, *52*, 1–9. [CrossRef]

22. Jantalia, C.P.; Resck, D.V.S.; Alves, B.J.R.; Zotarelli, L.; Urquiaga, S.; Boddey, R.M. Tillage effect on C stocks of a clayey Oxisol under a soybean-based crop rotation in the Brazilian Cerrado region. *Soil Tillage Res.* **2007**, *95*, 97–109. [CrossRef]

23. Dong, W.X.; Hu, C.S.; Chen, S.Y.; Zhang, Y.M. Tillage and residue management effects on soil carbon and CO_2 emission in a wheat–corn double-cropping system. *Nutr. Cycl. Agroecosyst.* **2009**, *83*, 27–37. [CrossRef]

24. Poirier, V.; Angers, D.A.; Rochette, P.; Chantigny, M.H.; Ziadi, N.; Tremblay, G.; Fortin, J. Interactive effects of tillage and mineral fertilization on soil carbon profiles. *Soil Sci. Soc. Am. J.* **2009**, *73*, 255–261. [CrossRef]

25. Du, Z.L.; Ren, T.S.; Hu, C.S. Tillage and Residue Removal Effects on Soil Carbon and Nitrogen Storage in the North China Plain. *Soil Sci. Soc. Am. J.* **2010**, *74*, 196–202. [CrossRef]

26. Dikgwatlhe, S.B.; Chen, Z.D.; Lal, R.; Zhang, H.L.; Chen, F. Changes in soil organic carbon and nitrogen as affected by tillage and residue management under wheat–maize cropping system in the North China Plain. *Soil Tillage Res.* **2014**, *144*, 110–118. [CrossRef]

27. Laird, D.A.; Chang, C.W. Long-term impacts of residue harvesting on soil quality. *Soil Tillage Res.* **2013**, *134*, 33–40. [CrossRef]

28. Vanden Bygaart, A.J.; Angers, D.A. Towards accurate measurements of soil organic carbon stock change in agroecosystems. *Can. J. Soil Sci.* **2006**, *86*, 465–471. [CrossRef]

29. Xue, J.F.; Pu, C.; Liu, S.L.; Chen, Z.D.; Chen, F.; Xiao, X.P.; Lal, R.; Zhang, H.L. Effects of tillage systems on soil organic carbon and total nitrogen in a double paddy cropping system in Southern China. *Soil Tillage Res.* **2015**, *153*, 161–168. [CrossRef]

30. You, D.B.; Tian, P.; Sui, P.X.; Zhang, W.K.; Yang, B.; Qi, H. Short-term effects of tillage and residue on spring maize yield through regulating root-shoot ratio in Northeast China. *Sci. Rep.* **2017**, *7*, 13314. [CrossRef]

31. Jin, X.X.; An, T.T.; Aaron, R.G.; Li, S.Y.; Timothy, F.; Wang, J.K. Enhanced conversion of newly-added maize straw to soil microbial biomass C under plastic film mulching and organic manure management. *Geoderma* **2018**, *313*, 154–162. [CrossRef]

32. Varela, M.F.; Scianca, C.M.; Taboada, M.A.; Rubio, G. Cover crop effects on soybean residue decomposition and P release in no-tillage systems of Argentina. *Soil Tillage Res.* **2014**, *143*, 59–66. [CrossRef]

33. Mei, N.; Yang, B.; Tian, P.; Jiang, Y.; Sui, P.X.; Sun, D.Q.; Zhang, Z.P.; Qi, H. Using a modified soil quality index to evaluate densely tilled soils with different yields in Northeast China. *Environ. Sci. Pollut. Res.* **2019**, *26*, 13867–13877. [CrossRef] [PubMed]

34. Ellert, B.H.; Bettany, J.R. Calculation of organic matter and nutrients stored in soils under contrasting management regimes. *Can. J. Soil Sci.* **1995**, *75*, 529–538. [CrossRef]

35. Wang, X.Y.; Sun, B.; Mao, J.D.; Sui, Y.Y.; Cao, X.Y. Structural convergence of maize and wheat straw during two-year decomposition under different climate conditions. *Environ. Sci. Technol.* **2012**, *46*, 7159–7165. [CrossRef] [PubMed]

36. Grandy, A.S.; Salam, D.S.; Wickings, K.; McDaniel, M.D.; Culman, S.W.; Snapp, S.S. Soil respiration and litter decomposition responses to nitrogen fertilization rate in no-till corn systems. *Agric. Ecosyst. Environ.* **2013**, *179*, 35–40. [CrossRef]

37. Devêvre, O.C.; Horwáth, W.R. Decomposition of rice straw and microbial carbon use efficiency under different soil temperatures and moistures. *Soil Biol. Biochem.* **2000**, *32*, 1773–1785. [CrossRef]
38. Puget, P.; Lal, R. Soil organic carbon and nitrogen in a Mollisol in central Ohio as affected by tillage and land use. *Soil Tillage Res.* **2005**, *80*, 201–213. [CrossRef]
39. Hiel, M.P.; Barbieux, S.; Pierreux, J.; Olivier, C.; Lobet, G.; Roisin, C.; Garré, S.; Colinet, G.; Bodson, B.; Dumont, B. Impact of crop residue management on crop production and soil chemistry after seven years of crop rotation in temperate climate, loamy soils. *PeerJ* **2018**, *6*, e4836. [CrossRef]
40. Russell, A.E.; Laird, D.A.; Parkin, T.B.; Mallarino, A.P. Impact of Nitrogen Fertilization and Cropping System on Carbon Sequestration in Midwestern Mollisols. *Soil Sci. Soc. Am. J.* **2005**, *69*, 413–422. [CrossRef]
41. Tong, C.L.; Xiao, H.A.; Tang, G.R.; Wang, H.Q.; Huang, T.P.; Xia, H.A.; Keith, S.J.; Li, Y.; Liu, S.L.; Wu, J.S. Long-term fertilizer effects on organic carbon and total nitrogen and coupling relationships of C and N in paddy soils in subtropical China. *Soil Tillage Res.* **2009**, *106*, 8–14. [CrossRef]
42. Kramer, S.; Marhan, S.; Haslwimmer, H.; Ruess, L.; Kandeler, E. Temporal variation in surface and subsoil abundance and function of the soil microbial community in an arable soil. *Soil Biol. Biochem.* **2013**, *61*, 76–85. [CrossRef]
43. Lou, Y.L.; Xu, M.G.; Chen, X.N.; He, X.H.; Zhao, K. Stratification of soil organic C, N and C:N ratio as affected by conservation tillage in two maize fields of China. *Catena* **2012**, *95*, 124–130. [CrossRef]
44. Maillard, É.; McConkey, B.G.; Luce, M.S.; Angers, D.A.; Fan, J.L. Crop rotation tillage system and precipitation regime effects on soil carbon stocks over 1 to 30 years in Saskatchewan, Canada. *Soil Tillage Res.* **2018**, *177*, 97–104. [CrossRef]
45. Hao, M.D.; Fan, J.; Wang, X.R.; Pen, L.F.; Lai, L. Effect of Fertilization on Soil Fertility and Wheat Yield of Dryland in the Loess Plateau. *Pedosphere* **2005**, *15*, 189–195.
46. Sá, J.C.M.; Lal, R. Stratification ratio of soil organic matter pools as an indicator of carbon sequestration in a tillage chronosequence on a Brazilian Oxisol. *Soil Tillage Res.* **2009**, *103*, 46–56. [CrossRef]
47. Kuzyakov, Y.; Friedel, J.K.; Stahr, K. Review of mechanisms and quantification of priming effects. *Soil Biol. Biochem.* **2000**, *32*, 1485–1498. [CrossRef]
48. Tian, Q.X.; Yang, X.L.; Wang, X.G.; Liao, C.; Li, Q.X.; Wang, M.; Wu, Y.; Liu, F. Microbial community mediated response of organic carbon mineralization to labile carbon and nitrogen addition in topsoil and subsoil. *Biogeochemistry* **2016**, *128*, 125–139. [CrossRef]

 *agronomy*

Article

Soybean in No-Till Cover-Crop Systems

Mosab Halwani [1,*], Moritz Reckling [1,2], Johannes Schuler [1], Ralf Bloch [1,3] and Johann Bachinger [1]

[1] Leibniz Centre for Agricultural Landscape Research, 15374 Müncheberg, Germany;
    moritz.reckling@zalf.de (M.R.); schuler@zalf.de (J.S.); bloch@zalf.de (R.B.); Jbachinger@zalf.de (J.B.)

[2] Department of Crop Production Ecology, Swedish University of Agricultural Sciences,
    75007 Uppsala, Sweden

[3] Faculty of Landscape Management and Nature Conservation,
    Eberswalde University for Sustainable Development, 16225 Eberswalde, Germany

* Correspondence: Mosab.Halwani@zalf.de

Received: 31 October 2019; Accepted: 10 December 2019; Published: 13 December 2019

Abstract: Introducing agro-ecological techniques such as no-tillage systems with cover crops in rotations with soybean (*Glycine max* (L.) Merr.) could provide more resilience to changing climatic conditions and, at the same time, reduce soil erosion, nitrate leaching, and weed density in the main crop. However, there are challenges in introducing no-tillage techniques in crop systems in Europe as there is little quantitative knowledge about the agro-economic impact. The objectives of this study were to evaluate the agronomic and economic impacts of three soybean cropping systems involving a rye (*Secale cereal* L.) cover crop prior to soybean, i.e., two no-tillage systems; either herbicide-free with crimping the rye or herbicide-based without rye crimping and one plough-based in which rye was cut as green silage. The impacts of these cropping strategies were compared in a three-year cropping system experiment at a research station in north-eastern Germany with and without irrigation. The following parameters were measured: (1) cover crop biomass; (2) weed biomass; (3) soybean plant density; (4) soybean grain yield; and (5) gross margin of the cropping system. The results showed that all three soybean cropping systems can effectively suppress weeds. System (C), the no-tillage herbicide-based system, produced the lowest rye biomass and highest soybean yield; system (B), the no-tillage herbicide-free/crimped rye system, produced the highest rye biomass and lowest soybean yield compared to system (A), the standard cutting/plough-based system. The differences in rye biomass and soybean yield observed between the three systems could be mainly attributed to the timing of the cover crop termination and the soybean sowing date. The gross margin was highest in system (C), due to the high soybean grain yield. The low soybean grain yield in system (B) resulted in lower revenues and gross margins compared to systems (A) and (C), although system (B) could be economically attractive in organic farming with higher prices for organic soybean. In the particularly dry year 2016, gross margins were higher when soybean was irrigated compared to the rainfed cultivation, due to significantly higher grain yields. Before recommending the application of the no-tillage with cover crop technique for the conditions tested in north-eastern Germany, more investigations on the benefits and risks of this technique are needed. Further research needs to focus on maintaining a high rye biomass as well as on ensuring an early soybean planting date. Optimizing the crimping and drilling equipment is still required in order to develop good management practices for no-tillage herbicide-free systems in European conditions.

Keywords: no-tillage; cover crop; irrigation; weed suppression; gross margin

---

1. Introduction

Climate change has adverse effects on crop production in Europe [1]. Adaption to climate change is crucial especially for grain legumes [2] and other spring crops that tend to have low yield stability [3].

Adaptation options include adjusting the crop management, choice of cultivars and design of the cropping system [4,5]. Northern Germany is expected to be affected by climate change including an increase in average annual temperature, shifts in seasonality of precipitation (decreased in summer, increased in winter), shifts in potential evaporation, increasing incidence of milder winters, an increased frequency of periods of drought and heavy precipitation, and an extended vegetation period [6].

As a consequence, the cool-season grain legumes such as lupins, field peas and faba beans are likely to respond with a reduced yield stability that has been observed in long-term field experiments over the last 60 years [7] and will not be able to take advantage of the higher temperatures and increase in vegetation period. Consequently, soybean (*Glycine max* (L.) Merr.) as a warm-season grain legume could benefit from these higher temperatures and longer vegetation periods in central Europe and provide an alternative to the cool-season grain legumes [8,9]. The cultivation of soybean in Europe is done as a summer annual grain crop using short-day adapted varieties. Late sowing dates of soybean in May means that a cover crop could be included prior to soybean planting to cover the soil during winter. However, little is known on cover crops used before grain legume sowing.

The main driver for growing cover crops prior to grain legumes, besides reducing wind and water erosion and increasing soil organic matter, is weed suppression before and during the growing season [10]. This is important since soybean is very sensitive to weed infestation especially in low-input and organic farming [11]. The mechanisms of weed suppression by cover crops are diverse and depend, besides other factors, on the termination method. A common approach uses broad spectrum herbicides based on glyphosate and atrazine to terminate the cover crops. Mechanical cutting and grazing can also be used to kill cover crops. A novel approach for cover crop termination is crimping, using a roller-crimper [12]. With all these approaches, the cover crop mulch layer inhibits the weed physically (due to shading), chemically (allelopathy) or by direct competition for resources [13].

Prevailing production systems couple cover crops with conservation tillage (CT) to enhance the ecological benefits e.g., no-tillage cover-crop systems after legume crops is the most common conservation agriculture (CA) application known to optimize utilization of nitrogen, thereby reducing nitrate leaching [14]. The use of cover crops in CA adds also further environmental benefits such as soil erosion control, increases soil organic carbon and thus enhancing soil fertility [15] as well as improvement of soil properties, microflora and fauna [16].

While such no-tillage cover-crop systems are more common for growing soybean in the US [17,18], their agronomic and economic potentials and limitations have not been investigated under Central European conditions except for the study by Weber et al. [12] in organic farming. To investigate the agro-economic potential of the novel no-tillage cover-crop systems in soybean, a three year field experiment was established on sandy soils in north-eastern Germany. The objectives were to assess the agronomic effects of the different cover crop systems on (i) soybean plant density, (ii) cover crop biomass, (iii) weed suppression and (iv) soybean grain yield as well as (v) the related economic opportunities and limitations of growing soybean in no-tillage cover-crop systems.

2. Materials and Methods

2.1. Study Site

The study was implemented in north-eastern Germany in the federal state of Brandenburg. A field experiment was conducted at the research station of the Leibniz Centre for Agricultural Landscape Research (ZALF) in Müncheberg (52°31′ N, 14°7′ E, 60 m.; NE Germany). Soils in the research station are from glacial deposits and are predominantly sandy loams and loamy sands with a high spatial heterogeneity containing on average 61% sand, 27% silt and 12% clay. The soil pH (KCl) ranges between 6.1 and 6.9, the total soil carbon between 4 g kg^{-1} and 7 g kg^{-1} and the plant available water holding capacity is estimated at 150 mm in the rooting zone of 1 m. During the last 40 years the climate measured in the experimental field in Müncheberg showed that the mean annual temperature increased from 8.5 °C in the period 1970–2000 to 9.0 °C in the period of 1980–2010. There was also an increase in the annual precipitation rate in the same period from 521 mm p.a. to 565 mm p.a. However, this increase was combined with a shift in the precipitation distribution with more precipitation in the winter months (Jan., Feb.) and less in April and more heavy precipitation events in the summer months [19]. During the study period from 2015–2017, the annual precipitation varied from 451 mm in the dry year in 2016 to 635 mm in 2017. Monthly rainfall totals and average temperatures for each year are presented in the results section.

The study took place on three different fields over three successive growing seasons, starting in fall 2014 with the establishment of winter rye (*Secale cereal* L.) as a cover crop. The experiments were designed by scientists, influenced by farmers' interests and constraints in managing weeds effectively without additional weeding or herbicide applications, following Reckling et al. [5].

2.2. Design of the Field Experiments

All treatments were arranged in a split-plot design with four replications. Whole plots consisted of cropping systems while subplots were irrigation (with/without) that was omitted in 2015 and 2017, when precipitation was sufficient and irrigation was unnecessary. Individual plots were 8.5 m × 12 m in size.

Three cropping systems were compared:

- System (A)—a standard "cutting/plough-based system", in which the previous rye cover crop was cut and removed, and plots were tilled immediately to 23 cm deep by reversible plow combined with packer to level the seedbed in spring (9–17 May). Then soybean was sown with a standard drill at the optimal time (11–19 May).
- System (B)—a "no-tillage herbicide-free system", where the previous rye cover crop was crimped at rye flowering (18–27 May), left on the field and soybean sown with a direct seeding drill immediately, without any later chemical regulation of weeds.
- System (C)—a "no-tillage herbicide-based system", in which the previous rye cover crop was killed with a broad-spectrum systemic herbicide (glyphosate) at the end of stem extension period (11–29 April) when the most mature rye was at booting stage (BBCH: 31); with no further soil tillage, soybean sown with a direct seeding drill at the optimal sowing time (11–17 May, similar to system (A)) and with subsequent application of pre-emergent and post-emergent herbicides.

All soybean crops followed a winter rye cover crop that was established in autumn (27 Sept–10 Oct) using a standard drill in 10 cm rows at 250 seeds m^{-2}. In 2017, the seeding rate was increased to 350 seeds m^{-2}. For weed management a single broadcast application for pre-emergence herbicides was applied (Table 1).

The soybean cultivar Merlin (maturity group 000) was planted at a rate of 65 seeds m^{-2} in 2015 and 80 seeds m^{-2} in 2016 and 2017. Adjustments were needed to increase the rate of emergence, especially in the system where the cover crop was crimped. All three systems were tested with and without irrigation. Irrigation was applied as needed, depending on rainfall events. Irrigation amounts and timing were modelled with the Web-BEREST irrigation software. Web-BEREST calculates the irrigation water based on the crop demand using the coefficient of actual to potential evapotranspiration. Detailed information on the model assumptions and equations are provided by Mirschel et al. [20].

Table 1. Schedule of field operations, in Müncheberg, from 2015 to 2017.

Field Operation		2015	2016	2017
Rye Sowing Date				
All cropping systems	Date	10.10.2014	29.09.2015	27.09.2016
	Rate (seeds m^{-2})	250	250	350
Rye termination				
System (A)	Date	11 May	17 May	09 May
	Forage harvester		Fortschritt E 281	
System (B)	Date	27 May	24 May	18 May
	Crimper		MaxiCut 600	
System (C)	Date	29 April	22 April	11 April
	Sprayer		Hardi Ranger Pro VHP	
Soybean inoculation				
All cropping systems	Preparation	HISTICK® Soy (*Bradyrhizobium japonicum*)		
	Rate (g 100kg^{-1})	400		
Fertilization				
Systems (A) & (C)	Date	20 March	01 April	27 March
	Triple Super Phosphate (46%)	100 kg ha^{-1}	100 kg ha^{-1}	100 kg ha^{-1}
	K$_2$O as potassium chloride (60%)	100 kg ha^{-1}	100 kg ha^{-1}	100 kg ha^{-1}
	Urea Ammonium nitrate and ammonium thiosulphate	230 L ha^{-1}	–	183 L ha^{-1}
Soybean planting				
System (A)	Date	13 May	19 May	11 May
	Rate (seeds m^{-2})	65	80	80
System (B)	Date	28 May	24 May	22 May
	Rate (seeds m^{-2})	65	80	80
System (C)	Date	13 May	17 May	11 May
	Rate (seeds m^{-2})	65	80	80
Pre-emergence herbicide application (soybean)				
Systems (A) & (C)	Date	18 May	20 May	12 May
	Metribuzin (700 g kg^{-1})		400 g ha^{-1}	
	Rate (L ha^{-1})	300	400	400
Soybean harvest				
	System (A)	05 Oct	05 Oct	21 Sep
	System (B)	22 Oct	05 Oct	17 Oct
	System (C)	05 Oct	22 Sep	21 Sep

2.3. Data Collection

Biomass of cover crop was measured in April for system (C) and May for systems (A) and (B), prior to termination, by clipping plants at the soil level from a single 0.5 m^2 quadrat per replication.

Plants were dried at 65 °C for 48 h and weighed to determine dry weight per plot. Soybean density was measured 12 days after planting, and soybean/weed stand was measured in late July, by clipping all plants within a single 0.5 m^2 quadrat for each replication. Soybean and weeds plants were separated to determine the fresh and dry weight. Soybean yield was determined between 5–22 October through the machine-harvesting operation of mature plants within 15–17 m^2. System (B) was omitted in 2015 when soybean density data were analyzed, due to difficulty in seeding through thick rye residue and poor plant stand.

A simple economic calculation was performed. Revenues were calculated for each cropping system i.e., the rye biomass and soybean grain yield and producer prices for both rye biomass and soybean grain in Germany from 2018. The variable costs included all inputs i.e., seeds, inoculum, fertilizers, herbicides and operations for soil tillage, sowing, fertilizer application, rye termination, harvesting, irrigation and/or harvesting of rye green biomass. The price for conventional soybean grain was 350 € t^{-1} and the price of organic soybean grain was 800 € t^{-1} in 2018, with 71% higher seed costs for organic than for conventional seeds. The price for rye biomass was valued at 78.6 € t^{-1}.

The rye residues that were crimped or killed by herbicide and which remained on the field after the termination were not assigned any economic value.

2.4. Statistical Analysis

Data were subject to two-way ANOVA, using the SAS® 9.2 PROC MIXED procedure (SAS Institute Inc., Cary, NC, USA), after satisfying the assumptions for normality and homogeneity of variance. When an interaction was not significant, it was removed from the model, and the data were reanalyzed without the interaction.

Two-way-ANOVA, followed by Tukey's post hoc test, was performed to compare the main effects of cropping system and year for the plots without irrigation and the interaction effect between cropping system and year on soybean yield, rye and weed biomass.

Additionally in 2016, Two-way-ANOVA, followed by classical t-tests between different treatments within irrigation and followed by Tukey's post hoc test between different treatments within cropping system, was performed to compare the main effects of cropping system and irrigation and the interaction effect between cropping system and irrigation on soybean yield and weed biomass. The cropping systems included three types (systems (A), (B), and (C)), data were collected for three years (2015, 2016, and 2017), and two irrigation conditions were considered (with and without irrigation).

Linear regression was performed to identify potential relationships between soybean grain yields and weed biomass under all three cropping systems. Additionally, regressions were used to identify potential effects of rye biomass in system (B) on soybean yield and weed biomass. The Pearson's correlation coefficients were calculated using the software SAS® 9.2 PROC REG procedures for multivariate data analysis (SAS Institute Inc., Cary, NC, USA).

3. Results

3.1. Climate

Cumulative precipitation during the 5-month growing season (May-September) was at least 12% higher than the 25-year average of 308 mm in 2017, but only 70% of the 25-year average in 2016, and only 64% of the 25-year average in 2015. Moreover, distribution of precipitation did not favor soybean germination in 2016, since less than 5 mm was received during May, this lower amount was only 9% of precipitation received in 25-year average of 55 mm.

Distribution of precipitation during the growing season also favored soybean seed production in 2017, with greater-than-average amounts of precipitation occurring during June and July when the majority of soybean vegetative and reproductive growth typically occurs in northeastern Germany.

Monthly growing-season temperatures were near the 25-year average of 16.3 °C in 2015 and 2017, 1.3 °C warmer than the 25-year average in 2016. Temperatures in June and July were comparable to the 25-year average in three experimental years (Table 2).

3.2. Soybean Plant Density

There was a significant interactions for soybean plant density between cropping system and year ($p = 0.0003$) (Figure 1). In 2017, soybean densities differed with the cropping system, with significantly higher soybean densities at system (A) compared to system (B) and system (C). In 2016 and 2015, higher soybean densities were found in system (C) compared to system (A) and system (B), however, these differences were not significant. In all systems, soybean densities in 2016 were reduced on average by 50% from the target plant density and ranged from 30 to 40 plants m^{-2}. Soybean density in 2017 achieved consistent soybean densities between 60 and 75 plants m^{-2} (Figure 1).

Table 2. Mean temperature, precipitation and irrigation amounts (in parentheses) from 2015–2017.

	Mean Temperature (°C)				Precipitation (mm)			
	2015	2016	2017	25-Year Mean	2015	2016	2017	25-Year Mean
January	1.3	−0.8	−1.6	−0.1	67.0	32.0	25.7	40.3
February	5.4	3.5	1.5	1.1	4.7	39.4	26.8	29.4
March	8.6	4.2	6.8	4.0	39.3	22.7	41.5	36.4
April	12.7	8.4	7.8	9.1	17.4	18.8	21.7	29.7
May	16.1	15.5	14.3	13.8	21.9	4.9 (15)	41.8	55.4
June	19.3	18.4	17.6	16.6	42.1	91.2	110.1	56.3
July	21.5	19.3	18.4	18.9	47.7	52.3 (75)	114.6	81.4
August	13.9	17.7	18.6	18.2	18.0	45.1 (10)	55.6	64.9
September	7.9	16.9	13.3	14.0	66.7	20.4	24.0	49.8
October	6.9	8.4	11.2	9.2	55.1	48.7	72.7	40.1
November	6.5	3.4	5.44	4.3	66	32.5	67.5	39.0
December	1.3	2.4	3.2	1.2	27.0	42.9	32.9	36.7
Annual	10.2	9.8	9.7	9.2	472.9	450.9	634.9	559.6

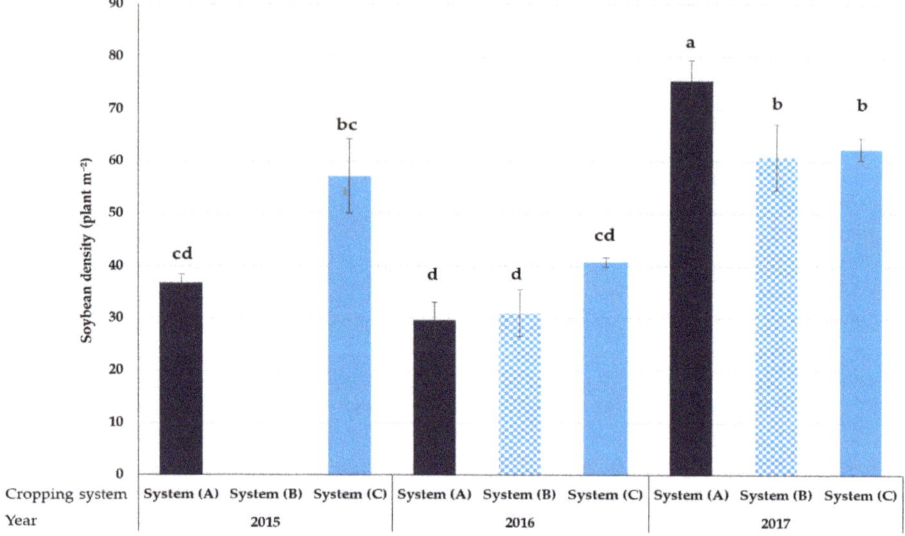

Figure 1. Interaction effect of cropping system on soybean density at Müncheberg, Germany, in 2015 to 2017. (p = 0.0003). Abbreviations: System (A), Cutting/plough-based system; System (B), No-tillage herbicide-free system; System (C), No-tillage herbicide-based system. Standard error bars are presented (n = 8). Means in the same year followed by the same letter are not significantly different at $p \leq 0.05$. The soybean density from system (B) in 2015 was excluded in data analysis because there were no plants to sample (see Section 2.3. Data collection).

3.3. Cover Crop Biomass

A two-way analysis of variance was conducted on the influence of cropping system and year on rye biomass. Significant interactions were detected between year and cropping systems ($p < 0.001$) (Figure 2).

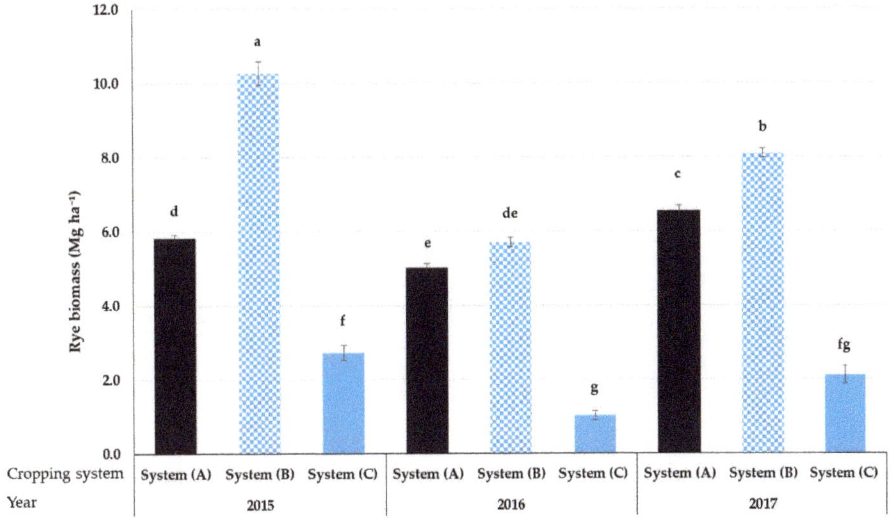

Figure 2. Interaction effect of cropping system on rye biomass at Müncheberg, Germany, in 2015 to 2017 ($p < 0.001$). Abbreviations: System (A), Cutting/plough-based system; System (B), No-tillage herbicide-free system; System (C), No-tillage herbicide-based system. Standard error bars are presented ($n = 8$). Means in the same year followed by the same letter are not significantly different at $p \leq 0.05$.

The dry weight of rye shoot biomass at the time of treatment with glyphosate at system (C) was significantly lowest in all years, whereas the dry weight of the mowed rye at the system (A) was intermediate. Biomass was the highest in no-tillage herbicide-free system in system (B) (Figure 2), with no significant difference between system (A) and system (B) in 2016.

3.4. Weed Suppression

There was a significant interaction between cropping system and year for weed biomass without irrigation ($p < 0.001$). Weed biomass at soybean flowering was not affected by the cropping system in 2017. In 2016, weed biomass in system (C) was significantly less than in systems (A) and (B). Thus, both conservation tillage systems (B) and (C) can effectively suppress weeds similarly to standard ploughed system (A) (Figure 3).

There was no significant interaction effects for weed biomass in 2016 between cropping system and irrigation ($p = 0.23$), while the main effects of cropping system and irrigation were significant. Irrigation in 2016 increased significantly weed biomass ($p = 0.006$) in all three cropping systems. There was a significant difference between systems ($p = 0.0003$), with system (C) having significant lower weed biomass than system (A) and system (B) (Figure 4).

Soybean grain yield was only weakly correlated with weed biomass in all cropping systems (Intercept = 30.25 ± 3.06 and slope = −0.74 ± 0.34 adjusted R^2 = 0.248, p =0.049). There was a strong negative correlation between rye cover crop biomass at the time of termination and weed biomass in the following soybean crop (Intercept = 58.17 ± 3.93 and slope = −0.69 ± 0.05 adjusted R^2 = 0.937, $p < 0.001$).

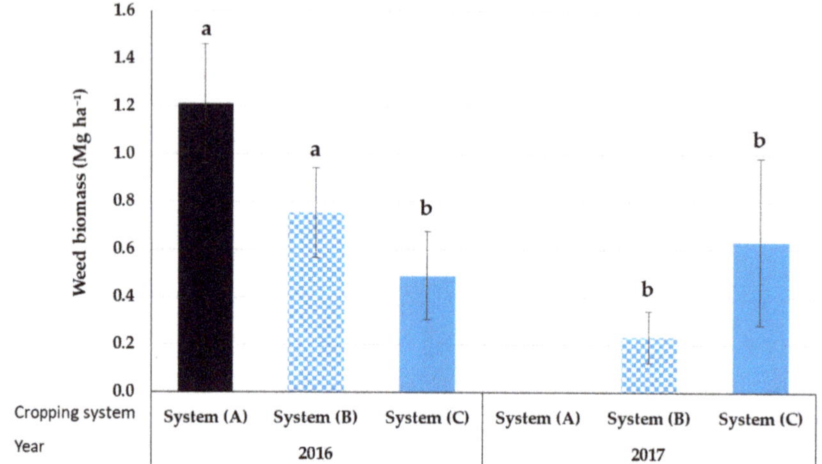

Figure 3. Interaction effect of cropping system on weed biomass without irrigation at Müncheberg, Germany, in 2016 and 2017. Abbreviations: System (A), Cutting/plough-based system; System (B), No-tillage herbicide-free system; System (C), No-tillage herbicide-based system. Standard error bars are presented ($n = 4$ in 2016 and $n = 8$ in 2017). Means in the same year followed by the same letter are not significantly different at $p \leq 0.05$. The weed biomass from system (A) in 2017 was excluded in data analysis because there were no weeds found.

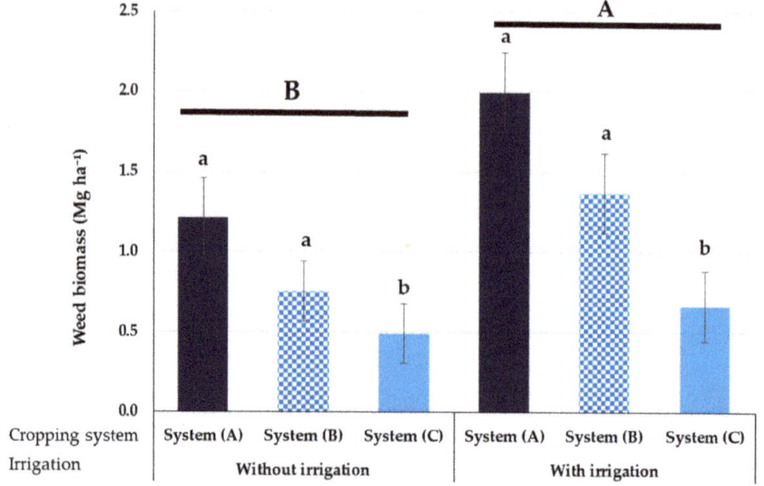

Figure 4. Effect of irrigation on weed biomass under three cropping systems at Müncheberg, Germany, in the dry year 2016. Abbreviations: System (A), Cutting/plough-based system; System (B), No-tillage herbicide-free system; System (C), No-tillage herbicide-based system. No interaction effect ($p = 0.23$). Standard error bars are presented ($n = 4$). Means followed by the same letter are not significantly different at P≤0.05. Lowercase letters allow for comparison among cropping system (P=0.0003). Uppercase letters allow for comparison between with and without irrigation ($p = 0.006$).

3.5. Soybean Yield

There was a significant interaction for soybean yield between cropping system and year under rainfall conditions ($p < 0.001$) (Figure 5). In 2015 no differences could be detected between cropping systems. While system (C) outperformed system (A) and (B) in 2016, system (C) was not statistically different from system (A) and system (B). Otherwise, a significant difference was observed between system (A) and system (B) in 2017 (Figure 5).

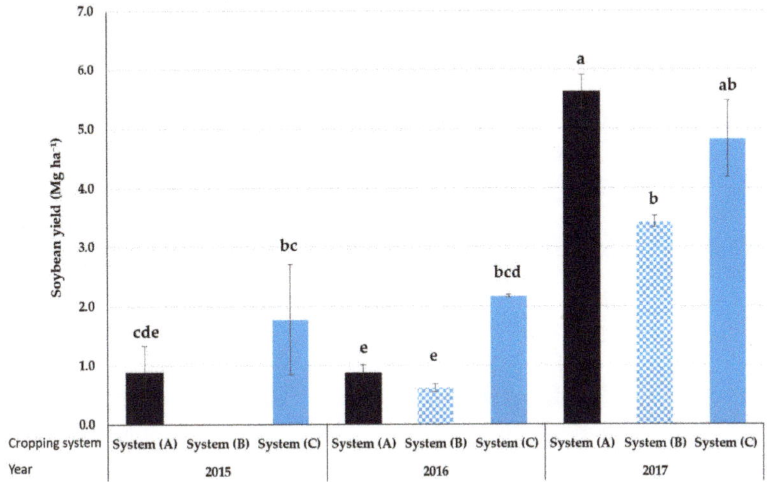

Figure 5. Interaction effect of cropping system on soybean yield without irrigation at Müncheberg, Germany, in 2015 to 2017. Abbreviations: System (A), Cutting/plough-based system; System (B), No-tillage herbicide-free system; System (C), No-tillage herbicide-based system. Soybean yield were analyzed separately each year, due to a significant interaction between years and cropping systems (p <0.001). Standard error bars are presented (2015 and 2017 n = 8, 2016 n = 4). Means in the same year followed by the same letter are not significantly different at $p \leq 0.05$. The soybean yield from system (B) in 2015 was excluded in data analysis because there were no plants to sample (see Section 2.3. Data collection).

In 2016, there were significant two-way interactions for soybean yield between cropping system and irrigation (p = 0.02). With and without irrigation, system (C) outperformed system (A) and (B) and there was no significant difference between system (A) and system (B) (Figure 6). Grain yield responded to irrigation in all systems. These responses to irrigation were highest in system (B) (264% higher yield with irrigation), compared to system (A) (105% higher yield with irrigation) and the lowest effect, but highest absolute yields were found in system (C) (61% higher yield with irrigation).

3.6. Economic Results

Economic results were affected by the cropping system, the presence of irrigation and differences in prices. System (A) achieved a gross margin of 380 € ha⁻¹ in the rainfed system, which was lower than system (C) with 463 € ha⁻¹. The gross margin for system (B) was negative with a conventional price (−172 € ha⁻¹) and positive with an organic price (447 € ha⁻¹). The main differences between the cropping systems resulted from the differences in soybean revenues, which were 975 € ha⁻¹ in the standard system (A), 1254 € ha⁻¹ in system (C), and 557 € ha⁻¹ in system (B) if the price of conventional soybean was used, 1176 € ha⁻¹ when the organic price was assumed. The second difference in the cropping system came from the variable costs, which were 1052 € ha⁻¹ when the rye biomass was harvested, and the field was ploughed (system (A)), 696 € ha⁻¹ when the rye was crimped (system (B)), and 792 € ha⁻¹ when rye was killed with the contact herbicide (system (C)). This amount was broken down into the cost for rye cultivation, which was equal for all systems, as well as the costs for soybean cultivation and rye termination, which were varied between the cropping systems (Figure 7).

In all three systems, the use of irrigation increased the variable costs by 270 €/ha on average in 2016 compared to rainfed cultivation. The largest effects on the additional revenues from irrigation in 2016 were achieved in system (B) which responded best to the addition water supply (Figure 8).

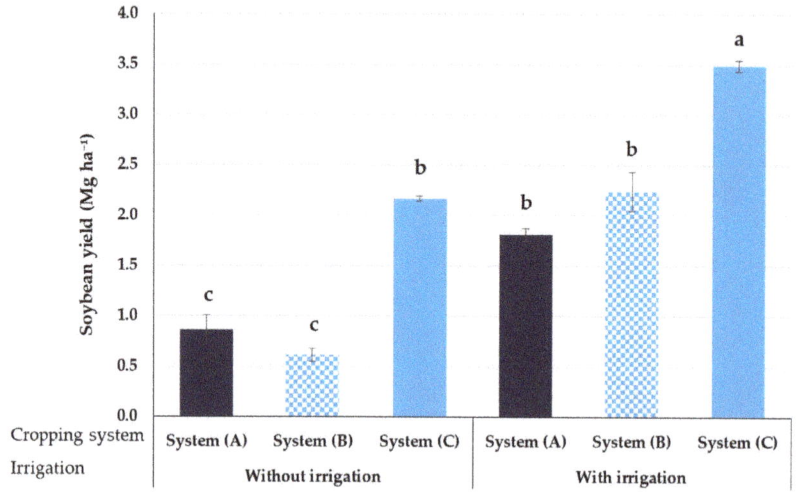

Figure 6. Interaction effect of irrigation on soybean yield under three cropping systems at Müncheberg, Germany, in dry year 2016. Abbreviations: System (A), Cutting/plough-based system; System (B), No-tillage herbicide-free system; System (C), No-tillage herbicide-based system. Interaction effect ($p = 0.02$). Standard error bars are presented ($n = 4$). Means followed by the same letter are not significantly different at $p \leq 0.05$.

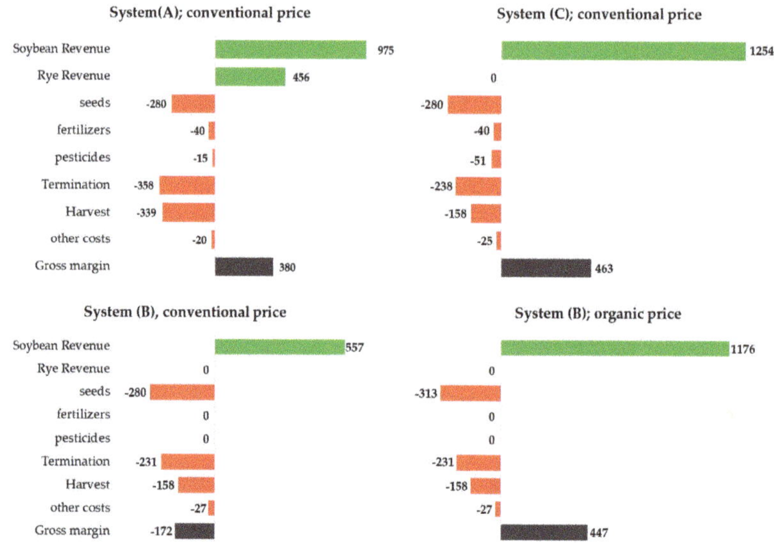

Figure 7. Comparison of average revenue, variable cost and gross margin for two no-tillage systems and tillage system from 2015 to 2017, all without irrigation. Abbreviations: System (A), Cutting/plough-based system; System (B), No-tillage herbicide-free system; System (C), No-tillage herbicide-based system.

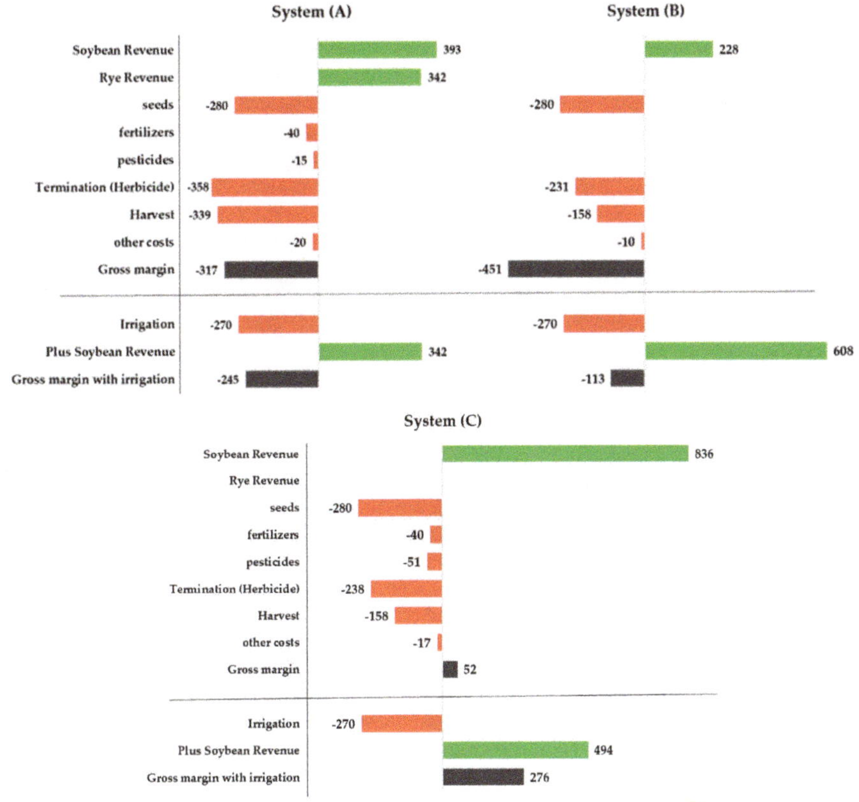

Figure 8. Comparison of gross margin for two no-tillage systems and tillage system under irrigated and without irrigation in the 2016 drought year. Abbreviations: System (A), Cutting/plough-based system; System (B), No-tillage herbicide-free system; System (C), No-tillage herbicide-based system.

4. Discussion

In central Europe, soybean production is only possible annually during summer, when the danger of frost ends. This poses a major challenge to sustainable agriculture, requiring to maintain ground cover over the winter months. For this reason, the interest in cover crops prior to soybean planting in central Europe is increasing [21]. Cover crops are preferred not just because they reduce wind and water erosion and increase soil organic matter but also because they have weed-suppressing effects before and during the growing season [22]. Terminating cover crops by plough is not suitable for sustainable agriculture, since soil disturbance should be minimized within this system.

In this trial, using two no-tillage cover-crop systems, weeds were managed during the soybean growing season equally or better than by using the cutting/plough-based system (system (A)). However, a higher soybean yield was obtained with the no-tillage herbicide-based system (System (C)) than with the cutting/plough-based system, which outperformed the no-tillage herbicide-free system (System (B)).

The date of termination differed among the cropping systems in this trial, and that could have a major effect on the sowing dates and finally, the yields. Rye in system (C) was terminated at the booting stage to maximize the effectiveness of glyphosate, which decreases when plants reach reproductive stages [23,24]. The cutting date of rye in system (A) was at the head emergence stage to maximize yield and feed quality traits [25,26]. In system (B), rye could only be terminated at anthesis (flowering) stage, following the recommendations of Keene et al. [18] in Canada, and Ashford and Reeves [27] in the northern USA, and thereby increasing the effectiveness of the roller-crimper for killing rye. The late cover crop termination in system (B) prevented an early soybean sowing date, which is needed for achieving high yields in no-tillage herbicide-free systems. Previous research suggested strategies to allow an earlier planting date, i.e., using earlier-flowering rye cultivars [17,28], planting rye earlier [10,26,29–31], planting soybean at the boot stage of rye cultivars with low regrowth potential [32,33], and planting soybean into standing rye prior to termination.

4.1. Soybean Productivity

The variation in soybean yield among cropping systems was largely influenced by the differences in soybean establishment. This makes it difficult to attribute this effect to the different features of the soil tillage systems. In our trial, the use of glyphosate to terminate the cover crop in the no-tillage system had the potential to achieve higher or similar soybean yields compared to those obtained using a standard cutting/plough-based system, whereas using a crimper to terminate the cover crop in the no-tillage system appeared to be less successful in this regard.

The typical soybean planting date in eastern Germany is from late April to mid-May [8]. However, soybean following crimped rye (system (B)) was planted during late May in this study to allow the cover crops to reach anthesis, thus facilitating better roller–crimper efficacy [17,27,34–39]. This delayed the planting date of soybean and subsequently limited the yield potential [40,41]. Soybean in the systems (A) and (C) was sown earlier than in system (B), providing the plants more time to establish and fill the pods, which is important considering that soybean is a short-day plant, sensitive to the photoperiod when grown in summer in Europe [42,43]. Conversely, a late planting date of soybean in system (B) resulted in a less vigorous vegetative growth, a lower number of branches able to produce pods, and a lower soybean yield potential per plant [41,43]. Additionally, the longer growing time of the cover crop often results in increased water use in the growing season. This, in turn, reduces the soil moisture available to soybean, which potentially increases the risk of a dry soil profile [28,35]. Moreover, Flower et al. [10] indicated that late killing of the cover crop reduces soil water storage. While cover crops are known to improve water retention (especially during winter), they are not able to offset crop yield declines that may occur because of temperature and water stress [19,30].

Soybean densities were lower for all systems in 2016 compared to 2017. Low soybean densities could be partially responsible for the relatively low soybean grain yields in 2016 for all systems compared to 2017. Higher plant densities result in canopy closure earlier in the season, which can be conducive to increased light interception, reduced soil moisture loss, prevention of the establishment of early-emerging weed, and increased crop yield [36,37]. Cumulative precipitation from May to September in the warmer year of 2016 was lower than in 2017, which could partly explain the yield differences between the two years. In dry years, the plants were often extremely short, and the lowest pods were very close to the ground, which increased losses during harvesting [37,38,40,42]. Mandić et al. [39] pointed out the importance of rainfall for soybean grain yield in north Serbia. Their study concluded that soybean grain yield had a strong positive relationship with the amount of rainfall during the growing season, especially in May, July, and August.

In 2016, in system (C), soybean produced higher yields than in systems (A) and (B), possibly because soil moisture was reduced after tillage during the harsh drought spring or was reduced by transpiration during cover crop growth until its flowering [24,33,34,36]. This theory is supported by the results of Bernstein et al. [33], who showed that after late rye termination in a dry year, soil moisture at depths below 20 cm was more reduced in a no-tillage system than in ploughed systems.

4.2. Cover Crop Biomass

Rye cover crop biomass was the lowest in system (C) compared to systems (A) and (B) in all years. Rye in the system (A) built intermediate rye biomass until the cutting date. The earlier termination dates of rye cover crops contributed to an apparent reduction in biomass of sprayed and mowed rye compared to the biomass of rolled-crimped rye. Rapid accumulation of rye biomass occurred in system (B) (roller-crimped) over the 28–37 days after the application of glyphosate in system (C) or 9–16 days after cutting in system (A). Previous research from Pennsylvania (USA) showed a37% increase in biomass with each 10-day delay in cover crop termination in spring [44].

In addition to the rye termination date, rainfall strongly influenced rye biomass accumulation. Rye biomass was lower in all systems in 2016 compared to 2015 and 2017. Differences in monthly precipitation from March to May during the study period explain why biomass results varied among these years. The rye cover crop was exposed to drought stress during the vegetative rye growth in 2016, which reduced the rye biomass production. This finding is in agreement with Alonso-Ayuso et al. [45], who showed that the above-ground biomass was double in a typical or humid year than in a dry year. Also, Idowu and Grover [46] found that a major challenge of cover crop adaptation in dry areas is the availability of water to grow cover crops.

4.3. Weed Suppression

Termination methods of rye suppressed weeds prior to soybean growing season, in system (A) during mechanical cutting and by ploughing the top soil layer, in system (B) during mechanical crimping and in system (C) by application of glyphosate that killed not only rye but also the weeds. Pre-emergence herbicide was applied in systems (A) and (C) to regulate weeds during the soybean growing season. Rye as a cover-crop in the crimped system (B) regulated weeds through mechanical (crimping) and biological suppression by inhibiting the germination of weed seeds (due to shadow and allelopathy) or by direct competition for resources [13]. In our trial, the use of both no-tillage cover-crop systems suppressed the weed and formed less or similar weed biomass compared to the standard cutting/plough-based system. Our results indicate efficiency of weed suppression in no-tillage systems with additional practices such as mechanical in system (B) or chemical system (C) including cover crops [10,47].

The weed biomass in 2017 was significantly lower than in 2016. This observation might be attributed to low precipitation in 2016, which had a greater effect on rye biomass at the termination date and the loss of rye residues than in the humid year 2017. Smith, et al. [48] found an increased weed competition within the row in conditions with low rye biomass. Furthermore, cover crop biomass was negatively correlated with weed density. This is in agreement with Witter, et al., [49] who showed that the higher the cover crop biomass, the lower the weed density. Insufficient soil moisture [28], inadequate seed placement [40], and poor seed-to-soil contact [31] can explain the influence of cover crop biomass on weed emergence in soybean.

4.4. Irrigation

Plots were only irrigated in 2016 because of insufficient rainfall from May to September. The use of irrigation increased soybean yields in our study (in 2016) and other investigations [50,51]. Drought stress at the critical phases, i.e., from flowering to pod filling, could lead to the defoliation of flower buds. However, the delay in the supply of irrigation water cannot always compensate for the damage [52].

The need for irrigation should be different among the systems, due to different sowing dates and water loss during termination. The negative effects of the rolled rye treatment on soybean yield reported by Davis [17] may have been caused by continued water use of cover crops crimped. In conservation tillage systems, the recommended rye termination date is 4 weeks prior to the soybean planting date [27], to minimize the problems of emergence as a result of soil water depletion.

The higher weed biomass on the irrigated plots can be attributed to the higher moisture in the topsoil, which stimulated the germination of weed seeds that could compete better with the soybeans. Ferdous et al. [53] reported a significant increase of weed dry matter at 40 days after planting in irrigated plots than in non-irrigated plots. In contrast, many previous studies reported that weed resurgence was not affected by irrigation significantly. This complete soybean canopy formed during the growing season can increase weed suppression and weed biomass better in the irrigated than non-irrigated plots [54].

4.5. Economic Assessment

The soybean revenue in system (C) were higher because of higher soybean yield. Although, low grain yields of soybean in system (B) had a negative effect on the total revenue, high prices for organic soybean of 800 € ha^{-1} versus 380 € ha^{-1} for conventional soybean could compensate for the lower yields resulting in an overall positive gross margin.

The highest variable cost of introducing rye as a cover crop in system (A) would be covered by selling the rye biomass as animal feed (at 456 € ha^{-1}). In general, conservation systems (B) and (C) appeared to perform better with irrigation, where competition for water reduced, resulting enhanced crop growth and increased revenues. This potentially enhanced returns on investment, as demonstrated by Snapp et al., [51]. An additional benefit of cover crops is that they can reduce nitrate leaching during winter [16], although, currently, this is not a sufficient incentive to overcome the additional costs from growing cover crops.

5. Conclusions

This study demonstrated the high efficiency of no-tillage cover-crop systems in controlling weeds during soybean growing season. Soybean yield was lower for the no-tillage herbicide-free system, which could be attributed to the late cover crop termination date in this system. The low soybean yield for the herbicide-free system resulted in low revenues and gross margins, which would only be economically attractive in organic farming, where the selling price of soybean is relatively higher.

Although irrigation promoted weed growth, it also increased soybean yields in all cropping systems. The additional costs of irrigation were covered by the higher revenues resulting from higher soybean yields in dry years, as observed in 2016.

We conclude that more investigations on the benefits and risks of no-tillage cover-crop management are needed before the technique can be recommended for the conditions tested in north-eastern Germany. Future research needs to focus on maintaining high rye biomass, ensuring an early rye termination and soybean planting date, optimizing the crimping and drilling equipment in order to develop good management practices in no-tillage herbicide-free systems for European conditions.

Author Contributions: R.B., J.B. and M.R. conceived and designed the experiments. M.H. and M.R. performed the agronomic and J.S. and R.B. performed the economic analysis. M.H. wrote the manuscript and all authors contributed.

Funding: This work was funded by the Federal Ministry of Education and Research (BMBF) Germany with the FACCE-JPI ERA-NET+ Project Climate-CAFE (Grant PTJ-031A544), the Innovation Network to Improve Soybean Production under the Global Change (INNISOY) through the European Interest Group CONCERT-Japan (01DR17011A) and the SusCrop- ERA-NET project LegumeGap (031B0807B). Moritz Reckling was funded by the Deutsche Forschungsgemeinschaft (DFG, German Research Foundation) – 420661662.

Acknowledgments: The authors would like to cordially thank Gunhild Rosner, Gerlinde Stange and the field research station in Müncheberg for their excellent technical support and establishment of the technically challenging field experiments.

Conflicts of Interest: The authors declare no competing financial interests in relation to the work described.

References

1. Olesen, J.E.; Trnka, M.; Kersebaum, K.C.; Skjelvåg, A.O.; Seguin, B.; Peltonen-Sainio, P.; Rossi, F.; Kozyra, J.; Micale, F. Impacts and adaptation of European crop production systems to climate change. *Eur. J. Agron.* **2011**, *34*, 96–112. [CrossRef]
2. Vadez, V.; Berger, J.D.; Warkentin, T.; Asseng, S.; Ratnakumar, P.; Rao, K.P.C.; Gaur, P.M.; Munier-Jolain, N.; Larmure, A.; Voisin, A.-S.; et al. Adaptation of grain legumes to climate change: A review. *Agron. Sustain. Dev.* **2011**, *32*, 31–44. [CrossRef]

3. Reckling, M.; Döring, T.F.; Bergkvist, G.; Stoddard, F.L.; Watson, C.A.; Seddig, S.; Chmielewski, F.-M.; Bachinger, J. Grain legume yields are as stable as other spring crops in long-term experiments across northern Europe. *Agron. Sustain. Dev.* **2018**, *38*, 63. [CrossRef] [PubMed]

4. Doré, T.; Makowski, D.; Malézieux, E.; Munier-Jolain, N.; Tchamitchian, M.; Tittonell, P. Facing up to the paradigm of ecological intensification in agronomy: Revisiting methods, concepts and knowledge. *Eur. J. Agron.* **2011**, *34*, 197–210. [CrossRef]

5. Reckling, M.; Bergkvist, G.; Watson, C.A.; Stoddard, F.L.; Bachinger, J. Re-designing organic grain legume cropping systems using systems agronomy. *Eur. J. Agron.* **2020**, *112*, 125951. [CrossRef]

6. Reyer, C.; Bachinger, J.; Bloch, R.; Hattermann, F.F.; Ibisch, P.L.; Kreft, S.; Lasch, P.; Lucht, W.; Nowicki, C.; Spathelf, P.; et al. Climate change adaptation and sustainable regional development: a case study for the Federal State of Brandenburg, Germany. *Reg. Environ. Chang.* **2012**, *12*, 523–542. [CrossRef]

7. Reckling, M.; Döring, T.F.; Bergkvist, G.; Chmielewski, F.-M.; Stoddard, F.L.; Watson, C.A.; Seddig, S.; Bachinger, J. Grain legume yield instability has increased over 60 years in long-term field experiments as measured by a scale-adjusted coefficient of variation. *Asp. Appl. Biol.* **2018**, *138*, 15–20.

8. Reckling, M.; Bachinger, J.; Bellingrath-Kimura, S.D. Taking Soybean Cultivation Further North: Agronomic Performance of Cultivars and Irrigation Treatments. In Proceedings of the XVe European Society for Agronomy Congress (ESA), Geneva, Switzerland, 27–31 August 2018; pp. 91–92.

9. Salamanca, C.; Cid, P.; Gomez-Macpherson, H. Conservation agriculture implications for soil water balance. In *A Modelling Approach, Proceedings of the 1st World Conference on Soil and Water Conservation under Global Change (CONSOWA), Lleida, Spain, 12–16 June 2017*; Simó, I., Poch, R.M., Pla, I., Eds.; Diputació de Lleida: Lleida, Spain; p. 2017.

10. Flower, K.; Cordingley, N.; Ward, P.; Weeks, C. Nitrogen, weed management and economics with cover crops in conservation agriculture in a Mediterranean climate. *Field Crops Res.* **2012**, *132*, 63–75. [CrossRef]

11. Davis, A.S.; Renner, K.A.; Gross, K.L. Weed Seedbank and Community Shifts in a Long-Term Cropping Systems Experiment. *Weed Sci.* **2005**, *53*, 296–306.

12. Weber, J.F.; Kunz, C.; Peteinatos, G.G.; Zikeli, S.; Gerhards, R. Weed Control Using Conventional Tillage, Reduced Tillage, No-Tillage, and Cover Crops in Organic Soybean. *Agriculture* **2017**, *7*, 43. [CrossRef]

13. Tabaglio, V.; Gavazzi, C.; Schulz, M.; Marocco, A. Alternative weed control using the allelopathic effect of natural benzoxazinoids from rye mulch. *Agron. Sustain. Dev.* **2008**, *28*, 397–401. [CrossRef]

14. Plaza-Bonilla, D.; Nolot, J.-M.; Raffaillac, D.; Justes, E. Cover crops mitigate nitrate leaching in cropping systems including grain legumes: Field evidence and model simulations. *Agric. Ecosyst. Environ.* **2015**, *212*, 1–12. [CrossRef]

15. Thapa, B.; Pande, K.R.; Khana, B.; Marahatta, S. Effect of Tillage, Residue Management and Cropping System on the Properties of Soil. *Int. J. Appl. Sci. Biotechnol.* **2018**, *6*, 164–168. [CrossRef]

16. Justes, E. *Cover Crops for Sustainable Farming*; Springer Nature: GX Dordrecht, The Netherlands, 2017. [CrossRef]

17. Davis, A.S. Cover-Crop Roller–Crimper Contributes to Weed Management in No-Till Soybean. *Weed Sci.* **2017**, *58*, 300–309. [CrossRef]

18. Keene, C.L.; Curran, W.S.; Wallace, J.M.; Ryan, M.R.; Mirsky, S.B.; VanGessel, M.J.; Barbercheck, M.E. Cover Crop Termination Timing is Critical in Organic Rotational No-Till Systems. *Agron. J.* **2017**, *109*, 272–282. [CrossRef]

19. Bloch, R. *The Vulnerability of Organic Farming to Climate Change Effects in the Federal State of Brandenburg, Germany*; Kassel University Press: Kassel, Germany, 2016.

20. Mirschel, W.; Klauss, H.; Berg, M.; Eisenhut, K.-U.; Ißbrücker, G.; Prochnow, A.; Schörling, B.; Wenkel, K.-O. Innovative Technologien für eine effiziente Bewässerung im Pflanzenbau. In *Land- und Ernährungswirtschaft im Klimawandel—Auswirkungen, Anpassungsstrategien und Entscheidungshilfen*; Bloch, R., Bachinger, J., Fohrmann, R., Pfriem, R., Eds.; oekom verlag: München, Germany, 2014; pp. 261–277.

21. Vincent-Caboud, L.; Casagrande, M.; David, C.; Ryan, M.R.; Silva, E.M.; Peigne, J. Using mulch from cover crops to facilitate organic no-till soybean and maize production. A review. *Agron. Sustain. Dev.* **2019**, *39*, 45. [CrossRef]

22. Carr, P. *Guest Editorial: Conservation Tillage for Organic Farming*; Multidisciplinary Digital Publishing Institute: Basel, Switzerland, 2017.

23. Werle, R.; Burr, C.; Blanco-Canqui, H. Cereal rye cover crop suppresses winter annual weeds. *Can. J. Plant Sci.* **2017**, *98*, 498–500. [CrossRef]

24. Clark, K.M.; Boardman, D.L.; Staples, J.S.; Easterby, S.; Reinbott, T.; Kremer, R.J.; Kitchen, N.R.; Veum, K.S. Crop yield and soil organic carbon in conventional and no-till organic systems on a claypan soil. *Agron. J.* **2017**, *109*, 588–599. [CrossRef]

25. Twidwell, E.K.; Beutler, M.; Kephart, K.; Boe, A. *Grass Seed Production in South Dakota Guidelines*; Agricultural Experiment Station Circulars: Logan, UT, USA, 1987.

26. De Bruin, J.L.; Porter, P.M.; Jordan, N.R. Use of a rye cover crop following corn in rotation with soybean in the upper Midwest. *Agron. J.* **2005**, *97*, 587–598. [CrossRef]

27. Ashford, D.L.; Reeves, D.W. Use of a mechanical roller-crimper as an alternative kill method for cover crops. *Am. J. Altern. Agric.* **2003**, *18*, 37–45. [CrossRef]

28. Wells, M.S.; Brinton, C.M.; Reberg-Horton, S.C. Weed suppression and soybean yield in a no-till cover-crop mulched system as influenced by six rye cultivars. *Renew. Agric. Food Syst.* **2015**, *31*, 429–440. [CrossRef]

29. Smith, A. Utilizing Rolled Rye Mulch for Weed Suppression in Organic No-Tillage Soybeans. Master's Thesis, Faculty of North Carolina State University, Raleigh, NC, USA, 2010.

30. Basche, A.D. *Climate-Smart Agriculture in Midwest Cropping Systems: Evaluating the Benefits and Tradeoffs of Cover Crops*; Iowa State University: Ames, IA, USA, 2015.

31. Williams, M.M.N.; Mortensen, D.A.; Doran, J.W. No-tillage soybean performance in cover crops for weed management in the western Corn Belt. *J. Soil Water Conserv.* **2000**, *55*, 79–84.

32. Silva, E.M.; Vereecke, L. Optimizing organic cover crop-based rotational tillage systems for early soybean growth. *Org. Agric.* **2019**, 1–11. [CrossRef]

33. Bernstein, E.R.; Posner, J.L.; Stoltenberg, D.E.; Hedtcke, J.L. Organically managed no-tillage rye–soybean systems: agronomic, economic, and environmental assessment. *Agron. J.* **2011**, *103*, 1169–1179. [CrossRef]

34. Mirsky, S.; Curran, W.S.; Mortensen, D.A.; Ryan, M.; Shumway, D. Control of cereal rye with a roller/crimper as influenced by cover crop phenology. *Agron. J.* **2009**, *101*, 1589–1596. [CrossRef]

35. Georgel, F. *Agroecological service crop termination with a roller-crimper in organic vegetable systems: A good alternative to conventional soil tillage?* Norwegian University of Life Sciences: Ås, Norway, 2017.

36. Liebert, J.; Ryan, M. High Planting Rates Improve Weed Suppression, Yield, and Profitability in Organically-Managed, No-till–Planted Soybean. *Weed Technol.* **2017**, *31*, 536–549. [CrossRef]

37. Aigner, A. Erkenntnisse aus den bayerischen Sojaversuchen. In Proceedings of the Fünf Jahre Soja-Netzwerk Wertschöpfungsketten und Impulse für die Zukunft, Würzburg, Germany, 23–24 October 2018; pp. 6–8.

38. Recknagel, J. Sojaanbau in Deutschland –Stand und Perspektiven. In Proceedings of the Fünf Jahre Soja-Netzwerk Wertschöpfungsketten und Impulse für die Zukunft, Würzburg, Germany, 23–24 October 2018; pp. 2–5.

39. Mandić, V.B.Z.; Krnjaja, V.; Simić, A.; Ružić-Muslić, D.; Dragičević, V.; Petričević, V. The rainfall use efficiency and soybean grain yield under rainfed conditions in Vojvodina. *Biotechnol. Anim. Husb.* **2017**, *33*, 475–486. [CrossRef]

40. Mirsky, S.B.; Ryan, M.R.; Teasdale, J.R.; Curran, W.S.; Reberg-Horton, C.S.; Spargo, J.T.; Wells, M.S.; Keene, C.L.; Moyer, J.W. Overcoming weed management challenges in cover crop–based organic rotational no-till soybean production in the eastern United States. *Weed Technol.* **2013**, *27*, 193–203. [CrossRef]

41. Frederick, J.R.; Bauer, P.J.; Busscher, W.J.; Mc Cutcheon, G.S. Tillage management for double cropped soybean grown in narrow and wide row width culture. *Crop Sci.* **1998**, *38*, 755–762. [CrossRef]

42. Kurasch, A.K.; Leiser, W.L.; Würschum, T.; Hahn, V. Identifizierung von Mega-Umwelten in Europa und Effekte der Reifegene auf die Anpassung europäischer Sojasorten. In Proceedings of the Soja-Tagung, Rastatt, Germany, 2017; pp. 72–73.

43. Linkemer, G.; Board, J.E.; Musgrave, M.E. Water logging effects on growth and yield components in lateplanted soybean. *Crop Sci.* **1998**, *38*, 1576–1584. [CrossRef]

44. Mirsky, S.B. Evaluating Constraints and Opportunities in Managing Weed Populations with Cover Crops. Ph.D. Thesis, Penn State University, Centre County, PA, USA, 2008.

45. Alonso-Ayuso, M.; Luis Gabriel, J.; Quemada, M. The Kill Date as a Management Tool for Cover Cropping Success. *PLoS ONE* **2014**, *9*, 12. [CrossRef] [PubMed]

46. Idowu, J.; Grover, K. *Principles of Cover Cropping for Arid and Semi-Arid Farming Systems*; New Mexico State University, College of Agricultural, Consumer and Environmental Sciences: Las Cruces, New Mexico, 2014.

47. Legleiter, T.; Johnson, B.; Young, B.G. *Successful Annual Ryegrass Termination with Herbicides*; Purdue University: West Lafayette, Indiana, 2015; p. 4.
48. Smith, A.N.; Chris Reberg-Horton, S.; Place, G.T.; Meijer, A.D.; Arellano, C.; Mueller, J.P. Rolled Rye Mulch for Weed Suppression in Organic No-Tillage Soybeans. *Weed Sci. Soc. Am.* **2011**, *59*, 224–231. [CrossRef]
49. Wittwer, R.A.; Dorn, B.; Jossi, W.; Van Der Heijden, M.G. Cover crops support ecological intensification of arable cropping systems. *Sci. Rep.* **2017**, *7*, 41911. [CrossRef]
50. Sadeghipour, O.; Abbasi, S. Soybean response to drought and seed inoculation. *World Appl. Sci. J.* **2012**, *17*, 55–60.
51. Snapp, S.; Swinton, S.; Labarta, R.; Mutch, D.; Black, J.; Leep, R.; Nyiraneza, J.; O'neil, K. Evaluating cover crops for benefits, costs and performance within cropping system niches. *Agron. J.* **2005**, *97*, 322–332.
52. Kokubun, M. Physiological mechanisms regulating flower abortion in soybean. In *Soybean-Biochemistry, Chemistry and Physiology*; IntechOpen: Rijeka, Croatia, 2011.
53. Ferdous, Z.; Datta, A.; Anwar, M. Plastic mulch and indigenous microorganism effects on yield and yield components of cauliflower and tomato in inland and coastal regions of Bangladesh. *J. Crop Improv.* **2017**, *31*, 261–279. [CrossRef]
54. Yelverton, F.H.; Coble, H.D. Narrow row spacing and canopy formation reduces weed resurgence in soybeans (Glycine max). *Weed Technol.* **1991**, *5*, 169–174. [CrossRef]

 *agronomy*

Article

Winter Rye Cover Crop with Liquid Manure Injection Reduces Spring Soil Nitrate but Not Maize Yield

Leslie A. Everett [1], Melissa L. Wilson [2], Randall J. Pepin [3] and Jeffrey A. Coulter [4,*]

[1]   Water Resources Center, University of Minnesota, St. Paul, MN 55108-6028, USA; evere003@umn.edu
[2]   Department of Soil, Water, and Climate, University of Minnesota, St. Paul, MN 55108-6028, USA;
       mlw@umn.edu
[3]   University of Minnesota Extension, University of Minnesota, St. Paul, MN 55108-6028, USA;
       pepin019@umn.edu
[4]   Department of Agronomy and Plant Genetics, University of Minnesota, St. Paul, MN 55108-6028, USA
*   Correspondence: jeffcoulter@umn.edu; Tel.: +1-612-625-1796

Received: 25 November 2019; Accepted: 3 December 2019; Published: 5 December 2019

Abstract: In maize-based cropping systems, leaching of nitrate-nitrogen (NO_3-N) to drainage tile and groundwater is a significant problem. The purpose of this study was to assess whether a winter rye cover crop planted after silage maize or soybean harvest and injected with liquid manure could decrease soil NO_3-N without reducing the yield of the following maize crop. An experiment was conducted at 19 sites with predominant occurrence of Mollisols (15 out of 19 sites) in the upper Midwest USA immediately after soybean or maize silage harvest to compare a drilled rye cover crop and a non-cover crop control. Later in the fall, liquid swine or dairy manure was injected into the cover crop and control plots. Rye was terminated the following spring using herbicide, usually before reaching 20 to 25 cm in height, and incorporated with tillage at most sites, after which maize was planted and harvested as silage or grain. Across sites, soil NO_3-N at rye termination was reduced by 36% (range = 4% to 67%) with rye compared to no rye. Nitrogen in aboveground rye biomass at termination ranged from 5 to 114 kg N ha^{-1} (mean = 51 kg N ha^{-1}). Across sites, there was no significant difference in yield of maize silage or grain between treatments. These results demonstrate in a Mollisol-dominated region the potential of a winter rye cover crop planted before manure application to effectively reduce soil NO_3-N without impacting yield of the following maize crop, thereby reducing risk of negative environmental impacts.

Keywords: cover crop; manure; nitrate; nitrogen; cereal rye; maize

1. Introduction

Agricultural intensification over the past few decades has led to degradation of aquatic systems, and the impact of downstream nutrient export from agricultural lands continues to be of concern. Nitrate-nitrogen (NO_3-N) is particularly troublesome as it leaches through the soil into subsurface drainage or groundwater, which ultimately leads to surface waters. There are now over 400 aquatic systems across the world experiencing hypoxia due to excess nutrients fueling algal blooms [1].

In the Midwest region of the United States, nutrient losses have been exacerbated by both intensified crop and livestock production. This has led to a large hypoxic, or dead, zone forming annually in the Gulf of Mexico [2,3]. In cropping systems, the use of commercial fertilizers along with increased mineralization of nitrogen from drained soils is problematic [4–6]. Where animal feeding operations have concentrated, the amount of manure often exceeds the nearby cropland nutrient needs and manure is treated as more of a waste rather than a resource [7,8]. As an example, the US Department of Agriculture's (USDA) Agricultural Resource Management Survey found that 95% of manure applications to maize (*Zea mays* L.) acres did not follow national recommendations for application rate, timing, and placement [9]. In Minnesota, USA, where the headwaters of the Mississippi River, which constitutes a major inflow to the Gulf of Mexico, are located, the 2-year maize-soybean (*Glycine max* (L.) Merr.) rotation dominates the agricultural landscape, similar to most places in the Midwest [10]. A survey of Minnesota growers found that those using manure applied on average 25.6 kg per hectare more N than recommended for maize [11], and overapplications most notably occurred where maize followed soybean [12].

In the upper Midwest USA, the cold climate adds further challenges to manure management. Due the short growing season and increasingly wet springtime conditions [13], liquid swine (*Sus scrofa* L.) and dairy cow (*Bos taurus* L.) manure are frequently applied in the fall prior to planting maize the following spring. However, maize does not begin taking up substantial amounts of nitrogen (N) until the mid-vegetative stages [14]. This creates risk of soil nitrate-nitrogen leaching below the maize rooting zone, since most of the N in liquid swine manure and about one-half in liquid dairy manure is ammonium [15], which is rapidly converted to NO_3-N when the soil temperature in the zone of application exceeds 10 °C [16]. This challenge is exacerbated when warm weather and excess precipitation occur between manure application and maize N uptake [5,17,18], which are expected to occur more frequently with climate change [19].

One strategy for managing excess N in soil between cash crops includes using cover crops to take up N and release it following termination. It has been well established that grasses are particularly effective at scavenging N and holding it in the biomass that is produced [20–24], although it is questionable whether N release from the cover crop biomass will synchronize with maize uptake the following growing season. Huntington et al. reported that in a no-till system, N release from hairy vetch (*Vicia villosa* Roth) and rye (*Secale cereale* L.) was typically highest after the maize silking stage and thus did not synchronize well with maize uptake [25]. Jahanzad et al. found that forage radish (*Raphanus sativus* L.) and winter pea (*Pisum sativum subsp. arvense* L.) decomposed much more quickly than rye and N release was more synchronized with early cash crop N demands, but they also reported that rye had faster decomposition and N release when buried in the soil as opposed to being left on the soil surface after being terminated [26]. In some cases, this lack of synchronization may cause losses in maize yield [27], but on the other hand, many studies have reported minimal or even positive impacts on yields following grass cover crops over the long term [28–31], particularly when fertilization regimes were optimized [32].

Most studies have evaluated cover crops in systems utilizing commercial fertilizers, but few studies have evaluated use with fall-applied manure. In Ontario Canada, fall cover crops of perennial ryegrass (*Lolium perenne* L.), oat (*Avena sativus* L.), and red clover (*Trifolium pratense* L.) reduced residual soil mineral N by 26% to 42% compared to no cover crop after fall-applied liquid swine manure, but the following year maize grain yield was reduced by up to 15% [33]. In Iowa, USA, Singer et al. found that a rye-oat mixture interseeded into soybean reduced soil NO_3-N in the 0–30 cm depth during the early growing season of maize without affecting maize grain yield after liquid swine manure was injected into the growing cover crop the previous fall [34]. In Pennsylvania, USA, maize yields were highest when manure was injected in late fall (November) into a rye cover crop compared with rye being planted after an early application of manure in September, and furthermore, manure injected into the cover crop resulted in better yields than when manure was simply broadcast onto the cover crop [35]. These findings have practical implications for growers because it is counterintuitive to plant a cover crop and then reduce biomass by injecting liquid manure into it.

While there have been some promising results when adding cover crops as a best management practice for fall manure application, this practice has not been evaluated in a cold climate, i.e., the upper Midwestern USA. It is unclear how results from other studies would transfer to this region when the short growing season adds additional constraints. Furthermore, practical applications for growers need to be considered because there has been little reported research conducted on working farms with large scale equipment. To convince growers of the merit of this practice and to increase adoption rates, more on-farm research is needed. The objective of this study was to assess whether a winter rye cover crop planted after silage maize or soybean harvest and injected with liquid swine or dairy manure in the fall could decrease soil NO_3-N without reducing yield of the following maize crop.

2. Materials and Methods

A field experiment was conducted across 15 on-farm and four research station sites in southern and central Minnesota during the 2016 and 2017 maize growing seasons. A composite soil sample from the 0–15 cm depth was collected from each site prior to applying treatments and analyzed for phosphorus (Olsen or Bray-1) [36,37], ammonium acetate extractable potassium [38], pH (1:1 soil:water) [39], and organic matter (combustion) [40]. Soil classification, texture, and initial soil-test levels are in Table S1. Soils were Mollisols with 27 to 83 g kg^{-1} organic matter in the 0–15 cm depth at 15 sites and Alfisols with 30 to 52 g kg^{-1} organic matter in the 0–15 cm depth at four sites.

The two treatments were winter rye cover crop and no cover crop, replicated three times within a randomized complete block design, and established immediately following maize silage or soybean harvest the year prior to the main crop growing season. All agronomic practices other than cover crop planting were consistent between treatments at all locations. At sites where the manure type (Table 1) was swine, the prior crop was soybean, and where the manure type was dairy, the prior crop was maize except for site 12 where it was soybean. Plots were equal to or greater in width than the cooperating growers' maize harvester, and were 4.6 to 6.7 m wide by 73.2 to 170.7 m long at the on-farm sites, and 1.5 m wide by 12.2 m long at the research station sites (sites 3, 8, 13, and 19). Rye was planted at 100 kg ha^{-1} by the cooperating growers using a grain drill. Rye was planted between late September and early November (Table 1). Aboveground biomass of rye in the fall was limited, especially with the later seeding dates, and therefore not measured.

Table 1. Dates of field operations, manure type, and maize harvest method at the 19 experimental sites.

Site	Rye Planting Date	Manure Type	Manure Application Date	Rye Sampling Date	Maize Planting Date	Maize Harvest Method
	2015				2016	
1	30 September	dairy	20 November	26 April	6 May	silage
2	26 September	dairy	9 October	18 April	4 May	silage
3	9 October	dairy	12 November	15 April	3 May	grain
4	25 September	dairy	2 November	21 April	9 May	silage
5	2 October	swine	12 November	25 April	13 May	grain
6	3 October	swine	10 November	22 April	6 May	grain
7	29 September	swine	12 November	22 April	25 April	grain
8	9 October	swine	12 November	15 April	3 May	grain
9	12 October	swine	10 November	26 April	7 May	grain
	2016				2017	
10	26 September	dairy	10 November	20 April	13 May	silage
11	24 October	dairy	27 October	28 April	9 May	silage
12	3 October	dairy	2 November	8 May	6 May	silage
13	26 October	dairy	5 December	24 April	15 May	grain
14	17 October	dairy	6 October	9 May	10 May	silage
15	7 October	dairy	22 October	25 April	11 May	silage
16	8 November	swine	25 October	8 May	9 May	grain
17	14 October	swine	18 October	20 April	6 May	grain
18	17 October	swine	27 November	17 April	9 May	grain
19	26 October	swine	5 December	24 April	15 May	grain

Liquid swine or dairy cow manure was injected into the soil of the cover crop and no cover crop treatments two to six weeks after the winter rye was planted, at most sites. However, in 2016, manure application occurred three to four days after rye planting at two sites, and 11 to 14 days before planting at two sites (Table 1). Manure application equipment included disk closures without knives (sites 6, 12), knives without terminal sweeps (sites 1, 2, 14), and knives with terminal sweeps narrower than 30 cm (all other sites). Manure was placed approximately 8 to 10 cm deep with the disk closures alone and 13 to 20 cm deep with knives. The manure application rate at each site was determined by the cooperating growers. A manure sample from each site was analyzed for total Kjeldahl N and inorganic N following methods of Peters et al. [41] on a continuous flow gas diffusion and conductivity cell analyzer (Timberline Instruments, Boulder, CO, USA). Total and inorganic N (NH_4-N) in manure and N from fertilizer are in Table 2. The forms, rates, and dates of application of mineral fertilizers applied are in Table S2. At all sites but three (sites 3, 8, 13), manure was the principal N source. At the three sites where it was not, manure had not been agitated and its N content was low (0.36–0.48 g L^{-1}), so fertilizer N applied in the spring after cover crop termination was the dominant N source.

Table 2. Total nitrogen (kg ha^{-1}) applied, plant nitrogen (N) measured in the aboveground rye biomass at termination and in the maize plants at harvest (silage or grain, plus dry stalks), and soil nitrate-N (0–60 cm, mg kg^{-1}) measured at cover crop termination.

Site	Total N Applied		Fertilizer	Plant N Uptake			Soil NO$_3$-N	
	Manure			Rye	Maize		w/Rye	w/o Rye
	TKN *	NH$_4$-N			w/Rye	w/o Rye		
				kg ha^{-1}			mg kg^{-1}	
2016 Maize Growing Season								
1	256	222	0	95 (14.2) †	85 (5.8)	120 (5.9)	4 (0.8)	12 (1.3)
2	179	142	0	98 (19.2)	193 (15.8)	188 (11.2)	24 (2.3)	26 (3.0)
3	30	7	157	15 (1.2)	139 (13.7)	135 (19.4)	6 (0.6)	9 (0.3)
4	186	177	28	107 (1.5)	151 (4.1)	149 (2.5)	16 (1.3)	21 (5.0)
5	320	232	36	114 (11.2)	186 (4.9)	177 (12.5)	7 (1.5)	21 (3.0)
6	226	209	0	84 (2.9)	177 (14.9)	177 (12.8)	20 (1.2)	38 (2.8)
7	219	217	0	71 (6.0)	152 (18.2)	136 (2.5)	12 (1.4)	25 (2.5)
8	22	20	112	14 (0.3)	148 (11.7)	142 (8.9)	7 (0.3)	11 (0.4)
9	75	73	0	43 (4.0)	91 (6.6)	99 (1.8)	4 (0.3)	10 (1.5)
2017 Maize Growing Season								
10	279	118	0	84 (4.2)	209 (5.2)	218 (6.2)	7 (1.8)	13 (1.5)
11	292	105	0	22 (2.8)	188 (7.4)	185 (15.6)	18 (2.2)	27 (0.8)
12	148	68	68	49 (0.3)	144 (1.5)	149 (3.5)	24 (3.0)	35 (4.4)
13	24	14	135	5 (0.3)	153 (9.2)	153 (6.0)	13 (1.5)	14 (1.2)
14	195	‡	68	54 (2.6)	144 (6.3)	170 (8.2)	9 (1.2)	15 (1.6)
15	374	169	35	25 (1.7)	166 (13.3)	167 (2.9)	27 (4.6)	28 (1.4)
16	226	137	86	5 (1.0)	156 (4.7)	165 (14.5)	23 (0.4)	35 (4.8)
17	140	100	0	65 (4.4)	141 (11.0)	144 (1.7)	10 (2.0)	20 (1.2)
18	186	127	67	8 (0.5)	174 (3.6)	163 (5.6)	44 (4.7)	64 (12.6)
19	206	153	0	5 (0.3)	181 (8.9)	180 (10.0)	22 (3.9)	56 (16.1)
Mean				49 (5.2)	157 (4.5)	159 (4.0)	16 (1.4)	25 (2.2)
Significance of difference with and without rye					n.s.		p < 0.001	

* Total Kjeldahl nitrogen. † Standard error is shown within parentheses. ‡ Not measured. n.s. = not significantly different.

Spring cover crop termination was targeted to be prior to rye reaching a 25 cm height to ensure that N uptake by rye did not reduce soil N below the level required by the subsequent maize crop during the early growing season [42]. Rye was terminated by herbicide followed by full-width tillage at all sites except three: at site 9, herbicide was followed by strip-tillage, and at sites 14 and 16, termination was by full-width tillage only. Agronomic practices for cover crop termination were also applied to plots of the no cover crop treatment at all sites. At rye termination, rye canopy height, plant density, aboveground biomass, and N concentration, as well as soil NO$_3$-N in the 0–60 cm layer, were measured. The average height of the rye canopy at each site was recorded. At three or four random locations per replication a 0.25 m^2 quadrat was delineated, plants were counted and cut at the soil surface, combined, weighed, subsampled, weighed, and then oven-dried at 60 °C until constant mass, weighed, ground to pass a 1-mm sieve, and analyzed for Kjeldahl N. Depending on plot length, 8 to 12 soil cores were randomly collected from the 0–60 cm depth in each plot, mixed, sub-sampled, oven-dried at 35 °C until constant mass, ground to pass a 2-mm sieve, and analyzed for NO_3-N (KCl extraction and cadmium reduction method) [43,44].

Maize was planted in 76-cm rows at 18 sites and in 56-cm rows at one site (site 16) and managed by cooperating growers. Maize hybrids and planting rates varied among sites and were selected by the cooperating growers. Some sites received supplemental N fertilizer at rates determined by the growers (Table 2). Maize was harvested by the growers as silage at eight sites and as grain at 11 sites. Silage yield was measured by weighing the silage wagon before and after chopping each plot. For each plot, a ~0.5-kg sample of the silage was weighed, oven-dried at 35 °C until constant mass, reweighed, ground to pass a 2-mm sieve, and analyzed for Kjeldahl N. At sites where maize was harvested for grain, grain yield was measured using a calibrated weigh wagon and samples were taken for laboratory determination of moisture and Kjeldahl N. Silage and grain yields were calculated at 650 and 155 mg kg^{-1} moisture content, respectively. Prior to machine harvest for grain, 20 plants were randomly selected from each plot, cut at the soil surface, ears removed, stalks and husk weighed, chopped, mixed, subsampled, weighed, oven-dried at 35 °C until constant mass, reweighed, ground to pass a 2-mm sieve, and analyzed for Kjeldahl N.

Monthly precipitation and average air temperature for August 2015 through November 2017 were collected from nine National Weather Service weather stations in Minnesota (Midwestern Regional Climate Center, Champaign, IL, USA) and compared to the 30-year average (1981–2010). Stations were chosen based on proximity to study sites and availability of 30-year data. To show the general trends in weather patterns over the study, data were averaged across weather stations.

Data were analyzed at $p \leq 0.05$ with JMP version 13 Pro (SAS Institute Inc., Cary, NC, USA) using standard least squares (restricted maximum likelihood method, REML) and linear regression. Treatment was considered a fixed effect and site and replication (nested within site), and site × treatment were considered random effects. Site means for aboveground rye dry matter yield (DM) at termination were regressed on rye planting day of the year and on canopy height. Nitrogen concentration in aboveground rye was regressed on rye aboveground DM and rye height. Total rye N was regressed on rye DM and rye N concentration. The difference in soil NO_3-N between cover cropped and bare plots was regressed on the number of days between rye planting and manure application. Rye plant density, rye DM, concentration of N in aboveground rye, and total N in aboveground rye were analyzed by site. An inspection of residuals indicated that the assumption of normality was met. Soil NO_3-N (mg kg^{-1}), maize grain yield, stalk DM, silage yield, grain N (kg ha^{-1}), stalk N, silage N, and total above ground plant N were each analyzed for rye and no rye treatments separately across sites. Those variables were analyzed individually across sites and treatments. Soil NO_3-N was also analyzed for rye and no rye treatments when maize was the forecrop versus soybean and also when swine manure was used versus dairy. An inspection of residuals indicated that the assumption of normality was met.

3. Results

3.1. Weather

Monthly precipitation and air temperature data from this study are compared to the 30-year average (1981–2010) in Figure 1. In 2015 when the cover crop was planted (September through October), the weather tended to be warmer and drier than the 30-year average. This allowed for good field conditions for harvest of the prior crop, and the cover crops were planted earlier than in 2016 (Table 1). While temperatures were warmer than the 30-year average in the fall of 2016, harvest of the crop prior to the cover crop was delayed in many fields due to excessively wet conditions, thus delaying planting of the cover crop.

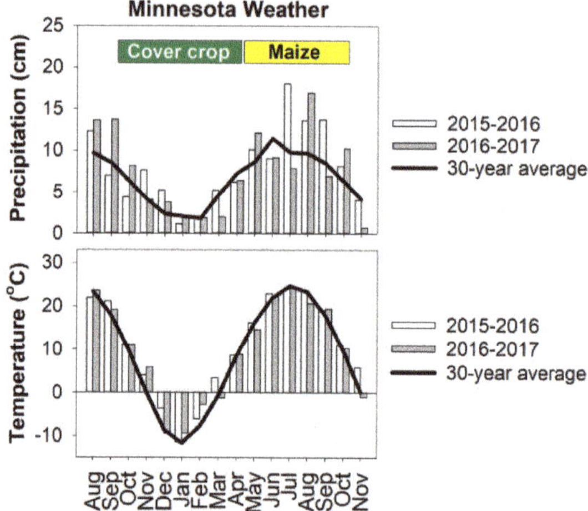

Figure 1. Average monthly precipitation (cm) and temperature (°C) during the study period compared with the 30-year mean for the period from 1981 to 2010. Data were averaged across nine National Weather Service weather stations in Minnesota.

Monthly average air temperature during winter (December through February) was normal or warmer than normal in both years. Precipitation was above average in both years in December, but near normal in January and February. On average, it was March when the 30-year temperature rose above the 3.3 °C threshold needed for rye vegetative growth to resume in the spring [45]. In 2016, it was warmer and wetter than usual, while in 2017 it was colder and drier than average. In April of both years, there was similar air temperature but lower precipitation than the 30-year average.

The maize growing season in both years had near-average temperatures, though precipitation was variable. In 2016, precipitation was above normal from May through October, with the exception of June. During this period, precipitation was 18.2 cm more than the 30-year average. In 2017, June, July, and September received below-normal precipitation while May, August, and October received more than average. From May through October, precipitation was 8.9 cm more than the 30-year average.

3.2. Rye Biomass, Rye N Uptake, and Soil NO₃-N

As indicated by spring biomass and density, recovery of rye following disturbance from manure injection was variable across sites (Figure 2, Table 3). The least disturbance occurred with smaller knives, with or without terminal sweeps. At most sites rye was terminated with herbicide and incorporated with tillage. Rye height at termination ranged from 5 to 30 cm among sites and was 25 to 30 cm at five sites (Table 3).

Aboveground rye DM at termination ranged from 102 to 3220 kg ha^{-1}. On average, rye growth was significantly greater ($p < 0.001$) in the 2016 maize growing season than in 2017 (mean of 1795 and 813 kg DM ha^{-1}, respectively), perhaps because rye planting dates were earlier in 2015. Rye DM at termination was greater when rye was planted earlier (Figure 3). Across sites, aboveground DM declined by 413 kg ha^{-1} with each one-week delay in rye planting. There was a significant positive linear relationship between rye DM (kg ha^{-1}) and height (cm) at termination (y = −444 + 99.88 × x, $R^2 = 0.76$, $p < 0.001$). At termination, N concentration in aboveground DM ranged from 29 to 53 g kg^{-1} and was not linearly related to rye aboveground DM ($R^2 = 0.001$, $p = 0.883$), total rye biomass N ($R^2 = 0.28$, $p = 0.490$), or rye height ($R^2 = 0.008$, $p = 0.712$). Total N uptake was primarily determined by rye DM (y = 2.25 + 0.038 × x, $R^2 = 0.92$, $p < 0.001$) and ranged from 5 to 114 kg ha^{-1} (Table 2).

(a)

(b)

(c)

Figure 2. An example of rye recovery after manure injection at site 4. Rye is shown at (**a**) manure injection in fall 2015, (**b**) two weeks after manure injection in fall 2015, and (**c**) at the time of termination in the spring of 2016.

Table 3. Rye aboveground biomass dry matter (DM), height, and density at termination.

Site	Biomass DM	Height *	Density
	kg ha^{-1}	cm	Plants m^{-2}
2016 Maize Growing Season			
1	3220 (572) †	30	116 (9)
2	1925 (224)	18	114 (5)
3	408 (34)	13	120 (8)
4	2526 (55)	25	128 (7)
5	2622 (301)	30	138 (7)
6	2002 (43)	28	74 (4)
7	1853 (160)	28	71 (6)
8	445 (10)	10	120 (7)
9	1153 (61)	23	84 (3)
2017 Maize Growing Season			
10	2141 (213)	15	160 (7)
11	590 (43)	10	107 (1)
12	1160 (27)	15	67 (7)
13	160 (5)	8	122 (6)
14	1760 (203)	20	178 (6)
15	609 (33)	20	79 (5)
16	102 (18)	5	51 (1)
17	1300 (92)	15	181 (2)
18	151 (5)	8	76 (9)
19	160 (10)	8	104 (4)

* Only one average measurement of height was made per site. † Standard error is shown within parentheses.

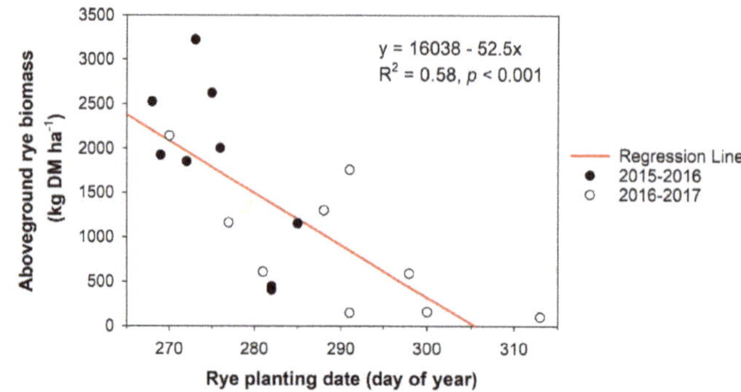

Figure 3. Aboveground rye biomass dry matter (DM) at termination in relation to rye planting date.

Across trials, the rye cover crop significantly reduced soil NO$_3$-N in the 0–60 cm layer at rye termination compared to no rye (Table 2, $p < 0.001$ for treatments, $p = 0.008$ for sites, $p = 0.159$ for treatment x site interaction). There were no differences within cover crop and no cover crop treatments when maize was the previous crop versus soybean ($p = 0.494$ and 0.056, respectively), nor when dairy manure was used versus swine manure ($p = 0.748$ and 0.113, respectively). There was not a significant linear relationship between rye biomass and soil NO$_3$-N. The amount of N supplied by manure the previous fall varied considerably among sites (Table 2) so the amount of NO$_3$-N available for uptake by the rye would have varied also. The amount of N taken up by the rye was not related to the difference between soil NO$_3$-N in the rye and no-rye treatment in each trial. When the difference in soil NO$_3$-N between cover cropped and bare plots was regressed on the number of days between rye planting and manure application, no significant relationship was found ($r = 0.204$, $p = 0.402$).

3.3. Maize Silage and Grain Yields

Across sites, maize silage and grain yields were not significantly influenced by cover cropping (p = 0.252 and p = 0.422, Tables 4 and 5). Similarly, there was no significant difference in maize N recovery in aboveground biomass with a winter rye cover crop compared to no cover crop (p = 0.412, Table 2).

Table 4. Maize silage yield (Mg ha^{-1} at 650 g kg^{-1} moisture) in plots following rye and no rye.

Site No.	Silage Yield	
	w/rye	w/o rye
	Mg ha^{-1}	
1	34.3 (0.7) *	39.8 (0.5)
2	59.1 (2.6)	57.9 (3.7)
4	48.1 (0.9)	46.5 (2.5)
10	46.5 (1.2)	49.8 (1.9)
11	43.6 (3.0)	43.4 (4.6)
12	39.0 (0.9)	41.5 (1.5)
14	48.6 (2.5)	48.6 (0.0)
15	48.1 (0.9)	48.4 (1.2)
Mean	45.7 (1.6)	46.9 (1.4)
Significance of difference with and without rye	n.s.	

* Standard error is shown within parentheses. n.s. = not significantly different.

Table 5. Maize grain yield (Mg ha^{-1} at 155 g kg^{-1} moisture) and stalk dry matter in plots following rye and no rye.

Site No.	Maize Grain Yield		Maize Stalk Dry Matter	
	w/rye	w/o rye	w/rye	w/o rye
	Mg ha^{-1}			
3	11.3 (0.6) *	11.1 (1.3)	5.5 (0.4)	5.7 (0.3)
5	12.4 (0.1)	11.8 (0.2)	8.4 (0.3)	7.9 (0.3)
6	14.0 (0.6)	14.5 (0.4)	8.7 (0.1)	8.0 (0.3)
7	13.8 (0.0)	14.1 (0.1)	7.2 (0.5)	6.6 (0.2)
8	12.9 (0.9)	13.1 (0.6)	6.5 (0.5)	7.0 (0.5)
9	10.0 (0.3)	10.5 (0.2)	5.1 (0.4)	6.1 (0.2)
13	12.6 (0.5)	13.3 (0.3)	7.5 (0.4)	7.2 (0.2)
16	10.2 (0.1)	10.3 (0.3)	7.2 (0.3)	8.0 (0.1)
17	11.7 (0.1)	12.5 (0.1)	7.9 (0.3)	7.9 (0.1)
18	14.4 (0.3)	13.7 (0.4)	7.5 (0.4)	7.3 (0.2)
19	13.8 (0.5)	13.8 (0.4)	7.7 (0.2)	7.6 (0.2)
Mean	12.5 (0.3)	12.6 (0.3)	7.2 (0.2)	7.2 (0.1)
Significance of difference with and without rye	n.s.		n.s.	

* Standard error is shown within parentheses. n.s. = not significantly different.

4. Discussion

A winter rye cover crop was established in 2015 and 2016 across 19 trials in southern and central Minnesota using drill-seeding after maize silage or soybean harvest. In both years, rye recovered after fall liquid manure injection; however, wide disk coverers that disturbed or covered most of the soil surface (sites 6 and 12) were associated with low rye density at termination. Singer et al. found that although manure injection reduced rye density in the disturbed zone when using 5.1-cm-wide chisel shanks, the rye biomass the following spring had fully recovered compared to the no-manure check [34]. Similarly, Milliron et al. reported that rye with manure injected with shallow disk injectors produced similar amounts of aboveground DM as when manure was applied prior to seeding the rye cover crop [35]. This illustrates the importance of using appropriate equipment that minimizes surface disturbance to reduce damage to a cover crop stand.

Rye establishment was more challenging in 2016 than in 2015 due to wide-spread precipitation which delayed harvest of the prior crop, indicating that post-harvest establishment of rye may not be successful in some years. An analysis of weather patterns in the upper Midwest USA by Strock et al. suggested that successful establishment post-harvest may only occur 25% of the time [46], although the paper only considered "successful" establishment to be the year when 2700 kg ha^{-1} of aboveground rye biomass was produced. Only one site of 19 in the current study reported greater than this amount. Others in the region have reported lower amounts of aboveground DM production that likely reflect more realistic production goals. For example, rye DM production ranged from 147 to 489 kg ha^{-1} by May 1 in southern Minnesota, USA [47]. In southwestern Minnesota, USA, Krueger et al. reported 680 to 872 kg ha^{-1} of rye DM produced by late April [48]. In northwestern Iowa, USA, 1480 to 2740 kg ha^{-1} of rye DM was produced by mid- to late-April when rye was drilled after harvest of the previous crop [49]. None of these researchers injected manure into the cover crop, potentially damaging the stand, however. In the current study which included manure injection, rye DM production tended to be similar to or higher than values in the literature in Minnesota, USA, and similar to those reported in Iowa, USA, which is further south and has a slightly warmer climate. This suggests the manure was beneficial for growth, despite possible damage to the rye stand during manure application.

One of the main goals of using a grass species for a cover crop is to capture nutrients that might otherwise be lost over the non-growing season. In this study, winter rye was successful in this regard. Not only did it hold N in the aboveground DM, differences were also found in spring soil N levels between the cover cropped and non-cover cropped treatments. Generally speaking, N uptake by the cover crop was related to DM production, which has been found in other studies as well [50,51]. This led to a wide range of total N uptake, from 4 to 114 kg N ha^{-1}, across the 19 sites. This variability is not uncommon in other cover crop studies. Rye N uptake ranged from 9 to 60 kg N ha^{-1} in Nebraska [52], 11 to 26 kg N ha^{-1} in Iowa [53], and 18.8 to 34.2 kg N ha^{-1} in Minnesota [48]. Nitrogen taken up by the rye is less likely to be lost through agricultural sub-surface drainage or by leaching from the soil [22,46]. Rye reduced soil NO_3-N in the 0–60 cm depth at the time of rye termination in the spring by an average of 36% across all 19 sites. There was considerable variability in the level of soil NO_3-N among sites ($p = 0.008$ for sites), however. Similarly, Krueger et al. found that soil nitrate was reduced by approximately 35% with a rye cover crop compared with the no-cover control in Minnesota, USA [48]. Cambardella et al. found a slightly higher reduction of 41% in Iowa, USA, which has a warmer climate, with a rye/oat mix [54]. In the current study, variability across sites was likely due to differences in rye biomass production as well as site-specific weather and soil conditions. For example, higher than expected soil NO_3-N concentration (based on the amount of N applied) in both cover cropped and bare plots may have been due to fall and spring mineralization of N from manure and soil organic matter. Where soil NO_3-N concentration was lower than expected, there are two possible scenarios. Nitrogen in the ammonium form had not yet mineralized to NO_3-N or the NO_3-N was leached or denitrified during wet fall and/or spring conditions and lost.

Across sites, maize grain and silage yields were not affected by cover cropping. This may have been due to N supply exceeding the N requirements of maize throughout its growth cycle, as maize aboveground N uptake with a winter rye cover crop was not significantly different from that with no cover crop. In addition to N supply from manure (and fertilizer at 10 of 19 sites), a large amount of N was likely supplied by soil N mineralization. Soils at 15 of 19 sites were Mollisols, which have high N mineralization capacity compared to other soil orders [55], and soil organic matter was relatively high (42 to 83 g kg^{-1} in the 0–15 cm soil layer) at 15 of 19 sites and moderate (27 to 32 g kg^{-1} in the 0–15 cm soil layer) at the remaining sites. It is the general trend in most cover cropping studies that yields of the following crop are positively or minimally impacted [31,46,56–58], but others have found negative impacts on yield dependent on management technique. For example, Acharya et al. suggested that the timing of when the cover crop is killed may influence plant disease and stand establishment of the following crop [59]. Crandall et al. found that when termination was delayed to a week prior to planting maize and fertilizer application was delayed until the V6 growth stage, maize yield was decreased [42]. Other studies have suggested that there may be an alleopathic effect of rye [60,61]. Few studies have evaluated the impact of integrating fall-applied manure with cover crops on the following maize. Krueger et al. found that in a system with fall-applied dairy manure, rye terminated a few days prior to planting maize for silage reduced yields compared with rye terminated approximately three to four weeks earlier [48]. Thilakarathna et al. reported that in a fall-applied swine manure system, non-legume cover crops did not impact maize yields, although cereal rye was not evaluated [33]. In the current study, the cover crop was terminated at or before reaching 25 cm height and 2.5 Mg ha^{-1} dry matter at most sites, and then incorporated into the soil. Tillage incorporation of the cover crop may have facilitated N mineralization for the following crop [26,62] since different cover crop termination methods and timing may affect N availability for and performance of the subsequent maize crop [42,59]. It is also possible that mineralization of organically bound N from the manure offset immobilization of N by the rye cover crop early in the growing season. More research is needed to understand the dynamics of nutrient release from cover crops in a manured system.

5. Conclusions

In these 19 trials conducted in the upper Midwest USA, 15 of which were carried out by commercial growers with their own equipment and management, a rye cover crop was successfully established by drill-seeding following harvest of maize for silage or soybean. Although aboveground fall biomass was limited both by the short growth period prior to freezing and by disturbance from the manure injection equipment, spring growth was sufficient to result in significant N uptake. This study was conducted at sites with predominant occurrence of Mollisols and relatively high soil organic matter levels, hence high soil N mineralization capacity. This, coupled with termination of the rye prior to the reproductive stage by herbicide and tillage and subsequent release of N, may have been partially responsible for the lack of yield reduction of the maize crop following rye compared to no rye. While removing N from exposure to leaching, as demonstrated in these trials as well as in earlier trials in Minnesota, USA [46], Iowa, USA [54] and Illinois, USA [57], the rye cover crop can also reduce soil erosion following the low residue crops of maize harvested for silage and soybean [63,64], an increasing threat as a changing climate in the Upper Midwest results in more intense rainstorms [65]. Reduced nitrate loss to groundwater and surface water and reduced soil erosion, increase the sustainability and reduce the environmental impact of the production system. Future research should focus on understanding the dynamics of nutrient release from grass legumes in a manured system.

Supplementary Materials: The following is available online at http://www.mdpi.com/2073-4395/9/12/852/s1, Table S1: Soil classification, texture, and initial soil-test levels (0–15 cm depth) at 19 field research locations in Minnesota, USA, Table S2: Forms, rates, and dates of fertilizer nitrogen (N) application at the 10 sites where applied.

Author Contributions: Conceptualization, L.A.E. and R.J.P.; data curation, L.A.E. and M.L.W.; formal analysis, L.A.E. and J.A.C.; funding acquisition, L.A.E.; investigation, L.A.E. and R.J.P.; methodology, L.A.E. and R.J.P.; project administration, L.A.E.; resources, L.A.E. and R.J.P.; supervision, L.A.E.; validation, L.A.E., R.J.P., and J.A.C.; visualization, L.A.E., J.A.C., and M.L.W.; writing—original draft, L.A.E., M.L.W. and J.A.C.; writing—review and editing, L.A.E., M.L.W. and J.A.C.

Funding: This research was funded by a Section 319 grant from the U.S. Environmental Protection Agency through the Minnesota Pollution Control Agency, grant number 103675/PO3000015229, and by funds from the Minnesota Agricultural Experiment Station, University of Minnesota.

Acknowledgments: On-farm trials were carried out with the assistance of University of Minnesota Extension Educators Karen Anderson, Daniel Martens, Emily Wilmes, Sarah Sheick, Brenda Miller, Melissa Runck, Lizabeth Stahl, Diane DeWitt, Brad Carlson, Gregory Klinger, Jason Ertl, and Rodney Greder, as well as University of Minnesota West Central Research and Outreach Center faculty and staff Bradley Heins and Curt Reese. Thanks also to the agricultural growers who hosted the on-farm trials.

Conflicts of Interest: The authors declare no conflicts of interest. The funders had no role in the design of the study; in the collection, analyses, or interpretation of data; in the writing of the manuscript, or in the decision to publish the results.

References

1. Diaz, R.J.; Rosenberg, R. Spreading Dead Zones and Consequences for Marine Ecosystems. *Science* **2008**, *321*, 926–929. [CrossRef] [PubMed]

2. Rabalais, N.N.; Turner, R.E.; Justić, D.; Dortch, Q.; Wiseman, W.J.; Gupta, B.K.S.; Justic, D. Nutrient Changes in the Mississippi River and System Responses on the Adjacent Continental Shelf. *Estuaries* **1996**, *19*, 386–407. [CrossRef]

3. Goolsby, D.A.; Battaglin, W.A.; Lawrence, G.B.; Artz, R.S.; Aulenbach, B.T.; Hooper, R.P.; Keeney, D.R.; Stensland, G.J. *Flux and Sources of Nutrients in the Mississippi-Atchafalaya River Basin: Topic 3 Report for the Integrated Assessment on Hypoxia in the Gulf of Mexico*; NOAA/National Centers for Coastal Ocean Science: Silver Spring, MD, USA, 1999.

4. David, M.B.; Gentry, L.E.; Kovacic, D.A.; Smith, K.M. Nitrogen Balance in and Export from an Agricultural Watershed. *J. Environ. Qual.* **1997**, *26*, 1038–1048. [CrossRef]

5. Randall, G.W.; Mulla, D.J. Nitrate Nitrogen in Surface Waters as Influenced by Climatic Conditions and Agricultural Practices. *J. Environ. Qual.* **2001**, *30*, 337–344. [CrossRef] [PubMed]

6. Lawlor, P.A.; Helmers, M.J.; Baker, J.L.; Melvin, S.W.; Lemke, D.W. Nitrogen Application Rate Effect on Nitrate-Nitrogen Concentration and Loss in Subsurface Drainage for a Corn-Soybean Rotation. *Trans. ASABE* **2008**, *51*, 83–94. [CrossRef]

7. Sharpley, A.; Foy, B.; Withers, P. Practical and Innovative Measures for the Control of Agricultural Phosphorus Losses to Water: An Overview. *J. Environ. Qual.* **2000**, *29*, 1–9. [CrossRef]

8. McLellan, E.; Robertson, D.; Schilling, K.; Tomer, M.; Kostel, J.; Smith, D.; King, K. Reducing Nitrogen Export from the Corn Belt to the Gulf of Mexico: Agricultural Strategies for Remediating Hypoxia. *J. Am. Water Resour. Assoc.* **2015**, *51*, 263–289. [CrossRef]

9. Ribaudo, M.; Delgado, J.; Hansen, L.; Livingston, M.; Mosheim, R.; Williamson, J. *Nitrogen in Agricultural Systems: Implications for Conservation Policy*; United Statesd Department of Agriculture-Economic Research Service: Washington, DC, USA, 2011.

10. USDA-NASS. *Crop. Production 2018 Summary*; United Statesd Department of Agriculture-National Agricultural Statistics Service: Washington, DC, USA, 2019.

11. Minnesota Department of Agriculture. *Fertilizer and Manure Selection and Management Practices Associated with Minnesota's 2010 Corn and Wheat Production*; Minnesota Department of Agriculture: Saint Paul, MN, USA, 2014.

12. Minnesota Department of Agriculture. *Commercial Nitrogen and Manure Applications on Minnesota's 2014 Corn Crop Compared to the University of Minnesota Nitrogen Guidelines*; Minnesota Department of Agriculture: Saint Paul, MN, USA, 2014.

13. Andresen, J.; Hilberg, S.; Kunkel, K. Historical Climate and Climate Trends in the Midwestern USA. In *U.S. National Climate Assessment Midwest Technical Input Report.*; Winkler, J., Andresen, J., Hatfield, J., Bidwell, D., Brown, D., Eds.; Great Lakes Integrated Sciences and Assessments (GLISA) Center: Ann Arbor, MI, USA, 2012.

14. Abendroth, L.J.; Elmore, R.W.; Boyer, M.J.; Marlay, S.K. *Corn Growth and Development*; Iowa State University Extension: Ames, IA, USA, 2011.

15. American Society of Agricultural and Biological Engineers (ASABE). *ASAE D384.2 Manure Production and Characteristics*; American Society of Agricultural and Biological Engineers: St. Joseph, MI, USA, 2005; (R2019); pp. 1–32.

16. Sabey, B.R.; Bartholomew, W.V.; Shaw, R.; Pesek, J. Influence of Temperature on Nitrification in Soils. *Soil Sci. Soc. Am. J.* **1956**, *20*, 357–360. [CrossRef]

17. Chang, C.; Entz, T. Nitrate Leaching Losses Under Repeated Cattle Feedlot Manure Applications in Southern Alberta. *J. Environ. Qual.* **1996**, *25*, 145–153. [CrossRef]

18. Donner, S.D.; Kucharik, C.J.; Foley, J.A. Impact of Changing Land Use Practices on Nitrate Export by the Mississippi River. *Glob. Biogeochem. Cycles* **2004**, *18*. [CrossRef]

19. Novotny, E.V.; Stefan, H.G. Stream Flow in Minnesota: Indicator of Climate Change. *J. Hydrol.* **2007**, *334*, 319–333. [CrossRef]

20. Ditsch, D.C.; Alley, M.M.; Kelley, K.R.; Lei, Y.Z. Effectiveness of Winter Rye for Accumulating Residual Fertilizer N Following Corn. *J. Soil Water Conserv.* **1993**, *48*, 125–132.

21. Coale, F.J.; Costa, J.M.; Bollero, G.A.; Schlosnagle, S.P. Small Grain Winter Cover Crops for Conservation of Residual Soil Nitrogen in the Mid-Atlantic Coastal Plain. *Am. J. Altern. Agric.* **2001**, *16*, 66–72. [CrossRef]

22. McCracken, D.V.; Smith, M.S.; Grove, J.H.; Blevins, R.L.; MacKown, C.T. Nitrate Leaching as Influenced by Cover Cropping and Nitrogen Source. *Soil Sci. Soc. Am. J.* **1994**, *58*, 1476–1483. [CrossRef]

23. Shah, S.; Hookway, S.; Pullen, H.; Clarke, T.; Wilkinson, S.; Reeve, V.; Fletcher, J.M. The Role of Cover Crops in Reducing Nitrate Leaching and Increasing Soil Organic Matter. *Asp. Appl. Biol.* **2017**, *134*, 243–251.

24. Meisinger, J.J.; Ricigliano, K.A. Nitrate Leaching from Winter Cereal Cover Crops Using Undisturbed Soil-Column Lysimeters. *J. Environ. Qual.* **2017**, *46*, 576–584. [CrossRef]

25. Huntington, T.G.; Grove, J.H.; Frye, W.W. Release and Recovery of Nitrogen from Winter Annual Cover Crops in No-till Corn Production. *Commun. Soil Sci. Plant. Anal.* **1985**, *16*, 193–211. [CrossRef]

26. Jahanzad, E.; Barker, A.V.; Hashemi, M.; Eaton, T.; Sadeghpour, A.; Weis, S.A. Nitrogen Release Dynamics and Decomposition of Buried and Surface Cover Crop Residues. *Agron. J.* **2016**, *108*, 1735–1741. [CrossRef]

27. Finney, D.M.; White, C.M.; Kaye, J.P. Biomass Production and Carbon/Nitrogen Ratio Influence Ecosystem Services from Cover Crop Mixtures. *Agron. J.* **2016**, *108*, 39–52. [CrossRef]

28. Miguez, F.E.; Bollero, G.A. Review of Corn Yield Response under Winter Cover Cropping Systems Using Meta-Analytic Methods. *Crop Sci.* **2005**, *45*, 2318–2329. [CrossRef]

29. Gabriel, J.L.; Quemada, M. Replacing Bare Fallow with Cover Crops in a Maize Cropping System: Yield, N Uptake and Fertiliser Fate. *Eur. J. Agron.* **2011**, *34*, 133–143. [CrossRef]

30. Ball-Coelho, B.R.; Roy, R.C. Overseeding Rye into Corn Reduces NO_3 Leaching and Increases Yields. *Can. J. Soil Sci.* **1997**, *77*, 443–451. [CrossRef]

31. Ball Coelho, B.R.; Roy, R.C.; Bruin, A.J. Long-Term Effects of Late-Summer Overseeding of Winter Rye on Corn Grain Yield and Nitrogen Balance. *Can. J. Plant. Sci.* **2005**, *85*, 543–554. [CrossRef]

32. Komainda, M.; Taube, F.; Kluß, C.; Herrmann, A. Effects of Catch Crops on Silage Maize (*Zea mays L.*): Yield, Nitrogen Uptake Efficiency and Losses. *Nutr. Cycl. Agroecosyst.* **2018**, *110*, 51–69. [CrossRef]

33. Thilakarathna, M.S.; Serran, S.; Lauzon, J.; Janovicek, K.; Deen, B. Management of Manure Nitrogen Using Cover Crops. *Agron. J.* **2015**, *107*, 1595–10607. [CrossRef]

34. Singer, J.W.; Cambardella, C.A.; Moorman, T.B. Enhancing Nutrient Cycling by Coupling Cover Crops with Manure Injection. *Agron. J.* **2008**, *100*, 1735–1739. [CrossRef]

35. Milliron, R.A.; Karsten, H.D.; Beegle, D.B. Influence of Dairy Slurry Manure Application Method, Fall Application-Timing, and Winter Rye Management on Nitrogen Conservation. *Agron. J.* **2019**, *111*, 1–15. [CrossRef]

36. Laverty, J.C. A Modified Procedure for Determination of Phosphorus in Soil Extracts. *Soil Sci. Soc. Am. Proc.* **1963**, *27*, 360–361. [CrossRef]

37. Frank, K.; Beegle, D.; Denning, J. Phosphorus. In *Recommended Chemical Soil Test Procedures for the North Central Region*; Brown, J.R., Ed.; Missouri Agricultural Experiment Station SB 1001: Columbia, MO, USA, 1998.

38. Warncke, D.; Brown, J.R. Potassium and Other Basic Cations. In *Recommended Chemical Soil Test Procedures for the North Central Region*; Brown, J.R., Ed.; Missouri Agricultural Experiment Station SB 1001: Columbia, MO, USA, 1998.

39. Peters, J.B.; Nathan, M.V.; Laboski, C.A.M. PH and Lime Requirement. In *Recommended Chemical Soil Test Procedures for the North Central Region*; Brown, J.R., Ed.; Missouri Agricultural Experiment Station SB 1001: Columbia, MO, USA, 1998.

40. Combs, S.M.; Nathan, M.V. Soil Organic Matter. In *Recommended Chemical Soil Test Procedures for the North Central Region*; Brown, J.R., Ed.; Missouri Agricultural Experiment Station SB 1001: Columbia, MO, USA, 1998.

41. Peters, J.; Combs, S.; Hoskins, B.; Jarman, J.; Kovar, J.; Watson, M.; Wolf, A.; Wolf, N. *Recommended Methods of Manure Analysis*; Peters, J., Ed.; Cooperative Extension Publishing: Madison, WI, USA, 2003.

42. Crandall, S.M.; Ruffo, M.L.; Bollero, G.A. Cropping System and Nitrogen Dynamics under a Cereal Winter Cover Crop Preceding Corn. *Plant. Soil* **2005**, *268*, 209–219. [CrossRef]

43. Gelderman, R.H.; Beegle, D. Nitrate-Nitrogen. In *Recommended Chemical Soil Test Procedures for the North Central Region*; Brown, J.R., Ed.; Missouri Agricultural Experiment Station SB 1001: Columbia, MO, USA, 1998.

44. Huffman, S.A.; Barbarick, K.A. Soil Nitrate Analysis by Cadmium Reduction. *Commun. Soil Sci. Plant. Anal.* **1981**, *12*, 79–89. [CrossRef]

45. SARE. *Managing Cover Crops Profitably*, 3rd ed.; Sustainable Agriculture Research and Education (SARE): College Park, MD, USA, 2012.

46. Strock, J.S.; Porter, P.M.; Russelle, M.P. Cover Cropping to Reduce Nitrate Loss through Subsurface Drainage in the Northern U.S. Corn Belt. *J. Environ. Qual.* **2004**, *33*, 1010–1016. [CrossRef] [PubMed]

47. De Bruin, J.L.; Porter, P.M.; Jordan, N.R. Use of a Rye Cover Crop Following Corn in Rotation with Soybean in the Upper Midwest. *Agron. J.* **2005**, *97*, 587–598. [CrossRef]

48. Krueger, E.S.; Ochsner, T.E.; Porter, P.M.; Baker, J.M. Winter Rye Cover Crop Management Influences on Soil Water, Soil Nitrate, and Corn Development. *Agron. J.* **2011**, *103*, 316–323. [CrossRef]

49. Kaspar, T.C.; Jaynes, D.B.; Parkin, T.B.; Moorman, T.B. Rye Cover Crop and Gamagrass Strip Effects on NO3 Concentration and Load in Tile Drainage. *J. Environ. Qual.* **2007**, *36*, 1503–1511. [CrossRef] [PubMed]

50. Wilson, M.L.; Baker, J.M.; Allan, D.L. Factors Affecting Successful Establishment of Aerially Seeded Winter Rye. *Agron. J.* **2013**, *105*, 1868–1877. [CrossRef]

51. Clark, A.J.; Decker, A.M.; Meisinger, J.J.; McIntosh, M.S. Kill Date of Vetch, Rye, and a Vetch-Rye Mixture: I. Cover Crop and Corn Nitrogen. *Agron. J.* **1997**, *89*, 427–434. [CrossRef]

52. Kessavalou, A.; Walters, D.T. Winter Rye Cover Crop Following Soybean Under Conservation Tillage. *Agron. J.* **1999**, *91*, 643–649. [CrossRef]

53. Pantoja, J.L.; Woli, K.P.; Sawyer, J.E.; Barker, D.W. Corn Nitrogen Fertilization Requirement and Corn–Soybean Productivity with a Rye Cover Crop. *Soil Sci. Soc. Am. J.* **2015**, *79*, 1482–1895. [CrossRef]

54. Cambardella, C.A.; Moorman, T.B.; Singer, J.W. Soil Nitrogen Response to Coupling Cover Crops with Manure Injection. *Nutr. Cycl. Agroecosyst* **2010**, *87*, 383–393. [CrossRef]

55. Brady, N.; Weil, R. *The Nature and Properties of Soils*, 13th ed.; Pearson/Prentice Hall: Upper Saddle River, NJ, USA, 2002.

56. Basche, A.D.; Kaspar, T.C.; Archontoulis, S.V.; Jaynes, D.B.; Sauer, T.J.; Parkin, T.B.; Miguez, F.E. Soil Water Improvements with the Long-Term Use of a Winter Rye Cover Crop. *Agric. Water Manag.* **2016**, *172*, 40–50. [CrossRef]

57. Ruffo, M.L.; Bullock, D.G.; Bollero, G.A. Soybean Yield as Affected by Biomass and Nitrogen Uptake of Cereal Rye in Winter Cover Crop Rotations. *Agron. J.* **2004**, *96*, 800–805. [CrossRef]

58. Duiker, S.W.; Curran, W.S. Rye Cover Crop Management for Corn Production in the Northern Mid-Atlantic Region. *Agron. J.* **2005**, *97*, 1413–1418. [CrossRef]

59. Acharya, J.; Bakker, M.G.; Moorman, T.B.; Kaspar, T.C.; Lenssen, A.W.; Robertson, A.E. Time Interval Between Cover Crop Termination and Planting Influences Corn Seedling Disease, Plant Growth, and Yield. *Plant. Dis.* **2017**, *101*, 591–600. [CrossRef] [PubMed]

60. Raimbault, B.A.; Vyn, T.J.; Tollenaar, M. Corn Response to Rye Cover Crop Management and Spring Tillage Systems. *Agron. J.* **1990**, *82*, 1088–1093. [CrossRef]

61. Tollenaar, M.; Mihajlovic, M.; Vyn, T.J. Annual Phytomass Production of a Rye-Corn Double-Cropping System in Ontario. *Agron. J.* **1992**, *84*, 963–967. [CrossRef]

62. Wilson, D.O.; Hargrove, W.L. Release of Nitrogen from Crimson Clover Residue under Two Tillage Systems1. *Soil Sci. Soc. Am. J.* **1986**, *50*, 1251–1254. [CrossRef]

63. Kaspar, T.C.; Radke, J.K.; Laflen, J.M. Small Grain Cover Crops and Wheel Traffic Effects on Infiltration, Runoff, and Erosion. *J. Soil Water Conserv.* **2001**, *56*, 160–164.

64. De Baets, S.; Poesen, J.; Meersmans, J.; Serlet, L. Cover Crops and Their Erosion-Reducing Effects during Concentrated Flow Erosion. *Catena* **2011**, *85*, 237–244. [CrossRef]

65. Easterling, D.R.; Kunkel, K.E.; Arnold, J.R.; Knutson, T.; Legrande, A.N.; Leung, L.R.; Vose, R.S.; Wal-Iser, D.E.; Wehner, M.F.; Fahey, D.J.W.; et al. *Precipitation Change in the United States*; Wuebbles, D.J., Fahey, D.W., Hibbard, K.A., Dokken, D.J., Stewart, B.C., Maycock, T.K., Eds.; U.S. Global Change Research Program: Washington, DC, USA, 2017. [CrossRef]

 *agronomy*

Article

Faba Bean and Pea Can Provide Late-Fall Forage Grazing without Affecting Maize Yield the Following Season

Bryce J. Andersen [1], Dulan P. Samarappuli [2], Abbey Wick [1] and Marisol T. Berti [2,*

[1] Department of Soil Sciences, North Dakota State University, Fargo, ND 58108, USA; bryce.j.andersen@ndsu.edu (B.J.A.); abbey.wick@ndsu.edu (A.W.)

[2] Department of Plant Sciences, North Dakota State University, PO Box 6050, Fargo, ND 58108, USA; dulan.samarappuli@ndsu.edu

* Correspondence: marisol.berti@ndsu.edu; Tel.: +1-701-231-6110

Received: 19 November 2019; Accepted: 31 December 2019; Published: 6 January 2020

Abstract: Faba bean (*Vicia faba* Roth) and pea (*Pisum sativum* L.) are grown worldwide as protein sources for food and feed and can be used as cover crops after wheat (*Triticum aestivum* L.). However, faba bean is underutilized in upper Midwest farming systems. This study was conducted to determine how faba bean relates to pea as a forage, cover crop, and in cycling of nutrients to maize (*Zea mays* L.) in the following season. Five faba bean cultivars and two pea cultivars, a forage pea and a field pea, were established after wheat harvest in North Dakota, in 2017 and 2018. Faba bean and pea cultivars averaged 1.3 Mg ha^{-1} of biomass, enough to support 1.5 animal unit month (AUM) ha^{-1} for a 450 kg cow (*Bos taurus* L.) with calf, at 50% harvest efficiency. Crude protein content was highest in faba bean cv. Boxer (304 g kg^{-1}), with faba bean cv. Laura and forage pea cv. Arvika having similar content, and field pea having the least (264 g kg^{-1}). Cover crop treatments did not affect maize in the following year, indicating no nutrient cycling from faba bean and pea to maize. Both cover crop species tested provided high protein forage, suitable for late grazing, with a more fibrous crop residue. Faba bean has potential as a cover crop in the upper Midwest while providing greater quality forage than pea.

Keywords: faba bean; forage pea; fall grazing; cover crop; catch crop; nutrient cycling

---

1. Introduction

The Natural Resources Conservation Service (NRCS) defines cover crops as grasses, legumes, and forbs sown for seasonal vegetative cover [1]. Cover crops may be established between successive production crops, companion-sown, or relay-sown into production crops. Species and sowing dates that will not compete with the production crop yield or harvest may be used. According to the NRCS [1], cover crops may be used to reduce water and wind erosion, maintain or increase soil health and organic matter, increase water quality, suppress weeds and break pest cycles, enhance soil water conservation, and minimize soil compaction. Cover crops may be grazed as long as the conservation purposes are not compromised. Different species of cover crops are also often sown together in mixtures to fulfill multiple purposes at the same time [1].

Cover crops are gaining popularity throughout the USA, rising from 4 million ha in 2012 to 6 million ha in 2017 [2]. North Dakota's cover crop acreage increased 89%, from 86,500 ha in 2012 to 163,600 ha in 2017 [2]. The use of cover crops is an important addition to upper Midwest farming systems in order to rebuild soil and reduce soil erosion [3].

One important use of legume cover crops is fall grazing, which can complement crop residue grazing. For cattle, crop residue grazing in the fall saves stored forage resources for winter feeding. In beef cattle production, feed is the most expensive part of the operation [4]. Wheat harvested in

August leaves ample time for grazing of residue, and cover crop growth throughout the fall and into winter. Distributing cattle to graze, and effectively reducing the manure concentration in an area, could help ensure cattle health by reducing the spread of disease, along with reducing the need for manure management [5].

Pea is used for grazing and hay in mixtures with barley (*Hordeum vulgare* L.) or oat (*Avena sativa* L.) in the upper Midwest. Pea alone yields less than 2 Mg ha^{-1}, which can increase to 5.7 Mg ha^{-1} when mixed with oat or forage barley [6]. Although differences in crude protein (CP) and biomass are not always seen when comparing forage and semi-leafless pea [7,8], depending on cultivar, forage pea can achieve greater biomass and crude protein yield than semi-leafless pea [9]. According to Anderson and Ilse [6], freshly weaned calves preferred pea and pea-barley hay to grass hay. Calves on pea and pea-barley hay, respectively, gained 304 g d^{-1} and 240 g d^{-1} more than calves on rations with grass hay alone, making pea hay 230% greater in value than the grass hay [6]. Due to the high CP content of some legumes leading to possible digestion issues, Amiri and Shariff [10] concluded that a combination of legume and grass species is safe a way to provide the needed feeding and protein requirements of grazing livestock.

Faba bean is not common in the upper Midwest, compared with other countries, where it is usually used in mixtures with oat, triticale (*Triticosecale* x Witt.), or barley [11–14]. Forage yield of oat-faba bean and triticale-faba bean mixtures fluctuated between 10 and 22 Mg ha^{-1} and total CP yield ranged between 1.0 and 3.3 Mg ha^{-1} in studies conducted in Greece [11,12]. In Canada, a faba bean–barley mixture had greater CP than a barley and barley–pea mixture [13], and silage maize–faba bean than maize [14]. Furthermore, faba bean silage had the highest CP content (220 g kg^{-1}) compared with pea (178 g kg^{-1}), and soybean [*Glycine max* (L.) Merr.] (197 g kg^{-1}) [15]. Lambs (*Ovis aries* L.) grazing on faba bean grew significantly faster (220 g head^{-1} day^{-1}) than lambs grazing on field pea (186 g d^{-1}) [16].

There are significant differences in both biomass quantity and quality between faba bean cultivars. Wegi et al. [17] found biomass production between five cultivars ranged from 3.3 to 5.1 Mg ha^{-1}. When comparing faba bean with forage pea, Iglesias and Lloveras [18] found that faba bean produced more biomass at pod fill than at initial flowering, but pea produced more biomass than most other cool-season legumes at earlier stages of development. Wichmann et al. [19] also found that pea had faster biomass accumulation than faba bean or blue lupin (*Lupinus angustifolius* L.).

Faba bean is grown as a winter annual in warm, temperate and subtropical areas. The hardiest European cultivars are able to tolerate temperatures down to -15 °C in vegetative stage without serious injury, but optimum temperatures for production are typically from 18 to 27 °C [20]. Field pea grown in the upper Midwest have shown ability to grow at temperatures down to -3 °C [21]. When compared with field pea, faba bean had slightly less winter-kill.

Faba bean has the highest reliance on N$_2$ fixation for growth in comparison with other cool-season legumes, such as pea and lupin, which leads to high N credit for the following crops [22–24]. Faba bean has been shown to attain high amounts of N derived from the atmosphere (Ndfa), ranging between 75% and 90% of its total shoot N [22–24], whereas pea Ndfa ranged between 50% and 70% [20,23,24]. Jensen et al. [20] reported that, on a global average, faba bean fixes 154 kg N ha^{-1} and pea fixes 86 kg N ha^{-1}. In addition, annual legumes can reduce NO$_3$-N leaching by scavenging residual soil NO$_3$-N, with shoot uptake values ranging from 92 to 276 kg N ha^{-1} for faba bean [20] and 104 kg N ha^{-1} for pea [23].

The possible legume benefits to the following crop are well known, but the data are inconsistent. Stevenson and van Kessell [25] concluded that wheat yield consistently was 43% greater when preceded by pea compared with wheat, with up to 14 of the extra 27 kg N ha^{-1} (Ndfa -total N content) accumulated in the wheat attributed to fixed N$_2$ from the pea and the rest to non-N-related benefits. Beckie et al. [26] had similar results, finding that the benefit of pea residue to the following crop was 25 kg N ha^{-1}, whereas Lupwayi and Soon [27] found that only 7.5 kg N ha^{-1} was released by pea residue to the subsequent crop. Faba bean was the only crop with a positive N balance after harvest when compared with lupin, pea, and oat [23], indicating it would be the only one providing soil mineral N to following crops [23]. Cupina et al. [28] found that field pea contributed 165 kg N ha^{-1} to the following crop after it was used as a cover crop over a mild winter. Couedel et al. [29] found that legume cover crops grown in the fall provided 35 to 54 kg N ha^{-1} as green manure to the following crop. These differences can be attributed to differences in soil type, moisture content, management, weather, especially rainfall, and how these factors affect the mineralization of plant tissue [20,27].

The North Dakota State University Extension [30] has evaluated faba bean as grain for food, however, very limited information and research has been done on faba bean as a cover crop and forage for grazing. In 2017, 3367 ha of faba bean for grain were reported in North Dakota [31]. Preliminary studies in North Dakota indicate that faba bean produces up to 680 kg of biomass yield with high crude protein and high digestibility for ruminants when sown in August after wheat in the northern Great Plains [3]. Faba bean has the potential to become an important cover crop in wheat-based or maize–soybean-based cropping systems.

The aim of this study was to evaluate forage yield and quality, and the effect on the following spring crop, along with other legume cover crop advantages of faba bean grown in the fall after wheat harvest compared with field and forage pea. This information could be used to provide information on fall cover crop use, late-season grazing opportunity, and the effects of late-season legume cover crops on soil NO_3–N in the fall and spring before maize sowing.

2. Materials and Methods

2.1. Field Establishment and Experimental Design

Experiments were conducted in 2017 and 2018 at two North Dakota State University (NDSU) research locations at Prosper, ND (−97°1143′ W, 46°9997′ N; 281 m elevation), and Hickson, ND (−96°8259′ W, 46°6335′ N; 281 m elevation). Location–year combinations will be referred to as environments henceforward. The soil type in Prosper is a Kindred–Bearden silty clay loam (Fine-silty, mixed, superactive frigid Typic Endoaquolls; Bearden: Fine-silty, mixed, superactive, frigid Aeric Calciaquolls), and the soil type in Hickson is a Hagne–Fargo silty clay loam (Hagne: Fine, smectitic, frigid Typic Calciaquerts; Fargo: Fine, smectitic, frigid Typic Epiaquerts) [32] Daily temperature and rainfall were monitored by the North Dakota Agricultural Weather Network (NDAWN) stations nearest to each experimental site.

The experimental design was a randomized complete-block design with four replicates, sown at two environments in August of 2017 and 2018 after the harvest of 'Glenn' cv. wheat. Wheat was grown during the season and cover crops were sown after wheat harvest. Wheat was drilled using a Great Plains 15-cm row spacing planter (Great Plains, Salinas, KS, USA) at 4,450,000 pure live seeds (PLS) ha^{-1} on 25 April 2017 and 2 May 2018 in Hickson, and on 20 April 2017 and 15 May 2018 in Prosper. Wheat in Hickson was fertilized with 88 kg N ha^{-1} and 24 kg P_2O_5 ha^{-1} both years, and Prosper was fertilized with 90 kg N ha^{-1} of N and 17 kg P_2O_5 ha^{-1} both years. Wheat was harvested on 8 August 2017 and 9 August 2018 in Hickson, and on 5 August 2017 and 8 August 2018 in Prosper. Cover crop treatments included five faba bean cultivars (Fanfare, Boxer, Laura, Snowdrop, and Tabasco), two pea cultivars (Arvika forage pea and Nette semi-leafless field pea), and one control plot without cover crop. After wheat harvest, a leaf blower (BR 200, Stihl, Waiblingen, Germany) was used to clear extra wheat stover from the plots to ensure an even sowing depth of 4 cm. All seeds were treated with inoculant (*Penicillium bilaiae, Rhizobium leguminosarum*) (TagTeam, Monsanto Company, St. Louis, MO, USA) at 6.1 kg ha^{-1} shortly before sowing. Cover crops were directly (no-till) sown with a plot drill (XL Plot seeder, Wintersteiger, Austria) into the wheat stubble on 22 August 2017 and 13 August 2018 in Hickson, and on 14 August 2017 and 16 August 2018 in Prosper (Table 1).

Each experimental unit had eight cover crop rows 15-cm apart and was 7.6-m in length. Prior to sowing, cover crop cultivars were tested for germination to calculate pure live seed (PLS). Cover crops were sown at 150,000 PLS ha^{-1} and 67 kg ha^{-1} PLS for faba bean and pea, respectively. Seeding rates were set at 75% of the recommended seeding rates for field production, as farmers using them for a cover crop would reduce sowing rates to save money. For faba bean, seeds for each plot were counted because of the variability in seed size among cultivars (Table 2). No herbicides or fertilizers were used on the cover crops, with the exception of a burndown application of glyphosate (*N*-(phosphonomethyl) glycine) (1.4 kg a.i. ha^{-1}) to kill volunteer wheat before cover crop sowing.

Table 1. Sowing and harvest of cover crops and weather data for each location and year (environment).

Environment	Sowing Date	Biomass Harvest	Total Rainfall [†]	Average Temp	Coldest Temp	Coldest Temp Date	GDD [‡]
			mm		**°C**		**°C**
Hickson 2017	22 August	26 October	81	14	−4	10 October	440
Hickson 2018	13 August	15 October	176	13	−7	11 October	436
Prosper 2017	14 August	25 October	161	14	−4	10 October	526
Prosper 2018	16 August	16 October	201	12	−11	12 October	409

[†] Total rainfall, temperature (temp), and growing degree days (GDD) measured from cover crop sowing to biomass harvest dates at each environment (mid-late August to mid-late October). [‡] Growing degree days calculated with 7 °C as the base temperature.

Table 2. Seed weight and sowing rate of cover crop cultivars sown after wheat harvest in Prosper and Hickson, ND, in 2017 and 2018.

Crop	Cultivar	100-Seed Weight	Sowing Rate
		g	**PLS ha^{-1}**
Faba bean	Fanfare	60.65	150,000
	Boxer	57.43	150,000
	Laura	57.73	150,000
	Snowdrop	35.28	150,000
	Tabasco	47.25	150,000
Field pea	Nette	19.17	350,000
Forage pea	Arvika	15.79	424,000

The following spring, maize was sown on the experiments. Peterson Farm Seed 75K85 VT2PRO (85-day maturity) maize was sown in 56-cm rows at a population of 79,262 plants ha^{-1} (MaxEmerge XP, John Deere, Moline, IL, USA) on 10 May 2018 in Hickson and 15 May 2018 in Prosper on the previous year's cover crops' plots. Each experimental unit was the same as the cover crop plots, consisting of three 56-cm maize rows 7.6-m in length. Maize was left unfertilized to allow for the determination of the difference in mineralization and release of N from the winter-killed cover crop biomass.

2.2. Plant Sampling and Analysis

Cover crop stand was recorded by counting emerged plants in 1-m^2 in each plot two weeks after emergence. The percent area of ground coverage for all cover crops was determined before the first killing frost with the Canopeo© application (Canopeo, Oklahoma State University, Stillwater, OK, USA) with pictures taken from 1-m above the canopy. Pictures were taken on 30 October 2017 in Prosper and 2 October 2018 in both Hickson and Prosper in 2018.

Shortly before the first expected killing frost, biomass samples were collected by clipping all aboveground biomass 0.2-m^2 from each cover crop plot. This was done on 26 October 2017 and 15 October 2018 in Hickson, and 25 October 2017 and 16 October 2018 in Prosper (Figure 1). The carrying capacity of cover crop biomass was calculated assuming 50% harvest efficiency using the NDSU Grazing Calculator Application (NDSU Grazing Calculator, North Dakota State University, Fargo, ND, USA). The leaves and stems of the collected plants were separated, dried at 70 °C until constant weight, and weighed to determine dry weight and leaf to stem ratio. For this ratio calculation, tendrils on the pea plants were counted as stems and stipules were counted as leaves. The leaf:stem ratio was calculated this way because tendril composition is more similar to stem than leaf, while stipules are similar to leaves. After this was calculated, dried samples were ground to pass through a 1-mm sieve with a mill (E3703.00, Eberbach Corporation, Bellville, MI, USA). Ground cover crop samples were analyzed using near-infrared reflectance spectroscopy (NIRS) with an XDS analyzer (Foss, Denmark) for N, P, and ash. Biomass N accumulation was calculated by multiplying the dry matter biomass yield by the total N content.

Maize was harvested on 24 October 2018 in Hickson and 18 October 2018 in Prosper (HP 5 combine, Almaco, Nevada, IA, USA). A 7.6-m row from the center of each plot was harvested for grain yield. A 56-cm row maize harvester was not available, so cobs were removed by hand and placed into the 76-cm row maize harvester. Grain from each plot was tested for water content and test weight (Mini GAC 2500, Dickey-John, Auburn, IL, USA). Before this, maize biomass yield was determined by cutting and weighing 1-m of maize row, 10-cm above the ground. After recording weights, plants were harvested for grain, but two were saved from each plot to determine harvest index. The two maize plants were dried at 70 °C until constant weight. Maize cobs were separated by hand and shelled (SCS-2, Agriculex, ON, Canada). Dry grain and stover were weighed separately. Harvest index was calculated using the equation

$$Harvest\ index = \frac{dry\ grain\ weight}{total\ biomass\ weight} \times 100$$

2.3. Soil Sampling and Analysis

Soil samples were taken at 0- to 15-cm and 0- to 60-cm across each replicate after wheat harvest. Wheat residue was removed from the surface before taking the sample (Table 3). Soil samples taken at the 0- to 15-cm depth were tested for soil pH, organic matter, P [33], and K with the ammonium acetate method [34], with a Buck Scientific Model 210 VGP Atomic Absorption Spectrophotometer (Buck Scientific, East Norwalk, CT, USA). Soil samples taken from 0- to 60-cm depth were analyzed for NO_3–N using the Vendrell and Zupancic [35] method. Soil samples were taken on 29 August in Hickson and 21 August in Prosper in 2017, and on 17 August in Hickson and 7 September in Prosper in 2018.

Soil samples were collected from each experimental unit after cover crop death in late fall on 3 November in both Hickson and Prosper in 2017, and 13 November in Hickson and 6 November in Prosper in 2018 (Figure 1). These soil samples were analyzed for NO_3–N, from 0- to 60-cm depth. The difference between the soil NO_3–N in the sample after wheat harvest and soil NO_3–N after cover crop harvest was considered as the change in NO_3–N.

Table 3. Soil sample results taken after wheat harvest and before cover crop sowing at each environment.

Environment	pH [†]	Organic Matter	Phosphorus	Potassium	NO_3-N [‡]
		g kg^{-1}	mg kg^{-1}		kg ha^{-1}
Hickson 2017	7.6	55	12	310	54.8
Hickson 2018	7.5	52	10	337	21.3
Prosper 2017	7.4	36	25	234	93.8
Prosper 2018	7.3	42	28	185	12.5

[†] pH, organic matter, P, and K, all sampled from 0–15 cm depth. [‡] NO_3–N sampled from 0–60 cm depth.

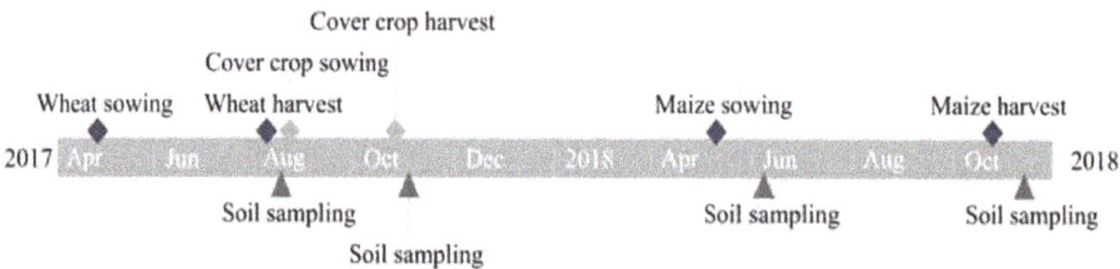

Figure 1. General timeline of sowing and harvesting of various crops and soil sampling throughout the experiment.

Soil samples were taken from each maize plot in the spring following cover crops at 0–60 cm and analyzed for NO_3-N on 8 June 2018 in both environments. In addition, soil samples were taken from each experimental unit at 0–60 cm and analyzed for NO_3-N after maize harvest on 13 November in Hickson and 6 November in Prosper in 2018. The difference between these spring and post-harvest soil NO_3–N samples in each plot was considered the change in NO_3–N. All evaluations were recorded for each plot of each replicate in all four environments.

2.4. Statistical Analysis

Statistical analysis was conducted using standard procedures for a randomized complete-block design. Biomass yield and forage quality data was analyzed using analysis of variance with the GLM procedure of SAS [36] Each location–year was analyzed separately and tested for homogeneity of variances before combining them. Each location–year combination was considered an environment and a random effect, while cover crops and grain crops were considered fixed effects in the analysis. All interactions of fixed effects with the environment were considered random in the analysis. The mean separation test was an *F*-protected least significant differences (LSD) ($p \leq 0.05$).

3. Results and Discussion

3.1. Weather

Both 2017 locations received less rainfall than the 30-yr average throughout the growing season except the month of September (Table 4). Hickson received 94 mm less rainfall than the 30-yr average in late summer and fall from July through October, and Prosper received 21 mm less rainfall than the 30-yr average from July through October (Table 4). Prosper 2017 received 86 mm more rainfall than the 30-yr average in September, relieving some possible water deficiency that had built up throughout the growing season. Hickson, in 2017, received only 6 mm more rainfall than average in September, making its total rainfall below average, much greater than Prosper 2017. Despite low rainfall, both locations had average temperatures throughout the growing season.

Both 2018 locations also started the growing season with drier-than-average rainfall. Hickson in 2018 received slightly above average rainfall in June through August, whereas Prosper 2018 received below average rainfall until August, where it received 12 mm above average, followed by 5 mm above average in both September and October. Along with a dry spring, both 2018 environments were 6 °C below average throughout April. In May and June, both environments had slightly above average temperatures, followed by slightly below average temperatures for the rest of the season.

Table 4. Total monthly rainfall, temperature, and difference from the 30-yr average for four environments at Hickson and Prosper, ND, USA, in 2017 and 2018.

Environment	Month	Rainfall		Temperature			
		Total	±30 yr [†]	Max	Min	Avg	±30 yr
		mm		°C			
Hickson 2017	July	37 [‡]	−46	29	14	21	0
	August	50	−13	25	12	18	−2
	September	69	6	22	9	15	0
	October	12	−41	15	0	8	0
Hickson 2018	April	2	−37	6	−5	1	−6
	May	22	−55	25	9	17	+3
	June	95	2	27	14	21	+2
	July	107	25	27	14	21	−1
	August	96	33	26	12	19	−1
	September	39	−25	21	7	14	−1
	October	46	−7	10	−2	4	−4
Prosper 2017	July	50	−38	28	14	21	0
	August	53	−14	25	11	18	−2
	September	152	86	22	8	15	0
	October	7	−55	15	0	8	0
Prosper 2018	April	4	−33	6	−6	0	−6
	May	54	−24	25	9	17	+3
	June	79	−21	27	14	20	+2
	July	65	−23	27	14	20	−1
	August	79	12	27	12	19	−1
	September	71	5	21	7	14	−1
	October	67	5	9	−1	4	−3

[†] 30 y average temperatures based on 1981–2010 long-term averages from NDAWN. [‡] Weather data obtained from: https://ndawn.ndsu.nodak.edu/weather-data-monthly.html.

3.2. Cover Crop Growth

Stand count of both pea cultivars averaged across environments was greater ($p \leq 0.05$) than the faba bean cultivars (Table 5) due to the higher seeding rate of the peas. There was no difference among faba bean cultivars, which showed similar emergence and growth; the same was true for the pea cultivars. There was a ground coverage by environment interaction across three environments ($p \leq 0.05$); data were not available for ground coverage in the Hickson 2017 environment. Forage pea cv. Arvika had greater ($p \leq 0.05$) ground coverage than any other treatment across each of the three environments (Table 5). Field pea cv. Nette had the next most coverage but was always less ($p \leq 0.05$) than 'Arvika'.

Prostrate growth, along with greater sowing density of pea cultivars, led to their greater ground coverage. Faba bean cv. Snowdrop and Boxer had consistently low ground coverage (Table 5). The control plots had weeds, explaining coverage observed in control plots in both 2018 environments. In 2017, there was uncontrolled volunteer wheat in the plots, explaining the 14.1% average coverage in the control plot. Volunteer wheat, along with the pictures being taken later, and faba beans beginning to turn a darker color with frost, led to skewed coverage readings of cover crops in that environment. Ground cover from cover crops could help significantly curtail soil losses due to wind erosion, which, according to Fryrear [37], can be reduced by 58% with just 20% ground cover. However, no-till management can greatly reduce soil erosion on its own; with low residue crops in no-till there is still chance for inter-rill erosion, which can be reduced by adding cover crops [38].

Table 5. Plant stand count averaged across four environments and ground coverage for each cultivar in each environment at Hickson and Prosper, ND, USA, in 2017 and 2018.

		Hickson	Prosper	
Cultivar	**Stand Count**	**2018**	**2017**	**2018**
	Plants m^{-2}	**Ground Coverage (%)**		
Fanfare	19 b [†]	23.0 def	12.1 k	19.1 ghi
Boxer	23 b	19.3 fgh	13.2 k	15.0 jk
Laura	22 b	21.4 efg	12.5 k	17.4 hij
Snowdrop	21 b	19.4 fgh	12.3 k	12.0 k
Tabasco	22 b	26.1 d	14.2 jk	15.3 ijk
Nette	40 a	40.9 c	15.2 jk	23.7 de
Arvika	41 a	61.9 a	21.1 efgh	49.4 b
No cover control	0 c	1.11 [‡]	14.1 jk	1.3 l
LSD (p = 0.05)	7	3.8		

[†] Means followed by the same letters within a column for each factor are not significantly different ($p \leq 0.05$) according to the least significant differences (LSD) test. [‡] Coverage in the no cover control corresponds to volunteer wheat in 2017 and weeds in 2018.

Cover crop biomass averaged across four environments was similar among all cultivars (Table 6), showing faba bean produced similar amounts of aboveground biomass as forage and field pea. This is in contrast to Iglesias and Lloveras [18], who reported significantly greater biomass production with faba bean than forage pea. This is likely because of the short growing period of the cover crops in this study (August–October), whereas, in Iglesias' study, cover crops were grown to maturity through a mild winter from late October to May. Cover crop biomass averaged 1207 kg ha^{-1} and ranged from 957 to 1630 kg ha^{-1} (Table 6) across cultivars, but were not significantly different. 'Arvika' averaged 1630 kg ha^{-1}, which was similar to the biomass yield reported by Iglesias and Lloveras [18] and Wichmann et al. [19]. They also indicated that, in earlier harvests, pea yielded greater biomass than other cold season legumes such as lupin, faba bean, and hairy vetch (*Vicia villosa* Roth). The differences in forage and field pea biomass align with findings from Uzun et al. [8] and Turk and Albayrak [9], who reported that leaved pea cultivars averaged greater biomass than semi-leafless cultivars. The carrying capacity per hectare of the average biomass produced by cover crops in this study (1207 kg ha^{-1}) resulted in 1.5 AUM ha^{-1} for a 450 kg cow and calf.

Table 6. Cover crop biomass yield and leaf:stem ratio averaged across four environments at Hickson and Prosper, ND, USA, in 2017 and 2018.

Cultivar	**Biomass Yield** **kg ha^{-1}**	**Leaf:Stem Ratio**
Faba bean		
Fanfare	1021 [†]	1.69 a
Boxer	1184	1.85 a
Laura	1190	1.82 a
Snowdrop	957	1.71 a
Tabasco	1320	1.69 a
Pea		
Nette	1149	0.49 c
Arvika	1630	1.12 b
LSD (p = 0.05)	NS	0.31

[†] Means followed by the same letters (or no letter) within a column for each factor are not significantly different ($p \leq 0.05$) according to the LSD test.

All faba bean cultivars, averaged across environments, showed greater ($p \leq 0.05$) leaf:stem ratios than forage pea, which had a greater ($p \leq 0.05$) ratio than field pea (Table 6). Alkhtib et al. [39] found that faba bean has a greater concentration of CP in the leaves than in the stems. The high leaf:stem ratio of faba bean gives them a high nutritive value forage. Forage pea had a greater number of leaves and stipules, but they were much smaller than faba bean leaves, giving them a lesser leaf:stem ratio. Field pea is largely a vine plant and had only stipules that were counted as leaves, contributing to its lesser ($p \leq 0.05$) leaf:stem ratio.

3.3. Cover Crop Biomass Chemical Composition

All biomass chemical composition parameters evaluated were significantly different among cultivars ($p \leq 0.05$) (Table 7). Faba bean cv. Boxer had the highest CP concentration, but not different from faba bean cv. Laura and forage pea cv. Arvika. Field pea cv. Nette had lesser ($p \leq 0.05$) CP concentration than any other cover crop. Wichmann et al. [19] also reported that faba bean contained more CP than pea. Faba bean cv. Fanfare, Tabasco, and Snowdrop all had less ($p \leq 0.05$) CP than 'Boxer'. Soto-Navarro et al. [7], Uzun et al. [8], and Strydhorst et al. [13] reported that forage pea types averaged a greater CP forage yield than semi-leafless pea, and that faba bean had more ($p \leq 0.05$) CP content than field pea. This could be related to the leaf:stem ratio, because Alkhtib et al. [39] found that leaves have a greater ($p \leq 0.05$) concentration of CP than stems. Legumes in this study were in a vegetative stage at harvest, which explains their high CP concentration. The crude protein concentration of typical legume forages, such as alfalfa (*Medicago sativa* L.), averaged 212 g kg^{-1} when cut at early bud stage [40].

Table 7. Cover crop crude protein, *p* and content averaged across four environments at Hickson and Prosper, ND, USA, in 2017 and 2018.

Cultivar	Crude Protein	Phosphorus	Ash
	g kg^{-1}		
Fanfare	289 b [†]	4.73 bc	74 a
Boxer	304 a	4.83 a	77 a
Laura	297 ab	4.76 ab	75 a
Snowdrop	284 b	4.66 cd	73 a
Tabasco	285 b	4.63 de	67 a
Nette	264 c	4.24 f	44 b
Arvika	298 ab	4.57 e	42 b
LSD ($p = 0.05$)	14	0.08	14

[†] Means followed by the same letters within a column for each factor are not significantly different ($p \leq 0.05$) according to the LSD test.

Forage pea harvested in vegetative stage usually have greater forage nutritive value than more mature plants (seed fill), with earlier harvests having greater CP, total digestible nutrients, and relative feed value [9]. Crude protein concentration in all cover crops in this study were much above those needed by beef cattle. A gestating cow in the mid-1/3 of pregnancy and weighing 540 kg requires 10 kg of dry matter intake with 71 g kg^{-1} of crude protein [41]. It is important to graze pea and faba bean along with an alternative source of fiber to increase the amount of DM intake, reduce CP concentration of feed intake to maintain rumen stability, and avoid bloating [41]. This could be provided by wheat stubble or wheat volunteers, along with other low-CP, high-fiber, dry hay.

Amiri and Shariff [10] found that combining legumes and grass species provided proper nutrition requirements for grazing livestock. Pea grain and forage are known to cause bloat in ruminant animals because of their high protein content [42]. Faba bean contains condensed tannins, which are attributed to reduced bloating typical of other legume forages [43]. Thus, faba bean as a sole feed are less likely to cause bloating than pea. However, it is important to note that tannin-free faba bean cultivars are available [13] to increase digestibility of the seed for feed. Although there is abundant information about the effects of tannin-free faba bean seed, there is little information on whether the amount of tannin in the plant biomass is affected. More research should be done on whether tannin-free faba bean seeds may produce biomass with reduced tannin content as well, increasing the probability of bloating.

Phosphorus concentration in the biomass was different among cultivars ($p \leq 0.05$), especially between faba bean and pea cultivars, with the greatest being faba bean 'Boxer' with 4.83 g kg^{-1} and the least being field pea 'Nette' with 4.24 g kg^{-1} (Table 7). Forage pea had less ($p \leq 0.05$) P content than all faba bean cultivars except for 'Tabasco', and field pea had less ($p \leq 0.05$) P content than forage pea. Both legumes have more than enough P for cattle, which need 1.4 g P kg^{-1} of feed [43].

Ash was greater ($p \leq 0.05$) in all faba bean cultivars than the pea cultivars (Table 7). This was likely due to the greater ($p \leq 0.05$) leaf:stem ratio of faba bean and the fact that leaves of faba bean have greater ash content than the stem [39]. Overall, ash concentrations (42–77 g kg^{-1}) were lesser compared with the average ash concentration of a grass–legume mixture (90 g kg^{-1}) [44].

No significant differences ($p \leq 0.05$) in nitrogen accumulation were found between cover crops averaged across environments (Table 8). Average nitrogen accumulation of all cover crop treatments was 55.3 kg N ha^{-1}, and ranged from 43.4–76.8 kg N ha^{-1}. Analysis indicated an interaction between N accumulation and environment ($p \leq 0.05$) (Table 8). Biomass and N accumulation averages follow the trend of available NO_3–N in the soil at cover crop sowing in each environment (Table 3). The fact that there was no significant interaction between biomass and environment, but there was between nitrogen accumulation and environment, suggests that available NO_3–N in the soil is not essential for legume biomass production, but they can accumulate excess soil NO_3–N when it is available. This shows that legumes can be scavengers as well when there are excessive nutrients in the soil, in alignment with findings from Jensen et al. [20] and Hauggard-Nielsen et al. [23]. The Prosper 2017 environment likely had the highest average N accumulation because it had the most growing degree days accumulated between cover crop sowing and harvest (Table 1), along with the most NO_3–N in the soil at cover crop sowing (Table 3), allowing for the most crop growth. The greatest amount of N was accumulated by forage pea cv. Arvika with 105.1 kg ha^{-1} in Hickson in 2017 (Table 8). In general, the least N accumulation was seen in most cultivars at Prosper in 2018 (Table 8).

Table 8. Environment by cover crop interaction and mean across all environments of N accumulation in the shoot biomass of each cultivar in Hickson and Prosper, ND, USA, in 2017 and 2018.

	Hickson		Prosper		Mean
Cultivar	**2017**	**2018**	**2017**	**2018**	
			kg ha^{-1}		
Fanfare	34.0 g [†]	41.7 efg	64.5 bcdef	47.6 defg	46.9
Boxer	55.1 cdefg	42.8 efg	90.2 ab	39.2 fg	56.9
Laura	38.8 fg	49.7 defg	91.2 ab	49.1 defg	57.2
Snowdrop	38.2 fg	50.2 defg	55.4 cdefg	29.6 g	43.4
Tabasco	73.0 bcd	47.4 defg	84.6 abc	29.0 g	58.5
Nette	57.3 cdefg	46.4 defg	55.8 cdefg	29.7 g	47.3
Arvika	105.1 a	71.3 bcde	66.1 bcdef	64.8 bcdef	76.8
LSD ($p = 0.05$)			30.0		NS

[†] Means followed by the same letters (or no letter) within a column for each factor are not significantly different ($p \leq 0.05$) according to the LSD test.

3.4. Soil

Soil residual NO_3–N was not significantly different among treatments after biomass harvest, October/November, due to the short time for the uptake of soil NO_3–N (Table 9). This could also be due to the N_2 fixation by legumes. Control plots without cover crop averaged slightly, but not significantly, greater residual NO_3–N in the soil than the cover crop plots. This shows that the cover crops in this study didn't accumulate more N from the soil than would have been leached away or lost by other means, alleviating the need to increase fertilization in the following year. Similarly, Hauggaard-Nielsen et al. [23] found that excess nutrients accumulate in cover crop biomass and can reduce nutrient leaching or loss through soil erosion throughout the fall, winter, and spring. However, Couedel et al. [29] found a decrease in soil NO_3–N from legume cover crops grown during a fallow period, which ranged from 25% to 56%, with pea causing the most reduction. This is likely due to the longer growing period of the cover crops in the Couedel et al. [29] study.

Table 9. End of season soil NO_3–N and change in soil NO_3–N throughout the life of the cover crop averaged across four environments at Hickson and Prosper, ND, in 2017 and 2018.

Cultivar	Soil NO_3–N	Change in NO_3–N [†]
	kg ha^{-1}	
Fanfare	24.0	21.6
Boxer	23.8	21.7
Laura	22.1	23.4
Snowdrop	22.7	22.8
Tabasco	23.7	21.8
Nette	25.9	19.7
Arvika	24.7	20.9
Control	26.4	19.1
LSD ($p = 0.05$)	NS	NS

[†] Change in soil NO_3-N from cover crop sowing to biomass harvest.

Change in soil NO_3–N during the life of the cover crops (August to October) was not different between treatments (Table 9). Although not significant, the absence of a cover crop resulted in a lesser change in soil NO_3–N from planting to harvest of cover crops than those with cover. Soil NO_3–N changes are mainly due to mineralization, leaching, and immobilization and N_2 fixation.

3.5. Maize Crop in Following Season

No differences were found between the previous year's cover crop treatments in any of the maize parameters that were tested in the following season. This neutral result indicates that N stored in the cover crop biomass did not mineralize fast enough, or in large enough amounts, for the following crop, maize, to utilize. This low response is likely due to low temperatures throughout April in 2018, delaying the start of the mineralization process (Table 4). Low rainfall throughout the spring of 2018 in both environments, and through July of 2018 in the Prosper environment likely decreased mineralization rate in the soil to less than expected (Table 4). Similarly, Hauggaard-Nielsen et al. [23] reported that production of wheat in the spring was unaffected after fall-sown grass/clover.

When analyzing the maize grain yield response to residual soil NO_3–N after each cover crop in the spring, there was a slight response ($r^2 = 0.30$) (Figure 2). Pea cultivars and the control plot had a lesser maize grain yield compared with the maize following faba bean cultivars. Likewise, Lupwayi and Soon [45] indicated that faba bean might have released more N in the first year of decomposition than both field and forage pea.

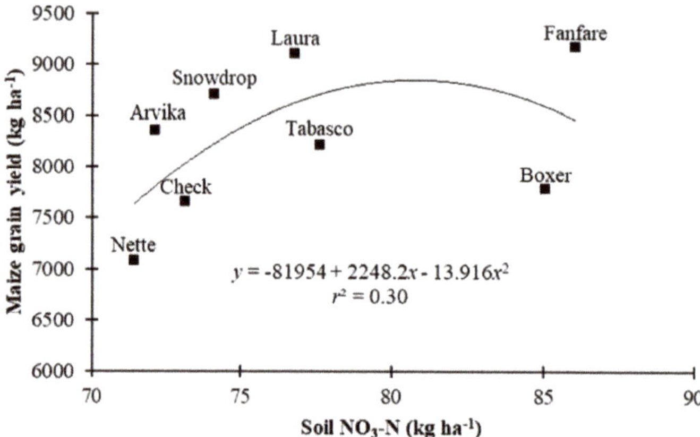

Figure 2. Interaction between spring soil NO$_3$–N and maize grain yield in 2018, following fall cover crops.

Slightly increased maize yield in the year following forage pea compared with the field pea was likely due to its greater leaf:stem ratio and greater N concentration; a response also reported by Lupwayi and Soon [45]. A greater N concentration in the faba bean biomass could allow for faster mineralization of the nutrients, leading to a slight boost in maize yield, but this has not been determined by this study. This trial was not fertilized with N; thus, a slightly greater availability of soil NO$_3$–N could have made a difference in grain yield.

4. Conclusions

Faba bean, forage pea, and field pea all show potential as late-season cover crops and forage grazing, due to their ability to produce biomass when sown after wheat harvest and their high CP content. All cover crops tested attained similar biomass accumulation and enough ground coverage to reduce erosion. Peas provided more ground coverage than faba bean. Increasing the faba bean sowing rate would likely increase biomass production and ground coverage. On just these cover crops, 1.5 AUM would be able to graze per hectare. Forage pea and faba bean had a better nutritive value than field pea because of their greater CP concentration. Faba bean cultivars had a greater leaf:stem ratio than peas, which could lead to increased intake of faba bean over peas for grazing cattle. Faba bean and forage pea also showed a slightly greater N benefit than field pea to the succeeding crop due to their high N content and superior leaf:stem ratio. There are multiple advantages to using these leguminous cover crops after wheat, though more research may need to be done to find conclusive results of their residual nutrient effects on the following crops.

Author Contributions: Conceptualization, M.T.B.; Data curation, B.J.A.; Formal analysis, B.J.A. and M.T.B.; Funding acquisition, M.T.B.; Investigation, B.J.A. and M.T.B.; Methodology, B.J.A. and M.T.B.; Project administration, M.T.B.; Resources, A.W. and M.T.B.; Supervision, M.T.B.; Validation, B.J.A. and M.T.B.; Writing original draft, B.J.A., D.P.S. and M.T.B.; Writing—review and editing, B.J.A., D.P.S., A.W. and M.T.B. All authors have read and agreed to the published version of the manuscript.

Funding: This research was funded by USDA-NIFA, Coordinated Agricultural Program, Award no. 2016-69004-24784 "CropSys—A novel management approach to increase productivity, resilience, and long-term sustainability of cropping systems in the northern Great Plains".

Acknowledgments: Our most sincere thanks to the forages and cover crops team, Alan Peterson, Alex Wittenberg, Sergio Cabello, and Swarup Podder, for their assistance in experiments establishment, data collection, harvest, and sample processing.

Conflicts of Interest: The authors declare no conflict of interest. The funders had no role in the design of the study; in the collection, analyses, or interpretation of data; in the writing of the manuscript, or in the decision to publish the results.

References

1. Natural Resources Conservation Service (NRCS). National Handbook of Conservation Practices. Conservation Practice Standards Code 340; 2014; Available online: https://www.nrcs.usda.gov/Internet/FSE_DOCUMENTS/stelprdb1263176.pdf (accessed on 4 November 2019).
2. Myers, R. A Preliminary Look at State Rankings for Cover Crop Acreage Based on Census of Agriculture Information. Sustainable Agriculture Network, 2019. Available online: https://files.constantcontact.com/c7de6aab001/914daacf-ca6d-4549-a57f-a2b3f861ec15.pdf (accessed on 4 November 2019).
3. Berti, M.T.; Li, H.; Samarappuli, D.; Peterson, A.; Cabello, S.; Andersen, B.; Wittenberg, A.; Johnson, B.; Steffl, N.; Aasand, K.; et al. MCCC-North Dakota Annual Report. 2019. Available online: http://mccc.msu.edu/wp-content/uploads/2018/04/ND_2017_MCCC_North-Dakota-Report.pdf (accessed on 4 November 2019).
4. Lawrence, J.D.; Strohbehn, D.R. Understanding and managing costs in beef cow-calf herds. In *Integrated Resource Management Committee. National Cattlemen's Beef Association Convention*; Iowa Beef Center, Iowa State University: Charlotte, NC, USA, 1999.
5. Strauch, D. Survival of pathogenic micro-organisms and parasites in excreta, manure and sewage sludge. *Rev. Sci. Tech. Off. Int. Epiz.* **1991**, *10*, 813–846. [CrossRef]
6. Anderson, V.L.; Ilse, B. Field Peas Make Excellent Quality Forage for Beef Cattle. 2012, pp. 1–2. Available online: https://northernpulse.com/uploads/resources/883/field-peas-make-excellenct-quality-forage-for-beef-cattle.jpg.pdf (accessed on 4 November 2019).
7. Soto-Navarro, S.A.; Encinias, A.M.; Bauer, M.L.; Lardy, G.P.; Caton, J.S. Feeding value of field pea as a protein source in forage-based diets fed to beef cattle. *J. Anim. Sci.* **2012**, *90*, 585–591. [CrossRef] [PubMed]
8. Uzun, A.; Bilgili, U.; Sincik, M.; Filya, I.; Acikgoz, E. Yield and quality of forage type pea lines of contrasting leaf types. *Eur. J. Agron.* **2005**, *22*, 85–94. [CrossRef]
9. Turk, M.; Albayrak, S. Effect of harvesting stages on forage yield and quality of different leaf types pea cultivar. *Turk. Field Crops* **2012**, *17*, 111–114.
10. Amiri, F.; Shariff, A.R.M. Comparison of nutritive values of grasses and legume species using forage quality index. *Songklanakarin J. Sci. Technol.* **2012**, *34*, 577–586.
11. Dordas, C.A.; Lithourgidis, A.S. Growth, yield, and nitrogen performance of faba bean intercrops with oat and triticale at varying seeding ratios. *Grass Forage Sci.* **2011**, *66*, 569–577. [CrossRef]
12. Dhima, K.V.; Vasilakoglou, I.B.; Keco, R.X.; Dima, A.K.; Paschalidis, K.A.; Gatsis, T.D. Forage yield and competition indices of faba bean intercropped with oat. *Grass Forage Sci.* **2014**, *69*, 376–383. [CrossRef]
13. Strydhorst, S.M.; King, J.R.; Lopetinsky, K.J.; Harker, K.N. Forage potential of intercropping barley with faba bean, lupin, or field pea. *Agron. J.* **2008**, *100*, 182–190. [CrossRef]
14. Stoltz, E.; Nadeau, E. Effects of intercropping on yield, weed incidence, forage quality and soil residual N in organically grown forage maize (*Zea mays* L.) and faba bean (*Vicia faba* L.). *Field Crops Res.* **2014**, *169*, 21–29. [CrossRef]
15. Mustafa, A.F.; Seguin, F. Characteristics and in situ degradability of whole crop faba bean, pea, and soybean silages. *Can. J. Anim. Sci.* **2003**, *83*, 793–799. [CrossRef]
16. Warner, K.S.A.; Hempworth, G.W.; Davidson, R.H.; Milton, J.T.B. Value of mature grain legume crops for out of season prime lamb production. *Anim. Prod. Aust.* **1998**, *22*, 217–220.
17. Wegi, T.; Tolera, A.; Wamatu, J.; Animut, G.; Rischkowsky, B. Effects of feeding different varieties of faba bean (*Vicia faba* L.) straws with concentrate supplement on feed intake, digestibility, body weight gain and carcass characteristics of Arsi-Bale sheep. *Asian Austral. J. Anim. Sci.* **2018**, *31*, 1221. [CrossRef] [PubMed]
18. Iglesias, I.; Lloveras, J. Annual cool-season legumes for forage production in mild winter areas. *Grass Forage Sci.* **1998**, *53*, 318–325. [CrossRef]
19. Wichmann, S.; Loges, R.; Taube, F. Development of whole crop yield and quality of different grain legumes in monocrop and intercropping with cereals. *Pflanzenbauwissenschaften* **2005**, *9*, 61–74.

20. Jensen, E.S.; Peoples, M.B.; Hauggaard-Nielsen, H. Faba bean in cropping systems. *Field Crops Res.* **2010**, *115*, 203–216. [CrossRef]
21. Schatz, B.G. Benefits of Growing Field Peas. 2012, p. 3. Available online: https://northernpulse.com/uploads/resources/883/field-peas-make-excellenct-quality-forage-for-beef-cattle.jpg.pdf (accessed on 4 November 2019).
22. Hardarson, G.; Danso, S.; Zapata, F.; Reichardt, K. Measurements of nitrogen fixation in faba bean at different N fertilizer rates using the 15N isotope dilution and 'A-value' methods. *Plant Soil* **1991**, *131*, 161–168. [CrossRef]
23. Hauggaard-Nielsen, H.; Mundus, S.; Jensen, E.S. Nitrogen dynamics following grain legumes and subsequent catch crops and the effects on succeeding cereal crops. *Nutr. Cycl. Agroecosyst.* **2009**, *84*, 281–291. [CrossRef]
24. Peoples, M.B.; Brockwell, J.; Herridge, D.F.; Rochester, I.J.; Alves, B.J.R.; Urquiaga, S.; Boddey, R.M.; Dakora, F.D.; Bhattarai, S.; Maskey, S.L.; et al. The contributions of nitrogen-fixing crop legumes to the productivity of agricultural systems. *Symbiosis* **2009**, *48*, 1–17. [CrossRef]
25. Stevenson, F.C.; van Kessel, C. The nitrogen and non-nitrogen rotation benefits of pea to succeeding crops. *Can. J. Plant Sci.* **1996**, *76*, 735–745. [CrossRef]
26. Beckie, H.J.; Brandt, S.A.; Schoenau, J.J.; Campbell, C.A.; Henry, J.L.; Janzen, H.H. Nitrogen contribution of field pea in annual cropping systems. 2. Total nitrogen benefit. *Can. J. Plant Sci.* **1997**, *77*, 323–331. [CrossRef]
27. Lupwayi, N.Z.; Soon, Y.K. Nitrogen release from field pea residues and soil inorganic N in a pea-wheat crop rotation in northwestern Canada. *Can. J. Plant Sci.* **2009**, *89*, 239–246. [CrossRef]
28. Cupina, B.; Manojlovic, M.; Krstic, D.; Cabilovski, R.; Mikic, A.; Ignjatovic-Cupina, A.; Eric, P. Effect of winter cover crops on the dynamics of soil mineral nitrogen and yield and quality of sudan grass *Sorghum bicolor* (L.) Moench. *Aust. J. Crop Sci.* **2011**, *5*, 839–845.
29. Couedel, A.; Alletto, L.; Tribouillois, H.; Justes, E. Cover crop crucifer-legume mixtures provide effective nitrate catch crop and nitrogen green manure ecosystem services. *Agric. Ecosyst. Environ.* **2018**, *254*, 50–59. [CrossRef]
30. North Dakota State University Extension. *Langdon 2016 Variety Trial Report*; North Dakota State Univ. Ext. Serv.: Fargo, ND, USA, 2016.
31. National Agricultural Statistics Service (NASS); Farm Service Agency (FSA). 2017 North Dakota Acreage Summary Report. 2017. Available online: https://www.fsa.usda.gov/Assets/USDA-FSA-Public/usdafiles/State-Offices/North-Dakota/pdfs/2017AcreageReportingSummary.pdf (accessed on 4 November 2019).
32. Web Soil Survey. *National Resources Conservation Service*; United States Dep. of Agric.: Washington, DC, USA, 2009; Available online: http://websoilsurvey.nrcs.usda.gov/app/WebSoilSurvey.aspx (accessed on 12 April 2018).
33. Olsen, R.S.; Cole, C.V.; Watanabe, F.S.; Dean, L.A. *Estimation of Available Phosphorus in Soils by Extraction with Sodium Bicarbonate. Circular 939*; United States Department of Agriculture: Washington, DC, USA, 1954.
34. Warncke, D.; Brown, J.R. Potassium and other basic cations. *Recomm. Chem. Soil Test Proced. N. Cent. Reg.* **1998**, *1001*, 31.
35. Vendrell, P.F.; Zupancic, J. Determination of soil nitrate by transnitration of salicylic-acid. In Proceedings of the International Symposium on Soil Testing and Plant Analysis, Fresno, CA, USA, 14–18 August 1989.
36. SAS Institute. *SAS User's Guide: Statistics*; SAS Institute: Cary, NC, USA, 2014.
37. Fryrear, D.W. Soil cover and wind erosion. *Trans ASAE* **1985**, *28*, 781–784. [CrossRef]
38. Kaspar, T.C.; Radke, J.K.; Laflen, J.M. Small grain cover crops and wheel traffic effects on infiltration, runoff, and erosion. *J. Soil Water Conserv.* **2001**, *56*, 160–164.
39. Alkhtib, A.S.; Warnatu, J.A.; Wegi, T.; Rischkowsky, B.A. Variation in the straw traits of morphological fractions of faba bean (*Vicia faba* L.) and implications for selecting for food-feed varieties. *Anim. Feed Sci. Technol.* **2016**, *222*, 122–131. [CrossRef]
40. Lloveras, J.; Ferran, J.; Alvarez, A.; Torres, L. Harvest management effects an alfalfa (*Medicago sativa* L.) production and quality in Mediterranean areas. *Grass Forage Sci.* **1998**, *53*, 88–92. [CrossRef]
41. Lalman, D.; Richards, C. Nutrient Requirements of Beef Cattle. Oklahoma Cooperative Extension Service, 2017. Available online: http://pods.dasnr.okstate.edu/docushare/dsweb/Get/Document-1921/E-974web.pdf (accessed on 4 November 2019).
42. Radostits, O.M.; Gay, C.G.; Hinchcliff, K.W.; Constable, P.D. *Veterinary Medicine: A Textbook of the Diseases of Cattle, Horses, Sheep, Pigs and Goats*, 10th ed.; Saunders Ltd.: Philadelphia, PA, USA, 2007.

43. Lees, G.L. Condensed tannins in some forage legumes: Their role in the prevention of ruminant pasture bloat. *Basic Life Sci.* **1992**, *59*, 915–934.

44. Hoffman, P.C. Ash Content in Forages. 2005. Available online: https://fyi.extension.wisc.edu/forage/ash-content-of-forages/ (accessed on 4 November 2019).

45. Lupwayi, N.Z.; Soon, Y.K. Carbon and nitrogen release from legume crop residues for three subsequent crops. *Soil Sci. Soc. Am. J.* **2015**, *79*, 1650–1659. [CrossRef]

Article

Leguminous Cover Crop *Astragalus sinicus* Enhances Grain Yields and Nitrogen Use Efficiency through Increased Tillering in an Intensive Double-Cropping Rice System in Southern China

Jiangwen Nie [1,3], Lixia Yi [1], Heshui Xu [2], Zhangyong Liu [1], Zhaohai Zeng [3], Paul Dijkstra [4], George W. Koch [4], Bruce A. Hungate [4] and Bo Zhu [1,*]

1 Hubei Collaborative Innovation Center for Grain Industry, Yangtze University, Jingzhou 434025, China; b20183010007@cau.edu.cn (J.N.); 501068@yangtzeu.edu.cn (L.Y.); lyz1331@hotmail.com (Z.L.)
2 Shanghai Academy of Agricultural Sciences, Shanghai 201403, China; xhsh0112@outlook.com
3 College of Agronomy and Biotechnology, China Agricultural University, Beijing 100193, China; zengzhaohai@cau.edu.cn
4 Center for Ecosystem Science and Society, Northern Arizona University, Flagstaff, AZ 86011, USA; Paul.Dijkstra@nau.edu (P.D.); George.Koch@nau.edu (G.W.K.); Bruce.Hungate@nau.edu (B.A.H.)
* Correspondence: 501045@yangtzeu.edu.cn; Tel.: +86-0716-8066314; Fax: +86-0716-8066314

Received: 24 July 2019; Accepted: 10 September 2019; Published: 16 September 2019

Abstract: Chinese milk vetch (*Astragalus sinicus* L., vetch), a leguminous winter cover crop, has been widely adopted by farmers in southern China to boost yield of the succeeding rice crop. However, the effects of vetch on rice grain yield and nitrogen (N) use efficiency have not yet been well studied in the intensive double-cropped rice cropping systems. To fill this gap, we conducted a three-year field experiment to evaluate the impacts of the vetch crop on yields and N use efficiency in the subsequent early and late rice seasons. With moderate N input (100 kg N ha^{-1} for each rice crop), vetch cover significantly increased grain yields by 7.3–13.4% for early rice, by 8.2–10.4% for late rice, and by 8.6–11.5% for total annual rice production when compared with winter fallow. When rice crops received an N input of 200 kg N ha^{-1}, vetch cover increased grain yields by 5.9–18.4% for early rice, by 3.8–10.1% for late rice, and by 6.2–11.3% for annual rice production. Moreover, comparable grain yields (11.9 vs. 12.0 Mg ha^{-1} for annual rice production) were observed between vetch cover with moderate N and fallow with added N fertilizer. Yield components analysis indicated that the increased tillering number was the main factor for the enhanced grain yields by vetch cover. Vetch cover with moderate and higher N input resulted in higher agronomic N use efficiency and applied N recovery efficiency compared with the fallow treatments. Here, our results showed that vetch as a winter cover crop can be combined with reduced N fertilizer input while maintaining high grain yields, thus gaining a more sustainable rice production system.

Keywords: leguminous cover crop; vetch; double cropping; grain yield; N uptake; N use efficiency; rice

---

1. Introduction

Rice (*Oryza sativa* L.) is one of the most important food crops in China, with a planted area of 30 million ha and a gross grain production of 206 million tons in 2014. Chemical nitrogen (N) fertilizer, as high as 193 kg N ha^{-1}, is regularly applied for each cropping season, with less than 36% of it recovered by the rice crop [1]. Sustainable rice production in this region is hampered by excessive N fertilizer application since: (1) rice grain yield increase is lower than N rate increase, leading to decreased partial factor productivity from applied N [2], (2) high N fertilizer input increases farmers' costs, and (3) decreased N use efficiency may be associated with high risk of N losses into the environment, resulting in environmental problems, such as nitrogen leaching [3], N_2O and NH_3 emissions, and subsequent high atmospheric N deposition rates [4].

Some strategies have been developed to balance the N effects on crop yield and environmental risks [1]. Improved N use efficiency can be achieved by using slow-release sulfur- or polymer-coated N fertilizer [5], multisplit topdressing, deep placement of fertilizer, and integrated soil–crop management [6]. These management technologies were introduced to help to enhance crop N uptake and reduce N losses into the environment. However, most of these fertilizer products or fertilizer management strategies require high cost or additional labor input, which limit their adaptation by farmers.

Biologically fixed N may be an alternative, low-cost N source for farmers in China. Cover crops are commonly sown during the fallow period and can be used as an effective tool to improve N management in annual crop rotations [7]. Nonleguminous cover crops such as turnip rape (*Brassica rapa*) and Italian ryegrass (*Lolium multiflorum* Lam.) have been used to intercept postharvest surplus N that would otherwise be leached into rivers and groundwater [8,9]. Leguminous cover crops such as Chinese milk vetch (*Astragalus sinicus* L.), crimson clover (*Trifolium incarnatum* L.), and purple vetch (*Vicia benghalensis* L.) can fix atmospheric N_2, reduce nitrate leaching, and increase soil N availability for the succeeding main crops [10–12], thus decreasing the reliance on synthetic fertilizer N and reducing the associated costs and environmental risks.

Chinese milk vetch is a widely used leguminous cover crop in the double-cropping rice systems in southern China. It is usually broadcast into the paddy field approximately two weeks before the late rice harvest and incorporated into the soil at blooming stage, about one or two weeks before the early season rice planting. It is hypothesized that there should be a synergistic interaction between vetch incorporation and chemical fertilizer N on rice grain yield and N use efficiency. However, the evidence is limited. Knowledge of the interaction between milk vetch incorporation and chemical fertilizer N is important for the optimal use of both cover crop and chemical fertilizers.

In the present study, we measured rice grain yield and N use efficiency for three years in a field experiment. The study has two main objectives: to evaluate the combined effects of vetch as a leguminous winter cover crop and fertilizer N application rate on (1) rice grain yields and (2) N use efficiency of the double-cropping rice system.

2. Materials and Methods

2.1. Experimental Site

A three-year field experiment (2015 to 2017) was conducted in a farmer's field in Huarong County (29°52′ N, 12°55′ E) in Dongting Lake Plain, Hunan province, China. The climate in this region is subtropical, monsoonal, and humid. The most widely practiced cropping system consists of growing two rice crops followed by winter fallow or a cover crop. The soil was a clay loam that contained 24.7% sand, 61.1% silt, and 14.2% clay, and pH (H_2O) was 6.1. Soil organic matter was 3.02%, while soil total N was 0.12%; Olsen extractable phosphorus was 12.7 mg kg^{-1} and available potassium was 107.5 mg kg^{-1} soil.

2.2. Treatments and Crop Management

The experimental treatments were: (1) winter fallow without N application during the main crop rice season as a control (FN_0); (2) winter fallow and a moderate 100 kg N ha^{-1} for each rice crop (FN_{100}); (3) vetch cover crop in winter and 100 kg N ha^{-1} for each rice crop (MN_{100}); (4) winter fallow and 200 kg N ha^{-1} input for each rice crop (FN_{200}), and (5) vetch cover cropping in winter and 200 kg N ha^{-1} for each rice crop (MN_{200}). The experimental design was a completely randomized block design with three replications. Blocks were separated by 1 m wide irrigation ditches and each treatment was randomly placed in 5 m × 6 m plots within each block. Plots were separated by 0.5 m wide ridges covered with plastic film to avoid water and nutrients leaking. Treatments were assigned to the same experimental plots for three years.

Vetch seeds (cv. *Xiangfei 3*) were hand broadcast approximately 15 days before late rice harvest in early October. The vetch received no fertilizer during the winter growing season. The aboveground biomass of the vetch was incorporated into the soil at the blooming stage by plowing and puddling approximately 10 days before early rice planting. In the winter fallow plots, the soil was bare without straw or grass cover (Figure 1).

Figure 1. Planting patterns of the winter fallow (**a**) and the winter cover cropping with milk vetch (**b**) in a double-rice cropping system.

For the early rice, seedlings (cv. Luliangyou 996, 30 d old, hybrid cultivar) were planted at a spacing of 15 cm × 20 cm with 2–3 plants per hill after field preparation and basal fertilization in late April and harvested in middle July. For the late rice, seedlings (cv. Yueyou 712, 30 d old, hybrid cultivar) were transplanted after early rice harvest and soil preparation, and harvested in late October. In each year, both early and late rice seasons received a basal fertilizer application in the form of 75 kg ha^{-1} P_2O_5 (calcium super phosphate), 100 kg ha^{-1} K_2O (potassium chloride), and 70% N (urea or urea plus vetch residue in vetch plots), with the remaining N applied as top dressing at the tillering stage for both early and late rice seasons. For plots with vetch cover, basal N application in the early rice crop was reduced by the estimated N content in the vetch aboveground biomass residue. The field was flooded after early rice and late rice transplanting, and a floodwater depth of 3–5 cm was maintained until 10 days before maturity except that the water was drained at maximum tillering stage to reduce unproductive tillers. Herbicides and pesticides were used to avoid yield loss according to the actual demands. All plots received a similar field management for irrigation and crop protection during rice growth season.

2.3. Sample Collection and Analysis

Rice grain yield was determined at maturity by randomly harvesting two 4 m^2 areas from each plot, while six representative hills of rice plants were collected to measure yield components and N uptake. Plants were separated into leaves, stems, and panicles. The dry weight of straw and grain was measured after oven drying at 75 °C to constant weight. Yield components included number of panicles m^{-2}, number of spikelets panicle^{-1}, grain-filling percentage, and 1000 grains weight. Rice plant sample and vetch plant sample were weighed fresh, dried at 60 °C to a constant weight. Tissue N concentration was determined by micro-Kjeldahl [13] and used to calculate total N uptake. Applied N recovery efficiency (RE$_N$) and agronomic N use efficiency (AE$_N$) were calculated following Xu et al. [14] as:

$$RE_N = \frac{NF - N0}{Applied\ N} \times 100 \tag{1}$$

where N_F and N_0 are plant total N content (kg N ha^{-1}) of N addition treatments and unfertilized control, respectively, and *Applied N* is the amount of N from urea and vetch.

$$AE_N = \frac{GF - G0}{Applied\ N} \times 100 \tag{2}$$

where G_F and G_0 are grain yield (kg ha^{-1}) of N addition treatments and unfertilized control, respectively.

2.4. Statistical Analyses

Data were analyzed using analysis of variance (SAS Institute, 2003) and means of rice grain yields, yield components, N uptake, and N use efficiency were compared based on the least significant difference (LSD) test at the 0.05 probability level.

3. Results

3.1. Grain Yield and Yield Components

Rice grain yield of the early rice, late rice, and annual rice production were significantly affected by year and treatment (Table 1). The interaction effect of year and treatment was significant for the early rice crop, but not significant for late rice and annual production. Vetch cover stimulated grain yield for the succeeding early and late rice compared with fallow at the same N application dose.

Compared with 100 kg ha^{-1} chemical N fertilizer addition, vetch cover increased grain yields by 7.3–13.4% for early rice, by 8.2–10.4% for late rice, and by 8.6–11.5% for the total annual production (Table 2). This was similar for the 200 kg N ha^{-1} treatments: 5.9–18.4% for early rice, 3.8–10.1% for late rice, and 6.2–11.3% for the total annual production higher with vetch cover than with winter fallow. With the exception of the late rice season in the year 2015, all differences between vetch and fallow were significant within the same N addition dose ($p < 0.05$).

Table 1. Analysis of variance (F values) for grain yield, total N uptake, agronomic N use efficiency (AE$_N$), and fertilizer N recovery efficiency (RE$_N$) of the early rice, late rice, and annual cycle rice in 2015, 2016, and 2017 in Huarong County.

Source of Variation	Early Rice				Late Rice				Annual Cycle Rice			
	Grain Yield	Total N Uptake	AE$_N$	RE$_N$	Grain Yield	Total N Uptake	AE$_N$	RE$_N$	Grain Yield	Total N Uptake	AE$_N$	RE$_N$
Year (Y)	217.2 **	58.6 **	6.5 **	28.7 **	191.9 **	173.8 **	7.1 **	7.8 **	100.7 **	144.1 **	7.1 **	22.3 **
Treatment (T)	359.8 **	144.8 **	78.0 **	21.3 **	89.0 **	92.0 **	5.1 **	4.7 *	272.8 **	164.0 **	21.8 **	14.9 **
Y × T	4.3 **	5.0 **	NS	NS	NS	NS	NS	NS	NS	4.0 **	NS	NS

Note: * and **, significant at $p \leq 0.05$ and $p \leq 0.01$, respectively; NS, nonsignificant.

Table 2. Grain yields for the early rice, late rice, and annual rice cycle in the years 2015–2017 influenced by vetch covering.

Treatment	2015			2016			2017		
	Early Rice (Mg ha^{-1})	Late Rice (Mg ha^{-1})	Annual Rice Cycle [a] (Mg ha^{-1})	Early Rice (Mg ha^{-1})	Late Rice (Mg ha^{-1})	Annual Rice Cycle (Mg ha^{-1})	Early Rice (Mg ha^{-1})	Late Rice (Mg ha^{-1})	Annual Rice Cycle (Mg ha^{-1})
FN$_0$	4.90 e	4.52 d	9.42 d	4.34 d	4.25 d	8.59 d	3.96 d	5.24 d	9.19 d
FN$_{100}$	6.26 d	5.06 c	11.32 c	5.30 c	4.64 c	9.94 c	5.09 c	6.07 c	11.16 c
MN$_{100}$	6.72 b	5.58 b	12.29 b	6.02 b	5.021 b	11.03 b	5.74 b	6.70 b	12.44 b
FN$_{200}$	6.38 c	5.92 a	12.31 b	5.83 b	5.17 b	11.00 b	5.81 b	6.91 b	12.72 b
MN$_{200}$	7.56 a	6.15 a	13.70 a	6.51 a	5.69 a	12.20 a	6.15 a	7.35 a	13.50 a

Note: Different letters in the same column mean there is a significant difference at the 0.05 level (LSD). The data shown in the panels are averages of the three replications for individual treatments. [a] The grain yield for annual rice cycle is the sum of the early and late rice grain yields from the same plot within a year.

Yield components analysis indicated (Table 3) that rice grain yield increases associated with vetch cover were mostly the result of the increased panicle m^{-2} number, while the other three yield components, spikelets panicle^{-1}, grain filling, and grain weight, were not significantly affected by treatments. Treatments with vetch cover showed higher panicle numbers for early rice in all of the years when compared with those with fallow. For panicle numbers of late rice, vetch cover increased panicle numbers compared to fallow at 100 kg N ha^{-1} in three years, while increased panicle numbers were observed only in 2017 for the 200 kg N ha^{-1} (Table 3).

Table 3. Yield components for the early rice and late rice in the years 2015–2017 influenced by vetch.

Treatment	Early Rice				Late Rice			
	Panicles (m^{-2})	Spikelets Panicle^{-1}	Grain Filling (%)	Grain Weight (mg)	Panicles (m^{-2})	Spikelets Panicle^{-1}	Grain Filling (%)	Grain Weight (mg)
2015								
FN$_0$	192 d	108 a	79 a	27 a	177 b	132 a	87 a	24 a
FN$_{100}$	249 c	120 a	77 a	27 a	178 b	133 a	89 a	24 a
MN$_{100}$	261 b	101 a	79 a	29 a	205 a	131 a	89 a	24 a
FN$_{200}$	276 b	114 a	76 a	26 a	210 a	136 a	87 a	24 a
MN$_{200}$	337 a	116 a	80 a	25 a	213 a	135 a	85 a	24 a
2016								
FN$_0$	151 d	121 a	79 a	29 a	228 c	102 a	65 a	24 a
FN$_{100}$	192 c	123 a	83 a	28 a	252 b	102 a	65 a	24 a
MN$_{100}$	222 b	129 a	80 a	28 a	286 a	98 a	66 a	25 a
FN$_{200}$	214 b	141 a	83 a	27 a	284 a	114 a	64 a	23 a
MN$_{200}$	250 a	133 a	78 a	27 a	298 a	112 a	63 a	23 a
2017								
FN$_0$	304 e	67 a	87 a	25 a	171 d	129 a	80 a	24 a
FN$_{100}$	354 d	76 a	84 a	25 a	198 c	137 a	79 a	24 a
MN$_{100}$	386 c	69 a	86 a	26 a	219 b	143 a	79 a	25 a
FN$_{200}$	415 b	78 a	84 a	25 a	246 a	153 a	73 a	24 a
MN$_{200}$	457 a	64 a	84 a	26 a	223 b	144 a	77 a	25 a

Note: Different letters in the same column in the same year mean there is a significant difference at the 0.05 level (LSD).

3.2. N Uptake

Year and treatment significantly affected total N uptake for the early and late rice crops and annual rice production. The interaction effect of year and treatment was significant for early rice and annual production, but not significant for late rice (Table 1).

The N uptake by early rice and late rice and annual N uptake increased significantly in response to N fertilizer (Figure 2). Consistent with the positive effect of vetch cover on grain yields, N uptake by rice plants was also significantly higher in the vetch treatments than in the fallow treatments (Figure 2). At 100 kg N ha^{-1}, vetch cover increased N uptake by 8.3–18.1% for early rice, by 6.0–8.8% for late rice, and by 7.1–13.2% for annual N uptake. At 200 kg N ha^{-1}, vetch cover enhanced N uptake by 9.2–26.6% for early rice, by 9.5–13.2% for late rice, and by 9.3–18.1% for annual N uptake (Figure 2).

3.3. N Use Efficiency

Both year and treatment had significant effects on AE$_N$ and RE$_N$ for early rice, late rice, and annual rice production, whereas the interaction between year and treatment on AE$_N$ and RE$_N$ was not significant (Table 1).

During the three-year study, AE$_N$ exhibited the same trend across treatments for early rice, late rice, and annual cycle rice crops (Figure 3). For the same N dose, treatments with the vetch cover demonstrated significantly greater AE$_N$ values than corresponding fallow treatments for early rice, late rice, and annual rice production ($p < 0.05$), with the exception of the late rice season in the years 2016 and 2017 (Figure 3).

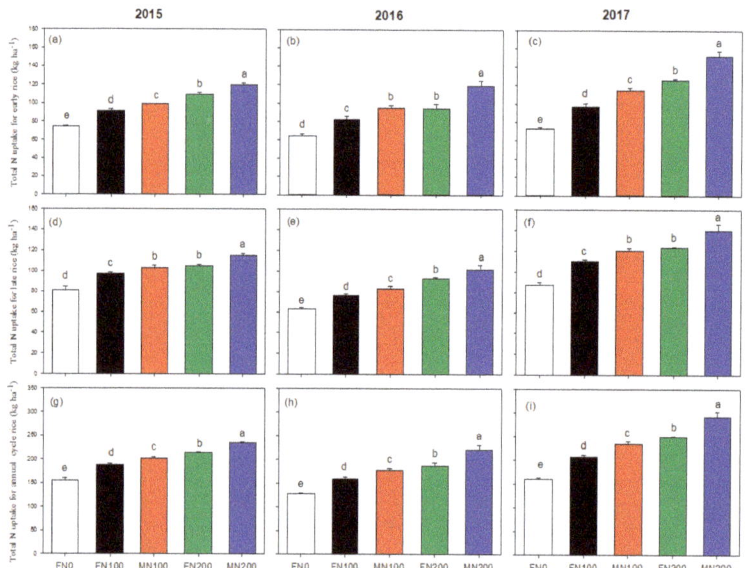

Figure 2. Total N uptake for early (**a–c**), late (**d–f**), and annual cycle (**g–i**) rice from 2015 to 2017. FN_0 represents winter fallow without N application in rice seasons as a check, MN_{100} represents vetch cover cropping with 100 kg N ha^{-1} each rice season, FN_{100} represents winter fallow with 100 kg N ha^{-1} each rice season, MN_{200} represents vetch cover cropping with 200 kg N ha^{-1} each rice season, and FN_{200} represents winter fallow with 200 kg N ha^{-1} each rice season. The data shown in the panels are averages of the three replications for individual treatments. Vertical bars represent the standard errors. Different lowercase letters represent the significant differences at $p < 0.05$.

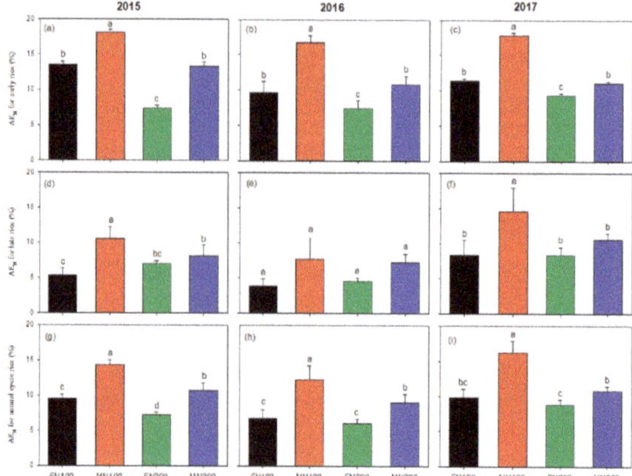

Figure 3. Agronomic N use efficiency for early (**a–c**), late (**d–f**), and annual cycle (**g–i**) rice from 2015 to 2017. FN_0 represents winter fallow without N application in rice seasons as a check, MN_{100} represents vetch cover cropping with 100 kg N ha^{-1} each rice season, FN_{100} represents winter fallow with 100 kg N ha^{-1} each rice season, MN_{200} represents vetch cover cropping with 200 kg N ha^{-1} each rice season, and FN_{200} represents winter fallow with 200 kg N ha^{-1} each rice season. The data shown in the panels are averages of the three replications for individual treatments. Vertical bars represent the standard errors. Different lowercase letters represent the significant differences at $p < 0.05$.

Similar to AE_N, RE_N was stimulated by vetch cover across the three experimental years (Figure 4). RE_N values for early rice, late rice, and annual production were highest in MN_{100} plots (24.6–41.8% for early rice, 19.6–32.8% for late rice, and 23.3–37.2% for annual cycle), followed by MN_{200}, FN_{100}, and FN_{200} (Figure 4). Under the same N dose, MN_{100} and MN_{200} also showed significantly greater RE_N values than FN_{100} and FN_{200}, respectively, for early rice, late rice, and annual cycle rice crops ($p < 0.05$).

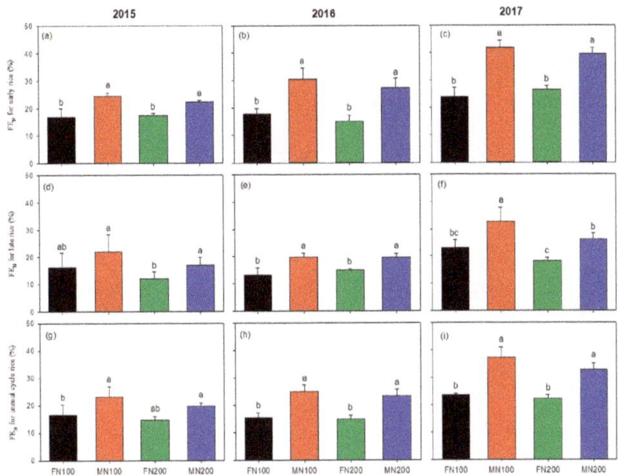

Figure 4. Fertilizer N recovery efficiency for early (**a–c**), late (**d–f**), and annual cycle (**g–i**) rice from 2015 to 2017. FN_0 represents winter fallow without N application in rice seasons as a check, MN_{100} represents vetch cover cropping with 100 kg N ha^{-1} each rice season, FN_{100} represents winter fallow with 100 kg N ha^{-1} each rice season, MN_{200} represents vetch cover cropping with 200 kg N ha^{-1} each rice season, and FN_{200} represents winter fallow with 200 kg N ha^{-1} each rice season. The data shown in the panels are averages of the three replications for individual treatments. Vertical bars represent the standard errors. Different lowercase letters represent the significant differences at $p < 0.05$.

4. Discussion

4.1. Effect of Vetch Winter Cropping on Rice Grain Yield

Effects of cover crops on the succeeding crop yields have been studied in various cropping systems around the world [9,15–17]. In most cases, leguminous cover crop species, such as alfalfa, red clover, and crimson clover, exhibit a positive effect on grain yield of subsequent corn or winter wheat in corn–cover crop and winter wheat–cover crop systems [18–20]. When it comes to rice, some studies reported grain yield increased following cover crops such as milk vetch, *Sesbania rostrate*, and *Aeschynomene afraspera* [17,21,22]. It is widely thought that the yield increase is the result of a better synchrony of N supply from the cover crop residues with main crop N demand. In addition to its ability to fix N, leguminous cover crops are more likely to give a higher N return after incorporation, suggesting a possibility for replacing fertilizer N during main crop seasons.

In this study, the positive effect of milk vetch on the subsequent rice crops is in agreement with results from Yu et al. [23] and Zhu et al. [17], which showed a 5–10% yield increase by including legumes such as bean or vetch in a rice–winter crop system. Similar positive effects on rice yield have been observed in multiple cropping systems such as rice–*Azolla* [24] and rice–fish [25]. In our study, milk vetch and subsequent early and late rice crops constituted a temporal multicropping system that resulted in enhanced rice crop yields. As reviewed by a number of reports [9,12,15,16], biologically-fixed N from the leguminous cover crops is likely associated with the increase in the rice crop yields. Interestingly, we observed comparable rice yields for the treatments MN_{100} and FN_{200} (6.1 vs. 6.0 Mg ha^{-1}, 5.8 vs. 6.0 Mg ha^{-1}, and 11.9 vs. 12.0 Mg ha^{-1} for early rice, late rice, and total annual rice production, respectively). This implies a considerable potential for vetch in double-rice systems to reduce fertilizer N requirement while still meeting the goal of high rice grain yields [23]. The positive effect of milk vetch on yield was not limited to the early rice crop, but was also observed in the late rice crop. The possible explanation for this late rice effect is the greater N return from early rice straw in milk-vetch-treated plots (Figure 2).

The results of yield components analysis indicated that a greater rice plant panicle number was the only factor affected by milk vetch and fertilizer rate, rather than number of spikelets panicle^{-1}, grain filling, and grain weight (Table 3). This implies that rice grain yield increase by milk vetch was a result of more vigorous growth of rice plants at or before the tillering stage [17].

4.2. Effect of Vetch on Rice N Uptake and N Use Efficiency

Another approach to managing the N economy of rice crops is to promote N retention by adding plant residues, such as straw, during the fallow period, but results from studies of the effects of straw application on the subsequent crop growth and N uptake are inconsistent. Some reported an insignificant increase or even a decrease in grain yield and N uptake after straw incorporation caused by short-term microbial immobilization associated with a high straw C/N ratio [26,27]. Anaerobic decomposition of incorporated residues in a flooded paddy will release low-molecular-weight organic acids, some of which can inhibit root growth [1,14]. Others observed enhanced crop yield and N uptake with residue incorporation of a suitable C/N ratio and when applied at the right time or in a right way [12,28,29]. In this study, the low C/N ratio of the vetch biomass residue (20–25) and 10 days between vetch incorporation and rice transplanting allowed enough time for N release coinciding with early rice growth resulting in increased grain yield and N uptake. The increase in late rice grain yields and N uptake in vetch covering treatments could be explained by a higher N return from early rice straw biomass incorporation before late rice transplanting.

To better evaluate the sustainability of the vetch–double-rice system, it is important to consider options that have the potential both to improve rice production and to decrease chemical N demand. AE_N and RE_N are commonly used to measure N cycling efficiency in ago-ecosystems [14,16,28]. As shown in Figure 3, the highest AE_N values for early rice, late rice, and annual production were observed in vetch with 100 kg N ha^{-1} (16.7–18.2% for early rice, 7.7–14.6% for late rice, and 12.2–16.3% for annual cycle). The higher AE_N in the vetch cropping treatments was the main explanation for the enhanced rice yields (Table 2). As reported by Bijay-Singh et al. [30] and Xu et al. [14], AE_N was not significantly increased by N return from incorporated straw unless the incorporated residue or aboveground biomass was applied with a lower N rate. However, our findings suggested an increased AE_N after the vetch winter crop and residue incorporation both under a moderate and high N rate. Yao et al. [31] also observed a higher AE_N in rice with a green manure duckweed (*Spirodela polyrhiza*) at a total N application of 225 kg N ha^{-1}, which was even higher than in our study. The effects of cover crop/green manure or crop straw incorporation on the subsequent crop AE_N may depend on the quality of the incorporated organic amendments, such as C/N ratio [14], nitrogen release rate [10], and toxic material produced during residue decomposition [1,14].

Values for RE_N in this three-year observation followed the range reported by other studies conducted in paddy soils [12,28,29]. The increase of RE_N (Figure 4) in the vetch treatments was the result of a better synchronization between N supply and crop N demand, which was also observed in previous studies [31–33]. As reported by Zhu et al. [12] and Xie et al. [33] from a greenhouse pot trail, vetch combined with urea application resulted in a higher and more persistent soil inorganic N supply than urea applied alone. Furthermore, vetch cropping helped to increase soil microbial activities because vetch covering and incorporation may enhance vetch–microbe interaction by organic C and N input from vetch biomass and root rhizodeposition [23,32]. According to Yuan et al. [34], biological N fixation by vetch could supply 54.1 to 70.8 kg N ha^{-1} to rice paddies in southern China.

Reduced N loss may be another essential factor explaining the increased N recovery efficiency in association with vetch cropping, because a cover crop could retain soil native N which otherwise might be lost through leaching during winter fallow [23] and decrease N loss in the form of ammonia gas during rice growing seasons [31].

5. Conclusions

In an intensive double-cropped rice system in southern China, maintaining high grain yields and decreasing chemical N input is a double-win target for farmers and the environment. The present study indicated that winter cover cropping with leguminous vetch increased grain yields, N uptake, and N use efficiency of the subsequent early and late rice crops when compared with a winter fallow. The interaction between vetch residue incorporation and urea application significantly increased rice production while reducing chemical N input. Comparable grain yields (11.9 vs. 12.0 Mg ha^{-1} for annual rice cycle) between vetch with 100 kg N ha^{-1} and fallow with 200 kg N ha^{-1} suggested that winter cropping of vetch, combined with reduced N fertilizer, is a possible way to help reach the double-win target. The enhanced N use efficiency when vetch is supplied with N fertilizer would reduce N loss into the environment. These results show that winter covering of vetch helps to gain a more sustainable rice production system.

Author Contributions: B.Z., Z.Z., and Z.L. designed the experiments. J.N., L.Y., and H.X. performed the experiments. J.N., P.D., G.W.K., and B.A.H. analyzed the data. J.N. wrote the manuscript. All authors approved the submission.

Funding: This study was funded by the National Key Program of Research & Development of China (2016YFD0300208-04), National Natural Science Foundation of China (No. 31870424, 31501274), Young Elite Scientists Sponsorship Program by CAST (No. YESS20160040), and China Scholarship Council (No. 201708420112).

Acknowledgments: Special thanks go to the anonymous reviewers for their constructive comments in improving this manuscript.

Conflicts of Interest: The authors declare no conflict of interest.

References

1. Peng, S.; Buresg, R.J.; Huang, J.; Zou, Y.; Zhong, X.; Wang, G.; Zhang, F. Strategies for overcoming low agronomic nitrogen use efficiency in irrigated rice systems in China. *Field Crops Res.* **2006**, *96*, 37–47. [CrossRef]

2. Ju, X.; Xing, G.; Chen, X.; Zhang, S.; Zhang, L.; Liu, X.; Cui, Z.; Yin, B.; Christie, P.; Zhu, Z.; et al. Reducing environmental risk by improving N management in intensive Chinese agricultural systems. *Proc. Natl. Acad. Sci. USA* **2009**, *106*, 3041–3046. [CrossRef] [PubMed]

3. Galloway, J.N.; Townsend, A.R.; Erisman, J.W.; Bekunda, M.; Cai, Z.; Freney, J.R.; Martinelli, L.A.; Seitzinger, S.P.; Sutton, M.A. Transformation of the nitrogen cycle: Recent trends, questions, and potential solutions. *Science* **2008**, *320*, 889–892. [CrossRef] [PubMed]

4. Liu, X.; Zhang, Y.; Han, W.; Tang, A.; Shen, J.; Cui, Z.; Vitousek, P.; Erisman, J.W.; Goulding, K.; Christie, P.; et al. Enhanced nitrogen deposition over China. *Nature* **2013**, *494*, 459–462. [CrossRef] [PubMed]

5. Geng, J.; Ma, Q.; Chen, J.; Zhang, M.; Li, C.; Yang, Y.; Yang, X.; Zhang, W.; Liu, Z. Effects of polymer coated urea and sulfur fertilization on yield, nitrogen use efficiency and leaf senescence of cotton. *Field Crops Res.* **2016**, *187*, 87–95. [CrossRef]

6. Chen, Y.; Peng, J.; Wang, J.; Fu, P.; Hou, Y.; Zhang, C.; Fahad, S.; Peng, S.; Cui, K.; Nie, L.; et al. Crop management based on multi-split topdressing enhances grain yield and nitrogen use efficiency in irrigated rice in China. *Field Crops Res.* **2015**, *184*, 50–57. [CrossRef]

7. Ramírez-García, J.; Carrillo, J.M.; Ruiz, M.; Alonso-Ayuso, M.; Quemada, M. Multicriteria decision analysis applied to cover crop species and cultivars selection. *Field Crops Res.* **2015**, *175*, 106–115.

8. Campiglia, E.; Mancinelli, R.; Radicetti, E.; Marinari, S. Legume cover crops and mulches: Effects on nitrate leaching and nitrogen input in a pepper crop (*Capsicum annuum* L.). *Nutr. Cycl. Agroecosyst.* **2011**, *89*, 399–412. [CrossRef]

9.   Gabriel, J.L.; Quemada, M. Replacing bare fallow with cover crops in a maize cropping system: Yield, N uptake and fertilizer fate. *Europ. J. Agron.* **2011**, *34*, 133–143. [CrossRef]

10. Baijukya, F.P.; de Ridder, N.; Giller, K.E. Nitrogen release from decomposing residue of leguminous cover crops and their effect on maize yield on depleted soil of Bukoba District, Tanzania. *Plant Soil* **2006**, *279*, 77–93. [CrossRef]

11. Tribouillois, H.; Cohan, J.P.; Justes, E. Cover crop mixtures including legume produce ecosystem services of nitrate capture and green manuring: Assessment combining experimentation and modeling. *Plant Soil* **2016**, *401*, 347–364. [CrossRef]

12. Zhu, B.; Yi, L.; Hu, Y.; Zeng, Z.; Lin, C.; Tang, H.; Yang, G.; Xiao, X. Nitrogen released from incorporated ^{15}N-labelled Chinese milk vetch (*Astragalus sinicus* L.) residue and its dynamics in a double rice cropping system. *Plant Soil* **2014**, *374*, 331–344. [CrossRef]

13. Bremner, J.M.; Mulvaney, C.S. N-Total. In *Methods of Soil Analysis, Part 2*; Page, A.L., Miller, R.H., Keeney, D.R., Eds.; American Society of Agronomy: Madison, WI, USA, 1982; pp. 595–624.

14. Xu, Y.; Nie, L.; Buresh, R.; Huang, J.; Cui, K.; Xu, B.; Gong, W.; Peng, S. Agronomic performance of late-season rice under different tillage, straw, and nitrogen management. *Field Crops Res.* **2010**, *115*, 79–84. [CrossRef]

15. Constantin, J.; Mary, B.; Laurent, F.; Aubrion, G.; Fontaine, A.; Kerverillant, P.; Beaudoin, N. Effects of catch crops, no till and reduced nitrogen fertilization on nitrogen leaching and balance in three long-term experiments. *Agric. Ecosyst. Environ.* **2010**, *135*, 268–278. [CrossRef]

16. Deligios, P.A.; Tiloca, M.T.; Sulas, L.; Buffa, M.; Caraffini, S.; Doro, L.; Sanna, G.; Spanu, E.; Spissu, E.; Urracci, G.R.; et al. Stable nutrient flows in sustainable and alternative cropping systems of globe artichoke. *Agron. Sustain. Dev.* **2017**, *37*, 54–65. [CrossRef]

17. Zhu, B.; Yi, L.; Guo, L.; Chen, G.; Hu, Y.; Tang, H.; Xiao, X.; Yang, G.; Surya, N.A.; Zeng, Z. Performance of two winter cover crops and their impacts on soil properties and two subsequent rice crops in Dongting Lake Plain, Hunan, China. *Soil Tillage Res.* **2012**, *124*, 95–101. [CrossRef]

18. Claire, C.; John, D.L.; Bill, D.; Laura, L.V.E. Legume cover crop management on nitrogen dynamics and yield in grain corn systems. *Field Crops Res.* **2017**, *201*, 75–85.

19. Gaudin, A.; Janovicek, K.; Martin, R.C.; Deen, W. Approaches to optimizing nitrogen fertilization in a winter wheat-red clover (*Trifolium pratense* L.) relay cropping system. *Field Crops Res.* **2014**, *155*, 192–201. [CrossRef]

20. Ketterings, Q.M.; Swink, S.N.; Duikr, S.W.; Czymmek, K.J.; Beegle, D.B.; Cox, W.J. Integrating cover crops for nitrogen management in corn systems on Northeastern U.S. dairies. *Agron. J.* **2015**, *107*, 1365–1376. [CrossRef]

21. Diekmann, K.K.; Ottow, J.C.G.; De Datta, S.K. Yield and nitrogen response of lowland rice (*Oryza sativa* L.) to *Sebania rostrata* and *Aeschynomeme afraspera* green manure in different marginally productive soils in Philippines. *Biol. Fertil. Soils* **1996**, *21*, 103–108. [CrossRef]

22. Lee, C.H.; Park, K.D.; Jung, K.Y.; Ali, M.A.; Lee, D.; Gutierrez, J.; Kim, P.J. Effects of Chinese milk vetch (*Astragalus sinicus* L.) as a green manure on rice productivity and methane emission in paddy soil. *Agric. Ecosyst. Environ.* **2010**, *138*, 343–347. [CrossRef]

23. Yu, Y.; Xue, L.; Yang, L. Winter legumes in rice crop rotations reduces nitrogen loss, and improves rice yield and soil nitrogen supply. *Agron. Sustain. Dev.* **2014**, *34*, 633–640. [CrossRef]

24. Xu, H.; Zhu, B.; Liu, J.; Li, D.; Yang, Y.; Zhang, K.; Jiang, Y.; Hu, Y.; Zeng, Z. *Azolla* planting reduces methane emission and nitrogen fertilizer application in double rice cropping system in southern China. *Agron. Sustain. Dev.* **2017**, *37*, 29–38. [CrossRef]

25. Xie, J.; Hu, L.; Tang, J.; Wu, X.; Li, N.; Yuan, Y.; Yang, H.; Zhang, J.; Luo, S.; Chen, X. Ecological mechanisms underlying the sustainability of the agricultural heritage rice-fish coculture system. *Proc. Natl. Acad. Sci. USA* **2011**, *108*, 1381–1387. [CrossRef] [PubMed]

26. Buresh, R.J.; Reddy, K.R.; van Kessel, C. Nitrogen transformations in submerged soils. In *Nitrogen in Agricultural Systems, Agronomy Monograph 49*; Schepers, J.S., Raun, W.R., Eds.; ASA, CSSA and SSSA: Madison, WI, USA, 2008; pp. 401–436.

27. Witt, C.; Cassman, K.G.; Olk, D.C.; Biker, U.; Liboon, S.P.; Sanson, M.I.; Ottow, J.C.G. Crop rotation and residue management effects on carbon sequestration, nitrogen cycling and productivity of irrigated rice systems. *Plant Soil* **2000**, *225*, 263–278. [CrossRef]

28. Asagi, N.; Ueno, H. Nitrogen dynamics in paddy soil applied with various ^{15}N-labelled green manures. *Plant Soil* **2009**, *322*, 251–262. [CrossRef]

29. Ashraf, M.; Mahnood, T.; Azam, F.; Qureshi, R.M. Comparative effects of applying leguminous and non-leguminous green manures and inorganic N on biomass yield and nitrogen uptake in flooded rice (*Oryza sativa* L.). *Biol. Fertil. Soils* **2004**, *40*, 147–152. [CrossRef]

30. Shan, Y.H.; Johnson-Beebout, S.E.; Buresh, R.J. Crop residue management for lowland rice-based cropping systems in Asia. *Adv. Agron.* **2008**, *98*, 117–199.

31. Yao, Y.; Zhang, M.; Tian, Y.; Zhao, M.; Zhang, B.; Zhao, M.; Zeng, K.; Yin, B. Duckweed (*Spirodela polyrhiza*) as green manure for increasing yield and reducing nitrogen loss in rice production. *Field Crops Res.* **2017**, *214*, 273–282. [CrossRef]

32. Srivastava, J.K.; Chandra, H.; Kalra, S.J.S.; Mishra, P.; Khan, H.; Yadav, P. Plant-microbe interaction in aquatic system and their role in the management of water quality: A review. *Appl. Water Sci.* **2017**, *7*, 1079–1090. [CrossRef]

33. Xie, Z.; He, Y.; Tu, S.; Xu, C.; Liu, G.; Wang, H.; Cao, W.; Liu, H. Chinese Milk Vetch improves plant growth, development and ^{15}N recovery in the rice-based rotation system of South China. *Sci. Rep.* **2017**, *7*, 3577. [CrossRef] [PubMed]

34. Yuan, M.; Liu, Q.; Zhang, S. Characteristics of nitrogen fixation of winter green manure in paddy soil in the Taihu Lake area. *Chin. J. Soil Sci.* **2010**, *41*, 1115–1119. (In Chinese)

Article

Kura Clover Living Mulch Reduces Fertilizer N Requirements and Increases Profitability of Maize

Jonathan R. Alexander [1],*, John M. Baker [1,2], Rodney T. Venterea [1,2] and Jeffrey A. Coulter [3]

1   Department of Soil, Water and Climate, University of Minnesota, St. Paul, MN 55108-6028, USA
2   United States Department of Agriculture—Agricultural Research Service, St. Paul, MN 55108-6028, USA
3   Department of Agronomy and Plant Genetics, University of Minnesota, St. Paul, MN 55108-6028, USA
*   Correspondence: alexa564@umn.edu; Tel.: +1-507-430-3239; Fax: +1-(621)-625-2208

Received: 20 June 2019; Accepted: 3 August 2019; Published: 6 August 2019

Abstract: Kura clover living mulch (KCLM) systems have been previously investigated for their incorporation into upper Midwestern row crop rotations to provide ecosystem services through continuous living cover. Reductions in soil erosion and nitrate loss to surface and groundwater have been reported, but factors affecting agronomic performance and nutrient management are not well defined. To achieve realized environmental benefits, research must develop agronomic management techniques, determine economic opportunities, and provide management recommendations for row crop production in KCLM systems. Two experiments were conducted in 2017 and 2018 to determine the response to N fertilizer application for maize production in KCLM. The first-year maize experiment followed forage management, and the second-year maize experiment followed maize after forage management. Eight fertilizer N treatments ranging from 0–250 kg N ha^{-1} were applied to each experiment and grain and stover yields were compared to conventionally managed maize hybrid trials that were conducted nearby. First-year maize did not need fertilizer N to maximize yield and profitability in either growing season, and second-year maize required a fertilizer N rate near local University guidelines for maize following soybean. The net economic return from maize grain and stover in the KCLM averaged over first and second-year maize experiments and 2017 and 2018 growing seasons were \$138 ha^{-1} greater than the conventional comparison.

Keywords: kura clover; living mulch; cover crop; perennial; conservation; nitrogen; forage; economics

---

1. Introduction

Kura clover (*Trifolium ambiguum* M. bieb), a rhizomous perennial legume forage, is well suited for incorporation into upper Midwestern row-cropping systems as a perennial cover crop or living mulch [1]. Kura clover's dense rhizome system holds large stores of metabolite energy that allow for perennial persistence and rapid reestablishment after intensive agronomic management [2]. In the U.S. Midwest, maize (*Zea mays*) and soybean (*Glycine max*) have been successfully grown in a kura clover living mulch (KCLM), and clover forage productivity recovered in the following growing season [1,3,4]. Living cover and active root uptake during the fall and spring months reduce soil erosion and nitrate leaching from maize production by up to three-quarters in the KCLM system compared to conventional management [5,6].

Maize production in KCLM requires stover harvest to prevent smothering of kura clover by crop residues [1]. Maize silage and stover are important forage and bedding materials in livestock operations and were harvested from 7% of maize acres in the central and eastern U.S. in 2010 [7,8]. While harvest of maize residue in conventional production systems can increase soil erosion and negatively affects soil carbon, structure, and fertility [9,10], soil physical and chemical properties were unaffected by 5–7 years of continuous maize stover removal under KCLM management in the upper Midwest [11]. Improved protection from soil erosion and increased carbon input to soils from living cover maintains soil quality and reduces environmental and economic costs of stover harvest by increasing the sustainable stover removal rate [12]. Higher stover removal rates will reduce the land area impacted by stover harvest, thus reducing harvest cost and increasing the sustainability of stover removal in upper Midwestern livestock production systems.

Disparities in the literature exist regarding the agronomic productivity between KCLM and conventional maize systems, where grain and forage yields are either reduced [3–6,13] or maintained [1,2,14,15]. Previous research comparing maize grain yield in KCLM and conventional management systems are sometimes confounded by nitrogen (N) management, where living mulch treatments were granted legume N credits of 67–146 kg N ha^{-1} [5,6,13]. Limited understanding of biological N fixation and cycling in KCLM systems has left researchers with little baseline information on N management guidelines, limiting the quantification and understanding of agronomic, environmental, and economic attributes of KCLM systems. Defining more robust N management recommendations for KCLM-row cropping systems requires identification of environmental and agronomic factors that affect in-season N contributions and availability. Recent work to isolate factors affecting in-season N contributions from KCLM identified rotary zone tillage as an important factor in promoting N mineralization from disturbed and incorporated clover residues [16]. This aligns with previous studies that identified rotary zone tillage as a promising strategy to reduce living mulch competition with the emerging row crop [14,17,18].

It is necessary to develop N management guidelines based on crop rotations and scenarios that may be utilized by growers to quantify the economic and environmental potentials of KCLM-maize management systems. The potential reduction of fertilizer N requirements for maize through in-season N contributions from KCLM may reduce management costs and improve economic competitiveness with other cropping systems. The objectives of this research were to determine the effect of N fertilizer management on the productivity of maize and kura clover in a KLCM-maize system and assess the economic performance of this system. Greater understanding of N cycling in KCLM systems will facilitate innovation and adoption of perennial-annual intercropping systems that reduce environmental impacts of agriculture while providing economic benefits to crop and livestock growers.

2. Materials and Methods

2.1. Site Description

Field experiments were conducted in 2017 and 2018 at the University of Minnesota Research and Outreach Center in Rosemont, MN (44.73° N, 93.09° W) on a Waukegan silt loam (Fine-silty over sandy or sandy-skeletal, mixed, super active, mesic Typic Hapludolls). Soils at the site contained 20.5 g kg^{-1} organic carbon and were 5.7 pH in the 0–30 cm layer. *Endura* kura clover was established at the site in 2006–2007 and used as a living mulch for maize and soybean from 2008–2009. Rhizomes were dug from the experimental site in 2010 using a potato (*Solanum tuberosum*) harvester for a vegetative repropagation study [19] and kura clover recovered before resuming as a living mulch for maize in 2011 and 2013 and soybean in 2012 and 2014. Phosphorous (P) and potassium (K) fertilizers were applied according to soil test values and University of Minnesota guidelines in 2015 [20] and clover was harvested for hay once each year in 2015 and 2016.

Daily precipitation and minimum and maximum air temperature from 1 April to 31 October were obtained from National Weather Service Cooperative Observer Station No. 217107. Daily weather observations were averaged over 1981–2010 to represent the climatic normal. Daily minimum and maximum temperature were used to calculate cumulative growing degree days with a base and upper limit of 10 and 30 °C, respectively, beginning on 12 May for the climatic average and the planting date of the corresponding growing seasons in these experiments.

2.2. Experimental Design

Two studies, each conducted over two growing seasons (2017 and 2018), investigated fertilizer N management for maize in KCLM (Table 1). The 'first-year' experiment was maize seeded into kura clover that was managed as forage since the 2015 growing season. The 'second-year' experiment was maize seeded into kura clover that had been planted to maize in the previous growing season and was previously managed as forage since the 2015 growing season. In the 2011–2014 growing seasons, the kura clover was managed as a living mulch in a maize-soybean rotation. First-year maize production in preparation for the second-year maize experiment received 150 kg N ha^{-1} as liquid urea-ammonium nitrate banded 10 cm from the center of the row at the six-leaf collar stage of maize phenological development. The second growing season of each study was placed adjacent to the first to maintain the crop rotation treatments; therefore, treatments were not applied to the same plots in both growing seasons. The studies investigated fertilizer N rate, where an unfertilized control was compared to plots that received a split application of urea containing urease and nitrification inhibitors (SuperU, Koch Agronomic Services, Wichita, KS, USA) at 40, 80, 120, 180, or 250 kg N ha^{-1}. SuperU was surface-banded at 40 kg N ha^{-1} onto the center of the tilled row and incorporated with a second pass of the rotary zone tillage tool prior to planting. The remainder of treatment fertilizer N was surface banded 5 cm from the center of the row at the four-leaf collar stage of maize phenological development. Two additional fertilizer N treatments applied urea instead of SuperU, either in a single application of 120 kg N ha^{-1} at planting or as a split application of 40 and 80 kg N ha^{-1} at planting and the four-leaf collar stage of maize phenological development, respectively. Plots were 4.7 m (6 rows) wide by 15.2 m long and the first- and second-year maize experiments were arranged as two 4 × 2 randomized complete blocks with four replications of the eight fertilizer N treatments.

Table 1. Crop rotations for first- and second-year maize experiments.

Experiment, Growing Season	Growing Season				
	2018	2017	2016	2015	2011–2014
Second-year maize, 2017		Maize †	Maize		
First-year maize, 2017		Maize †		Forage	Maize-Soybean rotation
Second-year maize, 2018	Maize †	Maize	Forage		
First-year maize, 2018	Maize †	Forage			

† Experimental data were collected from plots in the 2017 and 2018 growing seasons following corresponding crop rotations.

2.3. Agronomic Management

Prior to spring clover management, triple superphosphate, potash, and gypsum fertilizers were broadcast over the entire experimental area in both growing seasons based on soil test values and university guidelines [20]. Clover was cut with a flail mower to 5 cm prior to row establishment in 2017 and in the first-year maize experiment in 2018. Rows were established with a rotary zone tillage tool (Northwest Tillers, Yakima, WA, USA) that tilled 30-cm-wide rows every 76 cm on 11 May 2017 and 22 May 2018 [16,17]. Fertilizer N applications at planting were applied with a garden seeder (Earthway, Bristol, IN, USA) in the center of the tilled strip before incorporation with a second pass of the rotary zone tillage tool. Maize (Pioneer P0157AMX, Pioneer Dupont, Johnston, IA, USA) was planted in 76-cm rows in the center of the tilled strips at 86,000 seeds ha^{-1} with a John Deere 7000 planter (John Deere, Moline, IA, USA) on 12 May 2017 and 22 May 2018. An additional clover mowing 15 days after planting (DAP) was necessary in 2017 to reduce clover encroachment into maize rows, but mowing was not needed in the 2018 experiments. Herbicide suppression was used in both growing seasons to reduce clover encroachment into maize rows and was achieved with a broadcast application of 2.3 L a.e. ha^{-1} (N-(phosphonomethyl) glycine) (glyphosate) 39 and 31 DAP in 2017 and 2018, respectively [5,6,16].

2.4. Crop Sampling and Analysis

Clover biomass was sampled at 160 and 133 DAP maize in 2017 and 2018 by placing a 0.5-m^2 quadrat between the center two maize rows and cutting clover biomass to a 1 cm height [16]. Maize was harvested at physiological maturity by hand harvesting ears and cutting stalks to a 15 cm height from 4.6 m in two rows in each plot. Maize stover was weighed in the field before a six-stalk subsample was ground with a biomass chipper, collected, and weighed. Maize ears, stalks, and clover were dried at 60 °C for 3 d until reaching a constant mass. Dry stover subsamples were weighed to determine field moisture content to correspond with in-field measurements. Maize ears were shelled before dry grain and cobs were weighed. Subsamples of grain, cob, stover, and kura clover were ground to < 0.1 mm and analyzed for organic carbon and N with the Dumas dry combustion method in an elemental analyzer (Elementar, Langenselbold, DE) [21]. The N content of crop components were combined with corresponding dry biomass measurements to determine maize N uptake. Stover and cob yield and N content were summed and are herein referred to as stover yield and N content, respectively.

2.5. Residual Soil Nitrogen

Post-harvest soil samples were collected with a 41-mm i.d. hydraulic coring device. Soils were collected from the 0–30 and 30–60 cm soil layers in 2017 and 2018. Samples were homogenized and weighed before a 10-g subsample of wet soil was added to 38 mL of 2M KCl, shaken at 120 rpm for 1 h, and filtered through 11-µm filter paper. The filtered extractant was analyzed for NO_3-N (sum of NO_2-N and NO_3-N) and NH_4-N with the Greiss–Ilosvay with cadmium reduction and the sodium salicylate-nitroprusside methods, respectively, each modified for flow-through injection analysis [22] (Lachat, Loveland, CO, USA). A 5-g subsample of wet soil was dried at 105 °C until constant mass to determine gravimetric water content. Core volume and mass were used to determine soil bulk density, and soil N content was the product of soil N concentration and soil bulk density.

2.6. Economic Calculations

The partial net economic return was determined with a partial budget analysis of this study's KCLM experiments and conventionally managed maize hybrid trials, where maize hybrids followed soybean and were greater than 99-day relative maturity, similar to the 102-day relative maturity hybrid used in the KCLM experiments. Maize hybrid trials were conducted at Rosemount, MN, USA in 2017 ($n = 27$) and 2018 ($n = 28$) [23,24], and trial yields were compared to the corresponding growing season of the KCLM experiments. Agronomic management costs that differed between conventional and KCLM cropping systems included fertilizer N rate, spring tillage, spring mowing, fall tillage, and baling operations. Baling operations included raking, round baling, bale transportation, bale storage [25,26], and nutrient replacement costs associated with stover removal (1.65 kg P Mg^{-1} and 6.65 kg K Mg^{-1} dry stover) [27]. Management for conventional maize production was assumed as spring field cultivation and fall disk-chiseling, while management for the KCLM system was based on agronomic practices performed on experimental treatments with spring and in-season mowing, rotary zone tillage, and baling operations with the associated nutrient replacement cost [17,26,27]. All other costs, including land, P and K fertilization excluding stover nutrient removal replacement, seed, planting, pesticide and application of pesticide, harvest costs, and miscellaneous costs were assumed equal across treatments and conventional comparisons [26]. Net return from grain in the variety trials and the KCLM experiments were the product of grain yield at the economic optimum fertilizer N rate (EONR), which was determined from grain yield estimates of the fitted quadratic regression equations where N cost was $0.86 kg^{-1} and the grain valued at $133 Mg^{-1} at 155 g kg^{-1} moisture. When there was no grain yield response to fertilizer N, the EONR was set at 0 kg N ha^{-1}. Net return from stover in KCLM experiments was the product of stover yield at the EONR and stover value ($79.37 Mg^{-1}) at 200 g kg^{-1} moisture [28]. A comparison of system performance was based on parameter estimates of the fitted linear or quadratic regression of the response variables at the EONR when grain yield was affected by fertilizer N and the average response of the unfertilized treatment when grain yield was not affected by fertilizer N.

2.7. Data Analysis

Data were analyzed separately for first- and second-year maize experiments, where the first-year maize experiment was maize following forage management and the second-year maize experiment was maize following maize after forage management. Dependent variables were evaluated using restricted maximum likelihood in mixed effect models in the lme4 version 1.1-19 package of R version 3.5.2 [29], where growing season and N fertilizer treatment were considered fixed effects and block was considered a random effect. Scatter and quantile-quantile plots of predicted versus residual values were evaluated for homogeneity of variance and normality [30]. The requirements of normality and common variance were met for all dependent variables except residual NH_4-N, NO_3-N, and total inorganic N (NH_4-N + NO_3-N, TIN), which were therefore logarithm base 10 transformed and subsequently evaluated for normality and common variance using the aforementioned procedures. The requirements of homogeneity of variance and normality were met for NH_4-N, NO_3-N, and TIN following logarithm base 10 transformation. The significance of fixed effects were determined with analysis of variance at $p < 0.05$. When the main effect of growing season was significant, means were compared at $p < 0.05$ using pairwise comparisons with the emmeans version 1.3.3 package of R [31]. When the main effect of fertilizer N treatment or the growing season-by-fertilizer N treatment interaction was significant, means of the three treatments receiving a total of 120 kg N ha^{-1} were compared at $p < 0.05$ using pairwise comparisons with the emmeans package of R. Regression of the non-transformed parameter response to fertilizer N rate was conducted using the unfertilized treatment and the split SuperU N rate treatments with the lme4 package of R, where quadratic and linear regression functions were evaluated using analysis of variance and the quadratic function was used when $p < 0.05$.

3. Results

The 30-year (1981–2010) cumulative precipitation between 1 April and 31 October was 689 mm, and actual precipitation was 798 and 772 mm in 2017 and 2018, respectively (Figure 1a). Accumulated growing degree days were 1141, 1362, and 1435 for the climatic normal, 2017, and 2018, respectively (Figure 1b).

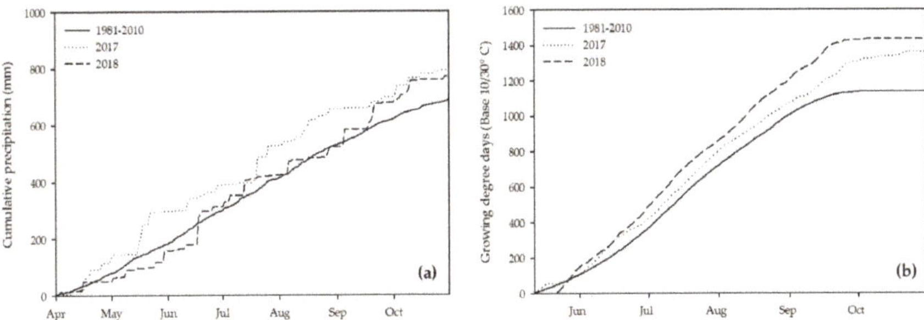

Figure 1. (**a**) cumulative precipitation between 1 April and 31 October and (**b**) accumulated growing degree days (base 10 °C and upper limit 30 °C) following planting for the 1981–2010 historic average and the 2017 and 2018 experiment growing seasons.

3.1. First-Year Maize

First-year maize grain yield was not affected by growing season or fertilizer N treatment (Table 2). Grain N yield was 9% greater in 2018 than in 2017. Stover and stover N yields were 25 and 34% greater in 2018 than in 2017, respectively, and late-season clover yield was reduced by 77% in 2018 compared to 2017.

Table 2. Growing season and treatment means, standard error, and statistical significance of dependent variables for first-year maize.

Fixed Effect	Grain Yield	Grain N [†]	Stover	Stover N	Kura Clover	NH_4-N	NO_3-N	TIN [‡]
	$Mg\ ha^{-1}$	$kg\ ha^{-1}$	$Mg\ ha^{-1}$	$kg\ ha^{-1}$	$kg\ ha^{-1}$		$kg\ ha^{-1}$ 0–0.6 m	
Growing Season								
2017	12.7 (0.2) *	140 (3) b #	6.75 (0.11) b	49.6 (1.6) b	616 (45) a	4.63 (0.40) b	14.99 (1.65)	19.62 (1.85)
2018	13.0 (0.2)	150 (3) a	8.52 (0.18) a	66.5 (2.4) a	140 (12) b	7.93 (0.86) a	12.35 (0.97)	20.28 (1.38)
Treatment								
0 ¶	12.8 (0.4)	134 (4)	7.37 (0.46)	59.2 (6.1)	444 (199)	6.61 (1.72)	7.96 (1.33)	14.57 (2.38)
40 ¶	12.7 (0.4)	137 (6)	7.41 (0.61)	54.4 (7.3)	330 (145)	5.56 (1.09)	9.31 (1.38)	14.87 (1.96)
80 ¶	13.1 (0.5)	146 (8)	7.38 (0.46)	54.7 (3.9)	439 (196)	7.98 (4.87)	10.54 (1.62)	18.52 (5.83)
120 §	12.6 (0.5)	140 (6)	7.70 (0.31)	60.4 (4.5)	407 (147)	6.03 (1.17)	15.15 (4.27)	21.18 (4.74)
120 ‖	12.7 (0.2)	143 (4)	7.83 (0.47)	58.7 (4.6)	401 (165)	4.75 (1.26)	10.28 (1.51)	15.03 (2.05)
120 ¶	13.2 (0.2)	158 (5)	7.58 (0.31)	58.1 (4.1)	361(157)	5.99 (1.50)	12.31 (2.22)	18.30 (2.17)
180 ¶	13.0 (0.3)	150 (5)	7.95 (0.53)	59.6 (6.4)	357 (119)	6.32 (1.10)	15.88 (3.62)	22.21 (3.91)
250 ¶	12.9 (0.8)	151 (11)	7.85 (0.33)	59.2 (4.1)	285 (118)	6.97 (1.35)	27.93 (2.85)	34.90 (2.95)
Significance								
Growing Season	0.311	0.020	<0.001	<0.001	<0.001	<0.001	0.203	0.163
Treatment	0.967	0.153	0.753	0.965	0.607	0.697	<0.001	<0.001
Growing Season × Treatment	0.959	0.741	0.565	0.954	0.723	0.527	0.002	0.003

[†] Nitrogen (N). [‡] Total inorganic N (NO_3-N + NH_4-N, TIN). § kg nitrogen (N) ha^{-1} fertilizer N as pre-plant urea. ‖ kg N ha^{-1} fertilizer N as split-applied urea. ¶ kg N ha^{-1} fertilizer N as split-applied SuperU. * Standard error is shown within parentheses. # Within a column for a given fixed effect, means followed by the same letter are not significantly different at $p < 0.05$.

Post-harvest soil NO_3-N and total inorganic N (NO_3-N + NH_4-N, TIN) in the 0–0.6 m layer were affected by the interaction between growing season and treatment (Table 2, Figure 2a,b). Significant differences among 120 kg N ha^{-1} treatments differing in N source and/or timing were found for NO_3-N and TIN. Residual NO_3-N and TIN did not differ between treatments in 2018 but were greater in the 120 kg N ha^{-1} SuperU and pre-plant urea treatments relative to the split-urea treatment in 2017.

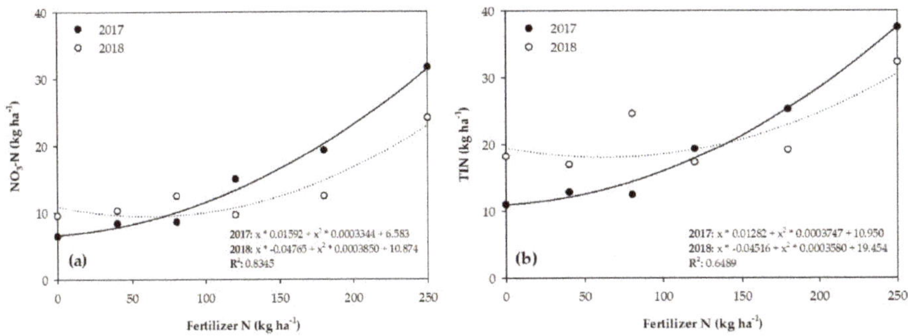

Figure 2. (**a**) response of post-harvest soil NO_3-N and (**b**) total inorganic nitrogen (NO_3-N + NH_4-N, TIN) to fertilizer nitrogen (N) rate for first-year maize in 2017 and 2018.

3.2. Second-Year Maize

There was a significant growing season-by-treatment interaction effect for maize grain, grain N, and stover yields (Table 3, Figure 3a,b, Figure 4), but N rate and timing treatments were not significantly different within a growing season. Late-season clover biomass responded negatively to fertilizer N application in 2017 and biomass was reduced in 2018 relative to 2017. Clover biomass was not significantly different between fertilizer N rate treatments in 2018 (Figure 5).

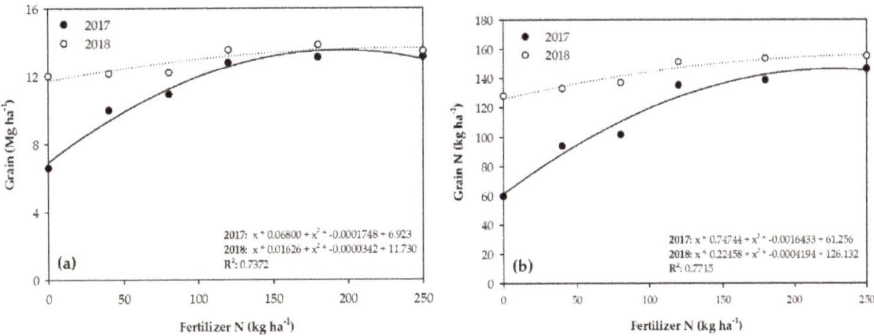

Figure 3. (**a**) response of grain yield and (**b**) grain N to fertilizer nitrogen (N) rate for second-year maize in 2017 and 2018.

Table 3. Growing season and treatment means, standard error, and statistical significance of dependent variables for second-year maize.

Fixed effect	Grain Yield	Grain N [†]	Stover	Stover N	Kura Clover	NH$_4$-N	NO$_3$-N	TIN [‡]
	Mg ha^{-1}	kg ha^{-1}	Mg ha^{-1}	kg ha^{-1}	kg ha^{-1}	kg ha^{-1} 0–0.6 m		
Growing Season								
2017	11.5 (0.4) *	118 (6)	6.10 (0.19)	41.9 (1.8) b #	593 (43)	4.48 (0.55) b	6.50 (0.93) b	10.98 (1.05) b
2018	12.9 (0.2)	143 (3)	8.34 (0.14)	64.2 (2.4) a	125 (18)	6.18 (0.39) a	9.02 (0.89) a	15.19 (0.96) a
Treatment								
0 ¶	9.3 (1.2)	94 (15)	6.18 (0.84)	46.6 (6.7)	558 (257)	4.37 (0.90)	4.27 (1.16)	8.64 (1.78)
40 ¶	11.1 (0.5)	113 (9)	6.98 (0.52)	56.1 (9.4)	353 (111)	6.03 (1.50)	5.40 (1.11)	11.43 (2.36)
80 ¶	11.6 (0.5)	119 (10)	7.18 (0.45)	48.4 (5.2)	327 (130)	4.91 (0.90)	5.93 (1.83)	10.85 (2.63)
120 §	12.3 (0.3)	132 (5)	7.46 (0.51)	54.9 (5.1)	381 (166)	4.79 (0.65)	6.44 (1.27)	11.24 (1.51)
120 ‖	13.2 (0.3)	146 (5)	7.29 (0.52)	47.1 (5.5)	253 (112)	4.89 (1.44)	7.96 (2.06)	12.85 (2.63)
120 ¶	13.2 (0.5)	143 (8)	7.74 (0.42)	52.9 (4.4)	346 (144)	6.17 (1.62)	5.94 (0.93)	12.11 (1.58)
180 ¶	13.5 (0.2)	146 (4)	7.63 (0.40)	56.9 (3.8)	265 (104)	5.53 (1.96)	9.22 (1.64)	14.75 (3.12)
250 ¶	13.3 (0.3)	151 (4)	7.30 (0.36)	61.4 (5.0)	389 (137)	5.92 (2.01)	16.91 (3.85)	22.83 (2.78)
Significance								
Growing Season	<0.001	<0.001	<0.001	<0.001	<0.001	<0.001	<0.001	<0.001
Treatment	<0.001	<0.001	0.014	0.124	0.009	0.944	<0.001	<0.001
Growing Season × Treatment	<0.001	0.006	0.032	0.448	0.004	0.982	0.120	0.242

† Nitrogen (N). ‡ Total inorganic N (NO$_3$-N + NH$_4$-N, TIN). § kg nitrogen (N) ha^{-1} fertilizer N as pre-plant urea. ‖ kg N ha^{-1} fertilizer N as split-applied urea. ¶ kg N ha^{-1} fertilizer N as split-applied SuperU. * Standard error is shown within parentheses. # Within a column for a given fixed effect, means followed by the same letter are not significantly different at $p < 0.05$.

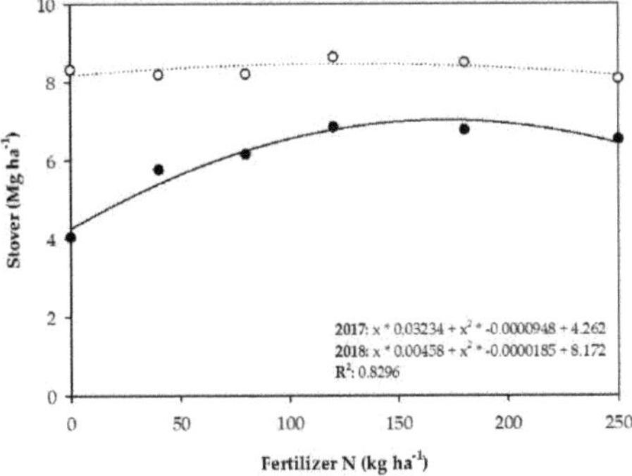

Figure 4. Response of stover yield to fertilizer nitrogen (N) rate for second-year maize in 2017 and 2018.

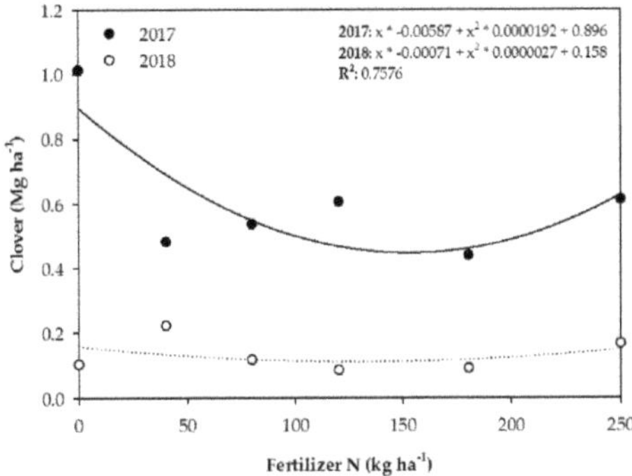

Figure 5. Response of late season clover biomass yield to fertilizer nitrogen (N) rate for second-year maize in 2017 and 2018.

Post-harvest soil NO_3-N and TIN were significantly affected by growing season and treatment, with greater soil N in 2018 than 2017 and increased residual N content with fertilizer N application (Figure 6a,b). Nitrate-N and TIN did not differ between source and timing treatments within a growing season.

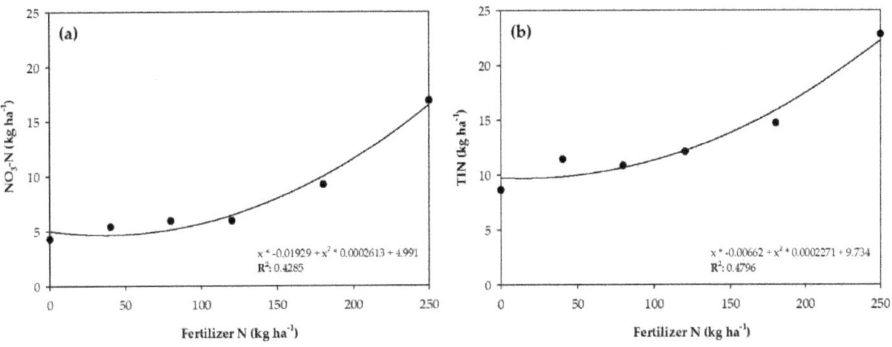

Figure 6. (**a**) response of post-harvest soil NO_3-N and (**b**) total inorganic nitrogen (NO_3-N + NH_4-N, TIN) to fertilizer nitrogen (N) rate for second-year maize, across 2017 and 2018.

3.3. Economic Performance

Fertilizer N was needed to maximize economic return from grain production in second-year maize (Table 3). The EONR was 0 kg N ha^{-1} for first-year maize in both growing seasons, and 177 and 146 kg N ha^{-1} for second-year maize in 2017 and 2018, respectively. At the EONR, crop and soil N variables were consistent, with approximately 13 Mg maize grain ha^{-1} and < 20 kg ha^{-1} of post-harvest TIN in the 0–0.6 m layer (Table 4).

Table 4. Level of dependent variables in first- and second-year maize at the economic optimum nitrogen rate (EONR) in 2017 and 2018.

Variable	Unit	2017		2018	
		First-Year Maize	Second-Year Maize	First-Year Maize	Second-Year Maize
EONR †	kg N ha^{-1}	0	177	0	146
Grain Yield	Mg ha^{-1}	12.3	13.5	13.2	13.4
Grain N ‡	kg ha^{-1}	126	142	141	150
Stover Yield	Mg ha^{-1}	6.2	7.0	8.5	8.4
Stover N	kg ha^{-1}	47.7	47.7	70.7	65.6
Clover Biomass	kg ha^{-1}	742	457	146	111
Residual TIN § 0–0.3 m	kg ha^{-1}	8.6	9.1	7.9	5.7
Residual TIN 0–0.6 m	kg ha^{-1}	2.4	4.5	10.3	10.3

† Economic optimum nitrogen rate (EONR). ‡ Nitrogen (N). § Total inorganic N (NO$_3$-N + NH$_4$-N, TIN).

Partial management cost, the sum of agronomic management costs that corresponded with treatment management and yield at the EONR for the KCLM experiments and the conventional comparison, were greater in the KCLM system than conventionally managed maize (Table 5). High spring tillage and stover harvest, handling, and nutrient replacement costs (1.65 kg P Mg^{-1} + 7.65 kg K Mg^{-1}, \$6 Mg^{-1} dry stover) [27] increased the partial management cost in the KCLM system by \$101–303 ha^{-1} relative to conventional management. Nitrogen costs were reduced by \$132 ha^{-1} in first-year maize, while the cost of fertilizer N for second-year maize was similar to the conventional comparison.

Table 5. Partial management cost (in U.S. \$ ha^{-1}) for maize produced conventionally and in KCLM in 2017 and 2018.

Management Practice	2017			2018		
	Conventional	First-Year KCLM	Second-Year KCLM	Conventional	First-Year KCLM	Second-Year KCLM
Spring Mowing §		26.93	26.93		26.93	
Spring Tillage	13.59 †	76.78 ‡	76.78 ‡	13.59 †	76.78 ‡	76.78 ‡
Fertilizer N	131.88	0.00	148.68	131.88	0.00	122.64
Mowing §		26.93	26.93			
Grain Handling and Storage §	231.10	170.53	186.58	215.88	182.48	185.09
Fall Tillage §	48.68			48.68		
Raking §		13.10	13.10		13.10	13.10
Baling §		33.36	33.36		33.36	33.36
Bale Handling and Storage §		141.16	158.73		192.33	191.08
Stover Nutrient Removal ‖		37.50	42.17		51.09	50.76
Partial Management Cost	**425.25**	**526.29**	**713.26**	**410.03**	**576.08**	**672.81**

Costs obtained from † [32], ‡ [17], § [26], ‖ [28]. Practices not listed are assumed to be equal between management systems (Table A1).

Net return from maize grain was reduced in the KCLM system compared to conventional management due to reduced yield and increased partial management cost. Net return was greater in 2018 than 2017 due to greater stover yields in first- and second-year maize. Additional revenue from stover harvest increased the partial economic net return of the KCLM system averaged between first- and second-year maize to \$-42 ha^{-1} and \$318 ha^{-1} in 2017 and 2018, respectively (Table 6).

Table 6. Economic return (in U.S. \$ ha^{-1}) for maize produced conventionally and in KCLM in 2017 and 2018.

Growing Season		2017			2018	
Management	Conventional	First-Year KCLM	Second-Year KCLM	Conventional	First-Year KCLM	Second-Year KCLM
Fixed Management Cost †	1235.47	1235.47	1235.47	1235.47	1235.47	1235.47
Partial Management Cost ‡	425.25	526.29	713.26	410.03	576.08	672.81
Grain Value §	2221.10	1638.95	1793.20	2074.80	1753.83	1778.87
Stover Value ‖	–	619.13	696.19	–	843.57	838.07
Net Return	**560.38**	**496.32**	**540.66**	**429.30**	**785.85**	**708.66**
Partial Economic Net Return #	–	**−64.06**	**−19.72**	–	**356.55**	**279.36**

† Fixed management cost: the sum of land rental, phosphorus and potassium fertilizer not associated with stover removal, fertilizer application, seed, planting, pesticide and application, harvest, labor, and miscellaneous (Table A1). ‡ Partial management cost (Table 5). § Grain value: \$133 Mg^{-1} at 155 g kg^{-1} moisture. ‖ Stover value: \$79.37 Mg^{-1} at 200 g kg^{-1} moisture [25]. # Partial economic net return is the net return of the treatment minus the net return of conventional management in the same growing season.

4. Discussion

Results from this study demonstrate that kura clover living mulch may be integrated into current cropping systems to provide economic and environmental benefits. Net economic return at the EONR in the KCLM system was similar or superior to the conventional comparison for first- and second-year maize in both growing seasons. Maize grain yield was reduced and management costs were increased in the KCLM system relative to the conventional comparison, but the added value from maize stover maintained or increased economic net return. With KCLM, residual soil TIN was <20 kg N ha^{-1} in the 0–0.6 m soil layer at the EONR in first- and second-year maize in both growing seasons. Additionally, KCLM has been shown to protect soils from erosion, decreasing the risk of degradation to soil and water resources during vulnerable fall and spring months when soils under conventional management are fallow [33,34].

Maintaining clover vigor in a KCLM system is challenging since spring clover regrowth is affected by variable weather conditions, maize production requires intensive clover suppression management, and shade from maize limits mid- and late-season clover growth [4]. Snowfall (25 cm) at the study site on 16 April 2018 delayed spring clover growth and development relative to 2017 and eliminated the need for mowing prior to row establishment in second-year maize. Following row establishment and maize planting, rapid growing degree day accumulation and maize development shaded the clover canopy before a second mowing was needed to manage competition between clover and maize. Less aggressive in-season suppression management in 2018 reduced mechanical clover disturbance, but late-season clover biomass was reduced by three-quarters relative to 2017 due to heavy shading. Management to mitigate reductions in clover biomass and vigor due to shading is limited, but may be important to reduce clover recovery time in the following spring [1,4].

Agronomic management preceding row crop production in KCLM influences N contributions during the maize year. Second-year maize grain yield was affected by the interaction between growing season and treatment, where grain yield for non-N-fertilized maize in 2018 was 4.8 Mg ha^{-1} greater and increased more gradually with applied N than in 2017. These differences may be attributed to differences in weather during the growing season or grain yield level. However, cumulative precipitation and growing degree day accumulation were similar in both growing seasons, and first-year maize did not respond to treatment, growing season, or the interaction of these factors. Thus, it is most likely that the number of years in forage management prior to treatment application was the main factor affecting maize yield response to fertilizer N in second-year maize experiment, confirming previous findings for maize following alfalfa (*Medicago sativa*) [35,36].

Forage legumes increase the soil labile N pool relative to fertilized maize systems [37,38]. Forage stands ≥3 yr. old at termination often accumulate enough labile organic N to eliminate the need for fertilizer N in first-year maize [35] and in many cases second-year maize [36,39]. While labile N accumulates under forage legume production, intensive grazing or harvest of sole kura clover reduces root and shoot biomass productivity over time [40,41]. The intensity of mechanical and chemical suppression of KCLM in the spring disturbs root and shoot tissues [17], while maize reduces late-season clover biomass due to competition for light and plant resources [42]. The translocation of metabolites from root biomass during the spring flush of clover growth and limited opportunity for biomass recovery in living mulch management is likely to exacerbate the decline of root biomass, spring vigor, and clover health. Rapid root accumulation after the establishment year has been observed in other forage legumes, where root biomass doubled between the first and second year of establishment [43]. The additional year of forage management preceding treatment application in 2018 relative to 2017 may have allowed for greater recovery of root biomass and accumulation of labile N. The magnitude of these accumulated N pools was large enough to reduce fertilizer N requirements for two years of maize production when the clover was managed as forage for ≥ 2 yr.

The relationships between clover forage production, maize production, and N contributions from the KCLM-maize cropping system add complexity to the current understanding of N management in these systems. Early studies found that KCLM systems supply most or all of the N requirements for maize [1], suggesting that N contributions from KCLM are supplied in the same year as biological N fixation. Although the response to fertilizer N in second-year maize may have been partially influenced by fall and spring growing conditions and maize development, both of which influence clover growth, early-season clover growth and vigor are closely linked to clover root biomass [40]. The number of years in forage preceding first-year maize is likely an important factor for re-accumulation of root biomass that is translocated to shoots in early spring, linking clover root biomass, spring vigor, and mineralized clover biomass N following suppression management. This study suggests that the N contribution from the living mulch is supplied in-season following row establishment and suppression management, and that mineralized N is sourced from labile and biomass N pools accumulated during forage management. First-year maize following $\geq$ 2yr of forage management does not need fertilizer N and second-year maize requires fertilizer N near University guidelines for maize following soybean. Additional research is needed to confirm these relationships with a greater number of rotation management variables, and to quantify the effect of KCLM management on root biomass pools. Optimization of crop rotation in KCLM systems may balance the health of the clover and the row crop to realize sustainable N contributions over a greater number of growing seasons.

5. Conclusions

Maize grown in KCLM was economically similar or superior to that produced conventionally due to additional income from maize stover harvest. Kura clover living mulch reduced the fertilizer N requirement for first-year maize in both growing seasons and second-year maize in 2018 relative to the conventional comparison. These benefits promote the use of KCLM systems for continuous maize production; however, further optimization is needed to reduce adoption barriers of KCLM-maize systems. Barriers to adoption of KCLM systems for maize production include slow clover establishment, which may take land out of production for a full growing season, the need for specialized row establishment equipment, suppression management operations that require multiple passes during the spring planting season and required maize stover harvest that can be challenging during wet fall conditions. These adoption barriers may be offset by agronomic benefits of KCLM systems, including the potential to reduce P and K application rates by up to one-half through band application with the rotary zone tillage tool relative to broadcasting [44,45]. Additionally, increased water infiltration in KCLM systems [11] may distribute precipitation over the landscape more evenly, reducing areas affected by flooding and reducing the time from rainfall until field conditions are suitable for field operations. Potential avenues for system optimization may include research to speed clover establishment, utilization of strip-tillage equipment that is more readily available than the rotary zone tillage tool, suppression techniques to reduce root and rhizome biomass disturbance, and alternative row-crop rotations. Alterations in agronomic management should consider potential impacts on mineralization of accumulated organic N pools and how this may affect the fertilizer N requirement of row crops. Research to address these constraints may improve the competitiveness of KCLM systems with conventional cropping systems, leading to increased adoption and realized environmental and economic benefits of KCLM systems. Kura clover living mulch-row crop systems may be an important strategy for reducing the negative impacts of agricultural production on water quality and soil health through improved water infiltration into soils, reduced residual soil N following row crops, increased protection against soil erosion, and increased gross crop productivity.

Author Contributions: Conceptualization, J.B., R.V. and J.C.; Data curation, J.A.; Formal analysis, J.A.; Funding acquisition, J.B. and R.V; Investigation, J.A., J.B., R.V. and J.C.; Methodology, J.A., J.B., R.V and J.C.; Project administration, J.B. and R.V; Resources, J.B. and R.V; Supervision, J.B., R.V. and J.C.; Validation, J.A and J.C.; Visualization, J.A; Writing—original draft, J.A; Writing—review and editing, J.A., J.B., R.V and J.C.

Funding: This research was funded by the Minnesota Corn Research and Promotion Council, Grant No. 4097-13SP.

Acknowledgments: The authors thank Michael Dolan, William Breiter, Cody Winker, Ryan Felton, and Seiya Wakahara for technical assistance with field sampling and laboratory analysis. The use of trade, firm, or corporation names in this publication is for the information and convenience of the reader. Such use does not constitute an official endorsement or approval by the United States Department of Agriculture or the Agricultural Research Service of any product or service to the exclusion of others that may be suitable.

Conflicts of Interest: The authors declare no conflict of interest.

Appendix A

Table A1. Fixed costs associated with grain production in conventional and KCLM systems.

Input (Unit)	\$ unit^{-1}	\$ ha^{-1} †
Land	-	500
Fertilizer P ‡ (kg)	0.86	57.84
Fertilizer K ‖ (kg)	0.60	32.28
Fertilizer Application	-	8.90
Seed (1000 seeds)	3.26	280.36
Planting	-	26.69
Herbicide	-	69.19
Insecticide	-	57.58
Spraying	-	10.38
Harvest	-	73.14
Labor	-	96.87
Miscellaneous	-	22.24
Total		**1235.47**

† Estimates were obtained from Plastina, 2018. ‡ Phosphorous (P). ‖ Potassium (K).

References

1. Zemenchik, R.A.; Albrecht, K.A.; Boerboom, C.M.; Lauer, J.G. Corn production with kura clover as a living mulch. *Agron. J.* **2000**, *92*, 698–705. [CrossRef]
2. Affeldt, R.P.; Albrecht, K.A.; Boerboom, C.M.; Bures, E.J. Integrating Herbicide-Resistant Corn Technology in a Kura Clover Living Mulch System. *Agron. J.* **2004**, *96*, 247–251. [CrossRef]
3. Pedersen, P.; Bures, E.J.; Albrecht, K.A. Soybean production in a kura clover living mulch system. *Agron. J.* **2009**, *101*, 653–656. [CrossRef]
4. Grabber, J.H.; Jokela, W.E.; Lauer, J.G. Soil nitrogen and forage yields of corn grown with clover or grass companion crops and manure. *Agron. J.* **2014**, *106*, 952–961. [CrossRef]
5. Siller, A.R.S.; Albrecht, K.A.; Jokela, W.E. Soil erosion and nutrient runoff in corn silage production with Kura clover living mulch and winter rye. *Agron. J.* **2016**, *108*, 989–999. [CrossRef]
6. Ochsner, T.E.; Albrecht, K.A.; Schumacher, T.W.; Baker, J.M.; Berkevich, R.J. Water balance and nitrate leaching under corn in kura clover living mulch. *Agron. J.* **2010**, *102*, 1169–1178. [CrossRef]

7. Schmer, M.R.; Brown, R.M.; Jin, V.L.; Mitchell, R.B.; Redfearn, D.D. Corn Residue Use by Livestock in the United States. *Agric. Environ. Lett.* **2017**, *2*, 1–4. [CrossRef]

8. USDA National Agricultural Statistics Service. *Quick Stats*; USDA: Washington, DC, USA, 2018.

9. Wilhelm, W.W.; Johnson, J.M.F.; Karlen, D.L.; Lightle, D.T. Corn stover to sustain soil organic carbon further constrains biomass supply. *Agron. J.* **2007**, *99*, 1665–1667. [CrossRef]

10. Johnson, J.M.F.; Strock, J.S.; Tallaksen, J.E.; Reese, M. Corn stover harvest changes soil hydrology and soil aggregation. *Soil Tillage Res.* **2016**, *161*, 106–115. [CrossRef]

11. Baker, J.M. *Long Term Affects of Kura Clover Living Mulch on Soil Physical Properties*; Working Paper; University of Minnesota: St. Paul, MN, USA, 2019.

12. Pratt, M.R.; Tyner, W.E.; Muth, D.J.; Kladivko, E.J. Synergies between cover crops and corn stover removal. *Agric. Syst.* **2014**, *130*, 67–76. [CrossRef]

13. Ochsner, T.E.; Schumacher, T.W.; Venterea, R.T.; Feyereisen, G.W.; Baker, J.M. Soil Water Dynamics and Nitrate Leaching Under Corn–Soybean Rotation, Continuous Corn, and Kura Clover. *Vadose Zo. J.* **2017**, *17*, 1. [CrossRef]

14. Pearson, C.H.; Brummer, J.E.; Beahm, A.T.; Hansen, N.C. Kura clover living mulch for furrow-irrigated corn in the intermountain west. *Agron. J.* **2014**, *106*, 1324–1328. [CrossRef]

15. Sawyer, J.E.; Pedersen, P.; Barker, D.W.; Ruiz Diaz, D.A.; Albrecht, K. Intercropping corn and kura clover: Response to nitrogen fertilization. *Agron. J.* **2010**, *102*, 568–574. [CrossRef]

16. Alexander, J.R.; Venterea, R.T.; Baker, J.M.; Coulter, J.A. Kura Clover Living Mulch: Spring Management Effects on Nitrogen. *Agronomy* **2019**, *9*, 69. [CrossRef]

17. Dobbratz, M.; Baker, J.M.; Grossman, J.; Wells, M.S.; Ginakes, P. Rotary zone tillage improves corn establishment in a kura clover living mulch. *Soil Tillage Res.* **2019**, *189*, 229–235. [CrossRef]

18. Ricks, N.R. Corn and Soybean Production on Irrigated Coarse-Textured Soils: Integrating Winter Rye and Kura Clover to Reduce Nitrate Leaching. Master's Thesis, University of Minnesota, Minneapolis, MN, USA, 2019.

19. Baker, J.M. Vegetative propagation of kura clover: A field-scale test. *Can. J. Pln. Sci.* **2012**, 1245–1251. [CrossRef]

20. Kaiser, D.E.; Lamb, J.A.; Eliason, R. *Fertilizer Guidelines for Agronomic Crops in Minnesota*; University of Minnesota Extension: St. Paul, MN, USA, 2011; Volume 1, pp. 1–44.

21. Bremner, J.M.; Mulvaney, C.S. Nitrogen—Total. In *Methods of Soil Analysis*; Sparks, D., Ed.; American Society of Agronomy: Madison, WI, USA, 1982; pp. 1085–1089.

22. Mulvaney, R.L. Nitrogen—Inorganic Forms. In *Methods of Soil Analysis Part 3: Chemical Methods*; Sparks, D.L., Ed.; American Society of Agronomy: Madison, WI, USA, 1996; pp. 1123–1184.

23. Hoverstad, T.; Ihlenfeld, W.; Coulter, J.; Reese, C.; Quiring, S.; Hanson, M. *Varietal Trial Results: Corn Grain*; University of Minnesota Extension: St. Paul, MN, USA, 2017; p. 6.

24. Hoverstad, T.; Ihlenfeld, W.; Coulter, J.; Reese, C.; Quiring, S.; Hanson, M. *Varietal Trial Results: Corn Grain*; University of Minnesota Extension: St. Paul, MN, USA, 2018; p. 6.

25. Edwards, W. *Estimating a Value for Corn Stover*; Ag Decision Maker—Iowa State University Extension and Outreach: Ames, IA, USA, 2014; Volume 1, pp. 1–4.

26. Plastina, A. *Estimating Costs of Crop Production in 2018*; Ag Decision Maker—Iowa State University Extension and Outreach: Ames, IA, USA, 2018; Volume 1, pp. 1–13.

27. Sawyer, J.E.; Mallarino, A.P. *Nutrient Considerations with Corn Stover Harvest*; Iowa State University (ISU) Extension and Outreach, ISU: Ames, IA, USA, 2014; pp. 1–3.

28. Battaglia, M.; Groover, G.; Thomason, W. *Harvesting and Nutrient Replacement Costs Associated with Corn Stover Removal in Virginia*; Virginia Cooperative Extension: Petersburg, VA, USA, 2018; pp. 1–9.

29. Bates, D.; Maechler, M.; Bolker, B.; Walker, S.; Christensen, R.H.B.; Singmann, H.; Dai, B.; Eigen, C. Package "lme4". *Compr. R Arch. Netw.* **2014**, 1–123.

30. Kutner, M.H.; Nachtsheim, C.J.; Neter, J. *Applied Linear Regression Models*, 4th ed.; McGraw-Hill: New York, NY, USA, 2004; pp. 107–152.

31. Lenth, R.; Singmann, H.; Love, J.; Buerkner, P.; Herve, M. Package "emmeans": Estimated Marginal Means, aka Least-Squares Means. *Compr. R Arch. Netw.* **2019**, 1–67.

32. Plastina, A.; Johanns, A.; Wynne, G. 2019 Iowa Farm Custom Rate Survey. *Ag Decis. Mak.* **2019**, 1–5.

33. Helmers, M.J.; Lawlor, P.A.; Baker, J.L.; Melvin, S.W.; Lemke, D.W. Temporal Subsurface Flow Patterns from Fifteen Years in North-Central Iowa. *Soc. Eng. Agri. Food Bio. Sys. Conf. Proc.* **2005**, 1–13.

34. Qi, Z.; Helmers, M.J.; Christianson, R.D.; Pederson, C.H. Nitrate-Nitrogen Losses through Subsurface Drainage under Various Agricultural Land Covers. *J. Environ. Qual.* **2011**, *40*, 1578. [CrossRef] [PubMed]

35. Yost, M.A.; Russelle, M.P.; Coulter, J.A.; Schmitt, M.A.; Sheaffer, C.C.; Randall, G.W. Stand age affects fertilizer nitrogen response in first-year corn following alfalfa. *Agron. J.* **2015**, *107*, 486–494. [CrossRef]

36. Yost, M.A.; Morris, T.F.; Russelle, M.P.; Coulter, J.A. Second-Year Corn after Alfalfa Often Requires No Fertilizer Nitrogen. *Agron. J.* **2014**, *106*, 659–669. [CrossRef]

37. Carpenter-Boggs, L.; Pikul, J.L.; Vigil, M.F.; Riedell, W.E. Soil Nitrogen Mineralization Influenced by Crop Rotation and Nitrogen Fertilization. *Soil Sci. Soc. Am. J.* **2000**, *64*, 2038. [CrossRef]

38. Singh, G.; Schoonover, J.E.; Williard, K.W.J.; Kaur, G.; Crim, J. Carbon and Nitrogen Pools in Deep Soil Horizons at Different Landscape Positions. *Soil Sci. Soc. Am. J.* **2018**, *82*, 1512. [CrossRef]

39. Yost, M.A.; Russelle, M.P.; Coulter, J.A. Field-specific fertilizer nitrogen requirements for first-year corn following alfalfa. *Agron. J.* **2014**, *106*, 645–658. [CrossRef]

40. Peterson, P.R.; Sheaffer, C.C.; Jordan, R.M.; Christians, C.J. Responses of kura clover to sheep grazing and clipping: II. Below-ground morphology, persistence, and total nonstructural carbohydrates. *Agron. J.* **1994**, *86*, 660–667. [CrossRef]

41. Peterson, P.R.; Sheaffer, C.C.; Jordan, R.M.; Christians, C.J. Responses of kura clover to sheep grazing and clipping: I. Yield and forage quality. *Agron. J.* **1994**, *86*, 655–660. [CrossRef]

42. Baributsa, D.N.; Foster, E.F.; Thelen, K.D.; Kravchenko, A.N.; Mutch, D.R.; Ngouajio, M. Corn and cover crop response to corn density in an interseeding system. *Agron. J.* **2008**, *100*, 981–987. [CrossRef]

43. Bolinder, M.A.; Angers, D.A.; Bélanger, G.; Michaud, R.; Laverdière, M.R. Root biomass and shoot to root ratios of perennial forage crops in eastern Canada. *Can. J. Plant Sci.* **2011**, *82*, 731–737. [CrossRef]

44. Kaiser, D.E.; Pagliari, P. Understanding phosphorus fertilizers|UMN Extension. Available online: https://extension.umn.edu/phosphorus-and-potassium/understanding-phosphorus-fertilizers#corn-624160 (accessed on 2 June 2019).

45. Kaiser, D.E.; Rosen, C.J. Potassium for crop production|UMN Extension. Available online: https://extension.umn.edu/phosphorus-and-potassium/potassium-crop-production#corn-and-small-grain-603411 (accessed on 2 June 2019).

 *agronomy*

Article

Agronomic and Economic Performance of Maize, Soybean, and Wheat in Different Rotations during the Transition to an Organic Cropping System

William J. Cox [1,*], John J. Hanchar [2] and Jerome Cherney [1]

[1] Unit of Soil and Crop Sciences, School of Integrated Plant Sciences, Cornell University, Ithaca, NY 14853, USA; jhc5@cornell.edu

[2] Northwestern New York Dairy, Livestock, and Field Crops Program, College of Agriculture and Life Sciences, Cornell University, Ithaca, NY 14853, USA; jjh6@cornell.edu

* Correspondence: wjc3@cornell.edu; Tel.: +1-607-255-1758

Received: 14 August 2018; Accepted: 14 September 2018; Published: 17 September 2018

Abstract: Crop producers transitioning to an organic cropping system must grow crops organically without price premiums for 36 months before certification. We evaluated red clover-maize, maize-soybean, and soybean-wheat/red clover rotations in organic and conventional cropping systems with recommended and high inputs in New York, USA to identify the best rotation and management practices during the transition. Organic compared with conventional maize with recommended inputs in the maize-soybean rotation (entry crop) averaged 32% lower yields, $878/ha higher production costs, and $1096/ha lower partial returns. Organic maize compared with conventional maize with recommended inputs in the red clover-maize rotation (second transition crop) had similar yields, production costs, and partial returns. Organic compared with conventional soybean with recommended inputs in soybean-wheat/red clover or maize-soybean rotations had similar yields, production costs, and partial returns. Organic compared with conventional wheat with recommended inputs in the soybean-wheat/clover rotation had similar yields, $416/ha higher production costs, and $491/ha lower partial returns. The organic compared with the conventional soybean-wheat/red clover rotation had the least negative impact on partial returns during the transition. Nevertheless, all organic rotations had similar partial returns ($434 to $495/ha) so transitioning immediately, regardless of entry crop, may be most prudent. High input management did not improve organic crop yields during the transition.

Keywords: organic cropping system; maize; soybean; wheat; partial returns

---

1. Introduction

Organically-produced maize (*Zea mays* L.), soybean {*Glycine max* (L.) Merr.} and wheat (*Triticum aestivum* L.) have substantial price premiums, providing market incentives for organic production. Downward trends in prices of all three crops have prompted some crop producers, who practice maize-soybean or maize-soybean-wheat/red clover (*Trifolium pretense* L.) rotations, to contemplate transitioning from conventional to an organic cropping system. The United States Department of Agriculture (USDA), however, requires a 36-month transition period that prohibits the use of genetically modified (GM) crops, synthetic fertilizer, pesticides, etc., before the land can be certified as organic and eligible for the organic price premium [1]. Furthermore, comprehensive survey data indicated that organic compared with conventional maize, soybean or wheat production, despite higher profits, had lower yields and higher/ha production costs [2]. Consequently, a major deterrent for conventional crop producers who wish to transition to an organic cropping system is the potential loss of significant profit during the transition. Identification of the best entry crop (first year transition crop) and subsequent rotation during the transition to an organic cropping system is essential for maintaining cash flow on the farm, especially given the relatively low cash receipts received by maize, soybean and wheat growers in recent years [3].

Numerous studies comparing organic and conventional cropping systems have been conducted. In a study established in 2002 near Morris, Minnesota, USA, organic compared with conventional maize yielded 34% lower, whereas organic compared with conventional soybean yielded statistically similar (but 15% numerically lower) from 2002–2005 when comparing 2-year conventional and organic maize-soybean rotations [4]. Organic maize yielded lower mostly due to lack of available soil N, associated with low N content of the solid dairy manure applied to organic maize. Despite $425/ha lower seed, fertilizer and pesticide costs, the 2-year organic compared with the 2-year conventional rotation had $128/ha higher production costs associated with higher labor, diesel, manure hauling, and machinery ownership costs. Consequently, the organic compared with the 2-year conventional rotation had $511/ha lower net present value during the transition because of lower yields, higher production costs, and the absence of an organic premium [4].

In this same study, the entry crop into an organic cropping system had a major impact on risks and returns during the transition phase [5]. Based on yield data and inputs from the same study, soybean as the entry crop provided a $283 advantage of net present value compared with maize in the maize-soybean organic rotation. In the 4-year organic rotation, wheat as the entry crop provided a $229 advantage over other entry crops [5]. Nevertheless, a simple dynamic adoption model indicated that transitioning to an organic cropping system as rapidly as possible, regardless of the entry crop, would result in the highest expected long-term profit [5].

In another Minnesota study established near Lamberton, organic maize in a maize-soybean-oat (*Avena sativa* L.)/alfalfa-alfalfa (*Medicago sativa* L.) rotation yielded similarly as conventional maize in a maize-soybean rotation from 1993–2009 [6]. Organic compared with conventional soybean, however, yielded 25% lower in their respective rotations over the same period. Organic compared with conventional maize, despite higher machinery costs, had $86/ha lower production costs because of lower seed costs as well as no herbicide costs [7]. Likewise, organic compared with conventional soybean had $101/ha lower production costs, primarily because of lower weed control costs. When factoring in the organic price premium (2.17 price ratio for maize and 2.27 for soybean), the 4-year organic rotation had $527 net revenue compared with $295 for the 2-year conventional rotation [7].

Machinery ownership costs, however, were not included in the first analyses of this study. When comparing the 4-year organic rotation with the 2-year conventional rotation, machinery ownership costs averaged $146/ha across organic farm sizes of 130, 225 and 325 ha compared with $183/ha across conventional farm sizes of 225, 455, and 630 ha [8]. The organic rotation had net returns of $114,000 compared with conventional net returns of $72,000 for a 225-hectare farm [8]. The organic rotation also had net returns of $296,000 for the largest farm size (325 ha), compared with conventional net returns of $220,000 for its largest farm size (630 ha), despite the farm-scale advantage for conventional production.

In a study established in 1990 in Wisconsin, USA, a no-till (NT) conventional maize-soybean rotation compared with an organic maize-soybean-wheat rotation averaged $130 and $408 higher economic mean returns, respectively, in the absence of organic premiums [9]. In the presence of government payments and organic premiums, the organic maize-soybean-wheat rotation had $321 and $165 higher economic mean returns, respectively, compared with the conventional NT maize-soybean rotation [9]. The conventional NT maize-soybean rotation yield trend, however, averaged 151 kg/ha/year compared with 101 kg/ha/year for the organic maize-soybean-wheat rotation from 2009 to 2012 [10], perhaps because of technological advances in the conventional cropping system and/or increased weed competition in the organic cropping system.

In a cropping system study established in 1996 in Maryland, USA, organic maize in a maize-soybean-wheat/vetch (*Vicia vilossa* Roth) rotation yielded 28% lower compared with conventional NT maize in a maize-soybean-wheat/soybean rotation during the transition years from 1996 to 1998 [11]. After the transition period, organic compared with conventional maize yielded 40% lower in their respective 3-crop rotations [11]. The lower organic maize yields were associated mostly with low soil N availability (73%) and weed competition (23%). After the transition period, organic soybean compared with NT conventional soybean yielded 24% lower in their respective 3-crop rotations because of greater weed competition [11]. In the 3-year period (2000–2002) following the transition, the organic compared with the conventional cropping system, despite lower maize and soybean yields, had $514/ha greater net returns [11]. Economic risk in the 3-year organic system, however, was 3.9 times greater compared with a 6-year organic rotation (maize/rye (*Secale cereale* L.)-soybean-wheat-alfalfa-alfalfa-alfalfa).

In a study, established in Iowa, USA, maize and soybean in an organic maize-soybean-oat/ alfalfa-alfalfa rotation compared with a conventional maize-soybean rotation yielded similarly during the transition [12], resulting in higher profitability for the organic cropping system because of lower production costs [13]. In the second phase of the study, maize and soybean again yielded similarly between cropping systems so the organic cropping system was far more profitable because of lower production costs in maize and higher prices received for organic maize and soybean [14].

Long-term cropping system experiments, though beneficial, are somewhat limited in the analyses of conventional vs. organic cropping systems because management practices are fixed, and the "human" management factor of organic production is missing [2]. Agricultural Resource Management Survey (ARMS) data from 2010 was used for maize (794 conventional and 451 organic farms); 2009 ARMS data for wheat (1641 conventional and 1458 organic farms); and 2006 ARMS data for soybean (748 conventional and 478 organic farms) to compare conventional and organic crop production [2]. Organic maize, soybean, and wheat had higher economic costs ($205 to $242/ha; $261 to $309/ha; and $135 to $153/ha higher, respectively) because increased costs for fuel, repair, capital, and labor offset lower seed, fertilizer and chemical costs. Furthermore, organic maize, soybean, and wheat compared with conventional crops yielded much lower (27%, 34%, and 32%, respectively). Consequently, organic compared with conventional producers had higher average economic costs per metric ton or Mg ($76 to $89/Mg, $143 to $164/Mg, and $243 to $287/Mg higher, respectively). Nevertheless, net economic returns were greater for organic compared with conventional maize and soybean producers ($126 to $163/ha, and $54 to $101/ha higher, respectively) because of the organic price premiums (~2.85 and ~2.25 ratios, respectively). Net economic returns for organic compared with conventional wheat, however, were slightly lower ($−5 to $−23/ha), despite the organic price premium (~2.4 price ratio). The survey data indicate that the price premium is crucial for profitability in organic maize, soybean, and wheat production because of lower yields and higher production costs.

A major deterrent to adoption of organic crop production is the uncertainty associated with selection of the best first year transition (entry) crop and subsequent rotation during the 36-month period when organic premiums do not exist [5]. Another deterrent is that novice organic crop producers are uncertain of the best organic management practices to use during the transition and beyond [5]. Two objectives of this study are: (a) to identify the best entry crop and subsequent organic rotation that results in the best partial economic returns to the organic cropping system during the transition, and (b) to evaluate recommended and high input organic management practices (organic seed treatment, and high seeding and high N rates) to determine if high input management increases weed competitiveness and improves soil N availability for organic crops.

2. Materials and Methods

We initiated a 4-year cropping system study at a Cornell University research farm near Aurora, NY, USA (42°44′ N, 76°40′ W) in 2015 to evaluate different sequences of the maize-soybean-wheat/red clover rotation. Winter wheat was not planted in the fall of 2014 before the onset of the study. Instead, red clover was seeded in mid-July of 2015 into bare soil to insure a green manure crop for the subsequent maize crop in 2016. Our three sequences during the transition thus included red clover-maize, soybean-wheat/red clover, and maize-soybean in 2015 and 2016 when organic crops were not eligible for an organic price premium. The 36-month transition period consisted of only two growing seasons in this study because the previous crops before the transition did not receive any prohibited inputs after June of 2014. Thus the 2017 crops were eligible for the organic price premium because they were harvested in July (wheat), October (soybean) and November (maize).

Three contiguous experimental fields (220 m × 40 m) with similar tile-drained silt loam soil (fine-loamy, mixed, mesic, Glossoboric Hapludalfs) but different previous crops in 2014 (spring barley, maize, and soybean) were used in the study. The experimental design is a split-split plot (four replications) with previous crops as whole plots, cropping systems (conventional and organic) as sub-plots, and management inputs (recommended and high inputs) as sub-sub plots. The entire 40 m lengths were planted to maize, soybean or winter wheat in each field, but plot length was shortened to 30 m to allow for 5 m borders on the north and south sides of the plots. Also, 3 m borders were inserted between sub-plots (cropping systems) to minimize spray drift or fertilizer movement from conventional into organic plots. Likewise, 3 m border plots were inserted between each sub-subplot to minimize border effects from each crop, which differed in height. To ensure that the 3 m border plots between sub-plots were adequate to minimize spray drift from conventional to organic plots, we only sprayed the conventional plots early in the morning when wind speed was less than 5 km/h. We did not plan to market the crops produced organically as organic in future years, so the 5 m borders were adequate for our study. Whole plot dimensions were 216 m wide and 30 m long, sub-plot dimensions were 27 m wide and 30 m long, and sub-subplot dimensions were 3 m wide and 30 m long.

Maize and soybean strips were moldboard plowed from 18–20 May in both years, followed by secondary tillage the following day. Maize and organic soybean were planted in 0.76 m row spacing immediately after secondary tillage in both years. Conventional soybean strips were also planted on the same day but in the typical 0.38 m row spacing. The maize and soybean planting dates, which were delayed so some early-season weeds could emerge before plowing in the organic cropping system, were just after the optimum planting dates for both crops (25 April–20 May for maize and 6 May–17 May for soybean) at this site.

Table 1 lists the management inputs for maize, soybean, and wheat for both years. Major differences between conventional and organic maize include (a) a treated (insecticide/fungicide seed treatment) GM (genetically modified) hybrid, P9675AMXT (Optimum®AcreMax®XTreme), with the AMXT, LL (Liberty Link) and RR2 (Roundup Ready 2) traits, versus the non-GM isoline, P9675 (no seed treatment in recommended input but an organic seed treatment, Sabrex, mixed in the seed hopper in the high input treatment), (b) 280 kg/ha of 10-20-20 (N-P-K analysis, respectively) versus 365 kg/ha of composted manure (5-4-3 analysis) as starter fertilizer, (c) 135 to 180 kg N/ha side-dressed in 2015 and 0 to 56 kg N/ha when following red clover in 2016 (recommended and high input treatments, respectively) with a liquid N source (32-0-0 analysis) versus the same N rates in organic maize with composted chicken manure applied pre-plant and (d) a single Glyphosate herbicide application for weed control in conventional versus tine weeding, followed by a close cultivation to the row, followed by two additional cultivations between the rows for organic maize. We estimated that 50% of the N from the composted poultry manure would be mineralized and available to organic maize. Seeding rates of 73,000 kernels/ha were used in recommended input and 86,500/ha in high input treatments of both cropping systems. We selected a non-GM isoline for organic maize instead of an organically developed and produced hybrid so we could determine how management practices (and not hybrid selection) affected yield and partial returns.

Table 1. Soil texture/drainage, planting rate, hybrid/cultivar, tillage, starter and N fertilizer practices, and weed control practices for maize, soybean, and wheat in conventional and organic cropping systems with two management treatments (recommended and high input) at a Cornell Research Farm near in central NY in 2015 and 2016.

Descriptor	Crop					
	Maize		Soybean		Wheat	
	Rec.	High	Rec.	High	Rec.	High
	Conventional					
Soil texture/Drainage	Well- drained silt loam					
Planting rate (seeds/ha)	73,100	87,700	370,500	494,000	2,964,000	4,200,000
Seed Treatment	Fungicide/insecticide		Fungicide/insecticide		Fungicide/insecticide	
Cultivar	GM hybrid	GM hybrid	GM variety	GM variety	Soft white (P24R46)	Soft white (P24R46)
Tillage	Moldboard Plow		Moldboard Plow		No-Till	
Starter Fertilizer (kg/ha)	280 kg/ha (10-20-20)		None		225 kg/ha (10-20-20)	
N fertilizer-side-dress (kg N/ha)	2015: 90–160 kg N/ha (liquid) 2016: none	2015: 135–200 kg N/ha (liquid) 2016: 56 kg N/ha	None	None	80 kg N/ha (33-0-0)	56 + 56 kg N/ha (33-0-0)
Herbicide application	Glyphosate	Glyphosate	Glyphosate	Glyphosate	None	Yes
Fungicide application	None	None	None	Yes	None	Yes
	Organic					
Soil texture/Drainage	Well- drained Honeoye silt loam					
Planting rate (kernels/acre)	73,100	87,700	370,500	494,000	2,964,000	4,200,000
Seed Treatment	None	Organic	None	Organic	None	Organic
Cultivar	Non-GM Isoline	Non-GM Isoline	Non-GM variety	Non-GM variety	Soft white (P24R46)	Soft white (P24R46)
Tillage	Moldboard Plow		Moldboard Plow		No-Till	
Starter Fertilizer	350 kg/ha composted chicken manure (5-4-3)		None		170 kg N/ha composted chicken manure (5-4-3)	
Pre-plant N fertilizer (kg N/ha)	2015: 90–160 kg N/ha composted manure 2016: none	2015: 56–200 kg N/ha composted manure 2016: 56 kg	None	None	80 kg N/ha composted manure	56 + 56 kg N/ha composted manure
Tine weeding	1×		1×		None	
Cultivate	3×		4×		None	

Maize was harvested with a small plot Almaco combine (Nevada, IA, USA) in November in both years when grain moistures were ~18%. An approximate 1000 g sample was collected from each sub-subplot to determine grain moisture and grain N% concentrations (by combustion with a LECO CN628 Nitrogen Analyzer, LECO Corporation, St. Joseph, MI, USA). Yields were adjusted to 15.5% moisture. Grain moisture differences were less than 1% between cropping systems so will not be reported.

Major differences between conventional and organic soybean include (a) treated (insecticide/ fungicide seed treatment) GM variety, P22T41R2 with the RR2Y and SCN traits versus a non-GMO variety, 92Y21 (organic seed treatment mixed in the seed hopper of the high input treatment), (b) 0.38 m versus 0.76 m row spacing (for cultivation of weeds in organic soybean), and c) a single Glyphosate herbicide application for weed control versus tine weeding, followed by close cultivation to the row, followed by three additional cultivations between the rows, respectively (Table 1). Seeding rates of 370,500 and 494,000 seeds/ha were used for recommended and high input treatments in both cropping systems. Conventional soybean in the high input treatment also received a fungicide (Fluxapyroxad + Pyraclostrobin at ~300 mL/ha) application at the early pod stage (R3 stage) in late July for potential disease problems and overall plant health. We did not fertilize soybean because conventional soybean growers typically do not use fertilizer on soybean.

Soybean was harvested on 23 September 2015 at ~11% moisture wheat was no-tilled into soybean stubble with a 1590 John Deere no-till drill (Molina, IL, USA) in 0.19 m rows the following day in 2015. We decided to no-till wheat because of the paucity of visible weeds, especially winter perennial weeds, in both cropping systems. Soybean developed green stem in 2016 and was not harvested until November 9, too late to plant wheat after soybean in this environment.

Major differences between conventional and organic wheat include (a) a treated (insecticide/ fungicide seed treatment) soft red winter wheat variety, 25R46, versus the untreated 25R46, (b) 225 kg/ha of 10-20-20 (N-P-K analysis) versus 175 kg/ha of composted chicken manure (5-4-3, N-P-K analysis) as starter fertilizer, (c) and top-dressing with 80 kg N/ha in late March or 56 kg N/ha in late March + 56 kg N/ha in late April in the recommended and high input treatments, respectively with ammonium nitrate (33-0-0) versus 80 kg N/ha (late March) or 56 kg N/ha (pre-plant) + 56 kg N/ha in late April in recommended and high input treatments, respectively with composted chicken manure (Table 1). We also applied an herbicide (thifensulfuron + tribenuron) in the fall and a fungicide (Prothioconazole + Tebuconazole) in the spring in high input conventional wheat. We frost-seeded red clover at ~30 kg/ha into all the wheat treatments in early March to provide N to the subsequent maize crop in 2017. Wheat was harvested on 7 July of 2016.

Maize densities were taken immediately before tine weeding (~1–2 days after 90% emergence) and again at the 9th leaf or V9 stage, after the completion of mechanical weed control practices, by counting all the plants along the 30 m plot length of the two harvest rows. The first maize density measurement was taken to determine if the treated GMO maize hybrid and non-treated non-GMO maize isoline differed in emergence rates and plant establishment. The second measurement was taken to determine the extent of maize damage by mechanical weed control practices (tine weeding, a close cultivation, and three in-row cultivations) in organic maize. Soybean densities were also taken immediately before tine weeding (~1–2 days after 90% emergence) to determine emergence rates and again a few days before harvesting to determine final soybean densities by counting all the soybean plants in three 1.52 m^2 regions along the 30-m harvest rows (2 center rows in organic soybean and 4 center rows in organic soybean). Weed densities were determined in maize by counting all the weeds taller than 5 cm in height along the 30 m length of the two harvest rows at the V14 stage, the end of the critical weed-free period for maize in this environment [15]. Weed densities were also determined in soybean by counting all visible weeds along the 30 m length of the entire 3.3 m wide soybean plot at the pod-filling stage (R4 stage), the end of the critical weed-free period for soybean in this environment. Predominant weed species in both crops included *Polygonum convovulus*, *Chenopodium album*, *Echinochloa crus-galli*, *Polygonum pensylvanicum* L., *Setaria vidis*, *Ambrosia artemisifolia* and *Amaranthus retroflexus*.

Wheat densities were taken about a week after emergence by counting all wheat plants in three 1.52 m^2 regions along the 30-m harvest rows (8 center rows). Weed densities were also determined in wheat by counting all visible weeds along the 30 m length of the entire 3.3 m wide wheat plot in early April, during the active spring tillering period in this environment. Predominant weed species, which not differ among previous crops or between cropping systems, included *Taraxacum officinale* F.H. Wigg, *Malva neglecta* Wallr., *Stellaria media* (L.) Vill., and *Lamium amplexicaule* L. Wheat was harvested with a small plot Almaco combine (Nevada, IA, USA) in early July. An approximate 1000 g sample was collected from each sub-subplot to determine grain moisture and grain N concentrations by combustion (LECO CN628 Nitrogen Analyzer, LECO Corporation, St. Joseph, MI, USA). Yields were adjusted to 13.5% moisture.

Costs for the different management inputs for the three crops in the two cropping systems are listed in Table 2. Production costs for organic maize and wheat will be somewhat inflated because of the use of composted chicken manure as the major N source (~13× higher cost/kg of N compared with liquid N in conventional maize and ammonium nitrate in wheat). We used composted chicken manure in organic maize and wheat because of its known analyses of N-P-K and its ease in calibration and application with a fertilizer spreader. We wished to avoid the problems with the use of solid animal manure in previous studies, which did not accurately estimate the N content [4,8]. Also, conventional maize received only a single application of Glyphosate compared with the typical two or more herbicide applications used by most growers because moldboard plowing reduced the need for supplemental weed control chemicals. Most maize growers use reduced tillage or no-till, which results in more chemical use and higher weed control costs. Our weed control costs were thus significantly lower than typical in conventional maize. Consequently, production costs are skewed in favor of conventional maize and wheat.

Table 2. Costs of variable inputs, including seed, hopper seed treatments, (inoculant for conventional soybean and Sabrex for organic crops), starter fertilizer, N fertilizer, herbicide, and fungicide in conventional and/or organic soybean, maize, and wheat.

Input	Conventional	Organic
	$	
	Soybean	
Seed/140,000	81.95 (including seed treatment)	50.95
Seed Treatment	48.80/g (Cell-Tech inoculant)	200/g (Sabrex)
Herbicide	280/L (Glyphosate)	-
Fungicide	2130/L (Fluxapyroxad + Pyraclostrobin)	-
	Maize	
Seed/80,000	330 (including seed treatment)	240
Seed Treatment	-	200/g (Sabrex)
Starter Fertilizer	448/tonne (Mg)	325/tonne (Mg)
Side-dress N	0.99/kg N	12.76/kg N
Herbicide	280/L (Glyphosate)	-
	Wheat	
Seed/bag	31 (including seed treatment)	24
Seed Treatment	-	200/g (Sabrex)
Starter Fertilizer	448/tonne (Mg)	325/tonne (Mg)
Herbicide	276/mL	-
Top-dress N	0.99/kg N	12.76/kg N
Fungicide	1325/L (Prothioconazole + Tebuconazole)	-

Soybean prices received by NY farmers averaged $0.345/kg in 2015 and 2016, maize prices averaged $0.156/kg in 2015 and 2016, and the wheat price averaged $0.149/kg in 2016 [16]. Economic analyses focused on enterprise budget items that differed among the treatments, namely the value of production associated with yield differences as well as cost differences for inputs for maize, soybean and wheat. Returns to variable and fixed inputs that do not differ between conventional and organic soybean production under recommended and high input management were calculated for the three crops. Our selected variable inputs include: Seed, fertilizer, and other inputs (inoculant, organic seed treatment, herbicide, and fungicide); labor and machinery operating inputs (repairs and maintenance, fuels and lubricants), excluding tillage, planting and harvesting tasks, except for hauling, where hauling cost is a function of yield [17]. Cost of production values reported for fixed inputs exclude farm machinery ownership costs for tillage, planting and harvest, land charges, and values of management inputs. Grain moistures did not differ between organic and conventional maize, and grain drying is not required for soybean and wheat, so we did not include those production costs in maize.

Previous crop (2014 crops), cropping systems (conventional and organic), and management inputs (recommended and high) were considered fixed and replications m (and years) random for statistical analyses for individual years and averaged across years using the REML function in the MIXED procedure of SAS (version 9.4; SAS Institute Inc., Cary, NC, USA). For statistical analyses of the partial returns data for the 2-year transition, rotations were considered a fixed variable and a sub-sub plot within cropping systems. Fields with different previous crops (2014 crops) had yield differences for maize and soybean in 2015 but did not have any interactions with cropping systems and rotations. Consequently, the data will be pooled across previous crops (the three contiguous fields) for each year. Least square means of the main effects (cropping system and management inputs) were computed and means separations were performed on significant effects using Tukey's HSD (Studentized Range) test, with statistical significance set at $p < 0.05$. Differences among least square means for cropping system interactions were calculated also using Tukey's HSD test. Two-way interactions (cropping system by management inputs) were detected for some variables so the interaction comparisons will be presented. Simple correlations (Pearson) among all measurements within each year were calculated using CORR in SAS.

3. Results and Discussion

3.1. Agronomic

Table 3 shows the agronomic maize data for 2015 and 2016. Organic compared with conventional maize in the maize-soybean rotation yielded 32% lower as an entry crop in 2015 when averaged across management input treatments. Organic compared with conventional maize had ~8% lower plant densities at the 9th leaf stage (V9), when all cultivations to organic maize had been completed, mostly due to cultivation damage [18]. Despite the close and repeated cultivations, organic compared with conventional maize had almost 5x higher weed densities. In addition, organic maize had very low grain N% concentrations (1.06%) compared with conventional maize (1.32%). Excessive precipitation (276 mm) from planting on 21 May to silking on 27 July probably leached or denitrified a considerable amount of the N in the pre-plow application of composted chicken manure [18].

In contrast, the experimental site received 98 mm of precipitation from the side-dressing N application (26 June) to silking, allowing for most of the side-dressed N to be available to conventional maize. Grain yield had a strong positive correlation with grain N% concentrations ($r = 0.81$, $n = 48$) and a strong negative correlation with weed densities ($r = -0.78$, $n = 48$). These results agree with findings that have reported lower organic maize yields during the transition because of limited soil N availability and weed competition [4,11].

In 2016, however, organic maize as the second-year transition crop in the soybean-maize rotation yielded similarly as conventional maize and input management did not influence yields. Maize yields were very low, however, because of exceedingly dry conditions from April through July (222 mm), which greatly reduces maize yields in northern latitudes of the USA [19]. It was also the driest (75 mm) 19 May (planting date) through 20 July (silking date) period in 61 years of record keeping at the experimental site [19]. Maize and weed densities were much lower in 2016 compared with 2015 undoubtedly because the exceedingly dry soil conditions reduced maize and weed emergence. Grain N% concentrations, however, were much greater because there was no leaching or denitrifying of applied N, as well as the concentration effect of grain N% associated with low yields. Grain yield did not correlate with weed densities nor grain N concentrations in 2016. Grain yield did correlate with maize densities ($r = 0.45$, $n = 48$) because plant densities were below threshold plant densities where yields decline, even in dry years [19].

Table 3. Maize densities at the 9th leaf stage (V9), weed densities at the V14 stage, yield, grain N content and revenue of maize in 2015 in 2016 under conventional and organic cropping systems at recommended and high input management in central NY.

| | Year | | |
Treatment	2015	2016	Mean
Maize densities-V9 stage (plants/ha)			
Conventional			
Recommended	72,158 b [+]	56,566 c	64,362 c
High Input	86,391 a	65,606 a	75,999 a
Organic			
Recommended	64,750 c	51,472 d	58,111 d
High Input	80,819 ab	60,648 b	70,734 b
Weed densities-V14 stage (weeds/m^2)			
Conventional			
Recommended	0.47 a	0.27 b	0.37 a
High Input	0.39 a	0.18 b	0.29 a
Organic			
Recommended	2.41 b	0.99 a	1.70 c
High Input	2.13 b	0.64 a	1.39 b
Yield (kg/ha)			
Conventional			
Recommended	10,321 a	7156 a	8739 a
High Input	10,545 a	7783 a	9164 a
Organic			
Recommended	6905 b	7093 a	6999 b
High Input	7281 b	7156 a	7219 b
Grain N (%)			
Conventional			
Recommended	1.32 a	1.56 ab	1.44 a
High Input	1.33 a	1.68 a	1.51 a
Organic			
Recommended	1.05 b	1.51 b	1.28 b
High Input	1.06 b	1.61 a	1.34 b
Revenue ($/ha)			
Conventional			
Recommended	1611 a	1116 a	1364 a
High Input	1645 a	1214 a	1430 a
Organic			
Recommended	1077 b	1107 a	1092 b
High Input	1136 b	1116 a	1127 b

[+] Treatment interaction means within the same column followed by the same letter are not significantly different according to Tukey's HSD (Studentized Range) test at the 0.05 level.

Table 4 shows the agronomic data for soybean in 2015 (soybean-wheat/red clover rotation) and 2016 (maize-soybean rotation). Organic and conventional soybean with recommended inputs yielded similarly in both years, but there was a cropping system × management input interaction in 2015.

Organic soybean did not respond to high inputs, but conventional soybean showed a 9% yield increase in 2015. Conventional soybean with recommended inputs had plant densities above the threshold for maximum yield in this environment [20] and very low weed densities so neither should have contributed greatly to the yield increase in the high input treatment in 2015. Nevertheless, seed yields in 2015 had a weak correlation with plant densities (r = 0.31, n = 48) and a weak negative correlation with weed densities (r = −0.36, n = 48). There was a cropping system × input management interaction for seed mass in 2015 with a 6 mg increase in seed mass for conventional soybean with high inputs, but no increase in seed mass for organic soybean with high inputs. The fungicide application to high input conventional soybean may have improved overall plant health resulting in greater seed mass and the 9% yield increase. Seed yield did not correlate with soybean densities or weed densities during the dry 2016 growing season. The organic soybean yield data agree with a Minnesota, USA study that showed that organic and conventional soybean yielded similarly during the transition [4].

Table 4. Harvest plant densities, weed densities at the full pod stage (R4), seed yield, and revenue of soybean in 2015 and 2016 under conventional and organic cropping systems at recommended and high input management in central NY.

	Year		
Treatment	**2015**	**2016**	**Mean**
Soybean densities (plants/ha)			
Conventional			
Recommended	307,967 c [+]	318,167 c	313,067 c
High Input	417,912 a	442,750 a	429,971 a
Organic			
Recommended	338,083 b	284,667 d	311,375 c
High Input	419,258 a	383,250 b	401,254 b
Weed densities-R4 stage (weeds/m²)			
Conventional			
Recommended	0.24 a	0.44 a	0.34 b
High Input	0.11 a	0.27 a	0.19 b
Organic			
Recommended	0.40 a	0.77 b	0.58 a
High Input	0.61 a	0.60 ab	0.60 a
Yield (kg/ha)			
Conventional			
Recommended	2977 b	2711 a	2844 a
High Input	3239 a	2806 a	3023 a
Organic			
Recommended	2851 b	2631 a	2741 a
High Input	2952 b	2655 a	2804 a
Revenue ($/ha)			
Conventional			
Recommended	1005 b	960 a	982 a
High Input	1093 a	994 a	1044 a
Organic			
Recommended	962 b	939 a	951 a
High Input	996 b	932 a	964 a

[+] Treatment means within the same column followed by the same letter are not significantly different according to Tukey's HSD (Studentized Range) test at the 0.05 level.

Table 5 shows the agronomic data for wheat in 2016. Wheat yields also had a cropping system x management input interaction. Organic wheat as the second-year transition crop in the soybean-wheat/red clover rotation yielded 11.5% lower than conventional wheat with recommended inputs. Conventional and organic wheat yielded the same with high inputs not because organic wheat responded to high inputs but because conventional wheat showed an 8.7% yield decrease with high inputs. It is not clear why conventional wheat yields actually declined with the use of high inputs. Yields were low because dry conditions (150 mm of precipitation, prevailed from the early tillering stage (1 April) until harvest (7 July). Surprisingly, organic compared with conventional wheat in the recommended input treatment had greater early plant establishment and fewer fall and spring weeds (Table 5). Conventional wheat, however, had an average grain N% of 2.03% compared to only 1.66% N in organic wheat, suggesting less available soil N for organic wheat. Grain yield, however, did not correlate with grain N% probably because dry soil conditions and not soil N availability was the major yield driver in 2016.

Table 5. Percent stand (early plant establishment), spikes/m^2 at harvest, weed densities in the early spring, grain yield, grain N%, and revenue of wheat in 2015–2016 under conventional and organic cropping systems at recommended and high input management in central NY.

	Wheat-2016		
Treatment	Stand/%	Spikes/m^2	Weeds/m^2
Conventional			
Recommended	88 b [+]	500 a	0.46 a
High Input	78 c	509 a	0.01 b
Organic			
Recommended	98 a	503 a	0.05 b
High Input	99 a	563 b	0.04 b
	Yield (kg/ha)	Grain N (%)	Revenue ($/ha)
Conventional			
Recommended	4314 a	1.95 b	642 a
High Input	3938 b	2.11 a	586 b
Organic			
Recommended	3817 b	1.65 c	568 b
High Input	3828 b	1.66 c	570 b

[+] Treatment means within the same column followed by the same letter are not significantly different according to Tukey's HSD (Studentized Range) test at the 0.05 level.

3.2. Economic

Maize revenue, a direct function of yield, had similar statistical relationships as yield so conventional compared with organic maize generated more revenue in 2015, but similar revenue in 2016 (Tables 3 and 6). Organic compared with conventional maize, averaged across the 2 years, had higher selected production costs when comparing their respective recommended and high input management treatments because of higher variable and fixed costs (Table 6). As expected, organic compared with conventional maize had lower seed costs because of the lack of seed treatment and GM traits. Organic compared with conventional maize, however, had higher fertilizer costs because of the much greater cost for composted chicken manure relative to conventional starter fertilizer and N fertilizer. A green manure crop was not in place for the 2015 maize crop so most of the composted chicken manure as an N source was applied in 2015 (none to the recommended input treatment and 56 kg N/acre in the high input treatment in 2016). Most organic crop producers in New York, USA do not use composted manure as an N source but rather use solid manure from their own livestock or from a neighbor's livestock, which is far less expensive. Organic compared with conventional maize also had higher labor, repair and maintenance, and fuel and lubricant costs because of the 4-time use of labor and equipment for mechanical weed control in organic maize compared with the 1-time use of labor and equipment for herbicide application in conventional maize. Organic compared with conventional maize thus had higher total selected variable costs ($190 to $687 in recommended and high input treatments, respectively).

Organic compared with conventional maize also had greater fixed costs because of greater wear and tear with the 4-time use of tractors and equipment for weed control purposes. Overall, organic compared with conventional maize had much higher total selected costs ($248 and $744/ha higher in recommended and high input treatments, respectively) compared to a Minnesota, USA study ($87/ha lower organic maize production costs) that used solid dairy manure [4]. In 2016, when composted chicken manure was not applied to organic maize with recommended inputs as an N source, organic compared to conventional maize had $75/ha lower total selected costs, similar to the Minnesota study.

Table 6. Income, selected costs, and partial returns for conventional maize with recommended management (M1) and high input management (M2), and organic maize with recommended management (M3) and high input management (M4) at a Cornell Research Farm in central NY averaged across the 2015 and 2016 growing seasons [1].

	Maize Treatments			
Production Value, Income	**M1**	**M2**	**M3**	**M4**
	$/ha			
Grain	1364	1430	1092	1127
Selected Production Costs [1]				
Variable Inputs				
Fertilizers	194.92	240.51	455.57	992.63
Seeds	301.07	360.84	219.07	262.49
Sprays & Other Crop Inputs	106.53	143.10	79.81	124.25
Labor	1.16	1.16	20.87	20.87
Repairs & Maintenance				
Tractor	0.22	0.22	5.53	5.53
Equipment	0.86	0.86	6.07	6.07
Fuels & Lubricants	0.73	0.73	12.09	12.09
Interest on Operating Capital	13.38	17.54	9.75	28.26
Total Selected Variable Input Costs	618.37	764.96	808.76	1452.1
Fixed Inputs				
Tractors	1.60	1.60	32.80	32.80
Equipment	4.47	4.47	30.23	30.23
Land charge	-	-	-	-
Value of management	-	-	-	-
Total Selected Fixed Input Costs	6.07	6.07	63.03	63.03
Total Selected Costs	624.44	771.03	871.79	1515.2
Partial Returns	739	659	220	−388

[1] This reporting of costs focused on those costs that differed among the four maize treatments. The land charge, and value of management input did not differ among treatments, so items are blank. Likewise, grain moistures did not differ among treatments so drying costs are not included. Seed costs differed among treatments due to price per unit differences between non-GMO and GMO hybrids, and seeding rate differences for recommended versus high input management. Spray and other crop inputs that differed included pest and disease management materials, and hauling as a function of yield. Labor costs reported included only those attributed to sprays for treatments C1 and C2, and those attributed to weeding tasks for C3 and C4. Labor costs reported do not include labor associated with tillage, planting and harvesting tasks considered constant, not differing among treatments. Similar explanations underlie estimates for the remaining cost items that differ. Costs for M3 and M4 were much higher in 2015 compared with 2016 because the use of composted chicken manure as an N source in 2015 vs. red clover in 2016.

Conventional compared with organic maize had much greater partial returns in 2015 because of higher yields and lower production costs (Table 7). If cash flow is of a major concern to the grower, maize should not be the entry crop in the transition to organic crop production unless there is animal manure on the farm (or close by) or a green manure crop in place. In 2016, when maize followed red clover, partial returns had a cropping system × management input interaction. Organic and conventional maize with recommended inputs had similar partial returns. In contrast, organic maize with high inputs (organic seed treatment, high seeding rates and 56 kg N/ha of composted manure), had lower partial returns compared with both conventional maize input treatments. Again, the 13× higher N/kg cost of composted chicken manure is almost solely responsible for the lower partial returns of organic maize with high input management. Organic maize with recommended inputs, which only received composted chicken manure as a starter fertilizer, had greater partial returns compared with conventional maize with high inputs, a management practice frequently used by conventional growers. If the grower wishes to plant maize during the transition, the partial returns data indicate that the grower should plant a green manure crop first, followed by maize with recommended inputs as the second crop. This strategy, however, would eliminate maize as the first crop eligible for the organic premium in 2017, which could reduce long-term economic benefits [6].

Table 7. Estimated partial returns of maize, soybean and wheat in conventional and organic cropping systems with recommended and high input management in 2015 (maize and soybean) and 2016 (all three crops) in central NY.

	Crop		
	Maize	**Soybean**	**Wheat**
Treatment	**2015 Estimated partial returns ($/ha)**		
Conventional			
Recommended	928 a [+]	706 a	-
High Input	844 a	664 a	-
Organic			
Recommended	−168 b	662 a	-
High Input	−562 c	630 a	-
Treatment	**2016 Estimated partial returns ($/ha)**		
Conventional			
Recommended	550 a	699 a	303 a
High Input	475 b	601 a	24 b
Organic			
Recommended	607 a	648 a	−188 c
High Input	−215 c	579 a	−588 d

[+] Treatment means within the same column followed by the same letter are not significantly different according to Tukey's HSD (Studentized Range) test at the 0.05 level.

Conventional soybean with high input management in 2015 had the highest revenue, but revenue did not differ between cropping systems nor management inputs in 2016 (Tables 4 and 8). Organic compared with conventional soybean had $50 to $105 lower total selected variable costs when comparing respective treatments (Table 8). Organic compared with conventional soybean had lower seed and other input costs (inoculant in conventional, organic seed treatment in organic high input, herbicide and fungicide in conventional high input), which offset higher remaining variable costs (labor, repairs and maintenance, and fuels and lubricants). As with maize, organic compared with conventional soybean had higher fixed input costs, associated with the greater use of tractor and equipment (tine weeder and cultivator) for repeated cultivations for weed control.

Organic compared with conventional soybean had slightly higher ($13/ha) total selected production costs with recommended input management but slightly lower ($47/ha) costs in high input management. Other cropping system studies have also reported similar or lower total production costs for organic soybean [7,8,13] mostly because of lower seed and pesticide costs. A Minnesota, USA study, however, reported $128/ha higher production costs in organic compared with conventional soybean because lower seed and pesticide costs did not offset higher labor, diesel, and machinery ownership costs [4]. Likewise, the USDA survey data [2] also reported that organic soybean producers had higher economic costs ($262 to $309/ha) compared with conventional producers.

Table 8. Income, selected costs, and partial returns for conventional soybean with recommended management (S1) and high input management (S2); and organic soybean with recommended management (S3) and high input management (S4) at a Research Farm in central NY averaged across the 2015 and 2016 growing seasons [1].

Production Value, Income	Soybean Treatments			
	S1	**S2**	**S3**	**S4**
	$/ha			
Seed	967	1028	932	953
Selected Production Costs [1]				
Variable Inputs				
Fertilizers	-	-	-	-
Seeds	216.94	289.25	134.90	179.88
Sprays & Other Crop Inputs	31.74	77.86	9.95	33.90
Labor	1.09	2.71	25.45	25.45
Repairs & Maintenance				
Tractor	0.22	0.45	6.08	6.08
Equipment	0.81	1.79	8.39	8.39
Fuels & Lubricants	0.75	1.52	17.69	17.69
Interest on Operating Capital	6.27	9.32	5.05	6.78
Total Selected Variable Input Costs	257.82	382.90	207.51	278.13
Fixed Inputs				
Tractors	1.58	3.18	35.40	35.40
Equipment	4.33	8.67	34.31	34.31
Land charge	-	-	-	-
Value of management	-	-	-	-
Total Selected Fixed Input Costs	5.91	11.85	69.71	69.71
Total Selected Costs	263.73	394.75	277.22	347.84
Partial Returns	703	633	655	605

[1] See Table 6 for an explanation of selected production costs.

Soybean partial returns in 2015 and 2016 did not differ among cropping systems nor management inputs because of mostly similar yields and production costs (Table 7). Organic soybean, especially with recommended inputs (no organic seed treatment to improve plant establishment or higher seeding rates to improve weed control) thus is an excellent entry or second year crop in the transition to an organic cropping system. Our economic data agree with another study that indicated that soybean is the preferred entry crop compared to maize [4]. A major advantage of using soybean as the entry crop is that soybean does not require N fertilizer so the prospective organic grower who does not own livestock will not have to find an organic N source, as in the case of maize or wheat.

Wheat revenue had a cropping system x management input interaction, similar to yield (Tables 5 and 9). Total selected production costs were more than 2-fold greater in organic compared with the respective conventional wheat management treatments (Table 9). The use of composted manure as starter fertilizer but more importantly as an N source is the major reason for the much greater variable and total production costs in organic wheat. As with maize, most organic growers in New York, USA would probably not use composted manure as an N source. Consequently, the $416 to $595/ha higher production costs for organic compared with conventional wheat in our study are much higher than the $243 to $257/ha higher production costs in the USDA survey report [2].

Table 9. Income, selected costs, and partial returns for conventional wheat with recommended management (W1) and high input management (W2), and organic wheat with recommended management (W3) and high input management (W4) at a Research Farm in central NY in 2015–2016 [1].

	Wheat Treatments			
Production Value, Income	**W1**	**W2**	**W3**	**W4**
	$/ha			
Grain	643	587	569	570
Selected Production Costs [1]				
Variable Inputs				
Fertilizers	165.49	198.84	601.15	891.61
Seeds	125.52	200.84	97.17	155.49
Sprays & Other Crop Inputs	41.74	131.03	40.01	82.25
Labor	0	2.42	0	0
Repairs & Maintenance				
Tractor	0	0.45	0	0
Equipment	0	0.88	0	0
Fuels & Lubricants	0	1.36	0	0
Interest on Operating Capital	8.35	13.41	18.45	28.26
Total Selected Variable Input Costs	341.08	550.24	756.78	1157.60
Fixed Inputs				
Tractors	0	3.29	0	0
Equipment	0	9.19	0	0
Land charge	-	-	-	-
Value of management	-	-	-	-
Total Selected Fixed Input Costs	0	12.47	0	0
Total Selected Costs	341.08	562.72	756.78	1157.60
Partial Returns	303	24	−188	−588

[1] See Table 6 for an explanation of selected production costs.

Organic compared with conventional wheat had much lower partial returns when comparing their respective management treatments because of similar or lower yields and much higher total selected production costs (Table 7). Many wheat growers in New York, USA, however, manage wheat with high inputs (high seeding rates, fall herbicide, high split-applied N rates, and late spring fungicide). Organic wheat with recommended inputs compared more favorably with typical conventional wheat management with high inputs ($212/ha lower partial returns). Organic wheat compared with organic maize and soybean as second-year crops in the transition had much lower partial returns. Conventional wheat compared with conventional maize or soybean, also had lower partial returns, which explains in part the record low hectares of wheat planted in the USA in 2017 [3]. Winter wheat, however is an ideal rotation crop that disrupts weed and insect cycles in maize and soybean [21] so must be evaluated in context of its benefits to an organic rotation.

When comparing partial returns of the three crop rotations (red clover-maize, maize-soybean, and soybean-wheat/red clover) during the transition, the organic red clover-maize rotation with recommended inputs had similar partial returns as the conventional red clover-maize rotation with recommended inputs and greater partial returns compared with the high input treatment (Table 10). Most conventional growers, however, who do not transition to organic production, would not practice such a rotation so comparisons should be made between the organic red clover-maize rotation with the conventional maize-soybean rotation. The organic red/clover-maize rotation with recommended inputs had $1127/ha lower partial returns compared with the conventional maize-soybean rotation with recommended inputs and $1024/ha lower partial returns compared with the high input treatment. We did not apply composted chicken manure as an N source to organic maize with recommended inputs in 2016, but rather utilized red clover as the N source. Therefore, production costs are not inflated and partial returns not deflated when comparing the organic clover-maize rotation with recommended inputs with the conventional maize-soybean rotation with recommended and high inputs.

The organic maize-soybean rotation with recommended inputs had similar partial returns ($434) as the organic red clover-maize rotation with recommended inputs ($441, Table 10). Consequently, partial returns of both organic rotations were similar when compared with the conventional maize-soybean rotation. The substitution of a green manure crop for maize as an entry crop instead of continuing a maize-soybean rotation during the transition thus did not improve partial returns.

The organic compared with the conventional soybean-wheat/red clover rotation with recommended inputs had $548/ha lower partial returns (Table 10). Many soybean and wheat growers in New York, USA, however, use high input management on both crops. The organic soybean-wheat/red clover rotation with recommended inputs compares more favorably with the conventional soybean-wheat red clover rotation with high inputs ($229/ha lower partial returns). If cash flow is of major concern to the grower during the transition, soybean was the best entry crop followed by wheat in this study. This agrees with the findings in a Minnesota USA study [5]. When comparing partial returns of all three organic rotations with recommended inputs, however, differences were only $54 to $61 /ha. Consequently, the red clover-maize, maize-soybean, and soybean-wheat/red clover rotations would essentially have the same cash flow impact on the farm during the transition. This agrees with the findings of Archer et al. [5] who reported that transitioning growers should begin the transition process immediately, regardless of the entry crop.

Table 10. Estimated partial returns of three rotations (red clover-maize, maize-soybean, and soybean-wheat) during the transition period (2015 and 2016) in conventional and organic cropping systems with recommended and high input management in central New York.

Treatment	Clover-Maize	Maize-Soybean	Soybean-Wheat
		Sequence during Transistion	
		Total Costs ($/ha)	
Conventional			
Recommended	741 [+]	958 [+]	605
High Input	909	1211	956
Organic			
Recommended	666	1556	1035
High Input	1503	2077	1505
		Total Revenue ($/ha)	
Conventional			
Recommended	1116 a	2526 a	1647 a
High Input	1214 a	2610 a	1679 a
Organic			
Recommended	1107 a	1990 b	1530 a
High Input	1116 a	2041 b	1567 a
		Total Partial Returns ($/ha)	
Conventional			
Recommended	375 a	1568 a	1043 a
High Input	305 b	1399 a	724 b
Organic			
Recommended	441 a	434 b	495 c
High Input	−387 c	−36 c	62 d

[+] Treatment means within the same column followed by the same letter are not significantly different according to Tukey's HSD (Studentized Range) test at the 0.05 level; [1] Maize costs in 2015 are much greater than costs in 2016 because of the use of composted chicken manure as the main N source in 2015 vs. red clover in 2016.

Organic maize and wheat had greater production costs than typical because of the use of composted chicken manure to ensure comparable N rates applied to organic and conventional maize and wheat. In addition, conventional weed control costs in maize are much lower than typical because most conventional maize growers in the USA do not moldboard plow their fields so they use more weed control chemicals than the single glyphosate application used in this study. Consequently, input costs of organic compared with conventional maize, soybean and wheat during the transition were much higher than typical, especially in maize and wheat. On the other hand, delaying the planting of conventional maize and soybean to just after the optimum planting dates probably reduced their yields and revenue somewhat. Likewise, the exceedingly dry 2016 growing season, probably contributed to the similar yields and revenue between organic and conventional maize and soybean and only an 11.5% lower organic wheat yield with recommended inputs during the second transition year [22]. In addition, we used untreated Pioneer varieties with no GMO traits in maize, soybean, and wheat instead of organic varieties, which probably also favored organic compared to conventional yields and revenue. Consequently, partial returns between organic and conventional cropping systems in our study during the transition did not differ greatly from other studies [4,8].

4. Conclusions

The two major constraints to organic field crop production are soil N availability and weed competition. Soybean was thus an excellent entry crop in the transition to organic production because it provided its own N, and tine weeding followed by four cultivations provided satisfactory weed control in this study. In contrast, maize as an entry crop, was more problematic because providing available soil N in the absence of a green manure crop was a challenge, and maize was less competitive with weeds (compared with wheat or soybean) in this study.

Organic wheat no-tilled into soybean stubble had very low weed densities in this study. Organic wheat in a maize-soybean-wheat/red clover rotation, however, must rely on an organic N source for its N uptake. Wheat takes up most of its N from late April through May in New York, USA when cool temperatures prevail, which may inhibit rapid mineralization of organic N. Nevertheless, organic wheat with recommended inputs compared with conventional wheat with high inputs, a typical management system for many growers in the Eastern USA, yielded similarly in 2016. Wheat was thus an excellent second year transition crop in this study. We recommend the organic soybean-wheat/red clover rotation during the transition for locations with similar environmental conditions of this study because this rotation compared most favorably with its conventional counterpart during the transition.

Author Contributions: W.J.C. designed and conducted the experiment, and wrote the initial draft of the manuscript. J.J.H. conducted the economic analyses for the study. J.C. conducted the statistical analyses and reviewed the manuscript.

Acknowledgments: This project was partially supported by the U.S. Department of Agriculture Cooperative State Research, Education, and Extension Service through New York Hatch Project 1257322. Pioneer Hi-Bred supplied all the seed for the study.

Conflicts of Interest: The authors declare no conflict of interests.

References

1. USDA, Organic. What Is Organic Certification? 2012. Available online: http://www.ams.usda.gov/sites/default/files/media/What%20is%20Organic%20Certification.pdf (accessed on 8 May 2018).
2. USDA, ERS. The Profit Potential of Certified Organic Field Crop Production. 2015. Available online: http://www.ers.usda.gov/media/1875176/err188_summary.pdf (accessed on 8 May 2018).
3. USDA, NASS. Prospective plantings. 2017b. Available online: http://usda.mannlib.cornell.edu/usda/current/ProsPlan/ProsPlan-03-31-2017.pdf (accessed on 8 May 2018).
4. Archer, D.W.; Jaradat, A.A.; Johnson, J.M.F.; Lachnicht-Weyers, S.; Gesch, R.W.; Forcella, F.; Kludze, H.K. Crop productivity and economics during the transition to alternative cropping systems. *Agron. J.* **2007**, *99*, 1538–1547. [CrossRef]
5. Archer, D.W.; Kludze, H. Transition to organic cropping systems under risk. In Proceedings of the American Agricultural Economics Assoc, Annual Meeting, Long Beach, CA, USA, 23–26 July 2006; p. 24.

6. Coulter, J.A.; Delbridge, T.A.; King, R.P.; Sheaffer, C.C. Productivity, economic, and soil quality in the minnesota variable-input cropping systems trial. *Crop Manag.* **2013**, *12*. [CrossRef]
7. Delbridge, T.A.; Coulter, J.A.; King, R.P.; Sheaffer, C.C. A profitability and risk analysis of organic and high-input cropping systems in southwestern Minnesota. In Proceedings of the Agricultural Applied Economics Assoc, Annual Meeting, Denver, CO, USA, 25–27 July 2010.
8. Delbridge, T.A.; Coulter, J.A.; King, R.P.; Sheaffer, C.C.; Wyse, L. Economic performance of long-term organic and conventional cropping systems in Minnesota. *Agron. J.* **2011**, *103*, 1372–1382. [CrossRef]
9. Chavas, J.P.; Posner, J.L.; Hedtcke, J.L. Organic and conventional production systems in the Wisconsin integrated cropping systems trial: II. Economic and risk analysis 1993–2006. *Agron. J.* **2009**, *101*, 288–295. [CrossRef]
10. Baldock, J.O.; Hedtcke, J.L.; Posner, J.L.; Hall, J.A. Organic and conventional production systems in the Wisconsin integrated cropping systems trial: III. Yield trends. *Agron. J.* **2014**, *106*, 1509–1522. [CrossRef]
11. Cavigelli, M.A.; Teasdale, J.R.; Conklin, A.E. Long-term agronomic performance of organic and conventional field crops in the mid-Atlantic region. *Agron. J.* **2008**, *100*, 785–794. [CrossRef]
12. Delate, K.; Cambardella, C.A. Agroecosystem performance during transition to certified organic grain production. *Agron. J.* **2004**, *96*, 1288–1298. [CrossRef]
13. Delate, K.M.; Duffy, C.; Chase, A.; Holste, H.; Friedrich, N.; Wantate, N. An economic comparison of organic and conventional grain crops in a long-term agroecological research (LTAR) site in Iowa. *Am. J. Altern. Agric.* **2003**, *18*, 59–69. [CrossRef]
14. Delate, K.; Cambardella, C.; Chase, C.; Johanns, A.; Turnbull, R. The long-term agroecological research (LTAR) experiment supports organic yields, soil quality, and economic performance in Iowa. *Crop Manag.* **2013**. [CrossRef]
15. Hall, M.R.; Swanton, C.L.; Anderson, G.W. The critical period of weed control in grain corn (*Zea mays*). *Weed Sci.* **1992**, *40*, 441–447.
16. USDA, NASS. Crop Values. 2016 Summary. 2017. Available online: http://usda.mannlib.cornell.edu/usda/current/CropValuSu/CropValuSu-02-24-2017_revision.pdf (accessed on 8 May 2018).
17. Lazarus, W.F. Agricultural Business Management News. Farm Machinery Cost Estimates Updated. Available online: http://blog-abm-news.extension.umn.edu/2017/05/farm-machinery-cost-estimates.html (accessed on 8 May 2018).
18. Cox, W.J.; Cherney, J.H. Agronomic comparisons of conventional and organic maize during the transition to an organic cropping system. *Agronomy* **2018**, *8*, 113. [CrossRef]
19. Kravchenko, A.N.; Snapp, S.S.; Robertson, G.P. Field-scale experiments reveal persistent yield gaps in low-input and organic cropping systems. *Proc. Natl. Acad. Sci. USA* **2017**, *114*, 926–931. [CrossRef] [PubMed]
20. Cox, W.J.; Cherney, J.H. Growth and yield responses of soybean to row spacing and seeding rate. *Agron. J.* **2011**, *103*, 123–128. [CrossRef]
21. Teasdale, J.R.; Mangum, R.W.; Radhakrishnan, J.; Cavigelli, M.A. Weed seedbank dynamics in three organic farming crop rotations. *Agron. J.* **2004**, *96*, 1429–1435. [CrossRef]
22. Lotter, D.; Seidel, R.; Liebhardt, W. The performance of organic and conventional cropping systems in an extreme climate year. *Am. J. Altern. Agric.* **2003**, *18*, 146–154. [CrossRef]

Article

Impact of Agroecological Practices on Greenhouse Vegetable Production: Comparison among Organic Production Systems

Corrado Ciaccia [1,*], Francesco Giovanni Ceglie [2], Giovanni Burgio [3], Suzana Madžarić [2], Elena Testani [1], Enrico Muzzi [3], Giancarlo Mimiola [2] and Fabio Tittarelli [1]

[1]   Council for Agricultural Research and Economics (CREA), Research Centre for Agriculture and Environment, Via della Navicella 2-4, 00184 Rome, Italy

[2]   Centre International De Hautes Études Agronomiques Méditerranéennes (CIHEAM)—Bari, Via Ceglie, 9, 70010 Valenzano (BA), Italy

[3]   Department of Agricultural and Food Sciences—DISTAL, University of Bologna, Viale G. Fanin 44, 40127 Bologna, Italy

*   Correspondence: corrado.ciaccia@crea.gov.it; Tel: +39-06-700-54-13

Received: 8 April 2019; Accepted: 6 July 2019; Published: 11 July 2019

Abstract: In greenhouses, where intensive systems are widely used for organic production, the differences between "conventionalized" and agroecological approaches are especially evident. Among the agronomic practices, green manure from agroecological service crops (ASCs) and organic amendments represent the main tools for soil fertility management with respect to the substitution of synthetic fertilizer with organic ones (the input substitution approach). Over a two-year organic rotation, we compared a conventionalized system (SB) and two agroecological systems, characterized by ASC introduction combined with the use of manure (AM) and compost (AC) amendments. A system approach was utilized for the comparison assessment. For this purpose, agronomic performance, soil fertility and the density of soil arthropod activity were monitored for the entire rotation. The comprehensive evaluation of the parameters measured provided evidence that clearly differentiated SB from AM and AC. The drivers of discrimination were soil parameters referring to long term fertility and soil arthropod dynamics. The study confirmed the higher productivity of SB but also no positive impact on soil fertility and soil arthropods, as highlighted by AM and AC. Based on the results, a trade-off between productivity and the promotion of long-term ecosystem diversity and functioning is needed for the assessment of systems of organic production.

Keywords: cover crops; agrobiodiversity; conventionalization; system approach

1. Introduction

The issue of intensive organic production systems has been the subject of debate during the last decade, since the concept of "conventionalization" of organic production was introduced [1,2]. "Conventionalized" organic production, also called as "input-substitution" system, is characterized by the substitution of synthetic inputs with those allowed by organic regulations [3]. Despite their conformity to organic regulations, these systems are, in general terms, very similar to conventional and integrated ones for the high yield per unit area and the potential environmentally negative externalities, such as degradation of soil, high greenhouse gases (GHG) emissions and biodiversity losses [4]. On the other hand, alternative organic production systems based on a balance between productivity, agrobiodiversity promotion and ecosystem service provision have been set up in different climatic conditions [5,6]. In the more recent scientific literature [7–9] the great majority of these organic systems are referred to open field conditions, where agronomic practices such as minimum tillage, organic amendments, long-term rotation and cover crop cultivation are more easily introduced. Only recently have such alternative organic systems of production been tested in protected conditions [10]. Among the many criteria involved in the definition of the price premium for organic vegetables, the potential environmental impact of the production system is considered one of the more influential among consumers of organic food [11–13].

These issues have led to the need to change the organic greenhouse production system, to meet the expected environmental sustainability (e.g., reduced water consumption and soil organic matter depletion), together with the market demand for organic production. In this context, an agroecological approach to organic production should be promoted, adopting a holistic perspective able to also encompass other aspects of agriculture (i.e., environmental, social), rather than just the production issues [14]. To implement such an approach, a deep comprehension of the complex interrelationships among the components of the agroecosystem is necessary, to maximize services and minimize negative externalities (e.g., biodiversity reduction due to external input use). Furthermore, practices aiming to increase the agrobiodiversity in space and time should be promoted, to foster the desired biodiversity components. Long-term rotation and introduction of agroecological service crops (ASCs)—cultivation of plant species not for yield purposes but for ecosystem service provision (e.g., cover crops, catch crops, living mulch, etc.) [15]—should also be adopted in greenhouse systems [10]. In this context, the real challenge for research on organic production is the comparison between different "certified" organic farming systems, which overcomes the mere comparison between conventional and organic agriculture. To contrast the widespread production intensification, different criteria for evaluating production systems must be identified. A holistic and agroecological based approach should reject the simplistic "conventional" criteria of maximizing yield per square meter or maximizing efficiency per input unit without any assessment of the negative impact on the environment [16]. Moreover, system studies in organic farming provide a sort of "knowledge package" including innovative solutions and multidisciplinary know-how, which go beyond the external inputs dependency of conventionalized systems of production. Since intensification of production systems implies simplification, and consequent biodiversity losses, which leads to reductions in ecosystem services, the use of arthropod biodiversity and dynamics can be a reliable indicator of system diversity, covering a wide range of ecological functions in the agroecosystem [17].

To investigate the effects of "agroecological" systems on agronomic performance, soil properties, and associated soil arthropod dynamics in greenhouse crops under a Mediterranean climate, a two-year field experiment was carried out at the Centre International De Hautes Études Agronomiques Méditerranéennes (CIHEAM)—Bari, in Valenzano (Puglia region, Southern Italy). The hypothesis underlying the experiment was that different agronomic management, within organic systems of production (organic fertilizers and ASC introduction and management, compared to a conventionalized-business as usual-system), would result in: (1) similar yield; (2) differences in soil fertility; and (3) differences in soil arthropods composition and their relative ecological services/disservices (e.g., pest control and organic matter decomposition). To test these hypotheses, the following aspects were studied: (1) crop performances over the two years of the experiment; (2) soil arthropod activity density in the management system tested; and (3) linkages between the cited biological community and the most important soil fertility parameters. A system approach was utilized for the assessment of different systems of production [18].

2. Materials and Methods

2.1. Experimental Site

The research was carried out at the long-term experiment on organic vegetable production systems in Mediterranean greenhouses (MOREGREEN LTE) located at the CIHEAM—Bari (MAIB), in Valenzano (Puglia region - Southern Italy). The location is about 72 meters above sea level (41°08′ N latitude and 16°51′ E longitude). The experimental greenhouse (300 m^2; 7.5 m × 40.0 m) was an un-heated tunnel (EUROPROGRESS s.r.l., Mirandola (MO), Italy) with galvanized steel frames covered by ethylene vinyl acetate (EVA) sheets. It was longitudinally divided into two areas/fields (field I and field II). The present research was based on the two-year rotation of field I, cultivated with kohlrabi (*Brassica oleracea* var. *gongyloides*, cv "Korist"), lettuce (*Lactuca sativa* L. var. *longifolia* Lam., cv Salad bowl), zucchini (*Cucurbita pepo* L., cv Striata di Puglia) and lamb's lettuce (*Valerianella locusta* L., cv Semegrosso d'Olanda) in rotation (2014–2015 and 2015–2016).

2.2. Experimental Design

The experimental layout was a completely randomized block (CRB) design with three replications (for a total of 9 plots, 3.0 m × 4.0 m each). The organic farming systems under comparison were as follows. (1) Substitution (SB), a business as usual organic production system (very diffused especially in greenhouse vegetable production), based on an "input substitution" approach. This system mimics conventional agriculture by substituting agrochemicals with products allowed in organic farming (e.g., substitution of synthetic fertilizers with organic ones). The following fertilizers were applied: commercial organic fertilizer based on guano (commercial name—"Guanito" by Italpollina); and liquid commercial organic fertilizer based on sugar beet molasses (commercial name—"Kappabios" by Serbios). (2) Agroman (AM), characterized by the use of a mixture of ASCs and mature organic cattle manure as a soil amendment. (3) Agrocom (AC), which utilizes a mixture of ASCs and on-farm made compost as a soil amendment. In the first year of rotation, before the kohlrabi crop, the same composition of ASC mixture was used in AM and AC, while two different mixtures of species were considered in the second cycle of ASCs, before the zucchini crop. The choice of mixtures, instead of the cultivation of a single ASC species, was made in order to better guarantee the provision of the ecological services in the long run. The ASC species mixture for each treatment and the ecological service corresponding to each botanical family are reported in Table 1. During the ASC cycles in AC and AM, soil was left bare in the SB plots.

Before ASC sowing, the soil was prepared using a rotary tiller. ASC mixture seeds were broadcasted by hand and gently covered with soil using a rake on 27 and 10 June in 2014 and 2015, respectively. In the first year of rotation, the ASCs were terminated differently at the start of the flowering stage (about 50 days after sowing); in AC, the ASCs were manually chopped using a sickle and ploughed into the soil as green manure using a rotary spader, whereas in AM, the ASCs were flattened using a roller crimper [19] to obtain a mulching layer made of the ASC biomass, in which kohlrabi was transplanted. After the kohlrabi cycle, before lettuce transplanting, ASC mulch residues were incorporated into the soil using a rotary spader. In the second experimental year, ASCs were instead terminated as green manure in both the AM and AC systems. Details on rotation, starting from the first year, were as follows: kohlrabi was transplanted on 22 October 2014, using seedlings from the nursery of CIHEAM—Bari, and harvesting was completed on 7 January 2015. Lettuce was transplanted on 9 March 2015, using seedlings from the nursery of CIHEAM—Bari, and harvested on 24 April 2015. The density of the kohlrabi and lettuce was 10 plants/m^2, (0.5 m between lines and 0.2 m within each line). Zucchini was transplanted at a density of 1.8 plants/m^2, (0.8 m between lines and 0.7 m within each line) on 8 September 2015, using seedlings from the nursery of CIHEAM—Bari, and harvesting was completed on 30 November 2015.

Finally, lamb's lettuce was sown at a rate of 50 seeds/m^2 on 14 March 2016, using a pre-sown blanket of biodegradable tissue (supplied by Virens, Padova, Italy), and harvesting was completed on 18 May 2016 [23]. All crops were harvested according to required market standards. Cumulative production (yield) per crop was calculated for each system. Air temperature at 2 m height was measured hourly by two probes during both ASC and cash crop cycles. Air humidity and temperature were kept under control (10–28 °C range) by manually opening/closing tunnel border sides when needed.

Table 1. Agroecological service crops (ASC) mixture composition and ecological services provided.

Year	System	Poaceae	Fabaceae	Brassicaceae	Seed Weight (%)
I (2014)	AM; AC	*Pennisetum glaucum* (L.) R. Br.			30.0
		Setaria italica (L.) Beauv.			30.0
			Lablab purpureus (L.) Sweet		20.0
			Vigna sinensis L.		20.0
II (2015)	AM		*Lablab purpureus* (L.) Sweet		20.0
			Vigna sinensis L.		20.0
			Crotolaria juncea L.		20.0
			Hedysarum coronarium L.		20.0
			Onebrychis vicifolia Scop.		20.0
	AC			*Brassica juncea* (L.) Czern.	33.3
				Raphanus sativus L.	33.3
				Sinapis alba L.	33.3
Ecological services provided		Nitrogen losses decrease (catch crops); soil organic matter increase, weed control [20]	Nitrogen fixation; soil organic matter increase; soil and water quality [21]	Nitrogen losses decrease; nematode suppression; soil-borne disease reduction; soil and water quality; root growth increase [22]	

2.3. Soil Preparation

The tunnel greenhouse was installed on May 2012, on a soil organically managed for ten years. Strawberry (*Fragaria* × *ananassa* var. *Duchesne*, cv Festival) was cultivated as previous crop in each system from September 2013 to May 2014. Before the first ASC cycle, the soil was ploughed using a rotary tiller (SICMA CS 105). In the SB system, for kohlrabi, lettuce and zucchini cultivation, the whole soil bed was covered by black polyethylene plastic mulch. Water was mainly supplied by drip irrigation, except for ASC and lamb's lettuce cultivation, where sprinkler irrigation was used.

2.4. Organic Amendment and Fertilizers

The total amount of amendments and fertilizers applied to each of the three systems, total nitrogen and carbon content, and distribution practices are reported in Tables 2 and 3. Compost and cattle manure samples were analysed in triplicate for dry matter, total organic carbon (TOC) and total nitrogen (TN). Dry matter was calculated by weight loss overnight in an oven at 105 °C. TOC was analyzed with a LECO analyzer (LECO RC-612; St. Joseph, MI, USA) using a dry combustion method [24]. TN was analyzed with the Dumas method using the elemental analyzer LECO FP 528.

Table 2. Characterization of amendments and fertilizers applied for each crop during the rotation.

Cash Crop	Fertilizer	Supplier	TOC	TN	C/N	Ptot
			(g kg^{-1})		(g kg^{-1})	
All	Commercial organic fertilizer based on guano, "Guanito".	Italpollina - Italy	320	60	5	6.6
All	Liquid commercial organic fertilizer based on Sugar beet molasses ("Kappabios")	Serbios - Italy	150	30	5	0.0
Before Kohlrabi	Compost	On-farm, CIHEAM—Bari facility	261	27	10	6.7
	Cattle manure	Organic farm "la Querceta", Bari - Italy	430	18	24	7.3
Before Zucchini	Compost	On-farm, CIHEAM—Bari facility	301	32	9	2.7
	Cattle manure	Organic farm "la Querceta", Bari - Italy	445	24	19	2.7

Table 3. Distribution of amendments and fertilizers to the three systems for each crop (on dry matter basis, except for "Kappabios", in liquid form).

Cash Crop	System	Fertilizers	Distribution
Kohlrabi	SB	Guanito	incorporated into the soil before transplanting at 1.67 Mg ha^{-1}
		Kappabios	fertigation, 7 times during crop cycle for a total of 0.32 Mg ha^{-1}
	AC	Compost	incorporated into the soil before transplanting at 7.41 Mg ha^{-1}
		Kappabios	fertigation, 3 times during crop cycle for a total of 0.14 Mg ha^{-1}
	AM	Cattle manure	incorporated into the soil before ASC sowing at 11.17 Mg ha^{-1}
		Kappabios	fertigation, 7 times during crop cycle for a total of 0.32 Mg ha^{-1}
Lettuce	SB	Kappabios	fertigation, 7 times during crop cycle for a total of 0.46 Mg ha^{-1}
	AC	Kappabios	fertigation, 3 times during crop cycle for a total of 0.20 Mg ha^{-1}
	AM	Kappabios	fertigation, 4 times during crop cycle for a total of 0.26 Mg ha^{-1}
Zucchini	SB	Guanito	incorporated into the soil before transplanting at 1.83 Mg ha^{-1}
		Kappabios	fertigation, 7 times during crop cycle for a total of 0.46 Mg ha^{-1}
	AC	Compost	incorporated into the soil before transplanting at 6.25 Mg ha^{-1}
		Kappabios	fertigation, 3 times during crop cycle for a total of 0.20 Mg ha^{-1}
	AM	Cattle manure	incorporated into the soil before transplanting at 8.33 Mg ha^{-1}
		Kappabios	fertigation, 4 times during crop cycle for a total of 0.26 Mg ha^{-1}
Lamb's lettuce	SB	Kappabios	fertigation, 7 times during crop cycle for a total of 0.46 Mg ha^{-1}
	AC	Kappabios	fertigation, 3 times during crop cycle for a total of 0.20 Mg ha^{-1}
	AM	Kappabios	fertigation, 4 times during crop cycle for a total of 0.26 Mg ha^{-1}

2.5. Plant Sampling and Analysis

Each year, at the end of the ASC cycle, three quadrats (0.25 m × 0.25 m) per plot were used to sample the fresh aboveground biomasses. At the harvest stage, cash crops were collected from the plots and partitioned into products and residues. Plant samples were divided into two subsets. One was dried at 105 °C for the determination of the dry matter content by gravimetric loss (total plant biomass, total yield and biomass of residues ploughed into the soil at the end of crop cycle—hereafter reported as Tot Biom, Tot Yield and Inc res, respectively), while the other was dried at 60 °C and stored for analyses. In SB, crop residues were removed from the field. TN was analyzed using samples of the ASCs, products and residues, with the Dumas method using the elemental analyzer LECO FP 528. N uptake of cultivated plants was then calculated by multiplying N content by the correspondent biomass per hectare values, in order to obtain the total N uptake (Tot Nu) and the yield N uptake (Yield Nu). All the organic biomass utilized as C input (ASC and cash crop residues—Inc res C) were analyzed for C content on a LECO analyzer (LECO RC-612) using a dry combustion method. To obtain the C input, the carbon content of plant materials was multiplied by the correspondent biomass per hectare values.

2.6. Soil Sampling and Analysis

For each crop in rotation, at different plant phenological phases (at transplanting, during the plant growth cycle—twice for kohlrabi and zucchini and once for lettuce and lamb's lettuce, and at harvest), elementary soil sub-samples were taken from each plot using an auger at a depth of 0–30 cm and mixed to form a composite sample for each plot. For each composite sample, total soil mineral nitrogen (SMN), as the sum of nitrate (NO_3^--N) and ammonium nitrogen (NH_4^+-N), was determined. Fresh soil samples were extracted by 2M KCl (1:10 w/v). Then NH_4^+-N and NO_3^--N were determined according to Henriksen and Selmer-Olsen [25] and Krom [26]. At the beginning of the experiment and at the end of each cash crop cycle, the soil samples were air dried, sieved at 2 mm and stored for the determination of total organic carbon (Soil TOC), total nitrogen (Soil TN) and available Phosphorus (Soil P). Soil TOC content was determined by combustion method on a LECO analyzer (LECO RC-612; St. Joseph, MI, USA) using a dry combustion method [24]. Soil TN was analyzed with the Dumas method using the elemental analyzer LECO FP 528. Soil P was extracted and measured according to the Olsen method [27].

2.7. Soil Arthropods Monitoring

Soil arthropods were monitored from July 2014, following the rotation plan, until May 2016. Depending on the crop cycle, monitoring periods were as follows: ASC '14 (49 days); kohlrabi (56 days); lettuce (43 days); ASC '15 (46 days); zucchini (93 days); and lamb's lettuce (58 days). Pitfall traps were used to collect arthropods (one per plot). Each trap consisted of a plastic cup (13 cm × 10 cm, 500 mL) half filled with 50% propylene glycol water solution. The cups were dug into the soil and the rim was levelled with the soil surface. A 15 cm × 15 cm plastic saucer was placed 4 cm above the cup to prevent irrigation water entering the cups. The traps were replaced every 15–25 days during the monitoring period (depending on the capture rate). The content of each cup was emptied in the field into plastic containers and stored in the fridge until soil arthropods were counted in the laboratory. In the present study, seven macrogroups were considered (Coleoptera Carabidae, Coleoptera Staphylinidae, Collembola, Araneae, Myriapoda, Isopoda and Opiliones). Data are presented as mean activity density for each macrogroup per crop and as mean for full rotation period (total number of individuals divided by days of trap activity and multiplied by 10; resulting in cumulative activity density for a 10 days period). Detailed elaboration of total soil arthropods and relative abundance is reported in [28], while the present paper reports the activity density of individual macrogroups and intends to explore their possible relations with the main soil parameters studied.

2.8. Statistical Analysis

Agronomic parameters and arthropod macrogroups were analysed with a mixed model, using YEAR (levels: 1°–2°) as random factor, SYSTEM (levels: AC, AM and SB) and CROP (levels: A, K, L, Z and LM, for ASC, kohlrabi, lettuce, zucchini and lamb's lettuce, respectively) as fixed factors. In the model, CROP was nested within YEARS. Mixed model analyses were performed by means of SAS 9.0.

Agronomic and arthropod data were also analyzed by means of an explorative multivariate analysis, using canonical discriminant analysis (CDA). CDA was used for understanding the temporal dynamics of the systems during the two-year rotation. In other words, to underpin the interactions evidenced by the mixed model (see previous analysis). The group variables of the CDA were System (SB, AC and AM) and Crop rotation (Year: 1 and 2), while the response variables (represented by arrows) were agronomic parameters and arthropod taxa. In the biplot, the multivariate 95% confidence coefficient of the canonical means of each group variable was represented by a circle. CDA was employed using R 3.3.2.

3. Results

The results of the mixed model analysis for agronomic parameters are reported in Table 4. The system x crop rotation (Year) showed significant interaction for most of the analyzed parameters, except for Soil P, Soil TN, Tot Biom, Inc res and Inc res C, while both System and Crop rotation (Year) factors highlighted significant differences, except for Soil TN in Crop rotation (Year). Tables 5 and 6 display means for soil chemical and plant parameters, respectively, for each system and crop in rotation, including the total means for the compared systems in the whole rotation. In particular, Soil TOC showed increasing trends during the rotation in AC and AM until ASC '15, whereas a slight variation around the average value was recorded for SB (Table 5). No difference due to the Year was recorded for Soil TN, while a significant effect of Systems was detected (Table 4), with lower value in SB than in agroecological systems (Table 5). Soil P showed the highest values for AC, followed by SB and then by AM, with higher values in the first year rotation than in the second one (36.6 and 31.0 ppm, respectively). Soil mineral nitrogen ($NO_3^--N + NH_4^+-N$ as SMN) parameters showed the lowest dynamics in SB and first crop year rotation, whereas the highest were recorded in the same year by AC, recording the highest NO_3^--N content during the lettuce cycle (179 ppm), with a similar trend also shown by AM. The lowest SMN values were instead recorded for the SB in the zucchini cycle (second year), corresponding to 23.3 ppm ($NO_3^--N + NH_4^+-N$) (Table 5). As far as the plant parameters are concerned (Table 6), SB showed the highest yield for all the cash crops, with the exception of lamb's lettuce when AM showed the highest yield. Similar results were recorded for the Nuptake parameters (Tot Nu and Yield Nu), whereas AM showed the highest Carbon input (Inc Res C) due to residues (0.23 t ha^{-1} vs 0.19 t ha^{-1} for AM and AC, respectively).

Crop rotation (System) showed significant differences, among the farming systems compared, in activity density for four out of the seven taxa (Araneae, Opiliones, Staphylinidae and Myriapoda) (Table 7). The significant interaction ($p < 0.01$) between systems and crops in the two-year rotation demonstrated that the effect of agroecological systems was strongly associated with the year of the rotation. Indeed, crop (Year) factor was strongly correlated with activity density for all the arthropod taxa. A detailed analysis of the mixed model interactions explains better the response of arthropod fauna to all the variables of the rotation (Table 8). For example, besides a general positive impact of agroecological systems on Staphylinidae in comparison with SB (mean value for the rotation), the AC system was positively associated with the activity density of this taxon mainly in the first year of rotation, and in particular during the kohlrabi and lettuce cultivation. Collembola activity density was higher in SB in comparison with agroecological systems, particularly for ASC during the first year.

This trend changed during the second year, when density of this taxon was higher in the agroecological systems (zucchini) or it had similar values for all systems. Araneae were positively correlated with AM in the ASC cycle during the second year of rotation, while the difference among the systems was not relevant in the other crops. Opiliones positively responded to AC and AM in the first year during lettuce cycle, while Myriapoda were more associated with the agroecological systems during the second year (ASC and zucchini). The mean activity density of Carabidae and Isopoda during the whole rotation was not correlated with the systems (Tables 7 and 8). Carabidae density showed a higher increase in AC than in SB and AM during the kohlrabi rotation (Table 8) and a partial response to AC in the ASC during the first year and in zucchini in the second year, even if the interaction is not significant (Table 7). Isopoda density was higher in agroecological systems in ASC during both the cultivation years.

Table 4. Results of the mixed model analysis on agronomic parameters (P levels). See material and methods for explanations.

Effect	NO₃⁻-N	NH₄⁺-N	Soil P	Soil TOC	Soil TN	Tot Biom	Tot Yield	Inc Res	Yield Nu	Tot Nu	Inc Res C
System	<0.001	<0.001	<0.001	<0.001	<0.05	<0.001	<0.001	<0.001	<0.001	<0.001	<0.001
Crop (year)	<0.001	<0.001	<0.001	<0.001	>0.05	<0.001	<0.001	<0.001	<0.001	<0.001	<0.001
System × Crop (year)	<0.001	<0.001	>0.05	<0.001	>0.05	>0.05	<0.001	>0.05	<0.001	<0.001	>0.05

Table 5. Means of soil chemical parameters for the three organic systems during the two-year rotation period.

	ASC '14			Kohlrabi			Lettuce			ASC '15			Zucchini			Lamb's Lettuce			Mean for Rotation (SE)		
	SB[1]	AC	AM	SB	AC	AM	SB	AC	AM	SB	AC	AM	SB	AC	AM	SB	AC	AM	SB	AC	AM
NO₃⁻-N ppm	78.9	81.7	10.3	77.6	58.7	19.3	65.9	179.	85.6	21.8	40.1	22.9	8.07	18.5	14.3	5.8	19.6	46.8	43.0 (8.94)	66.2 (13.6)	33.2 (7.29)
NH₄⁺-N ppm	15.2	21.6	16.2	23.4	27.9	18.4	22.1	22.8	19.8	29.1	30.1	30.9	15.2	21.6	16.2	23.4	27.9	18.4	22.8 (1.42)	23.4 (1.27)	19.8 (1.25)
Soil P ppm	36.5	44.4	33.1	31.4	40.9	28.3	40.8	39.0	35.3	25.1	18.6	21.1	36.5	44.4	33.1	31.4	40.9	28.3	34.6 (2.23)	36.7 (2.51)	33.5 (1.92)
Soil TOC %	1.37	1.54	1.50	1.50	1.78	1.65	1.67	2.01	1.89	1.53	2.16	2.10	1.37	1.54	1.50	1.50	1.78	1.65	1.42 (0.05)	1.69 (0.08)	1.63 (0.08)
Soil TN %	0.12	0.14	0.13	0.12	0.14	0.13	0.12	0.15	0.14	0.11	0.19	0.18	0.12	0.14	0.13	0.12	0.14	0.13	0.12 (0.01)	0.15 (0.01)	0.14 (0.01)

[1] SB–Substitution, AC–Agrocom; AM–Agroman.

Table 6. Means of plant parameters for the three organic systems during the two-year rotation period.

	ASC '14			Kohlrabi			Lettuce			ASC '15			Zucchini			Lamb's Lettuce			Mean for Rotation (SE)		
	SB[3]	AC	AM	SB	AC	AM	SB	AC	AM	SB	AC	AM	SB	AC	AM	SB	AC	AM	SB	AC	AM
Tot Biom [1]	-	22.6	26.9	2.30	1.91	1.31	1.59	1.15	1.40	-	4.16	4.73	3.87	2.43	2.74	0.50	0.75	0.93	1.38 (0.34)	5.50 (1.98)	6.34 (2.29)
Tot Yield [1]	-	-	-	1.75	1.51	1.03	1.59	1.15	1.40	-	-	-	0.79	0.49	0.54	0.50	0.75	0.93	0.77 (0.17)	0.65 (0.14)	0.65 (0.13)
Res inc [1]	-	22.6	26.9	-	0.39	0.29	-	-	-	-	4.16	4.73	-	1.94	2.20	-	-	-	- (-)	4.85 (2.06)	5.69 (2.38)
Tot Nu [2]	-	502.	588.	75.9	56.4	41.2	65.5	47.3	48.1	-	75.3	123.	189.	112.	117.	18.7	25.9	40.5	58.3 (16.1)	137 (42.7)	159 (48.3)
Yield Nu [2]	-	-	-	53.5	41.8	30.2	65.5	47.3	48.1	-	-	-	51.3	30.1	32.2	18.7	26.0	40.5	31.5 (6.65)	24.2 (4.63)	25.2 (4.84)
Inc Res C [1]	-	0.89	0.10	-	0.16	0.11	-	-	-	-	0.18	0.26	-	0.76	0.89	-	-	-	- (-)	0.19 (0.08)	0.23 (0.09)

[1] t ha⁻¹; [2] kg ha⁻¹; [3] SB—Substitution, AC—Agrocom; AM—Agroman.

Table 7. Results of the mixed model analysis on arthropod taxa (P levels). See material and methods for explanations.

Effect	Carabidae	Araneae	Opiliones	Staphylinidae	Isopoda	Myriapoda
System	>0.05	<0.01	<0.01	<0.01	>0.05	<0.01
Crop (year)	<0.01	<0.01	<0.01	<0.01	<0.01	<0.01
System × crop (year)	>0.05 (0.0711)	<0.01	<0.01	<0.01	>0.05 (0.0587)	<0.05

Table 8. Activity density of soil arthropods macrogroups for the three organic systems during the two-year rotation period.

Macrogroup	ASC '14			Kohlrabi			Lettuce			ASC '15			Zucchini			Lamb's Lettuce			Mean for Rotation (SE)		
	SB [1]	AC	AM	SB	AC	AM	SB	AC	AM	SB	AC	AM	SB	AC	AM	SB	AC	AM	SB	AC	AM
Carabidae	0.1	0.8	0.1	0.8	1.3	0.2	0.2	0.4	0.3	0.1	0.3	0.4	0.4	0.9	0.6	0.1	0.1	0.2	0.3 (0.1)	0.6 (0.2)	0.3 (0.1)
Araneae	1.6	1.1	2.7	1.3	1.2	1.0	2.6	2.3	2.2	5.7	8.4	11.5	3.0	3.1	1.9	0.7	0.9	0.7	2.5 (0.7)	2.8 (1.2)	3.3 (1.7)
Opiliones	0.0	0.2	0.0	0.1	0.7	0.2	1.1	2.9	4.0	0.1	0.3	0.1	0.2	1.0	1.3	0.8	1.3	1.1	0.4 (0.2)	1.1 (0.4)	1.1 (0.6)
Isopoda	1.0	3.8	4.6	2.5	1.1	0.7	1.3	1.8	1.7	2.6	4.8	10.9	3.7	2.7	2.3	3.2	2.9	2.8	2.4 (0.4)	2.9 (0.5)	3.8 (1.5)
Myriapoda	0.1	0.5	0.3	0.5	0.8	0.1	0.2	0.5	0.3	0.8	4.0	3.1	0.3	1.1	1.1	1.4	1.7	1.5	0.6 (0.2)	1.4 (0.5)	1.1 (0.5)
Staphylinidae	0.1	2.1	2.9	0.7	3.2	1.9	1.2	6.9	2.1	0.9	4.0	4.3	0.9	1.4	1.2	0.2	0.2	0.2	0.7 (0.2)	3.0 (1.0)	2.1 (0.6)

[1] SB–Substitution, AC–Agrocom; AM–Agroman.

Figures 1 and 2 refer to the system discrimination due to the agronomical parameters variability. In particular, in Figure 1, SB differentiates with respect to AC and AM for yield (higher in SB), for Yield N uptake (higher nutritional efficiency of SB respect to AM and AC) and partially for Soil P. On the other hand, AM and AC differentiate from SB mainly for their effects on Soil TOC and Soil TN.

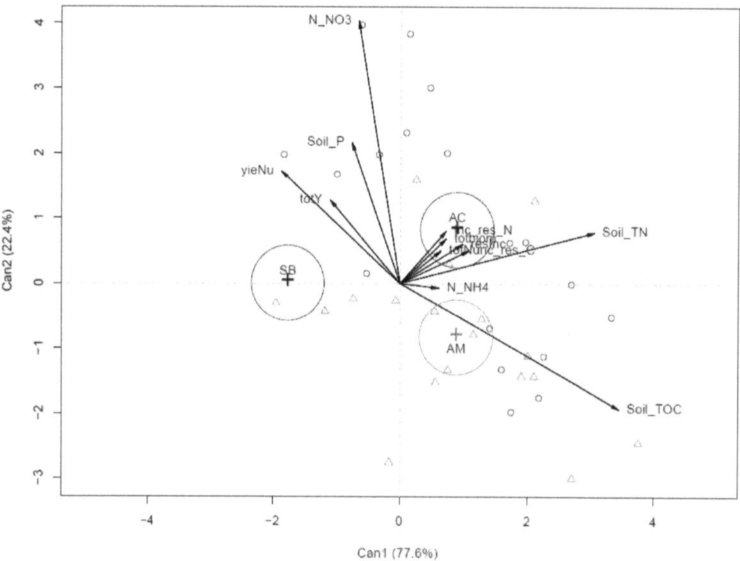

Figure 1. Canonical discriminant analysis (CDA) output for agronomic parameters in the three compared systems (SB, AC and AM) pooling the two-year rotation.

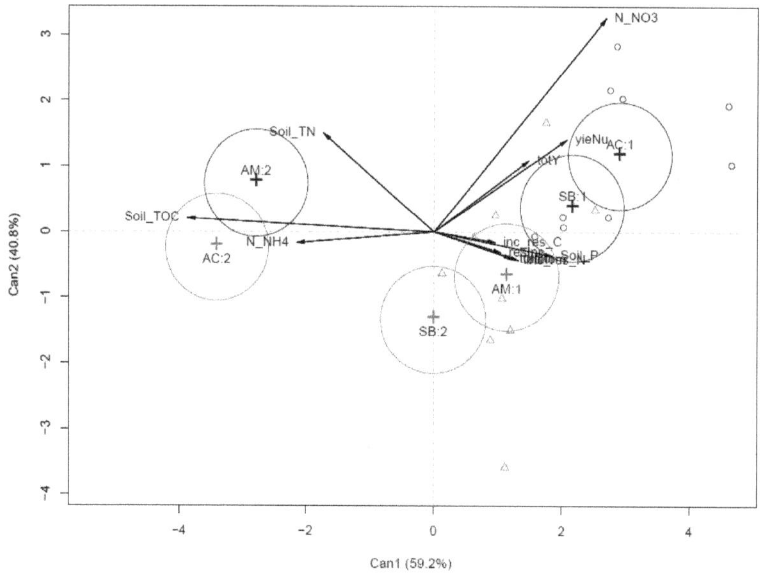

Figure 2. CDA output for agronomic parameters in the three compared systems (SB, AC and AM) over the two experimental years.

Figure 2 refers to the same dataset utilized in Figure 1, with the difference that the effect of single systems of production (SB, AM and AC) on soil and plant parameters are analyzed during the two years of the experiment (SB1 and SB2; AM1 and AM2; AC1 and AC2), giving to the production systems compared a dynamic effect on plant and soil parameters. In particular, it is worth noting how, during the first year of rotation, the three systems compared (SB1, AM1 and AC1) did not differentiate clearly. In the second year, the differences among SB2, AM2 and AC2 seem to strongly discriminate substitution from agroecological systems in terms of soil parameter vectors (Soil TOC, NH_4^+-N and Soil TN).

Figure 3 graphically represents the distribution of the systems based on soil arthropod macrogroup abundances, pooling the two-year rotation together. In this case, AM assumed an intermediate position between SB and AC, along the first axis. So, the discrimination among the systems did not follow the same trend reported for soil chemical parameters. In any case, both AC and AM seem to be associated with vectors representing most of the groups of arthropods (higher abundance). Conversely, SB was correlated only with Collembola.

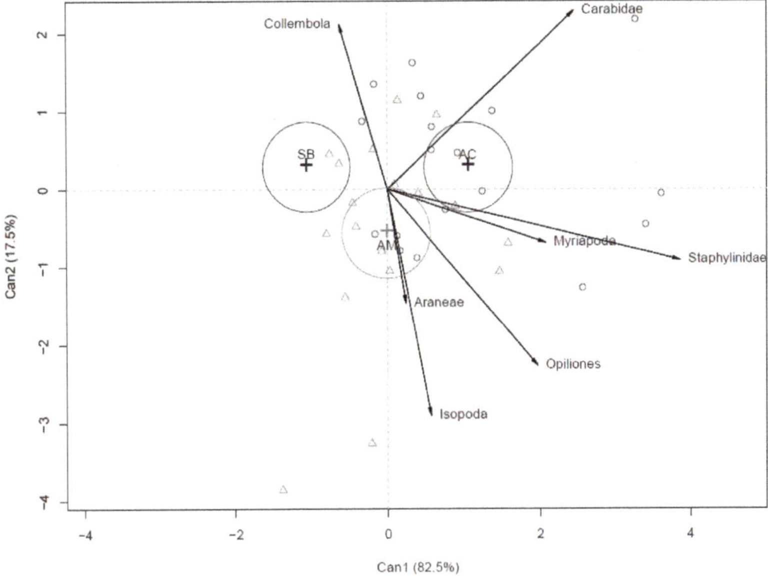

Figure 3. CDA output for arthropods abundance in the three compared systems (SB, AC and AM) pooling the two-year rotation.

In Figure 4, the dynamic effect of the production systems (SB, AM and AC) on soil arthropod abundance is represented. CDA corroborates mixed model analysis, and in particular the interaction between system and crop rotation (year). The almost complete overlap of SB1 and SB2 confirms that the influence of the input substitution system of production on soil arthropods did not change during the two-year rotation. Conversely, for the two agroecological treatments, crops cultivated in the two years of rotation affected the trends of soil arthropods. In particular, during the first year, Staphilinidae and Opiliones showed strong correlation with AC and AM, respectively. During the second year, an overlap of the effects of AM2 and AC2 was observed, with a strong correlation of the agroecological systems with Araneae, Collembola and Myriapoda. During the second year, Isopoda showed slight correlation with AC and AM, confirmed by the almost significant interaction between systems and crop in rotation (Year) of the mixed model (Table 7). Both agroecological systems in the second year (AC2 and AM2), were associated with groups (Araneae, Myriapoda, Collembola) which were different in comparison to those of the first year (AC1 and AM1) of the experiment (Opiliones, Staphylinidae).

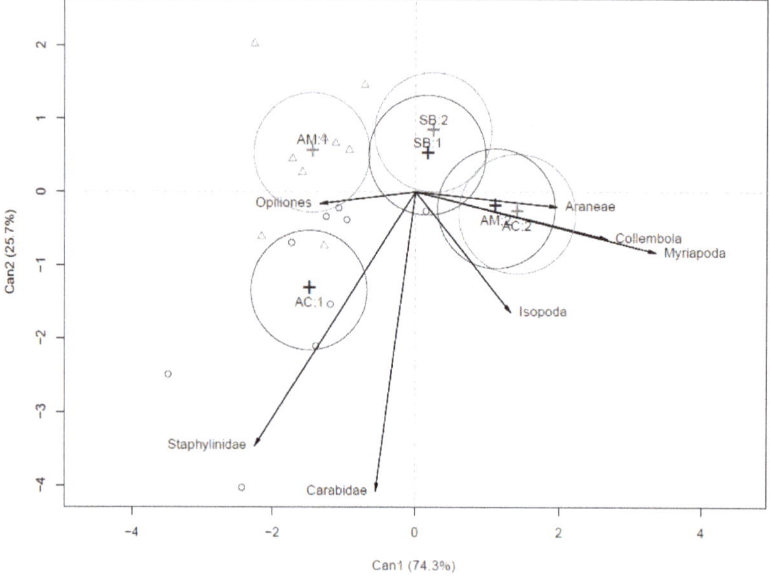

Figure 4. CDA output for arthropods abundance in the three compared systems (SB, AC and AM) over the two experimental years

4. Discussion

Yield comparisons between conventional and organic farming systems have been reported by several authors and show that marketable yields are usually higher in conventional systems compared to organic systems by 10% to 25% [20,29,30]. In our research, as expected, similar results were observed for SB and agroecological treatments. Indeed, organic fertilisers used in SB seemed to play exactly the role of the synthetic input in a conventional system, guaranteeing a better productive effect and more efficient nitrogen uptake of the cash crop compared to AM and AC. So, the hypothesis of similar results was not verified. As observed by Tittarelli et al. [10], agroecological systems have a lower efficiency in N utilization because the amount of N associated to organic biomass incorporated to soil (e.g., crop residues, green manure and organic amendments) is higher compared to the amount applied with organic fertilizer. These agronomic practices have the wide objective of guaranteeing not only nutrients availability to the crops but also to increase soil organic matter content [31]. The analysis of the results in terms of soil fertility are further supported by the evidence shown in Figure 2, where the increasing amount of organic matter incorporated into the soil in the AC and AM systems is graphically represented in CDA by the shifting of agroecological systems in the second year of the trial under the Soil TOC and Soil TN drivers. On the other side, SB slightly changed during the biennial rotation with respect to the soil parameters analyzed, and it was mainly driven by yield and available mineral nitrogen. This result is in accordance with those of Garcia-Franco et al. [32]. Agroecological practices, which are mainly characterized by the introduction in the system of higher amounts of organic carbon, differentiated with respect to the input substitution system during the second year, when the amount of organic amendments incorporated into the soil became significantly higher than in SB, where just organic fertilizers were broadcasted.

The spatio-temporal distribution of soil arthropods in the systems of production we compared was more complex and seemed to follow a multi-effect response. Besides a general positive influence of agroecological systems on arthropod activity density, responses of macrogroups do not allow us to identify precise drivers of their patterns. Indeed, the agroecological systems (AM and AC) positively affected the activity density of most of the arthropod taxa, but these effects were strongly influenced by the year of cultivation. The positive effect of the agroecological systems, and in particular AC, was evident for Staphylinidaee during the first year of rotation. Some positive effects of the agroecological systems were found also for Opiliones, Araneae and Myriapoda, but only in the second year of rotation (Table 8). On the contrary, the SB system did not affect activity density of soil arthropod taxa in the two years of the cultivation of cash crops, and this lack of effect was constant during the whole rotation. The partial responses of taxa only for few crops in rotation seems to be the reason for the interactions close to significant values (Table 7), whereas the CDAs carried out for evaluating the temporal performance of systems on agronomic parameters and arthropod activity density confirmed the interactions observed with the mixed model and allowed a more in-depth evaluation of the results obtained.

The macrogroup of Collembola demonstrated, in our study, how complex soil arthropods reactions to production systems are. Agricultural practices implemented can affect nutrients supply and can be a source of physical disturbance, both considered to be among the key factors determining biological diversity in aboveground systems [33,34]. Coulibaly et al. [35] found that during a 4-year period, tillage intensity (i.e., physical disturbance) had a significant effect on diversity and abundance of Collembola, while residue incorporation (resource supply) had no significant impact. Our results only partially confirmed these findings, since in the first year of our experiment, additional disturbance of soil due to the sowing of ASC in AM and AC resulted in much lower activity density of the Collembola macrogroup in these systems than in SB. While the beneficial effects of agroecological practices (e.g., resource supply) appeared to be significant for the following crop (kohlrabi) and with a residual effect for the subsequent one (lettuce). During the second year, the effects of rotation (crop) could be observed, since the AC and AM systems had similar or higher values of Collembola activity density than in SB. In other words, the direction of the Collembola vector changed toward the AC and AM systems in the second year of the study. These results should be attributed to systems stabilization, since after the ASC cultivation in 2014, Collembola activity density was not significantly different between the systems for the whole rotation.

According to the scientific literature, soil fauna plays a key role in soil functioning and in the maintenance of soil quality [36] and can reflect anthropogenic disturbances to soil ecosystems [37,38]. Our results demonstrated how the community structure of soil arthropods is strictly linked to changes in soil fertility parameters. Soil arthropods are recognized as reliable bioindicators, due to their important role in terrestrial ecosystem services delivery, especially because of their capability to affect organic matter decomposition processes [39] and to contribute to biological pest control [40]. When soil arthropods are divided according to ecosystem service delivery groups (functional subgroups), such as biological pest control subgroups (Carabidae, Aranea, Opiliones, Staphylinidae) and nutrient cycling subgroups (Isopoda, Myriapoda, Collembola), the results obtained indicate significantly higher contribution of AC and AM systems to biological pest control [28]. In the case of the nutrient cycling subgroup, the systems did not differ significantly; thus, the hypothesis related to diversified services among systems is only partially confirmed.

It is worth noting that the introduction of agroecological practices, such as the cultivation of ASC as green manure and the use of soil amendments instead of commercial organic fertilizers, determined a quick change in some soil characteristics, which were able to discriminate systems of production.

5. Conclusions

The results obtained during the two-year rotation generated evidence for how the input substitution system (SB) is different to agroecological systems, both in terms of agronomic parameters and soil arthropods abundance. Thus, SB showed effects on short term soil fertility but, as expected, did not have any effect on parameters related to the long run. These slight differences in agronomic parameters were in line with the results obtained for soil arthropods, since significant changes in abundance and dynamics were not observed. On the other hand, both agroecological systems showed a significant change in long-term soil fertility parameters, and also an evolution of soil arthropod groups. Actually, in accordance with previous findings, patterns analyzed can be considered to be bioindicators for a comprehensive assessment of management systems. In protected condition, where the systems are often very intensive, the cultivation of ASC which are not aimed to produce yield represents an agronomic mean which is not easily acceptable by growers. Recent publication of European regulation (EU 2018/848) on organic production reports the need of short-term green manure for certified organic production in protected conditions. It is a challenge which would only be accepted by growers if its environmental impact is demonstrated.

Our results demonstrated that the label "organic" does not represent the same impact on soil fertility and biodiversity per se. Since, in protected conditions the input substitution approach is the more widely diffused system of production, it could be worth investigating its potential impact also on other aspects of soil health and suppressiveness (e.g., nematode biodiversity, soil microbial activity, etc.) in comparison with organic systems characterized by the implementation of agroecological practices.

Author Contributions: Conceptualization, F.T., C.C.; methodology, F.T., C.C., E.T., F.G.C., and G.B.; formal analysis, C.C., E.T. and S.M.; investigation S.M., E.T., and G.B.; data curation, E.T., F.G.C. and S.M.; writing—original draft preparation, C.C. and F.T.; writing—review and editing, E.T., F.G.C., S.M., G.M., and G.B.; visualization, C.C. and G.B.; supervision, F.T.; project administration, F.T.; funding acquisition, F.T.

Funding: This study has been carried out in the framework of the activities of BIOSEMED project funded by the Italian Ministry of Agriculture, and its findings relative to biodiversity issues have been discussed and shared with Dr. Roberta Farina in the DIVERFARMING H2020 project.

Acknowledgments: We would like to thank Stefano Trotta for his help in plants and soil analysis.

Conflicts of Interest: The authors declare no conflict of interest. The funders had no role in the design of the study; in the collection, analyses, or interpretation of data; in the writing of the manuscript; or in the decision to publish the results.

References

1. Darnhofer, I.; Lindenthal, T.; Bartel-Kratochvil, R.; Zollitsch, W. Conventionalisation of organic farming practices: From structural criteria towards an assessment based on organic principles. A review. *Agron. Sustain. Dev.* **2010**, *30*, 67–81. [CrossRef]

2. Goldberger, J.R. Conventionalization, civic engagement, and the sustainability of organic agriculture. *J. Rural Stud.* **2011**, *27*, 288–296. [CrossRef]

3. European Commission. Commission Regulation (EC) No. 889/2008 laying down detailed rules for the implementation of Council Regulation (EC) No. 834/2007 on organic production and labelling of organic products with regard to organic production, labelling and control. *Off. J.* **2008**, *L250*, 1–84.

4. Frison, E.A. IPES-Food. In *From Uniformity to Diversity: A Paradigm Shift from Industrial Agriculture to Diversified Agroecological Systems*; IPES: Louvain-la-Neuve, Belgium, 2016.

5. Moonen, A.C.; Barberi, P. Functional biodiversity: An agroecosystem approach. *Agric. Ecosyst. Environ.* **2008**, *127*, 7–21. [CrossRef]

6. Wood, S.A.; Karp, D.S.; DeClerck, F.; Kremen, C.; Naeem, S.; Palm, C.A. Functional traits in agriculture: Agrobiodiversity and ecosystem services. *Trends Ecol. Evol.* **2015**, *30*, 531–539. [CrossRef] [PubMed]

7. Campanelli, G.; Canali, S. Crop Production and environmental effects in conventional and organic vegetable farming systems: The case of a long-term experiment in Mediterranean conditions (Central Italy). *J. Sustain. Agric.* **2012**, *36*, 599–619. [CrossRef]

8. Rahmann, G.; Reza Ardakani, M.; Bàrberi, P.; Boehm, H.; Canali, S.; Chander, M.; David, W.; Dengel, L.; Erisman, J.W.; Galvis-Martinez, A.C.; et al. Organic Agriculture 3.0 is innovation with research. *Org. Agric.* **2017**, *7*, 169–197. [CrossRef]

9. Raviv, M. Sustainability of Organic Horticulture. In *Horticultural Reviews, 36*; Janick, J., Ed.; Wiley-Blackwell: Hoboken, NJ, USA, 2010; pp. 289–333.

10. Tittarelli, F.; Ceglie, F.G.; Ciaccia, C.; Mimiola, G.; Amodio, M.L.; Colelli, G. Organic strawberry in Mediterranean greenhouse: Effect of different production systems on soil fertility and fruit quality. *Renew. Agric. Food Syst.* **2017**, *32*, 485–497. [CrossRef]

11. Pearson, D.; Henryks, J.; Jones, H. Organic food: What we know (and do not know) about consumers. *Renew. Agric. Food Syst.* **2010**, *26*, 171–177. [CrossRef]

12. Yiridoe, E.K.; Bonti-Ankomah, S.; Martin, R.C. Comparison of consumer perceptions and preference toward organic versus conventionally produced foods: A review and update of the literature. *Renew. Agric. Food Syst.* **2005**, *20*, 193–205. [CrossRef]

13. EGTOP (Expert Group for Technical Advice on Organic Production). Final Report on Greenhouse Production (Protected Cropping). 2013. Available online: https://ec.europa.eu/info/sites/info/files/food-farming-fisheries/farming/documents/final-report-etop-greenhouse-production.pdf (accessed on July 10 2019).

14. Niggli, U. Incorporating agroecology into organic research-an ongoing challenge. *Sustain. Agric. Res.* **2015**, *4*, 149–157. [CrossRef]

15. Magagnoli, S.; Depalo, L.; Masetti, A.; Campanelli, G.; Canali, S.; Leteo, F.; Burgio, G. Influence of agroecological service crop termination and synthetic biodegradable film covering on Aphis gossypii Glover (Rhynchota: Aphididae) infestation and natural enemy dynamics. *Renew. Agric. Food Syst.* **2018**, *33*, 386–392. [CrossRef]

16. Marliac, G.; Mazzia, C.; Pasquet, A.; Cornic, J.F.; Hedde, M.; Capowiez, Y. Management diversity within organic production influences epigeal spider communities in apple orchards. *Agric. Ecosyst. Environ.* **2016**, *216*, 73–81. [CrossRef]

17. Hendrickx, F.; Maelfait, J.P.; Van Wingerden, W.; Schweiger, O.; Speelmans, M.; Aviron, S.; Augenstein, I.; Billeter, R.; Bailey, D.; Bukacek, R.; et al. How landscape structure, land-use intensity and habitat diversity affect components of total arthropod diversity in agricultural landscapes. *J. Appl. Ecol.* **2007**, *44*, 340–351. [CrossRef]

18. Drinkwater, L.E. Cropping systems rsearch: Reconsidering agricultural experimental approaches. *HortTechnology* **2002**, *12*, 355–361. [CrossRef]

19. Canali, S.; Campanelli, G.; Ciaccia, C.; Leteo, F.; Testani, E.; Montemurro, F. Conservation tillage strategy based on the roller crimper technology for weed control in Mediterranean vegetable organic cropping systems. *Eur. J. Agron.* **2013**, *50*, 11–18. [CrossRef]

20. Canali, S.; Diacono, M.; Campanelli, G.; Montemurro, F. Organic no-till with roller crimpers: Agro-ecosystem services and applications in organic Mediterranean vegetable productions. *Sustain. Agric. Res.* **2015**, *4*, 70. [CrossRef]

21. Snapp, S.S.; Swinton, S.M.; Labarta, R.; Mutch, D.; Black, J.R.; Leep, R.; Nyiraneza, J.; O'neil, K. Evaluating cover crops for benefits, costs and performance within cropping system niches. *Agron. J.* **2005**, *97*, 322–332.

22. Nett, L.; Feller, C.; George, E.; Fink, M. Effect of winter catch crops on nitrogen surplus in intensive vegetable crop rotations. *Nutr. Cycl. Agroecosyst.* **2011**, *91*, 327–337. [CrossRef]

23. Ceglie, F.G.; Amodio, M.L.; de Chiara, M.L.V.; Madzaric, S.; Mimiola, G.; Testani, E.; Tittarelli, F.; Colelli, G. Effect of organic agronomic techniques and packaging on the quality of lamb's lettuce. *J. Sci. Food Agric.* **2018**, *98*, 4606–4615. [CrossRef] [PubMed]

24. Wang, D.; Anderson, D.W. Direct measurement of organic carbon content in soils by the Leco CR-12 carbon analyzer. *Commun. Soil Sci Plan.* **1988**, *29*, 15–21. [CrossRef]

25. Henriksen, A.; Selmer-Olsen, A.R. Automatic methods for determining nitrate and nitrite in water and soil extracts. *Analyst* **1970**, *95*, 514–518. [CrossRef]

26. Krom, M.D. Spectrophotometric determination of ammonia: A study of a modified Berthelot reaction using salicylate and dichloroisocyanurate. *Analyst* **1980**, *105*, 305–316. [CrossRef]

27. Olsen, S.R.; Sommers, L.E. Phosphorus. In *Methods of Soil Analysis*; Part 2. AL Page; X e Y Madison: Madison, WI, USA, 1982; pp. 403–430.

28. Madzaric, S.; Ceglie, F.G.; Depalo, L.; Al Bitar, L.; Mimiola, G.; Tittarelli, F.; Burgio, G. Organic versus organic—Soil arthropods as bioindicators of ecological sustainability in greenhouse system experiment under Mediterranean conditions. *Bull. Entomol. Res.* **2018**, *108*, 625–635. [CrossRef] [PubMed]

29. Reganold, J.P.; Wachter, J.M. Organic agriculture in the twenty-first century. *Nat. Plants* **2016**, *2*, 15221. [CrossRef]

30. De Ponti, T.; Rijk, B.; van Ittersum, M. The crop yield gap between organic and conventional agriculture. *Agric. Syst.* **2012**, *108*, 1–9. [CrossRef]

31. Mazzoncini, M.; Sapkota, T.B.; Bàrberi, P.; Antichi, D.; Risaliti, R. Long-term effect of tillage, nitrogen fertilisation and ASCs on soil organic carbon and total nitrogen content. *Soil Tillage Res.* **2011**, *114*, 165–174. [CrossRef]

32. Garcia-Franco, N.; Albaladejo, J.; Almagro, M.; Martínez-Mena, M. Beneficial effects of reduced tillage and green manure on soil aggregation and stabilization of organic carbon in a Mediterranean agroecosystem. *Soil Tillage Res.* **2015**, *153*, 66–75. [CrossRef]

33. Cole, L.; Buckland, S.M.; Bardgett, R.D. Influence of disturbance and nitrogen addition on plant and soil animal diversity in grassland. *Soil Biol. Biochem.* **2008**, *40*, 505–514. [CrossRef]

34. Scheunemann, N.; Digel, C.; Scheu, S.; Butenschoen, O. Roots rather than shoot residues drive soil arthropod communities of arable fields. *Oecologia* **2015**, *179*, 1135–1145. [CrossRef]

35. Coulibaly, F.M.; Coudrain, V.; Hedde, M.; Brunet, N.; Mary, B.; Recous, S.; Chauvat, M. Effect of different crop management practices on soil Collembola assemblages: A 4-year follow-up. *Appl. Soil Ecol.* **2017**, *119*, 354–366. [CrossRef]

36. De Vries, F.T.; Thébault, E.; Liiri, M.; Birkhofer, K.; Tsiafouli, M.A.; BjØrnlund, L.; JØrgensen, H.B.; Brady, M.V.; Christensen, S.; de Ruiter, P.C.; et al. Soil food web properties explain ecosystem services across European land use systems. *Proc. Natl. Acad. Sci. USA* **2013**, *110*, 14296–14301. [CrossRef] [PubMed]

37. Paoletti, M.G.; Favretto, M.R.; Stinner, B.R.; Purrington, F.F.; Bater, J.E. Invertebrates as bioindicators of soil use. *Agric. Ecosyst. Environ.* **1991**, *34*, 341–362. [CrossRef]

38. Van Straalen, N.M. Evaluation of bioindicator systems derived from soil arthropod communities. *Appl. Soil Ecol.* **1998**, *9*, 429–437. [CrossRef]

39. Sackett, T.E.; Classen, A.T.; Sanders, N.J. Linking soil food web structure to above- and belowground ecosystem processes: A meta-analysis. *Oikos* **2010**, *119*, 1984–1992. [CrossRef]

40. Coleman, D.C.; Wall, D.H. Soil Fauna: Occurrence, Biodiversity, and Roles in Ecosystem Function. In *Soil Microbiology, Ecology and Biochemistry*; Academic Press: London, UK, 2015; pp. 111–149.

 *agronomy*

Article

Modeling Water and Nitrogen Balance of Different Cropping Systems in the North China Plain

Shah Jahan Leghari [1], **Kelin Hu** [1,*], **Hao Liang** [1] **and Yichang Wei** [2]

[1] College of Land Science and Technology, China Agricultural University, Beijing 100193, China; leghari222@gmail.com (S.J.L.); haoliang@hhu.edu.cn (H.L.)

[2] College of Surveying and Geo-informatics, North China University of Water Resources and Electric Power, Zhengzhou 450046, China; weiyichang@ncwu.edu.cn

* Correspondence: hukel@cau.edu.cn; Tel.: +86-10-62732412; Fax: +86-10-62733596

Received: 2 October 2019; Accepted: 23 October 2019; Published: 30 October 2019

Abstract: The North China Plain (NCP) is experiencing serious groundwater level decline and groundwater nitrate contamination due to excessive water pumping and application of nitrogen (N) fertilizer. In this study, grain yield, water and N use efficiencies under different cropping systems including two harvests in 1 year (winter wheat–summer maize) based on farmer $(2H1Y)^{FP}$ and optimized practices $(2H1Y)^{OPT}$, three harvests in 2 years (winter wheat–summer maize–spring maize, 3H2Y), and one harvest in 1 year (spring maize, 1H1Y) were evaluated using the water-heat-carbon-nitrogen simulator (WHCNS) model. The $2H1Y^{FP}$ system was maintained with 100% irrigation and fertilizer, while crop water requirement and N demand for other cropping systems were optimized and managed by soil testing. In addition, a scenario analysis was also performed under the interaction of linearly increasing and decreasing N rates, and irrigation levels. Results showed that the model performed well with simulated soil water content, soil N concentration, leaf area index, dry matter, and grain yield. Statistically acceptable ranges of root mean square error, Nash–Sutcliffe model efficiency, index of agreement values close to 1, and strong correlation coefficients existed between simulated and observed values. We concluded that replacing the prevalent $2H1Y^{FP}$ with 1H1Y would be ecofriendly at the cost of some grain yield decline. This cropping system had the highest average water use (2.1 kg m^{-3}) and N use efficiencies (4.8 kg kg^{-1}) on reduced water (56.64%) and N (81.36%) inputs than $2H1Y^{FP}$. Whereas 3H2Y showed insignificant results in terms of grain yield, and $2H1Y^{FP}$ was unsustainable. The $2H1Y^{FP}$ system consumed a total of 745 mm irrigation and 1100 kg N ha^{-1} in two years. When farming practices were optimized for two harvests in 1 year system $(2H1Y)^{OPT}$, then grain yield improved and water (18.12%) plus N (61.82%) consumptions were minimized. There was an ample amount of N saved, but water conservation was still unsatisfactory. However, considering the results of scenario analyses, it is recommended that winter wheat would be cultivated at <200 mm irrigation by reducing one irrigation event.

Keywords: cropping systems; water; nitrogen; WHCNS; scenario analyses

---

1. Introduction

The North China Plain (NCP) is located in the eastern coastal region of China (34°46–40°25′ N and 112°30′–119°30′ E). The plain covers an area of about 409,500 km^2, most of which is <50 m above sea level and extends to Henan, Hebei, and Shandong provinces [1–3]. The region has a subtropical monsoon climate with highly variable rainfall and a typical temperature [4]. Cereal crops such as wheat, maize, sorghum, rice, barley, millet, and oats are grown here. While winter wheat and summer maize are staple crops, widely cultivated in rotations using the double cropping system [5], which accounts for >60% of the national wheat and >30% of the maize grain demand [6].

Since the 1970s, the annual use of the two-harvest cropping system greatly increased the total grain output. However, it has also caused severe groundwater decline [7,8], because both crops generally have high water requirements to complete their life cycle [9,10]. Crop water requirement does not meet the balance between groundwater recharge through rain and evapotranspiration (ET) loss from plant and soil [11]. An estimated amount of more than 450 mm yr^{-1} irrigation water is required to maintain high crop yield under the conventional double cropping system [12], which dominatingly depends on pumping of groundwater, and there is no access to irrigation from the river source. Thus, the groundwater is almost the only source of irrigation [13]. In the last 20 years, sustained water pumping has adversely affected aquifers. Consequently, the groundwater level fell rapidly, at the rate of approximately 1 m yr^{-1}. On the other hand, the double cropping system under conventional farming practices contaminated groundwater by NO_3^- leaching [14], surface water quality degradation [2], soil acidification [15,16], and air pollution [17] due to excessive and long-term use of N fertilizers under low efficiency of farmer cultivation practices. Around 500 to 600 kg N ha^{-1} yr^{-1} of fertilizer is normally applied to achieve maximum yield. Agronomically, it is an excessive application rate based on general crop N requirements, which range from 200 to 300 kg N ha^{-1} yr^{-1} [18].

A reduction in the application of N and proper irrigation scheduling can help to achieve sustainable yield goals by simultaneously protecting natural resources and atmosphere [19]. About 15–35% of water consumption can be lowered by improving water use efficiency [20]. It is possible through efficient management strategy and optimal cropping system [21]. However, there are numerous constraints associated with the optimal cropping system [22]. Since the field, dynamic observation of evapotranspiration, water drainage, N losses, and crop growing indices are time consuming and costly. Crop model-based improved simulation techniques with response to environmental conditions have been developed for making effective and proper use of input resources [23]. Simulation models define the systematic processes related to crop growth and development and are useful tools to assess the impact of soil extrinsic factors (fertilization), variability of climatic conditions, and crop management practices [24,25] by providing credible predictions [26]. Typically, crop models consider the time span in which a specific growth stage takes place and initiate biomass of crop components, e.g., roots, leaves, stems, and yield attributes, as they change time to time, and similarly, changes in the nutrients content and soil moisture status. Mostly, models use major information on crop growth factors. They are of various types, classified into three broad categories i.e., statistical, functional, and mechanistic models [27], and are widely used in agricultural research to solve and identify problems integrated in the complex farming system [28]. The crop modeling analyses showed that cereal crops are most vulnerable to increasing water scarcity in many countries of the world [29]. Recently, an integrated soil-crop model (WHCNS, soil water-heat-carbon-nitrogen simulator) was developed based on the Chinese climate, soil types, and field management and the model has been successfully applied to simulate water use, N loss, and N use efficiency in China [30–32]. Previous research showed that the model simulated soil and crop indicators agreed well with observed data [33]. However, there is still limited work on the simulation of water and nitrogen balance under diversified crop rotations using this model.

Therefore, the main aims of this study were: (1) To evaluate grain yield, water and N use efficiencies under different cropping systems using the WHCNS model, and (2) to optimize water and N management by scenario analyses.

2. Materials and Methods

2.1. Description of Study Site

Quzhou County (114.9° E, 36.7° N, 40 m above sea level) belongs to the prefecture-level city of Handan, located upstream of the Heilonggang River basin in southern Hebei province, which is one of the most affected area in terms of groundwater depletion in the world [34]. Fertilizers are highly used with inefficiency in the study area. A recent research of Cai et al. [35] and Ling et al. [36] reported an increasing trend of dependence on chemical fertilizers up to 36.3% in the current production system exceeding international standards for the sake of yield [37].

The county belongs to a continental monsoonal climate. The long-term annual mean air temperature is 13.1 °C and the mean annual precipitation is 534.9 mm. About >68% of precipitation occurs during the June to September months of the year. The soil of this site is alkaline in nature with a pH > 8.0, has a silty texture, nitrogen < 0.68 g kg^{-1}, phosphorus < 6 mg kg^{-1}, and potassium < 75 mg kg^{-1} [3]. The soil properties at different depths are presented in Table 1.

Table 1. Basic soil properties at different depths.

Depth (cm)	BD (g cm^{-3})	Particle Fraction (%)			Soil Texture (USDA)	pH	θ_s (cm^3 cm^{-3})	θ_{fc} (cm^3 cm^{-3})	θ_{wp} (cm^3 cm^{-3})	Ks (cm d^{-1})
		Sand	Silt	Clay						
0–30	1.42	33.1	46.6	20.3	loamy	8.4	0.47	0.275	0.125	17.27
30–60	1.42	24.4	53.1	22.5	Silty loam	8.7	0.39	0.295	0.132	15.88
60–90	1.33	29.6	45.5	24.9	Loamy	8.5	0.38	0.295	0.143	18.23
90–120	1.51	36.4	47.4	16.2	Loamy	8.6	0.39	0.260	0.111	5.11

Note: BD, Bulk density, θ_s, Soil saturated water content, θ_{fc}, Field capacity, θ_{wp}, Soil water content permanent wilting point and Ks, Saturated hydraulic conductivity.

2.2. Detail of Experimental Design and Management Practices

Four replicated randomized complete block design field experiments were conducted from 2004 to 2006 on different cropping systems including two harvests in 1 year (winter wheat–summer maize) based on farmer (2H1Y)^{FP} and optimized practices (2H1Y)^{OPT}, three harvests in 2 years (winter wheat–summer maize–spring maize, 3H2Y), and one harvest in 1 year (spring maize, 1H1Y). The plot size was 84 m^2 (10 m long × 8.4 m wide). Wheat variety Shijiazhuang-8 was sown (row spacing 15 cm) on 10 October and harvested on 15 June, then summer maize cultivar Zhengdan-958 was grown (row spacing 60 cm) after 2–5 days gap for two years. Nongda-1505 spring maize variety was employed for both monoculture (1H1Y) and triple cropping (3H2Y) systems. The sowing date of spring maize was on 27 April which was harvested on 8 October of the same year. Cropping systems were rotated for a two-year period by following these sowing dates. The seed rate for winter wheat was 385 kg ha^{-1}, and 24–30 kg for maize. The final plant density of maize was 6–7 plants m^2. Except for 2H1Y^{FP}, the other cropping systems were managed under optimized water and fertilizer practices based on soil testing. Irrigation was applied to keep soil moisture in the rooting zone between 50% and 85% of the available field capacity. The 2H1Y^{FP} system was fully irrigated based on local farmer practices. Winter wheat was irrigated 3–5 times and maize crop was irrigated 2 times through the flooding method. In 2H1Y^{FP} system, the N rates were 300 kg N ha^{-1} and 250 kg N ha^{-1} for wheat and maize cultivation in each season, respectively, through 2–3 split doses. The optimized cropping systems 2H1Y^{OPT}, 3H2Y, and 1H1Y received 420, 325, and 205 kg N ha^{-1}, respectively. Fertilizer application was through the broadcasting method. Precision land leveling was done for each cropping system to ensure uniform germination and growth, and 45 kg ha^{-1} of phosphorus was applied as a basal dose for vigorous seedling establishment. Other agronomical measurements such as weed, pest, and disease controlling were according to local farmer practices.

2.3. Data Collection, Methods, and Analyses

The soil water content was measured using TDR (MP917) probes (Environmental Sensors, Inc., Sidney; BC; Canada) [38] from shallow to deep soil layers of 0–30, 30–60, 60–90 and 90–120 cm, and soil samples were taken as well with help of auger for observation of soil N concentrations. Five core samples were properly mixed, sieved with 4 mm sieves, extracted with 0.01 mol L^{-1} CaCl$_2$ (1:10, soil/water), and then filtered. The ready samples were analyzed by continuous flow analyzer (TRAAC 2000, Bran and Luebbe, Norderstedt; SH; Germany). The soil water content (cm^3 cm^{-3}), and N concentrations (mg kg$^{-1)}$, dry matter (kg ha$^{-1)}$ and leaf area index (m^2 m^{-3}) were observed during key growth stages, and grain yield (kg ha$^{-1)}$ was measured after harvesting. Winter wheat was harvested by its full physiological maturity indicator (completely loss of green color from glumes), and maize crop was harvested when the grain moisture content was dropped <20%. Replicated cob samples were oven dried to keep grain moisture <15%, and then threshed. The N balance, NUE, water balance, and WUE were calculated using Equations (1)–(4) respectively. All the collected data were simulated using the WHCNS model and statistically analyzed using Statistix 8.1 software (Tallahassee; FL; USA) to determine significant differences at $p < 0.05$ probability level [36]. In addition, we established different scenarios to optimize water and N management based on the experimental results. Complete detail of scenarios is given in Section 2.4.1.

$$NB = \left(N_{fer} + N_{min}\right) - \left(N_{vol} + N_{den} + N_{lea} + N_{up}\right) \tag{1}$$

$$NUE = \frac{GY\ (kg\ h^{-1})}{Applied\ N\ (kg\ h^{-1})} \tag{2}$$

$$WB = (I + P) - (ET + D + R) \tag{3}$$

$$WUE = \frac{GY\ (kg\ h^{-1})}{ET\ (mm)} \tag{4}$$

$$WS\ and\ NF\ (\%) = \frac{Wu\ and\ Fu\ (\ system\ A\ -\ system\ B)}{Wu\ and\ Fu\ (system\ A)} \times 100 \tag{5}$$

where, N_{fer} is total N fertilizer rate (kg N ha^{-1}), N_{min} is net mineralization (kg N ha^{-1}), N_{vol} = volatilization (kg N ha^{-1}), N_{den} = denitrification (kg N ha^{-1}), N_{lea} is N leaching (kg N ha^{-1}), N_{up} is crop N uptake (kg N ha^{-1}), GY is grain yield (kg ha^{-1}), I is irrigation (mm), P is precipitation (mm), ET is evapotranspiration (mm) and D is drainage (mm), R is runoff (mm) [39,40]. WS and NF are water-saving and N fertilizer, respectively, Wu/Fu = Water use (mm) and Fertilizer use (kg) of cropping system [41].

2.4. WHCNS Model

The WHCNS model was used to simulate the effect of irrigation levels and N fertilizer rates on the growth and yield of different cropping systems. Model is efficient in the simulation of soil water movement, N transport, crop growth, development, and grain yield under different crop rotations and management practices. It can simulate complex and intensive cropping systems characteristic to NCP. It is comprised of 5 primary modules, (a) soil water, (b) heat transfer, (c) N transport, (d) soil organic carbon turnover, and (e) crop growth. The reference crop evapotranspiration (ETo) was calculated by using the Penman-Monteith equation [42] proposed by Allen et al. [43]. Water infiltration and their redistribution in the soil profile, runoff, and soil N concentrations (NH_4^+ and NO_3^-) were simulated by using the Green-Ampt model and Richards equation, SCS curve number and convection dispersion equations respectively [44–46]. Soil heat transfer simulation imported from the Hydrus-1D and soil organic matter dynamics were directly derived from the Daisy model. Crop growth and development processes, dry matter formation and grain yield simulations were performed using the improved PS123 model [47]. More detailed information is available in the paper of Liang et al. [30].

2.4.1. Scenarios Setting

Scenario setup is one of the important technique to set an unlimited number of treatments that could be not possible or difficult in field experiments due to space, time and financial limitations etc. Therefore considering successful prediction of model, best management practice, and latest research findings of Ling et al. [36], we further changed farming practices, where N rates and irrigation levels were linearly increased and decreased for winter wheat, summer and spring maize to extensively examine the effect of excessive and deficit irrigation, and N input managements on cropping systems. These scenarios were simulated using the WHCNS model (Table 2).

Table 2. Scenarios setting based on irrigation and N fertilizer management.

| | Irrigation (mm ha⁻¹) | | | | | | Fertilization (kg N ha⁻¹) | | | | | |
I	WW	Dist	SM	Dist	SPM	Dist	F	WW	Splits	SM	Splits	SPM	Splits
I_1	450	5	400	5	350	5	F_1	400	4	350	4	300	4
I_2	400	5	350	5	300	5	F_2	350	4	300	4	250	4
I_3	350	5	300	5	250	4	F_3	300	4	250	4	200	4
I_4	300	5	250	4	200	4	F_4	250	3	200	3	150	3
I_5	250	4	200	4	150	4	F_5	200	3	150	3	100	3
I_6	200	4	150	3	100	3	F_6	150	3	100	3	50	2
I_7	150	3	100	3	50	2	F_7	100	2	50	2	–	–
I_8	100	2	50	1	0	–	F_8	50	2	–	–	–	–
I_9	50	1	0	–	–	–	–	–	–	–	–	–	–
I_{10}	0	–	–	–	–	–	–	–	–	–	–	–	–

Note: WW, Winter wheat, SM, Summer maize, SPM, Spring maize, I, Irrigations, F, Fertilizers, Dist, Distribution of total irrigation amount into frequencies and Splits, Division of total N into a number of doses.

Hence, each crop received excessive and deficit N, and irrigation treatments as well as evaluated under rainfed conditions.

2.4.2. Model Input Data

Model input system requires information about crop growth stages with root, shoot and leaf weight (g), plant type C3 or C4, critical temperature ranges, sowing date, seed, and fertilizer rate, application timing and harvesting date, basic soil properties, carbon (kg h⁻¹⁾ and C/N ratio, latitude and longitude of the site, as well as daily rainfall, relative humidity, maximum, minimum and mean air temperature. The soil data for soil saturated water content (θ_s), water content at field capacity (θ_{fc}) and permanent wilting point (θ_{wp}), and saturated hydraulic conductivity (Ks) were calibrated from the soil analyses laboratory report, which is presented in Table 1 and basic crop parameters are given in Table 3. The meteorological data used in the model were acquired from a nearby automatic weather station, located approximately 5.5 km away from the research site.

Table 3. Crop parameters.

Parameters	Description	Crops		
		WW	SM	SPM
Tbase	Base temperature for crop growth (°C)	0	8	8
Tsum	Effective accumulated temperature (°C)	2100	1650	2250
Ke	Extinction coefficient (0–1)	0.6	0.6	0.6
Kc_ini	Crop coefficients in initial stage (0–1)	0.65	0.65	0.65
Kc_mid	Crop coefficients in middle stage (0–1)	1.25	1.35	1.25
Kc_end	Crop coefficients in final stage (0–1)	0.45	0.8	0.8
SLA_max	Maximum specific leaf area (m² kg⁻¹)	28	32	28
SLA_min	Minimum specific leaf area (m² kg⁻¹)	16	15	13
AMAX	Maximal assimilation rate (kg ha⁻¹ h⁻¹)	45	60	50
R_max	Maximal rooting depth (cm)	160	145	145

Note: WW, Winter wheat, SM, Summer maize, SPM, Spring maize.

2.4.3. Statistical Evaluation of the Model

The model performance was evaluated using: (i) Root mean square error (RMSE) to measure how much error there is between two data sets of predicted and observed, (ii) the Nash-Sutcliffe model efficiency coefficient (NSE) and (iii) index of agreement (*d*), which are standardized measures in the modeling, and (iv) Pearson correlation coefficient analyses was applied to quantify the magnitude of the collinearity between observed and predicted variables.

$$RMSE = \sqrt{\frac{1}{n}\sum_{i=1}^{n}(P_i - O_i)^2} \tag{6}$$

$$NSE = 1 - \frac{\sum_{i=1}^{n}(O_i - P_i)^2}{\sum_{i=1}^{n}(O_i - O)^2} \tag{7}$$

$$d = 1 - \frac{\sum_{i=1}^{n}(O_i - P_i)^2}{\sum_{i=1}^{n}(|O_i - O| + |P_i - O|)^2} \tag{8}$$

$$r = \frac{\sum_{i=1}^{n}(O_i - \overline{O}) + \left(P_i - \overline{P}_i\right)}{\sqrt{\sum_{i=1}^{n}(O_i - \overline{O})^2}\,\sqrt{\sum_{i=1}^{n}\left(P_i - \overline{P}_i\right)^2}} \tag{9}$$

where O_i = observed value, P_i = predicted value, and O is the mean of the measured values, n is the number of data values, GY = grain yield.

The lower RMSE, the good model performance [48]. The NSE ranges from $-\infty$ to 1.1 show the great efficiency, 0 indicates the predicted values are as accurate as of the mean of the observed data, and <0 result indicates that observed average value is a good predictor as compared to the model [49]. While d value varies from 0 to 1. A closer value to 1 show a perfect match of simulated results with observed data, and 0 specify no agreement at all [50]. In case of correlationship, the thumb rule for interpretation based on positive (+) or negatives (−) sizes is, 0.90–1.00 (Very high), 0.70–0.90 (High), 0.50–0.70 (Moderate), 0.30–0.50 (low), 0.0–0.30 (No) [51]. The r close to 1 illustrates model fit [52].

3. Results

3.1. WHCNS Model Calibration and Validation

The observed soil water content (cm^3 cm^{-3}), soil N concentration (mg kg^{-1}), leaf area index (m^2 m^{-3}), dry matter (kg ha^{-1}), and grain yield (kg ha^{-1}) data from 2H1Y^{FP} were used to calibrate the model, and then validated by the field experimental data obtained from other cropping systems (2H1Y^{OPT}, 3H2Y and 1H1Y). The results were statistically evaluated by RMSE and related statistical indices. The model simulation was satisfactory for all cropping systems. The observed data points fluctuated up-down close to the predicted curve which was in good agreement between model simulated and absolute values. Hence, RMSE values of soil water content at 0–120 cm soil depth for 2H1Y^{FP}, 2H1Y^{OPT}, 3H2Y and 1H1Y were 0.05, 0.68, 0.13 and 0.43, respectively (Figure 1). While, shallow to deep soil layer (0–30, 30–60, 60–90, 90–120 cm) wise RMSE values are given in Table 4. It has been found that RMSE of volumetric water contents decreased as soil depth increased. The minimum RMSE for upper soil (0–30 cm) was 0.06, and for deeper soil (90–120 cm) was 0.04. The RMSE values for soil N concentration were varied from 7.43–2.10, and for leaf area index, dry matter and grain yield ranged from 0.49–0.55, 1326.9–1608.3 and 280.91–609.96 respectively. From the yield point of view, the minimum RMSE 280.91 was showed from the 2H1Y^{FP} system. The cumulative simulated grain yield was only a 7.47% difference than observed. Predictions were accurate for each system with 3.42–1.02% variation (Figure 2). In addition, the correlation degree of strength varied from parameter-to-parameter. The sizes of correlation coefficient values were within very high, high and moderate. Figure 3 Illustrates integrated correlationships for soil water content, soil N concentration, leaf area index, dry matter, and grain yield of different cropping systems. While cropping system wise r values and other statistical indices are shown in Table 4. The simulated and observed data of crop growth and yield attributes had a very strong correlation coefficient strength (0.89–0.99). Whereas, soil N concentration had strong (0.61–0.88) and soil water content had moderate correlation (0.53–0.76) according to Mukaka [51], who briefly elaborated the correlation coefficient and provided guidance for correct use. The WHCNS model performance was valid season-to-season, because of better parameters assumption. Statistical indices well specified performance of the model. The model can be effectively used to simulate soil, crop growth, and yield indicators. There was a minor simulation error based on overall figures, which indicated a better sensitivity of the model. The slight fluctuation of RMSE values in all cropping systems could be attributed to the temporal change of soil moisture and nutrients status after irrigation and N fertilization [53–59]. In the current study, RMSE ranges were <8 at the top layer, and <3 at 60–90 cm soil depth. Statistically, there is no threshold limit. Generally, low RMSE is a sign of model efficiency [37,46].

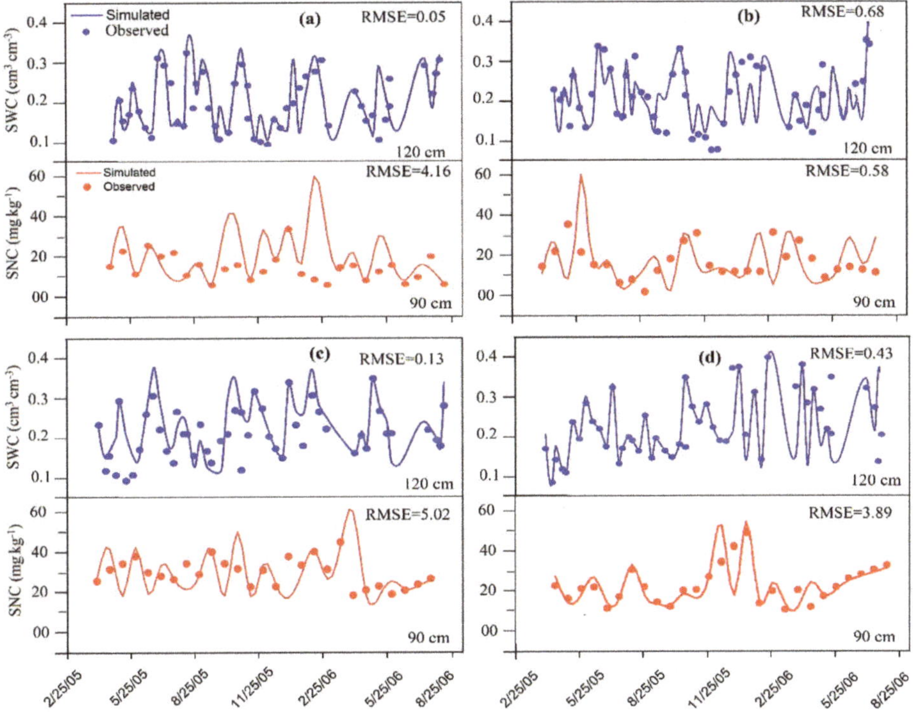

Figure 1. WHCNS model simulated (line) and observed (dots) soil water content (SWC, cm^3 cm^{-3}) and soil mineral N concentration (SNC, mg kg^{-1}) of different cropping systems. (**a**), Two harvests in 1 year based on farmer (2H1Y)^{FP}, (**b**), Optimized practices (2H1Y)^{OPT}, (**c**), Three harvests in 2 years (3H2Y), and (**d**), One harvest in 1 year (1H1Y).

Table 4. Statistical indices of WHCNS model accuracy for soil water content, soil mineral N concentrations, leaf area index, crop dry matter and yield of all cropping systems.

Cropping Systems	Soil and Crop Indicators															
	Soil Water Content ($cm^3\ cm^{-3}$)															
	0–30 cm				30–60 cm				60–90 cm				90–120 cm			
	RMSE	*NSE*	*d*	*r*	*RMSE*	*NSE*	*d*	*r*	*RMSE*	*NSE*	*d*	*r*	*RMSE*	*NSE*	*d*	*r*
(2H1Y)[FP]	0.07	0.92	0.99	0.53	0.04	0.96	0.96	0.70	0.05	0.96	0.97	0.64	0.07	0.98	0.98	0.57
(2H1Y)[OPT]	0.08	0.96	0.95	0.60	0.09	0.92	0.97	0.54	0.06	0.98	0.99	0.52	0.05	0.97	0.99	0.60
(3H2Y)	0.07	0.91	0.96	0.63	0.05	0.92	0.94	0.67	0.05	0.95	0.96	0.56	0.04	0.95	0.92	0.58
(1H1Y)	0.06	0.96	0.98	0.55	0.07	0.91	0.95	0.52	0.06	0.91	0.95	0.76	0.05	0.96	0.93	0.53

	Soil Nitrogen Concentration ($mg\ kg^{-1}$)											
	0–30 cm				30–60 cm				60–90 cm			
	RMSE	*NSE*	*d*	*r*	*RMSE*	*NSE*	*d*	*r*	*RMSE*	*NSE*	*d*	*r*
(2H1Y[FP])	5.41	0.76	0.97	0.70	3.67	0.75	0.97	0.79	2.98	0.68	0.98	0.84
(2H1Y)[OPT]	7.43	0.73	0.98	0.75	4.89	0.65	0.97	0.72	3.87	0.83	0.96	0.85
(3H2Y)	6.64	0.62	0.98	0.66	4.26	0.85	0.98	0.86	3.67	0.55	0.99	0.88
(1H1Y)	5.60	0.87	0.87	0.75	3.58	0.73	0.91	0.61	2.10	0.86	0.93	0.83

	Leaf Area Index ($m^2\ m^{-2}$)				Dry Matter ($kg\ ha^{-1}$)				Grain Yield ($kg\ ha^{-1}$)			
	RMSE	*NSE*	*d*	*r*	*RMSE*	*NSE*	*d*	*r*	*RMSE*	*NSE*	*d*	*r*
(2H1Y)[FP]	0.55	0.96	0.94	0.90	1326.9	0.92	0.93	0.96	280.91	0.99		0.98
(2H1Y)[OPT]	0.49	0.94	0.95	0.89	1534.6	0.95	0.92	0.94	609.96	0.98	0.90	0.92
(3H2Y)	0.51	0.95	0.96	0.94	1608.3	0.97	0.96	0.97	380.03	0.99		0.99
(1H1Y)	0.81	0.92	0.94	0.94	1493.8	0.97	0.95	0.99	290.31	0.99		−1.0

Note: (2H1Y)[FP], Tow harvests in 1 year based on farmer practices, (2H1Y)[OPT], Two harvests in 1 year based optimized, (3H2Y), Three harvests in 2 years, and (1H1Y), One harvest in 1 year.

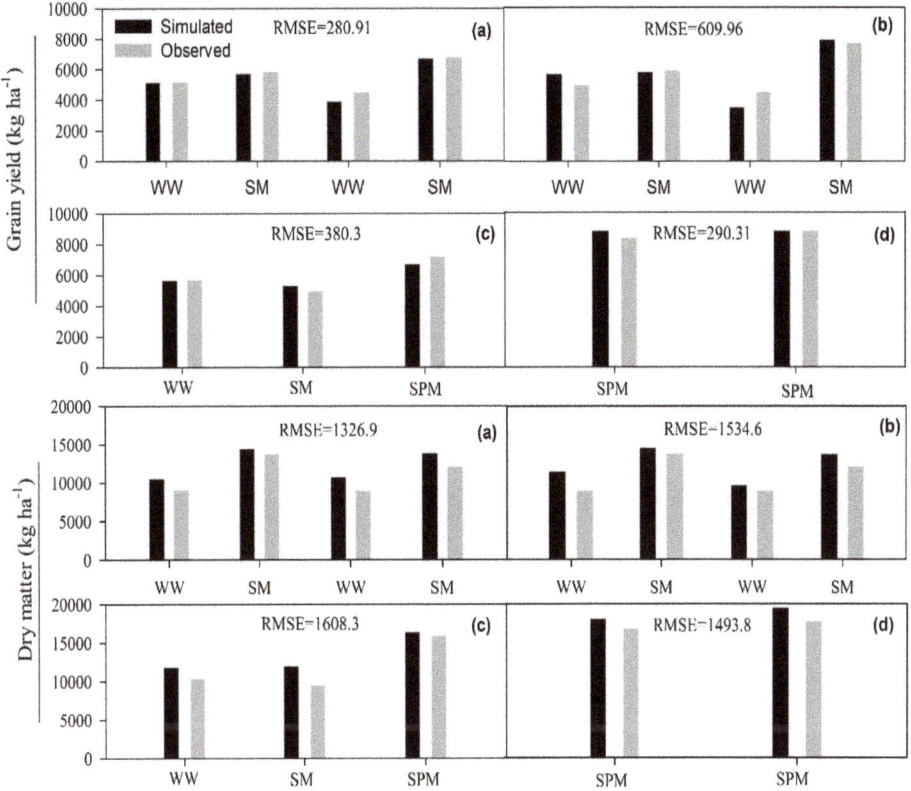

Figure 2. Simulated (black bars) and observed (gray bars) grain yield and dray matter of different cropping systems. (**a**), two harvests in 1 year based on farmer (2H1Y)^{FP} and (**b**), optimized practices (2H1Y)^{OPT}, (**c**), three harvests in 2 years (3H2Y), and (**d**), one harvest in 1 year (1H1Y). WW, Winter wheat, SM, Summer maize and SPM, Spring maize.

3.2. Grain Productivity

Results revealed that the 2H1Y^{FP} cropping system consumed excessive irrigation (745 mm) and N fertilizer (1100 kg) as compared to optimized management practices, and showed a low total yield of 21,474 kg as well (Figure 2a). Significantly ($p < 0.05$) high yield 22,758 kg obtained from 2H1Y^{OPT}, where 18.12% low irrigation (610 mm) and 61.82% deficit N rate were applied. In this system, winter wheat accounted for 5635 kg ha^{-1} and maize 5788 kg ha^{-1} in the first growing season, but in the subsequent year, winter wheat yield was 3488 kg ha^{-1} with 38.10% reduction and summer maize yield was 7847 kg ha^{-1} with 35.57% addition. The reason for the decrease in winter wheat yield was low rainfall, and the increase of maize yield in second year cycle was precipitation events. The heavy rainfall recorded during the late season, particularly in maize crop period throughout the experimental period. While the 3H2Y system showed total grain yield of 17,612 kg and monoculture 1H1Y had total grain yield of 17,471 kg. In 1H1Y system, spring maize yield was high in both two years. First year yield was 8331 kg ha^{-1} and second years was 8761 kg ha^{-1}. There was no significant difference between 3H2Y and 1H1Y in terms of total productivity (Figure 2). On an annual basis, the yield differences were very low among all cropping systems. The 2H1Y^{FP} had average annual yield range of 10737.5 kg. It was <5.64% than 2H1Y^{OPT}, and >17.98% to 18.64% than 3H2Y and 1H1Y cropping systems.

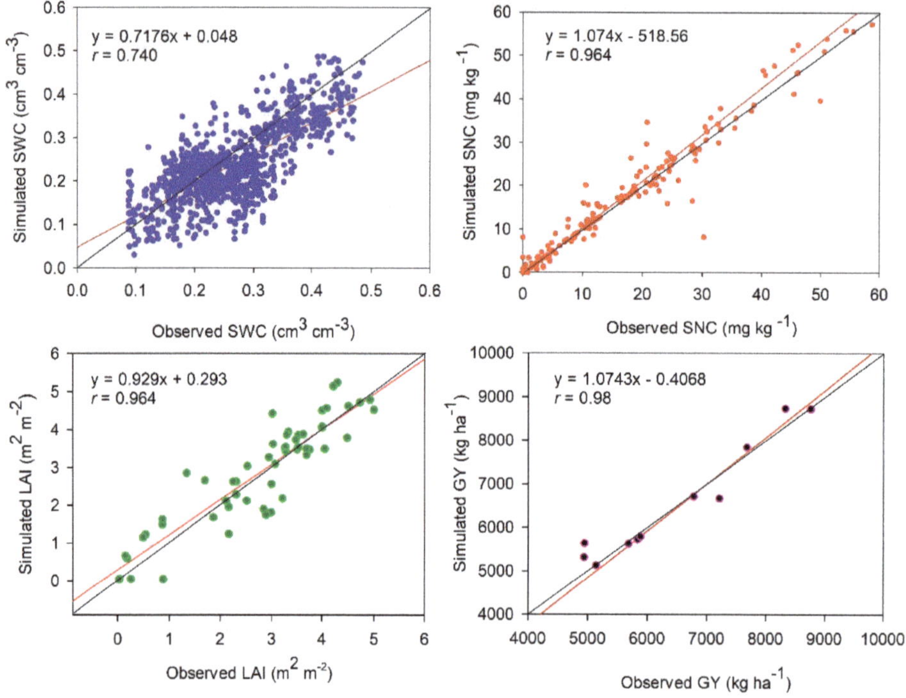

Figure 3. Relationship between simulated and observed soil water content (SWC, cm^3 cm^{-3}), soil mineral concentration (SNC, mg kg^{-1}), leaf area index (LAI, m^2 m^{-2}) and grain yield (GY, kg ha^{-1}) under different cropping systems.

3.3. Evapotranspiration, Drainage, Runoff and Water Use Efficiency

The model was applied to simulate runoff, drainage and evapotranspiration (Table 5). Results showed that drainage and water runoff were associated with rainfall intensity. The 2H1Y^{FP} system had 243 mm drainage (>2H1Y^{OPT}), and 3H2Y had 113 mm drainage (≥1H1Y). The commutative drainage was high, about 517 mm during maize growing seasons, and only 106 mm during wheat seasons of all cropping systems. The maximum runoff water loss occurred when rainfall was >300 mm. In case of total evapotranspiration water loss, the lowest evapotranspiration (812 mm) was simulated from 1H1Y, followed by 3H2Y (1013 mm). Double cropping system under farmer and optimized management practices had equal evapotranspiration rate (128 mm). Thus, there was a significantly high (351.7 mm) positive water balance calculated from spring maize cultivation in the second year. The average water balance order was 1H1Y > 3H2Y > 2H1Y^{FP} > 2H1Y^{OPT}.

Across all cropping systems, the average water use efficiency was low in the 2H1Y^{FP} (1.8 kg m^{-3}), 2H1Y^{OPT} (1.9 kg m^{-3}) and 3H2Y (1.8 kg m^{-3}) as compared to 1H1Y (2.1 kg m^{-3}). The summer and spring maize crops had maximum water use efficiency throughout the experimental trails.

3.4. Crop N Uptake, N Loss and N Use Efficiency

The summer and spring maize crops had higher crop N uptake than winter wheat in all cropping systems (Table 6). Crop N uptake was significantly high in 1H1Y monoculture rotation. First year spring maize had 216 kg N ha^{-1} uptake, and subsequent year had 234 kg N ha^{-1}. Average N uptakes were 184.25, 163.5, 177.6 and 225 kg N ha^{-1} in 2H1Y^{FP}, 2H1Y^{OPT}, 3H2Y and 1H1Y systems respectively. Thus the sequence was 1H1Y > 2H1Y^{FP} > 3H2Y > 2H1Y^{OPT}.

Table 5. Water balance summary of different cropping systems.

Cropping Systems	Rotations	Ir (mm)	Pr (mm)	Eta (mm)	D (mm)	R (mm)	Wbal (mm)	WUE (kg m^{-3})
(2H1Y)^{FP}	WW	240	109	328	13	0	8	1.6
	SM	75	224	287	33	21	−42	2.0
	WW	315	223	345	17	6	170	1.3
	SM	115	569	320	180	255	−71	2.1
(2H1Y)^{OPT}	WW	180	109	302	7	0	−20	1.6
	SM	75	224	286	7	21	−1.5	2.1
	WW	240	223	331	7	6	119	1.4
	SM	115	569	320	133	251	−20	2.4
(3H2Y)	WW	180	109	302	7	0	−20	1.9
	SM	75	224	285	8	21	−1 5	1.7
	WW	183	633	426	98	245	47	1.7
(1H1Y)	SPM	243	276	389	7	20	103	2.1
	SPM	80	638	423	106	245	−56	2.1

Note: Ir, Irrigation, Pr, Precipitation, Eta, Actual evapotranspiration, D, Drainage, R, Runoff, Wbal, Water balance and WUE, Water use efficiency.

Table 6. N balance summary of different cropping systems.

Cropping Systems	Rotations	N$_{fer}$	N$_{min}$	N$_{vol}$	N$_{den}$	N$_{lea}$	N$_{up}$	Nbal	NUE
		(kg ha^{-1})							(kg kg^{-1})
(2H1Y)^{FP}	WW	300	−8	30	2	2	144	114	28.9
	SM	250	−4	47	2	10	215	−28	21.3
	WW	300	−3	26	2	6	146	117	24.7
	SM	250	−6	55	4	79	232	−1.26	18.3
(2H1Y)^{OPT}	WW	120	11	2	9	1	134	−1.4	34.1
	SM	75	36	6	9	1	172	−77	31.3
	WW	145	79	13	11	0	129	71	29.3
	SM	80	84	33	14	19	219	−1.21	26.9
(3H2Y)	WW	120	11	2	9	0.5	137	−1.8	38.3
	SM	75	37	7	9	1.3	169	−74	26.5
	WW	130	100	25	10	3.0	227	−35	27.2
(1H1Y)	SPM	75	7	3	1	0.3	216	−1.38	37.8
	SPM	130	117	24	12	5.8	234	−29	31.8

Note: N$_{fer}$, Fertilization, N$_{min}$, Net mineralization, N$_{vol}$, Ammonia volatilization, N$_{den}$, Denitrification, N$_{lea}$, N leaching, N$_{up}$, Crop N uptake, Nbal, Nitrogen balance and NUE, Nitrogen use efficiency.

The N loss through volatilization, denitrification and leaching are not only yield affecting factor, but also the primary cause of maximum N deposition in the terrestrial and aquatic systems. The current study showed that the double cropping system under farmer (2H1Y^{FP}) and optimized practices (2H1Y^{OPT}) had higher N loss, total ranged from 118–265 kg. The lowest N loss was 46.1 kg from the 1H1Y, and 3H2Y had 66.8 kg N loss. Therefore, the 2H1Y^{FP} and 2H1Y^{OPT} had low average N use efficiency than 3H2Y and 1H1Y systems. The highest mean N use efficiency (34.8 kg kg^{-1}) was from 1H1Y monoculture. Generally, N use efficiency was higher for all crops grown in the first season of first rotation in each cropping system and subsequently dropped (Table 6).

3.5. Results of Scenario Analyses

Initially, an increasing yield response of winter wheat, summer and spring maize showed with the interaction of increasing N rates and irrigation levels (Figure 4a). When irrigation and N doses were increased to excessive, then yield was decreased or there was no significant increase. The grain yield was <3000, <4000 and <7000 kg ha^{-1}, respectively at 0–50 mm irrigation in combination with all N rates. The grain yield curves of winter wheat reached to peak from the 200 mm irrigation × 150–250 kg N ha^{-1} with the highest simulated yield of 5000–5750 kg ha^{-1}, summer maize from the 200 mm irrigation $\times$ 100–1 50 kg N ha^{-1} with yield of 6200–7000 kg ha^{-1}, and spring maize from the >150 mm irrigation ha^{-1} $\times$ 100–250 kg N ha^{-1} with yield of 8400–8540 kg ha^{-1}. In the case of N and water use efficiencies, which is the ratio of grain yield to the amount of N and irrigation depleted by crop during the growth period (Figure 4b,c). The N use efficiency was the lowest at >250, >200 and >150 kg N ha^{-1} for winter wheat, summer and spring maize, respectively due to low yield. The spring maize had the highest N use efficiency (37.5 kg kg^{-1}) from the 50 to100 kg N ha^{-1}. While water use efficiency also lowered as the irrigation increased. The spring maize had the highest water use efficiency (>2.8 kg m^{-3}). Furthermore, when irrigation level was set to 0 mm in the model as rainfed condition, then simulated yield of spring maize yield was >4000 kg ha^{-1} with the combination of 50–100 kg N ha^{-1}, and the yield response of winter wheat was <2500 kg ha^{-1} (Figure 4a). Going in depth, the N leaching and drainage prediction of WHCNS model showed that the N loss was correlated to N rate and drainage was closely associated to irrigation amount. As the N rates and irrigation amounts were increased, the N leaching and water drainages increased too. The rates of N leaching and drainage were notable when N rate was >250 kg ha^{-1} and irrigation level was >200 mm ha^{-1} for winter wheat with a maximum N leaching of 28.3 kg ha^{-1} and drainage 29.0 mm. Both, summer and spring maize crops had the higher N leaching when N rate was >150 and >100 kg N ha^{-1}, respectively (Figure 5b,c).

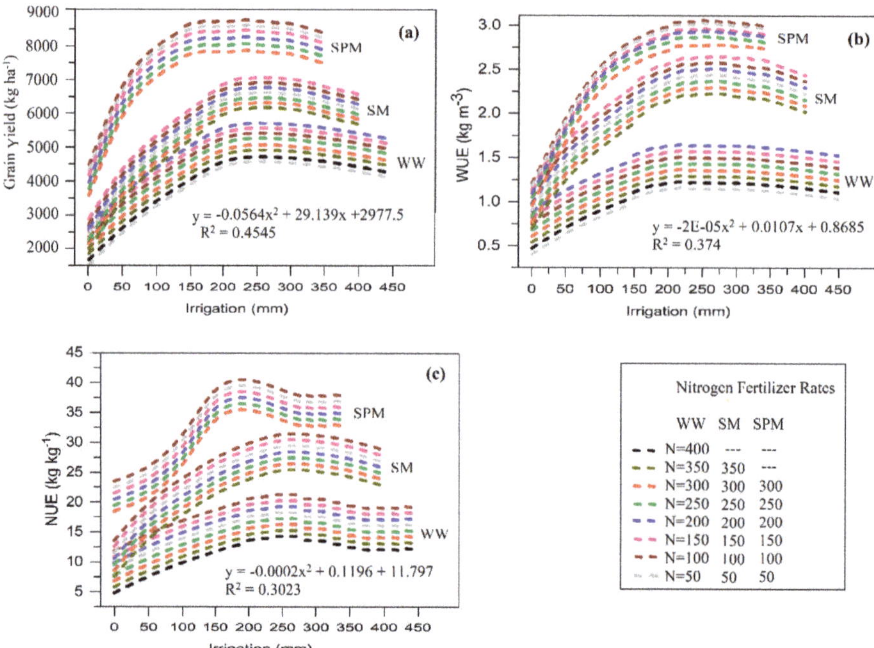

Figure 4. Scenarios based simulated grain yield (**a**), N use efficiency (**b**), and water use efficiency (**c**) under effect of deficit and excessive irrigation levels, and N rates. WW, winter wheat, SM, summer maize, SPM, spring maize.

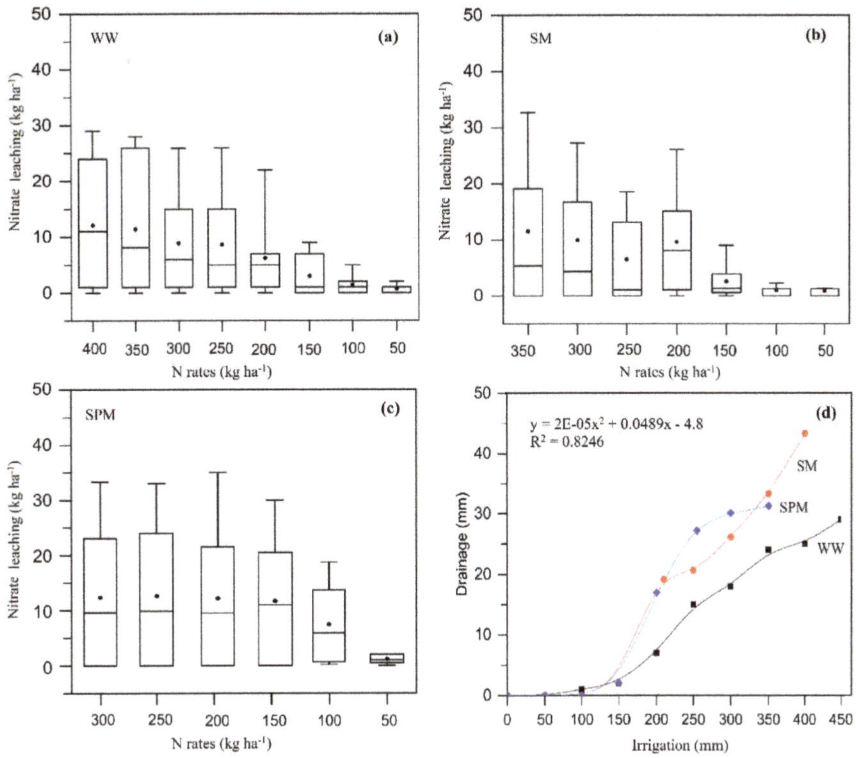

Figure 5. Simulated nitrate leaching (**a–c**) and drainage (**d**) under effect of excessive and deficit irrigation levels, and N rates. The boxes show the 25 50 and 75 percentiles, the line in the box indicates the median, black dots show mean, and the crosses indicate maximum and minimum. WW, winter wheat, SM, summer maize and SPM, spring maize.

4. Discussion

4.1. Effect of Different Cropping Systems on Grain Yield, Water and N Use Efficiencies

The modeling based study of crops is an effective method to optimize the farming practices and suggest possible solution for sustainable agriculture development. The predicted results for 2H1Y^{FP} cropping system under local farmer practices indicated that the yield is not related to excessive irrigation and N application. This is also confirmed from the observed data. The low yield was obtained from the double cropping system plots, despite the field soil had the high amendment of irrigation and N fertilizer. We found that, although 2H1Y^{OPT} increased grain yield (5.97%) and reduced N fertilizer (61.81%) as compared to 2H1Y^{FP}, the 610 mm irrigation consumption is still high. The high water use in the double cropping system also reported by many other researchers [60–62]. Therefore, only N saving is unsustainable, because irrigation water is quite important factor in the crop production system. About 420 mm irrigation was used by wheat crop in the 2H1Y^{OPT}. The rainfall was significantly low (<250 mm) throughout the winter wheat growing season of this study period, which limited grain yield. Water saving technology [63], proper plant density [64], minimum tillage [65], and other agronomic and breeding approaches would be advantageous to reduce water consumption of winter wheat. Furthermore, the evapotranspiration sequence was 1H1Y < 3H2Y < 2H1Y^{OPT} < 2H1Y^{FP}. The irrigation levels and N fertilizer rates of 3H2Y were <41.20% and <70.45%, respectively than 2H1Y^{FP}, but this 3H2Y system failed due to low grain yield output. Only monoculture 1H1Y system was dominant over all other systems in the light of both environmental protection and groundwater security at cost of some grain yield loss. In this system, irrigation level was <56.64% (323 mm) and N fertilizer was <81.36% (205 kg) than 2H1Y^{FP}. Similar results also reported by Meng et al. [66]. Yang et al., [37], studied five crops (wheat–maize–cotton–sweet potato –peanuts) in rotations using DSSAT crop model for a 12 years long term field experiment, and found that wheat accounted for >40% crop water use. The 2H1Y^{FP} had the highest annual evapotranspiration rate (734 mm) with 1.1 m groundwater decline. The monoculture of any crop used less water than prevalent 2H1Y^{FP}. Monoculture cultivation of maize reduced >60% irrigation and >70% N fertilizer [3]. There were two important reasons of high yield of continuous 1H1Y system to low inputs in the present study, one was high average N mineralization (62.23 kg ha^{-1}), N uptake (225 kg ha^{-1}), N use efficiency (35.65 kg ha^{-1}) and water use efficiency (2.15 kg m^{-3}). It can be explained as temporal match between spring maize growth period and available resources. The spring maize being a C4 efficient, do not exhibit photorespiration, uses CO_2, solar radiation, temperature, and water more efficiently than winter wheat [66,67], and other reasons of high yield of continuous 1H1Y system to low input treatments could be balanced N dose, which was <8.29% than 43–179 kg N ha^{-1} as recommended by Zhang et al. [68].

4.2. Effect of Cropping Systems on Water Drainage, and N Fates

Irrigation and N balancing as per need of normal plant growth by reducing the nitrate that transport to water bodies and gaseous loss to environment, is a big challenge in the crop production system [69]. When field water become excess, then it drains down from root zone carrying NO_3^- which contaminates groundwater. It has been identified that agricultural farming in China is causing more environmental pollution than any other source [70]. Our study revealed that the 1H1Y had the lowest N loss through leaching, denitrification, and volatilizations than the rest of the cropping systems (2H1Y^{FP}, 2H1Y^{OPT} and 3H2Y). The amounts of N leaching and volatilization were higher in 2H1Y^{FP} cropping system. Particularly, maximum N leaching and N volatilization loss measured from maize crop managed under farmer practices. The linking factor was high dose application of N fertilizer rates exceeding crop demand. The annual application of N fertilizer >550 kg ha^{-1} did not significantly increase grain yield, but led to twice time higher NO_3^- losses [71,72]. Bing et al. [73] observed the peak of N emission at >180 kg N ha^{-1}, because crop plant can only uptake required amount to complete life cycle, when soil and environmental condition are favorable. The maximum N loss from double cropping system under farmer practices was due to excessive N application (550 kg N ha^{-1}), and topdressing of N via broadcasting application method. This method of fertilizer application is inefficient. Beside greater N loss in volatilization due to increased time of N granules exposed to sun and air. There are also number of other disadvantages such as N fixation as granules cover large soil mass and wrong placement since plant cannot utilize N as N move latterly over some distance from the range of roots [73,74].

Based on the evidence of current research results, it was noticed that N leaching and drainage occurred due to spring rainfall events and excessive irrigation. The cause of volatilization was the raised temperature during the period from May to October in each year of entire experimental period. An average >20% NH_3 volatilization was observed by Cui et al. [18] due to higher temperature. Another N loss was from denitrification pathway. It is a process, where NO_3^- converts into N_2 by serial facilitation of soil microbes, depends on soil pH, aeration and energy [74]. The N denitrification was low in 2H1Y^{FP}, and higher in all other cropping systems. In case of the N mineralization, it was effective in 2H1Y^{OPT}, 3H2Y and 1H1Y systems, indicates stimulated soil microbial activity in optimized cropping systems, which increased plant available N. There was 210 kg N became available for 2H1Y^{OPT}, 148 kg N for 3H2Y and 124 kg for 1H1Y. We noticed that maize crop had bigger N net mineralization and N uptake than wheat crop. These results are in line with Ju et al. [75].

4.3. Best Management Practices (BMPs) for Different Cropping Systems

Based on experimental findings, which are briefly described above and deep analyses of management practices, showed that it is important to further optimize farming practices for reduction of nitrate loss while sustaining yield. The results of scenario analyses indicated that there is an opportunity to reduce some amount of N and irrigation from different cropping systems. In the $2H1Y^{OPT}$ cropping system, the cultivation of winter wheat using <200 mm irrigation and 100–150 kg N ha^{-1} and summer maize with inputs of <150 mm irrigation and <100 kg N ha^{-1} would be both environmentally and economically beneficial. Furthermore, irrigation ranging in between 100–150 mm ha^{-1} and <150 kg N ha^{-1} for spring maize cultivation would be a suitable management practice. These findings are in agreement with the conclusion of Zhao et al. [76] and Luo et al. [77]. Cultivation of crops under conventional practices will further increase groundwater decline. Keeping field moisture at optimum level and lowering N application would reduce the risk of environmental security [78].

5. Conclusions

It was a very critical decision to recommend an ideal cropping system considering simultaneously environmental and socioeconomic factors of the region. Therefore, grain yield, water and N use efficiencies under different cropping systems were compared through the WHCNS model, and optimized management practices were explored by scenario analyses. The results indicated that simulated soil water content, soil N concentration, leaf area index, dry matter and grain yield agreed well with observed values. So, the model could be used to quantitatively analyze the N fates, water and N use efficiencies under the different management practices. We concluded that replacing the prevalent $2H1Y^{FP}$ cropping system with 1H1Y would be ecofriendly at cost of some grain yield decline. Whereas, the 3H2Y showed insignificant results in terms of grain yield, and $2H1Y^{FP}$ was unsustainable. The $2H1Y^{FP}$ system consumed 745 mm water and 1100 kg N ha^{-1} in two years. When farming practices were optimized for two harvests in 1 year system $(2H1Y)^{OPT}$, then grain yield improved and water (18.12%) and N (61.82%) consumption were minimized. There was an ample amount of N saved, but water conservation was still unsatisfactory. However, considering the results of scenario analyses, it is recommended that winter wheat would be cultivated at <200 mm irrigation by reducing one irrigation event.

Author Contributions: Data analyses, S.J.L. and H.L.; wrote the paper, S.J.L., K.H. and Y.W.; supervision, K.H.

Funding: The work was supported by the National Key R&D Program of China (2016YFD0800102 and 2017YFD0301102-5).

Acknowledgments: We thank Liang Jin for providing soil data for our study.

Conflicts of Interest: The authors declare no conflict of interest.

References

1. Quan, J.; Zhang, Q.; He, H.; Liu, J.; Huang, M.; Jin, H. Analysis of the formation of fog and haze in North China Plain (NCP). *Atmos. Chem. Phys.* **2011**, *11*, 8205–8214. [CrossRef]
2. Zhang, Z.; Tao, F.L.; Du, J.; Shi, P.J.; Yu, D.Y.; Meng, Y.B.; Sun, Y. Surface water quality and its control in a river with intensive human impacts. A case study of the Xiangjiang River. *Chin. J. Environ. Manag.* **2010**, *91*, 2483–2490. [CrossRef] [PubMed]
3. Meng, Q.; Sun, Q.; Chen, X.; Cui, Z.; Yue, S.; Zhang, F.; Romheld, V. Alternative cropping systems for sustainable water and nitrogen use in the North China Plain. *Agric. Ecosyst. Environ.* **2012**, *146*, 93–102. [CrossRef]
4. Lv, L.; Yao, Y.; Zhang, L.; Dong, Z.; Jia, X.; Liang, S.; Ji, J. Winter wheat grain yield and its components in the North China Plain: Irrigation management, cultivation, and climate. *Chil. J. Agric. Res.* **2013**, *73*, 233–242. [CrossRef]
5. Xiao, D.P.; Tao, F.L.; Liu, Y.J.; Shi, W.J.; Wang, M.; Liu, F.S.; Zhang, S.; Zhu, Z. Observed changes in winter wheat phenology in the North China Plain for 1981-2009. *Int. J. Biometeorol.* **2013**, *57*, 275–285. [CrossRef]

6. NBS (National Bureau of Statistics of China). *China Agriculture Yearbook*; China Agriculture Press: Beijing, China, 2009. Available online: http://www.stats.gov.cn/tjsj/ndsj/2009/indexeh.htm (accessed on 1 January 2012). (In Chinese)

7. Gao, B.; Ju, X.T.; Meng, Q.F.; Cui, Z.L.; Christie, P.; Chen, X.P.; Zhang, F.S. The impact of alternative cropping systems on global warming potential, grain yield and groundwater use. *Agric. Ecosyst. Environ.* **2015**, *203*, 46–54. [CrossRef]

8. Van Oort, P.A.J.; Wang, G.; Vos, J.; Meinke, H.; Li, B.G.; Huang, J.K.; van der Werf, W. Towards groundwater neutral cropping systems in the alluvial fans of the North China Plain. *Agric. Water Manag.* **2016**, *165*, 131–140. [CrossRef]

9. An, P.; Ren, W.; Liu, X.; Song, M.; Li, X. Adjustment and optimization of the cropping systems under water constraint. *Sustainability* **2016**, *8*, 1207. [CrossRef]

10. Kukal, M.S.; Irmak, S. Characterization of water use and productivity dynamics across four C3 and C4 row crops under optimal growth conditions. *Agric. Water Manag.* **2020**, *227*, 105840. [CrossRef]

11. Yuan, Z.; Yan, D.H.; Yang, Z.Y.; Yin, J.; Breach, P.; Wang, D.Y. Impacts of climate change on winter wheat water requirement in Haihe River Basin. *Mitig. Adapt. Strateg. Glob. Chang.* **2016**, *21*, 677–697. [CrossRef]

12. Sun, H.Y.; Liu, C.M.; Zhang, X.Y.; Shen, Y.J.; Zhang, Y.Q. Effects of irrigation on water balance, yield and WUE of winter wheat in the North China Plain. *Agric. Water Manag.* **2006**, *85*, 211–218. [CrossRef]

13. Xiao, D.P.; Tao, F.L. Contributions of cultivars, management and climate change to winter wheat yield in the North China Plain in the past three decades. *Eur. J. Agron.* **2014**, *52*, 112–122. [CrossRef]

14. Ju, X.T.; Kou, C.L.; Zhang, F.S.; Christie, P. Nitrogen balance and groundwater nitrate contamination: Comparison among three intensive cropping systems on the North China Plain. *Environ. Pollut.* **2006**, *143*, 117–125. [CrossRef] [PubMed]

15. Guo, J.H.; Liu, X.J.; Zhang, Y.; Shen, J.L.; Han, W.X.; Zhang, W.F.; Christie, P.; Goulding, K.W.T.; Vitousek, P.M.; Zhang, F.S. Significant acidification in major Chinese croplands. *Science* **2010**, *327*, 1008–1010. [CrossRef]

16. Schroder, J.L.; Zhang, H.; Girma, K.; Raun, W.R.; Penn, C.J.; Payton, M.E. Soil acidification from long-term use of nitrogen fertilizers on winter wheat. *Soil Sci. Soc. Am. J.* **2011**, *75*, 957–964. [CrossRef]

17. Liu, C.; Wang, K.; Meng, S.; Zheng, X.; Zhou, Z.; Han, S.; Yang, Z. Effects of irrigation, fertilization and crop straw management on nitrous oxide and nitric oxide emissions from a wheat-maize rotation field in northern China. *Agric. Ecosyst. Environ.* **2011**, *140*, 226–233. [CrossRef]

18. Cui, Z.L.; Chen, X.P.; Zhang, F.S. Current Nitrogen management status and measures to improve the intensive wheat-maize system in China. *AMBIO* **2010**, *39*, 376–384. [CrossRef]

19. Liu, X.; Ju, X.; Zhang, F.; Pan, J.; Christie, P. Nitrogen dynamics and budgets in a winter wheat–maize cropping system in the North China Plain. *Field Crops Res.* **2003**, *83*, 111–124. [CrossRef]

20. Kendy, E.; Zhang, Y.Q.; Liu, C.M.; Wang, J.X.; Steenhuis, T. Groundwater recharge from irrigated cropland in the North China Plain: Case study of Luancheng County, Hebei province, 1949–2000. *Hydrol. Process.* **2004**, *18*, 2289–2302. [CrossRef]

21. Fang, Q.; Yu, Q.; Wang, E.; Chen, Y.; Zhang, G.; Wang, J.; Li, L. Soil nitrate accumulation, leaching and crop nitrogen use as influenced by fertilization and irrigation in an intensive wheat–maize double cropping system in the North China Plain. *Plant Soil* **2006**, *284*, 335–350. [CrossRef]

22. Srivastava, P.; Singh, R.M. Optimization of cropping pattern in a canal command area using fuzzy programming approach. *Water Resour. Manag.* **2015**, *29*, 4481–4500. [CrossRef]

23. Osama, S.; Elkholy, M.; Kansoh, R.M. Optimization of the cropping pattern in Egypt. *Alex. Eng. J.* **2017**, *56*, 557–566. [CrossRef]

24. Craufurd, P.Q.; Vadez, V.; Jagadish, S.K.; Prasad, P.V.; Zaman, A.M. Crop science experiments designed to inform crop modeling. *Agric. For. Meteorol.* **2013**, *170*, 8–18. [CrossRef]

25. Constantin, J.; Willaume, M.; Murgue, C.; Lacroix, B.; Therond, O. The soil-crop models STICS and AqYield predict yield and soil water content for irrigated crops equally well with limited data. *Agric. For. Meteorol.* **2015**, *206*, 55–68. [CrossRef]

26. Yin, X.; van der Linden, C.G.; Struik, P.C. Bringing genetics and biochemistry to crop modelling, and vice versa. *Eur. J. Agron.* **2018**, *100*, 132–140. [CrossRef]

27. Dourado, N.D.; Teruel, D.A.; Reichardt, K.; Nielsen, D.R.; Frizzone, J.A.; Bacchi, O.O.S. Principles of crop modeling and simulation: I. Uses of mathematical models in agricultural science. *Sci. Agric.* **1998**, *55*, 46–50. [CrossRef]

28. Salo, T.J.; Palosuo, T.; Kersebaum, K.C.; Nendel, C.; Angulo, C.; Ewert, F.; Ferrise, R. Comparing the performance of 11 crop simulation models in predicting yield response to nitrogen fertilization. *J. Agric. Sci.* **2016**, *154*, 1218–1240. [CrossRef]

29. Leng, G.; Hall, J. Crop yield sensitivity of global major agricultural countries to droughts and the projected changes in the future. *Sci. Total Environ.* **2018**, *654*, 811–821. [CrossRef]

30. Liang, H.; Hu, K.; Batchelor, W.D.; Qi, Z.; Li, B. An integrated soil-crop system model for water and nitrogen management in North China. *Sci. Rep.* **2016**, *6*, 25755. [CrossRef]

31. Li, Z.; Hu, K.; Li, B.; He, M.; Zhang, J. Evaluation of water and nitrogen use efficiencies in a double cropping system under different integrated management practices based on a model approach. *Agric. Water Manag.* **2015**, *159*, 19–34. [CrossRef]

32. Xu, Q.; Li, Z.; Hu, K.; Li, B. Optimal management of water and nitrogen for farmland in North China Plain based on osculating value method and WHCNS model. *Trans. Chin. Soc. Agric. Eng.* **2017**, *33*, 152–158. [CrossRef]

33. Zhang, H.; Hu, K.; Zhang, L.; Ji, Y.; Qin, W. Exploring optimal catch crops for reducing nitrate leaching in vegetable greenhouse in north China. *Agric. Water Manag.* **2019**, *212*, 273–282. [CrossRef]

34. Xu, T.; Dengming, Y.; Baisha, W.; Wuxia, B.; Pierre, D.; Fang, L.Y.W.; Jun, M. The Effect evaluation of comprehensive treatment for groundwater overdraft in Quzhou county, china. *Water* **2018**, *10*, 874. [CrossRef]

35. Cai, J.; Xia, X.; Chen, H.; Wang, T.; Zhang, H. Decomposition of fertilizer use intensity and its environmental risk in china's grain production process. *Sustainability* **2018**, *10*, 498. [CrossRef]

36. Du, L.; Xu, C.; Wu, Y.; Wang, M.; Chen, F. Water degradation footprint of crop production in Hebei province. *J. Agro-Environ. Sci.* **2018**, *37*, 286–293. [CrossRef]

37. Yang, X.; Chen, Y.; Pacenka, S.; Gao, W.; Zhang, M.; Sui, P.; Steenhuis, T.S. Recharge and groundwater use in the North China Plain for six irrigated crops for an eleven year period. *PLoS ONE* **2015**, *10*, e0115269. [CrossRef] [PubMed]

38. Chow, L.; Xing, Z.; Rees, H.; Meng, F.; Monteith, J.; Stevens, L. Field Performance of Nine Soil Water Content Sensors on a Sandy Loam Soil in New Brunswick, Maritime Region, Canada. *Sensors* **2009**, *9*, 9398–9413. [CrossRef]

39. Liang, H.; Hu, K.; Qin, W.; Zuo, Q.; Zhang, Y. Modelling the effect of mulching on soil heat transfer, water movement and crop growth for ground cover rice production system. *Field Crops Res.* **2017**, *201*, 97–107. [CrossRef]

40. Hussain, G.; Aljaloud, A.A.; Alshammary, S.F.; Karimulla, S. Effect of saline irrigation on the biomass yield and the protein, nitrogen, phosphorus and potassium composition of alfalfa in a pot experiment. *J. Plant Nutr.* **1995**, *18*, 2389–2408. [CrossRef]

41. Siyal, A.A.; Mashori, A.S.; Bristow, K.L.; Van Genuchten, M.T. Alternate furrow irrigation can radically improve water productivity of okra. *Agric. Water Manag.* **2016**, *173*, 55–60. [CrossRef]

42. Cai, J.; Liu, Y.; Lei, T.; Pereira, L.S. Estimating reference evapotranspiration with the FAO Penman-Monteith equation using daily weather forecast messages. *Agric. For. Meteorol.* **2007**, *145*, 22–35. [CrossRef]

43. Allen, R.G.; Pereira, L.S.; Raes, D.; Smith, M. *Crop Evapotranspiration-Guidelines for Computing Crop Water Requirements*; FAO Irrigation and Drainage Paper 56; FAO: Rome, Italy, 1998.

44. Green, W.H.; Ampt, G.A. Studies on soil physics: Part 1. The flow of air and water through soils. *J. Agric. Sci.* **1911**, *4*, 1–24.

45. Richards, L.A. Capillary conduction of liquids through porous mediums. *J. Appl. Phys.* **1931**, *1*, 318–333. [CrossRef]

46. Thomas, S. Introduction to climate modelling. In *Advances in Geophysical and Environmental Mechanics and Mathematics*, 1st ed.; Springer: Berlin/Heidelberg, Germany, 2011. [CrossRef]

47. Driessen, P.M.; Konijn, N.T. *Land-Use System Analysis*; Wageningen Agricultural University: Wageningen, The Netherlands, 1992.

48. Moriasi, D.N.; Arnold, J.G.; Van Liew, M.W.; Bingner, R.L.; Harmel, R.D.; Veith, T.L. Model evaluation guidelines for systematic quantification of accuracy in watershed simulations. *Trans. ASABE* **2007**, *50*, 885–900. [CrossRef]

49. Nash, J.E.; Sutcliffe, J.V. River flow forecasting through conceptual models. Part I—A discussion of principles. *J. Hydrol.* **1970**, *10*, 282–290. [CrossRef]

50. Willmott, C.J. On the validation of models. *Phys. Geogr.* **1981**, *2*, 184–194. [CrossRef]
51. Mukaka, M.M. A guide to appropriate use of correlation coefficient in medical research. *Malawi Med. J.* **2012**, *24*, 69–71.
52. Yang, Y.; Cui, Y.; Luo, Y.; Lyu, X.; Traore, S.; Khan, S.; Wang, W. Short term forecasting of daily reference evapotranspiration using the Penman-Monteith model and public weather forecasts. *Agric. Water Manag.* **2016**, *177*, 329–339. [CrossRef]
53. Fang, Q.; Ma, L.; Yu, Q.; Malone, R.W.; Saseendran, S.A.; Ahuja, L.R. Modeling nitrogen and water management effects in a wheat-maize double-cropping system. *J. Environ. Qual.* **2008**, *37*, 2232–2242. [CrossRef]
54. Hu, K.; Li, Y.; Chen, W.; Chen, D.; Wei, Y.; Edis, R.; Zhang, Y. Modeling nitrate leaching and optimizing water and nitrogen management under irrigated maize in desert oases in Northwestern China. *J. Environ. Qual.* **2010**, *39*, 667–677. [CrossRef]
55. Liang, H.; Qi, Z.; Hu, K.; Li, B.; Prasher, S.O. Modelling subsurface drainage and nitrogen losses from artificially drained cropland using coupled DRAINMOD and WHCNS models. *Agric. Water Manag.* **2018**, *195*, 201–210. [CrossRef]
56. Liang, H.; Hu, K.; Qin, W.; Zuo, Q.; Guo, L.; Tao, Y.; Lin, S. Ground cover rice production system reduces water consumption and nitrogen loss and increases water and nitrogen use efficiencies. *Field Crops Res.* **2019**, *233*, 70–79. [CrossRef]
57. He, Y.; Liang, H.; Hu, K.; Wang, H.; Hou, L. Modeling nitrogen leaching in a spring maize system under changing climate and genotype scenarios in arid Inner Mongolia, China. *Agric. Water Manag.* **2018**, *210*, 316–323. [CrossRef]
58. Wang, H.; Tong, W.; Zhao, X.; Wang, Z. Optimum management of water and fertilizer for potato in soft rock and sand compound soil based on WHCNS model: Scenario prediction. *IOP Conf. Series: Earth and Environmental Sci.* **2018**, *170*, 022171. [CrossRef]
59. Jego, G.; Anchez, P.; Jose, M.; Justes, E. Predicting soil water and mineral nitrogen contents with the STICS model for estimating nitrate leaching under agricultural fields. *Agric. Water Manag.* **2012**, *107*, 54–65. [CrossRef]
60. Zhang, X.; Qin, W.; Chen, S.; Shao, L.; Sun, H. Responses of yield and WUE of winter wheat to water stress during the past three decades: A case study in the North China Plain. *Agric. Water Manag.* **2017**, *179*, 47–54. [CrossRef]
61. Umair, M.; Shen, Y.; Qi, Y.; Zhang, Y.; Ahmad, A.; Pei, H.; Liu, M. Evaluation of the CropSyst Model during Wheat-Maize Rotations on the North China Plain for Identifying Soil Evaporation Losses. *Front. Plant Sci.* **2017**, *8*, 1667. [CrossRef]
62. Chu, Y.; Shen, Y.; Yuan, Z. Water footprint of crop production for different crop structures in the Hebei southern plain, North China. *Hydrol. Earth Syst. Sci.* **2017**, *21*, 3061–3069. [CrossRef]
63. Sidhu, H.S.; Jat, M.L.; Singh, Y.; Sidhu, R.K.; Gupta, N.; Singh, P.; Gerard, B. Sub-surface drip fertigation with conservation agriculture in a rice-wheat system: A breakthrough for addressing water and nitrogen use efficiency. *Agric. Water Manag.* **2019**, *216*, 273–283. [CrossRef]
64. Li, Q.; Bian, C.; Liu, X.; Ma, C.; Liu, Q. Winter wheat grain yield and water use efficiency in wide-precision planting pattern under deficit irrigation in North China Plain. *Agric. Water Manag.* **2015**, *153*, 71–76. [CrossRef]
65. Guan, D.; Zhang, Y.; Al-Kaisi, M.M.; Wang, Q.; Zhang, M.; Li, Z. Tillage practices effect on root distribution and water use efficiency of winter wheat under rain-fed condition in the North China Plain. *Soil Tillage Res.* **2015**, *146*, 286–295. [CrossRef]
66. Meng, Q.; Wang, H.; Yan, P.; Pan, J.; Lu, D.; Cui, Z.; Chen, X. Designing a new cropping system for high productivity and sustainable water usage under climate change. *Sci. Rep.* **2017**, *7*, 41587. [CrossRef] [PubMed]
67. Sun, H.; Zhang, X.; Liu, X.; Liu, X.; Shao, L.; Chen, S.; Dong, X. Impact of different cropping systems and irrigation schedules on evapotranspiration, grain yield and groundwater level in the North China Plain. *Agric. Water Manag.* **2019**, *211*, 202–209. [CrossRef]
68. Zhang, Y.; Wang, H.; Lei, Q.; Luo, J.; Lindsey, S.; Zhang, J.; Ren, T. Optimizing the nitrogen application rate for maize and wheat based on yield and environment on the Northern China Plain. *Sci. Total Environ.* **2018**, *618*, 1173–1183. [CrossRef]

69. Dinnes, D.L.; Karlen, D.L.; Jaynes, D.B.; Kaspar, T.C.; Hatfield, J.L.; Colvin, T.S.; Cambardella, C.A. Nitrogen management strategies to reduce nitrate leaching in tile-drained Midwestern soils. *Agron. J.* **2002**, *94*, 153–171. [CrossRef]

70. Watts, J. Chinese farms causes more pollution than factories, says official survey. *The Guardian.* 13 September 2010. Available online: https://www.theguardian.com/environment/2010/feb/09/china-farms-pollution (accessed on 20 June 2019).

71. Ju, X.T.; Xing, G.X.; Chen, X.P.; Zhang, S.L.; Zhang, L.J.; Liu, X.J.; Zhang, F.S. Reducing environmental risk by improving N management in intensive Chinese agricultural systems. *Proc. Natl. Acad. Sci. USA* **2009**, *106*, 3041–3046. [CrossRef]

72. Ji, B.; Yang, K.; Zhu, L.; Jiang, Y.; Wang, H.; Zhou, J.; Zhang, H. Aerobic denitrification: A review of important advances of the last 30 years. *Biotechnol. Bioprocess Eng.* **2015**, *20*, 643–651. [CrossRef]

73. Cao, B.; He, F.Y.; Xu, Q.M.; Yin, B.; Cai, G.-X. Denitrification losses and N2O emissions from nitrogen fertilizer applied to a vegetable field. *Pedosph* **2006**, *16*, 390–397. [CrossRef]

74. Leghari, S.J.; Wahocho, N.A.; Laghari, G.M.; Laghari, A.H.; Mustafa, B.G.; Hussain, T.K.; Lashari, A.A. Role of nitrogen for plant growth and development: A review. *Adv. Environ. Biol.* **2016**, *10*, 209–219.

75. Ju, X.T.; Liu, X.J.; Zhang, F.S. Soil nitrogen mineralization and its prediction in winter wheat-summer maize rotation system. *Chin. J. Appl. Ecol.* **2003**, *14*, 2241–2245.

76. Zhao, Z.; Qin, X.; Wang, Z.; Wang, E. Performance of different cropping systems across precipitation gradient in North China Plain. *Agric. For. Meteorol.* **2018**, *259*, 162–172. [CrossRef]

77. Luo, J.; Shen, Y.; Qi, Y.; Zhang, Y.; Xiao, D. Evaluating water conservation effects due to cropping system optimization on the Beijing-Tianjin-Hebei plain, China. *Agric. Syst.* **2018**, *159*, 32–41. [CrossRef]

78. Xiao, G.; Zhao, Z.; Liang, L.; Meng, F.; Wu, W.; Guo, Y. Improving nitrogen and water use efficiency in a wheat-maize rotation system in the North China Plain using optimized farming practices. *Agric. Water Manag.* **2019**, *212*, 172–180. [CrossRef]

 *agronomy*

Article

Harvesting Regimes Affect Brown Midrib Sorghum-Sudangrass and Brown Midrib Pearl Millet Forage Production and Quality

Joshua A. Machicek [1], Brock C. Blaser [1,*], Murali Darapuneni [2] and Marty B. Rhoades [1]

1   Department of Agricultural Sciences, West Texas A&M University, 2501 4th Ave, Canyon, TX 79016, USA
2   Agricultural Science Center at Tucumcari, New Mexico State University, Tucumcari, NM 88401, USA
*   Correspondence: bblaser@wtamu.edu; Tel.: +1-806-651-2555

Received: 30 June 2019; Accepted: 27 July 2019; Published: 30 July 2019

Abstract: As water levels in the Ogallala Aquifer continue to decline in the Texas High Plains, alternative forage crops that utilize less water must be identified to meet the forage demand of the livestock industry in this region. A two-year (2016 and 2017) study was conducted at West Texas A&M University Nance Ranch near Canyon, TX to evaluate the forage production and quality of brown midrib (BMR) sorghum-sudangrass (SS) (*Sorghum bicolor* (L.) Moench ssp. *Drummondii*) and BMR pearl millet (PM) (*Pennisetum glaucum* (L.) Leeke)) harvested under three regimes (three 30-d, two 45-d, and one 90-d harvests). Sorghum-sudangrass consistently out yielded PM in total DM production in both tested years (yield range 3.96 to 6.28 Mg DM ha^{-1} vs. 5.38 to 11.19 Mg DM ha^{-1} in 2016 and 6.00 to 9.87 Mg DM ha^{-1} vs. 6.53 to 15.51 Mg DM ha^{-1} in 2017). Water use efficiency was higher in PM compared to SS. The 90-d harvesting regime maximized the water use efficiency and DM production compared to other regimes in both crops; however, some forage quality may be sacrificed. In general, the higher forage quality was achieved in shorter interval harvesting regimes (frequent cuttings). The selection of suitable forage crop and harvesting regime based on this research can be extremely beneficial to the producers of Texas High Plains to meet their individual forage needs and demand.

Keywords: harvesting strategies; forage yield and quality; forage sorghum; pearl millet; Texas High Plains

---

1. Introduction

Alternative forage crops that utilize less water must be identified to meet the demands of the livestock industry in the Texas High Plains as water levels in the Ogallala Aquifer continue to decline [1–3]. Forage sorghums, including BMR SS, are widely utilized in the High Plains region because of their relative drought and heat tolerance [4,5]. These forage types have potential to produce large amounts of nutritious forage during summer months while the versatility of the crop allows for incorporation into many different types of cropping or livestock operations [6]. However, less is known about PM production in the region and the potential of the crop to meet some of these forage needs.

Pearl millet is one of the low-input crops primarily grown in semi-arid regions of Africa and southeast Asia [7]. This crop is grown in the southeastern US primarily for grain purposes ([7], although there were some attempts to highlight the forage purpose of PM [8,9]. Similar to sorghum, PM is also a drought and heat tolerant crop that survives well under limited rainfed conditions [10]. Regrowth of PM is affected by stubble height, cutting frequency, and stage of harvest [11]. Unlike many sorghums, PM contains no prussic acid [11,12]. Both species have varieties that contain the BMR trait; therefore, they have reduced lignin to increase forage quality and give producers more flexibility in harvest scheduling [13].

Therefore, PM may have a potential to be as productive as forage sorghums and provide the same quality. Cutting height and yield attributes of PM and SS have been evaluated in Kansas and New Mexico, but not in the Texas High Plains [6,9,11]. However, additional information is necessary to identify the ideal cutting intervals to optimize yield and quality for SS and PM. Such knowledge will help increase the production of these crops and provide more management versatility to producers as they seek to meet the forage demands of the livestock industry in the region. The objectives of this study were to (i) evaluate SS and PM forage production and regrowth patterns under three different harvest intervals, and (ii) evaluate the effects of harvest interval on feed nutritive components and value.

2. Materials and Methods

2.1. Study Site, Experimental Design, and Crop Management

This study was conducted during the 2016 and 2017 growing seasons at the West Texas A&M University Nance Ranch near Canyon (34°58′6″ N, 101°47′16″ W; 1097 m above sea level). The experiments were conducted on Olton clay loam soil (fine, mixed, superactive, thermic, Aridic Paleustoll). The plots were prepared for planting with two passes of a tandem disk followed by one pass with a rotary tiller. 'Bodacious' BMR sorghum-sudangrass (7272 seeds kg^{-1}, 85% germination, 98% purity) and 'Graze King' BMR pearl millet (176,211 seeds kg^{-1}, 85% germination, 98% purity) were planted at 75 and 85 seeds m^{-2}, respectively, on 17 June 2016 using a tractor mounted 150 cm wide Great Plains 3P500 grain drill (Great Plains Manufacturing, Salina, KS) with 19-cm row widths. In 2017, a PM seed lot (116,280 seeds kg^{-1}, 85% germination, 98% purity) was acquired from Winfield United. On 31 May 2017, both species were planted at 85 seeds m^{-2}. Main plot size was 24.4 by 18.2 m. The planted area for each subplot was 3 by 6.1 m in both years. The experimental design was a nested split plot with crop species being main plot and harvesting regime being sub-plot. The treatments were replicated four times.

The crops were irrigated with a flow metered, surface drip line system with two lines 150 cm apart and drip line emitters every 60 cm. The emitters applied 7.5 L h^{-1} and 25 mm of water was applied weekly for 10 weeks. Soil samples prior to planting were analyzed for a forage sorghum yield goal of 25 Mg ha^{-1}. Phosphorus and potassium levels were found to be sufficient, but nitrogen (N) was required. Urea N fertilizer was broadcast applied based on soil sampling recommendations on 12 July 2016 and 7 July 2017 at 84 kg N ha^{-1} and 78 kg N ha^{-1}, respectively.

2.2. Forage Management and Physiology Measurements

Forage DM was sampled in three harvest regimes: three 30-d, two 45-d and one 90-d harvest. Samples were cut at 15 cm cutting height within a one m quadrat. Two samples were taken per subplot and the dry weights were averaged. Ratoon harvests were taken from the same sampled area each time. Samples were oven dried at 60 °C for 120 h.

Leaf area index (LAI) was determined every 14 d and after each harvest beginning on 12 July 2016 and 21 June 2017, using Li-Cor 2200 plant canopy analyzer (Li-Cor Incorporated, Lincoln, NE, USA). Two LAI measurements were obtained in each plot. A LAI measurement was defined as one above canopy (incident) reading and four below canopy readings. The four below canopy readings were taken across three rows and averaged for one LAI value. Measurements were collected under low light at sunrise, sunset, or overcast conditions.

Photosynthetically active radiation (PAR) interception by the crop canopy was determined every 7 d beginning on 6 July 2016 and 21 June 2017 using AccuPAR Linear PAR Ceptometer, Model PAR-80 light measuring instrument (Decagon Devices, Pullman, WA, USA). Measurements were obtained by placing the ceptometer diagonally across three rows. Measurements were collected under full sunlight between 11:00 and 14:00 Percent light interception was calculated by dividing the average of two below canopy PAR readings by one above canopy reading and multiplying by 100.

2.3. Forage Analysis

Forage analysis samples were taken from aboveground biomass samples, ground with a wood chipper and sent to Servi-Tech Laboratories in Amarillo, TX. Samples were ground through a Wiley mill (Arthur H. Thomas Co., Philadelphia, PA, USA) to pass a 1-mm screen. Crude protein (CP), acid detergent fiber (ADF), neutral detergent fiber (NDF), total digestible nutrients (TDN), and relative feed value (RFV) were measured or calculated.

Crude protein was measured using the combustion method, an AOAC Official Method 990.03 [14]. The ADF analysis was measured using Ankom technology method 5 [15]. This method is a modification of AOAC Official Method 973.18. The NDF analysis was performed with Ankom technology method 6, a modification of AOAC Official Method 2002.04 [16]. Total digestible nutrients was calculated using the formula: TDN = (NFC × 0.98) + (CP × 0.87) + (FA × 0.97 × 2.25) + (NDFn × NDFD/100) − 10. Where NFC = non-fibrous carbohydrate, FA = fatty acid, NDFn = nitrogen free NDF and NDFD = in vitro NDF digestibility.

The RFV is a calculation of digestible dry matter (DDM) and dry matter intake (DMI) and a constant. The RFV was calculated using the formula: RFV = (DDM × DMI)/1.29. The DDM was calculated using the formula: DDM = 88.9 − (0.779 × %ADF). The DMI was calculated by using the formula: DMI = 120/%NDF.

2.4. Water Use Efficiency

Water use efficiency (WUE) measurements were taken in the 2017 growing season only. Plant available soil water (PAW) at planting was taken in adjacent plots less than 50 m away from four random sites to a depth of 75 cm using a tractor mounted Giddings hydraulic press (Giddings Machine Company Inc., Windsor, CO). One core sample per plot, divided into three depth sectors: 0–15, 15–45, and 45–75 cm, was taken at end-of-season harvest to determine PAW, weighed, then oven dried at 104 °C for 72 h. The PAW was found using the equation: PAW = [(volumetric water − permanent wilting point)/100) × depth in cm of the measured soil profile]. The WUE was calculated using the following formula: WUE = DM/[(PAW planting + total rainfall + total irrigation) − PAW harvest].

2.5. Weather Data

Weather conditions during the study in 2016 and climatic data were obtained from the National Weather Service station for Canyon, TX approximately 7 km from the research site. In 2017, daily maximum and minimum air temperature and rainfall were recorded from a weather station (Campbell Scientific, Logan, UT, USA) located 100 m from the experimental site (Table 1). Canyon, TX has a mean annual rainfall of 474 mm. Growing degree days (GDD) were calculated beginning June 2016 and May 2017 of each season using the formula: GDD = Σ {[(daily max. temp. + daily min. temp.)/2] − base temp.} with base temperature = 10 °C, and maximum temperature = 34 °C [17,18].

Table 1. Average monthly air temperature and total rainfall near Canyon, TX for 2016–2017. Thirty year averages (30-year) were calculated from data collected from the National Weather Service Forecast Office from 1985–2015.

Month	Air Temperature			Precipitation		
	2016 [†]	2017 [‡]	30-year	2016	2017	30-year
	°C			mm		
May	17.3	17.8	19.8	34	40	69
June	25.2	24.2	24.7	26	104	73
July	28.1	25.3	26.3	58	80	58
August	18.6	22.2	25.5	81	128	79
September	15.3	21.2	21.4	40	85	61

[†] 2016 weather data collected from the National Weather Service approx. 7 km from research site for Canyon, TX. [‡] 2017 weather data collected from onsite weather station (Campbell Scientific, Logan, UT) 100 m from the experimental site.

2.6. Statistical Design and Analysis

Statistical analysis was performed using the PROC MIXED model in Statistical Analysis System Version 9.4 (SAS Institute, Cary, NC, USA, 2017). The test years and replications were treated as random and appropriate error term was used as denominator to determine the F-test significance. Homogeneity of variance test was used to determine the appropriateness of combined test or 'by year' analysis. Based on this test, a separate analysis by year was carried out. The separate analysis was also justified by different weather and crop-growing conditions existed in the two tested years. When F-test for the treatment was significant, a LSD/PDIFF (α = 0.05) was used to test significant differences between treatment means unless otherwise noted.

3. Results and Discussion

3.1. Weather Conditions

The 2016 growing conditions were unfavorable due to warmer temperatures in June and July and cooler temperatures, 6.9 and 6.1 °C below the average in August and September (Table 1). For the month of June, in 2016, only 35% of the 30-year average rainfall accumulated. However, in 2017 growing conditions were similar to the 30-year average. June through September averaged 145% more precipitation than the 30-year average. In 2016, both crops accumulated 134 GDDs before the crops emerged, although in 2017, 161 GDDs were accumulated. The 2016 growing season accumulated 1461 GDDs while in the 2017 growing season, 1377 GDDs were accumulated.

3.2. Forage Production

The maximum amount of forage was produced in the 90-d harvest interval with 11.05 and 15.51 Mg DM ha^{-1} in SS and 6.29 and 9.87 Mg DM ha^{-1} in PM in 2016 and 2017, respectively (Figure 1, Table 2). When total DM harvest intervals were averaged, PM only produced 70% of the total SS DM. These results were contrary to the results reported by Bishnoi [8], where they reported that PM produced 1.5 to 2 times the amount of forage compared to SS and forage sorghum. In 2016, the SS yields for the 30-d and 45-d harvests ranged from 48–45% of the 90-d harvest. While in 2017, 30-d and 45-d harvest yields ranged from 42–45% of the 90-d harvest.

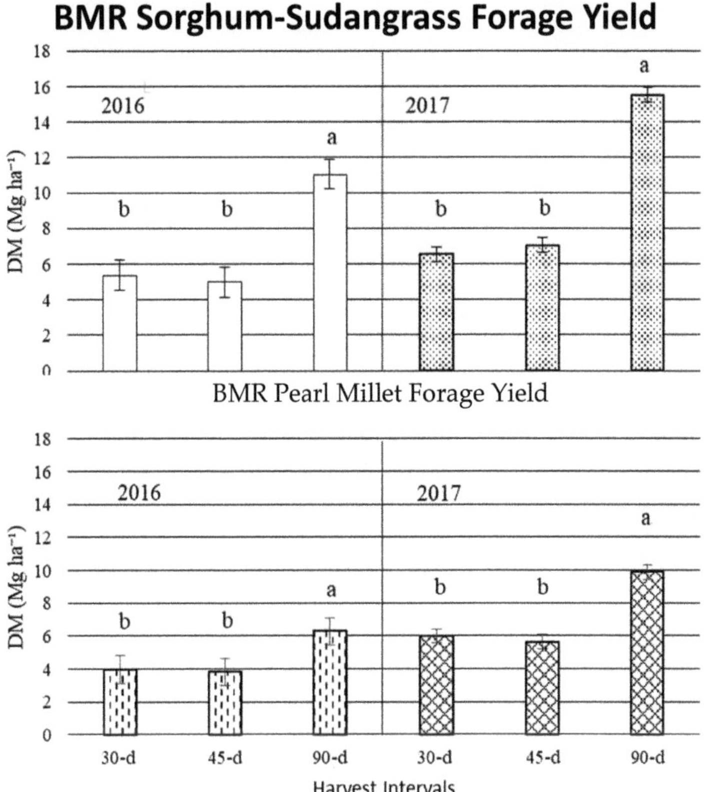

Figure 1. Average total (summed across harvest intervals) brown midrib (BMR) sorghum-sudangrass and BMR pearl millet aboveground dry matter (DM) in Canyon, TX during 2016–2017. The same lowercase letter represents similar means within year, within crop ($\alpha < 0.05$).

Table 2. Brown midrib (BMR) sorghum-sudangrass (SS) and BMR pearl millet (PM) aboveground dry matter (DM) and forage quality means for crude protein (CP), acid detergent fiber (ADF), neutral detergent fiber (NDF), total digestible nutrients (TDN), and relative feed value (RFV) near Canyon, TX, in 2016 and 2017. Harvest intervals included three 30-day, two 45-day, and one 90-day harvest for each crop.

Interval	Crop	Harvest	2016						2017					
			DM [†]	CP	ADF	NDF	TDN	RFV	DM	CP	ADF	NDF	TDN	RFV
			Mg ha^{-1}		%				Mg ha^{-1}		%			
30-day	PM	H$_{30}$	0.50*b*	14.6*a*	30.4*b*	56.5*b*	68.5*a*	107.8*a*	0.88*b*	11.4*a*	34.7*b*	61.2*ab*	63.5*a*	94.3*a*
		H$_{60}$	2.53*a*	11.0*b*	31.7*b*	58.6*b*	67.4*a*	102.0*a*	4.02*a*	10.5*b*	38.6*a*	63.0*a*	59.7*b*	87.0*b*
		H$_{90}$	0.93*b*	9.1*b*	37.7*a*	60.6*a*	60.4*b*	91.5*b*	1.09*b*	8.0*c*	39.0*a*	59.4*b*	59.0*b*	91.8*a*
		Total	3.96						6.00					
	SS	H$_{30}$	1.06*b*	11.0*a*	34.7*b*	62.1*a*	63.5*a*	92.8*a*	1.83*b*	10.6*a*	35.8*c*	58.3*a*	62.6*a*	97.3*b*
		H$_{60}$	3.53*a*	9.2*a*	34.5*b*	60.6*ab*	64.0*a*	95.3*a*	4.09*a*	10.1*ab*	40.2*a*	59.8*a*	57.5*c*	89.5*c*
		H$_{90}$	0.80*b*	10.5*a*	38.4*a*	57.8*b*	59.5*b*	95.3*a*	0.61*c*	9.4*b*	37.9*b*	53.9*b*	60.2*b*	102.8*a*
		Total	5.38						6.53					
		SE	0.239	0.83	0.89	1.16	1.01	2.94	0.224	0.27	0.46	0.67	0.47	1.46
45-day	PM	H$_{45}$	1.57*a*	14.8*a*	32.5*b*	60.4*b*	66.0*a*	98.0*a*	2.78	12.7*a*	36.8*b*	63.8*a*	61.5*a*	87.8*a*
		H$_{90}$	2.24*a*	12.0*b*	35.9*a*	61.5*a*	62.6*b*	92.5*b*	2.86	6.0*b*	38.8*a*	60.3*b*	59.0*b*	90.5*a*
		Total	3.81						5.64					
	SS	H$_{45}$	2.65a	5.8*b*	38.9*b*	63.9*a*	59.0*a*	85.3*b*	4.59*a*	9.9*a*	38.9*a*	62.1*a*	59.0*a*	87.8*b*
		H$_{90}$	2.34a	6.5*a*	39.9*a*	62.1*a*	57.9*b*	86.5*a*	2.46*b*	6.7*b*	38.2*a*	57.2*b*	59.9*a*	96.3*a*
		Total	4.99						7.05					
		SE	0.271	0.62	0.62	0.70	0.68	1.65	0.225	0.38	0.46	0.59	0.55	1.30
90-day	PM	H$_{90}$	6.29*B*	5.1*A*	38.0*A*	64.5*A*	59.9*A*	85.5*A*	9.87*B*	4.3*A*	39.3*A*	59.8*A*	58.6*A*	90.8*A*
	SS	H$_{90}$	11.05*A*	4.4*A*	38.6*A*	62.0*A*	59.5*A*	88.5*A*	15.51*A*	4.2*A*	39.9*A*	58.3*B*	57.9*A*	92.5*A*
		SE	1.306	0.43	0.97	1.03	1.12	2.64	0.429	0.30	0.22	0.40	0.30	0.69

[†] Columns with same lowercase letter are not different between harvests within harvest interval, within crop, within year. Columns with same uppercase letter are not different between harvests within harvest interval, between crops, within year ($\alpha < 0.05$).

In 2016, PM yields for the 30-d and 45-d harvests ranged from 63–60% of the 90-d harvests (Table 2). While in 2017, 30-d and 45-d harvests ranged from 60–57% of the 90-d PM yields. This is contrary to Stephenson [11] that found PM to produce 83% of total DM when comparing a three cut system to a two cut system when harvested at the boot stage.

In SS, 45-d harvest ratoon cut responded differently in both years (Table 2). There was no difference between the first harvest and the ratoon harvest in the first year, while in 2017, more forage was produced in the first harvest. This can be attributed to above average rainfall received in June and July of 2017. In 2016, the ratoon cut made up 47% while in 2017, the ratoon cut made up only 35% of the total DM yield. This is higher when compared to the results reported in Maughan [19], where they reported that the ratoon cut yielded 29% of the total DM produced. However, the current study results are similar to the results reported by Duncan [20], where they reported the ratoon efficiency ranged between 35 and 45%.

In both years, the PM 30-d harvest intervals responded similarly with 13–15% of the total DM yielding from the first cut and 64–67% produced from the second cut (Table 2). This differs from Stephenson [11] that found 33% of the total forage DM yielding from the first cut while the second cut yielded 44% of the total forage DM. However, the third cut responded similarly to Stephenson [11] with 23% of the total DM. Therefore, most of the growth in the current study was produced between 30 and 60 days.

3.3. Forage Quality

In the 30-d interval, when averaged across harvests, the SS CP content was 10.2 and 10% in 2016 and 2017 growing seasons, respectively (Table 2). This is similar to the results reported by Sanderson [21] with an average CP content of 10.9%. Crude protein did not differ significantly among three harvests in 2016. These results are slightly higher than the findings by El-Latif [22] that reported decreasing CP at each cut, 9.0, 7.8, and 7.5% CP for first, second, and third harvests, respectively. However, the 2017 CP results were consistent with the results of El-Latif [22], showing gradual decrease in CP content from first to third harvest. Sanderson [21] also reported an ADF content of 36.5% when harvests were averaged together. Their findings were similar to the current study, where SS ADF was 35.8 and 38%, when averaged across the three harvest intervals, in 2016 and 2017, respectively. In 2016, the ADF content in SS at third harvest was higher than first two harvests; whereas the NDF and TDN values were decreased gradually from first to third harvests. The 2017 results for ADF were slightly differed from 2016 results that showed higher ADF for second harvest.

In both years, the PM 30-d harvest interval %CP decreased with each harvest (Table 2). When averaged across the three harvests, the CP was 10.8% compared to 17.1% reported by Rostamza [23]. In 2016, both ADF and NDF increased with each harvest; however, only ADF increased in 2017 while NDF declined. This is contrary to Rostamza [23] that found as water is limited, ADF increases. Rostamza [23] reported TDN decreases as available water decreases; however, in this study, rainfall was less in 2016 but produced a TDN that was five percentage points higher than the TDN in 2017 when averaged across harvests. At the third harvest, in 2016, RFV declined but in 2017, RFV increased from the second harvest.

Sorghum-sudangrass, in the 45-d harvest, had higher CP and RFV in the ratoon cut but lower TDN (Table 2). However, in 2017, a higher CP was found in the first harvest. In the 90-d harvest, only the DM differed in both years between crops except for NDF in 2017. This study reported slightly better CP than Nasiyev [24] that reported 3.6, 2.4, and 3.0% CP in course millet, sorghum, and sudangrass, respectively.

In the 45-d harvest, PM had higher CP and TDN in the first harvest compared to the ratoon harvest (Table 2). However, only in 2016, the first harvest had a higher RFV. This study reported slightly lower NDF when compared to Cherney [25] that found BMR PM to contain 67.7 and 62.3% NDF in the first and second harvest, respectively. On the contrary, Cherney [25] reported lower ADF% in the first and second cut of 34.5 and 31.4%, respectively. Crude protein content conveyed by Cherney [25] is similar to 2016. The 2016 CP for each cut is 14.8 and 12.0% compared to 14.6 and 12.3% CP reported by Cherney [25] for the first and second harvest, respectively. The CP of the first and second harvest in 2017 were 2:1 and fifty percent lower than values reported by Cherney [25], respectively.

As both crops advanced in maturity, an overall decline in forage quality can be noted. This is attributed to a decline in the leaf:stem ratio in both crops (data not shown). When the leaf:stem ratio degenerates, more mass is allocated in the stem; thus, more lignin is produced and the forage quality is reduced.

3.4. Crop Canopy and Morphology

The LAI in the 90-d SS harvest responded similarly in both years, plateauing between 1078 and 1029 GDDs (Figures 2 and 3). However, 2016 had a 3% higher end of season LAI value, 4.65, when compared to 4.50 in 2017. This is similar to LAI values of 4.5 to 5.0 in the 2010 growing season reported by Maughan [19]. However, the 2016 and 2017 study are much lower than Singh [26] that reported a LAI value of 6.4.

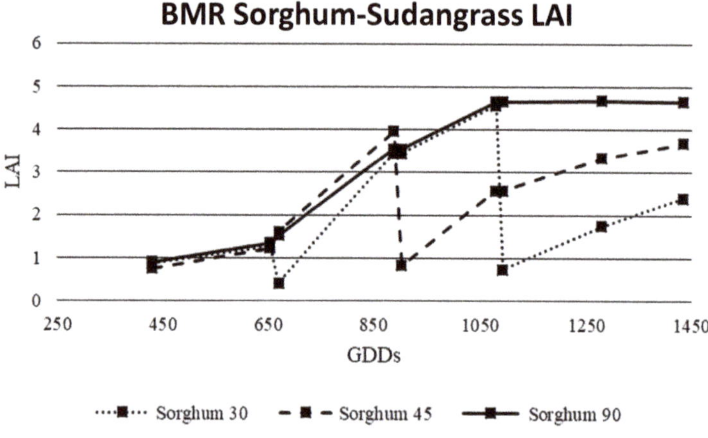

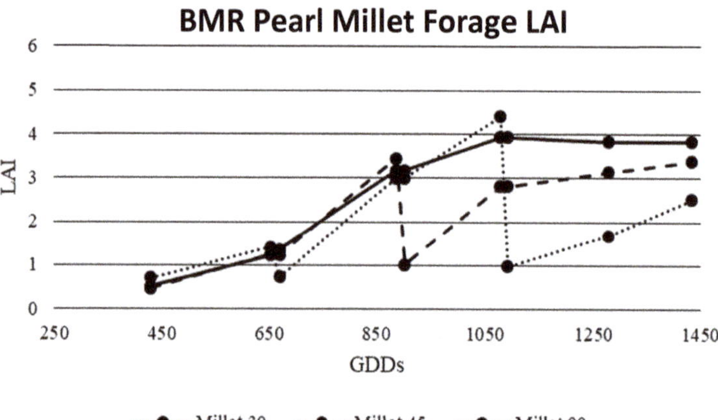

Figure 2. Brown midrib (BMR) sorghum-sudangrass and BMR pearl millet leaf area index (LAI) for three different harvest intervals (three 30-d, two 45-d, and one 90-d) in 2016 at Canyon, TX, USA.

In 2016, the 90-d PM harvest reached maximum LAI, 3.95, at 1078 GDDs and ended the season at a LAI of 3.83 (Figure 2). However, LAI in 2017 peaked at a value of 5.64 at 1029 GDDs and sloped off at the end of the season to a LAI of 4.61. This study is much lower than results found by Singh [26] that reported LAI values of 11.4 and 6.5 in unstressed and severely stressed PM, respectively. Singh [26] attributed this to the severely stressed plants having higher ground cover due to the death of the lower leaves and the incident interception of the upper profile of the crop canopy.

In both years and both crops, the 45-d harvest had a higher LAI value before harvest than at the end of the season (Figures 2 and 3). A similar trend occurred in the 30-d harvest interval where maximum LAI was attained before the second harvest. The SS, 30-d harvests recovered from the maximum LAI to the end of season LAI, 52% and 47% in 2016 and 2017, respectively. While 30-d harvest, PM recovered to 57% and 54% in both years after the second harvest. In 2016, both SS and PM recovered to 93 and 98% of the maximum LAI; however, in 2017, only 88 and 73% of the LAI value was reached in SS and PM, respectively.

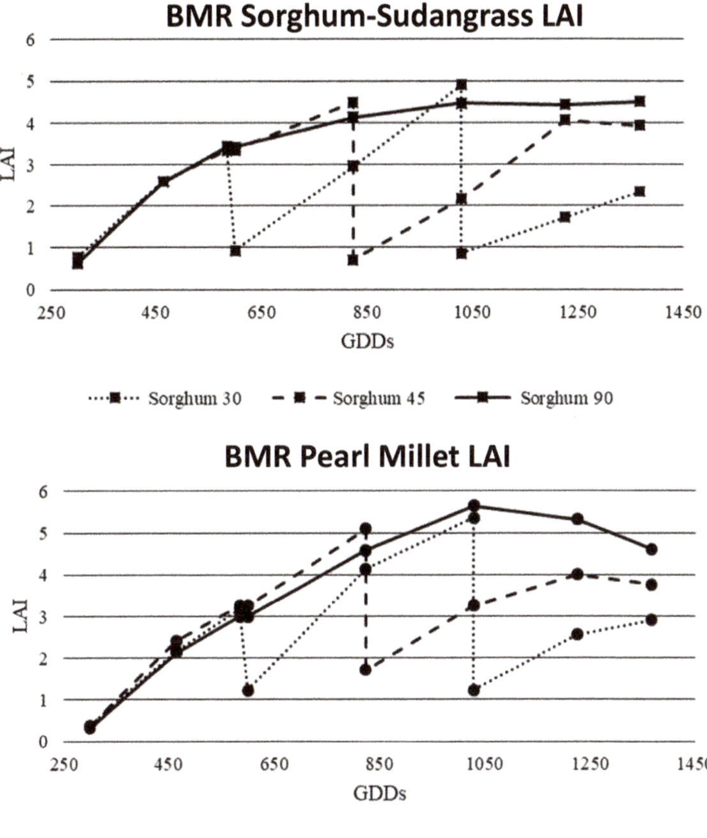

Figure 3. Brown midrib (BMR) sorghum-sudangrass and BMR pearl millet leaf area index (LAI) for three different harvest intervals (three 30-d, two 45-d, and one 90-d) in 2017 at Canyon, TX, USA.

In 2016, in PM, end of season light interception ranged 20% between the three cutting intervals (Figure 4). The 30, 45, and 90-d harvest intervals had end of season PAR interception at 67, 77, and 87% PAR, respectively. In both years, PM reached maximum PAR, 90 and 98%, 350 GDDs before the final harvest (Figures 4 and 5). However, SS in both years reached maximum PAR, 94 and 96%, 200 GDDs after PM reached maximum PAR interception. This is similar to Maughan [19] that found energy sorghum reached 95% PAR interception.

In 2016, both species in the 45-d harvest reached 82% PAR before harvest; while in 2017, both crops averaged 93% intercepted PAR before harvest (Figures 4 and 5). In 2017, PM seemed to have higher light interception after harvest. Across both species and years in the 30-d harvest interval, maximum PAR was achieved right before the second harvest. The SS peaked at 90 and 94% and PM peaked at 81 and 98% PAR in both years, respectively.

In both years after a harvest, PM had higher LAI and PAR than SS. This can be attributed to at 300 GDDs, SS produced 69 and 80% of PM leaf:stem ratios in 2016 and 2017, respectively (data not shown); and SS produced 39% of PM tillers per plant at 550 GDDs (data not shown). Because PM had higher values in all four of these attributes, the crop was better able to reduce soil moisture evaporative losses resulting in a higher WUE.

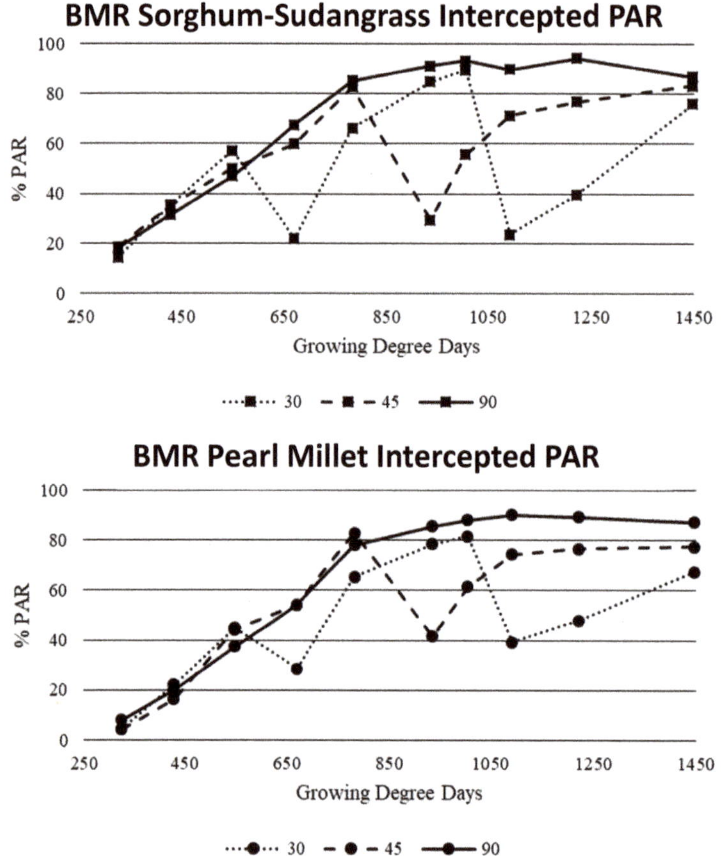

Figure 4. Brown midrib (BMR) sorghum-sudangrass and BMR pearl millet intercepted photosynthetically active radiation (PAR) for three different harvest intervals (three 30-d, two 45-d, and one 90-d) in 2016 at Canyon, TX, USA.

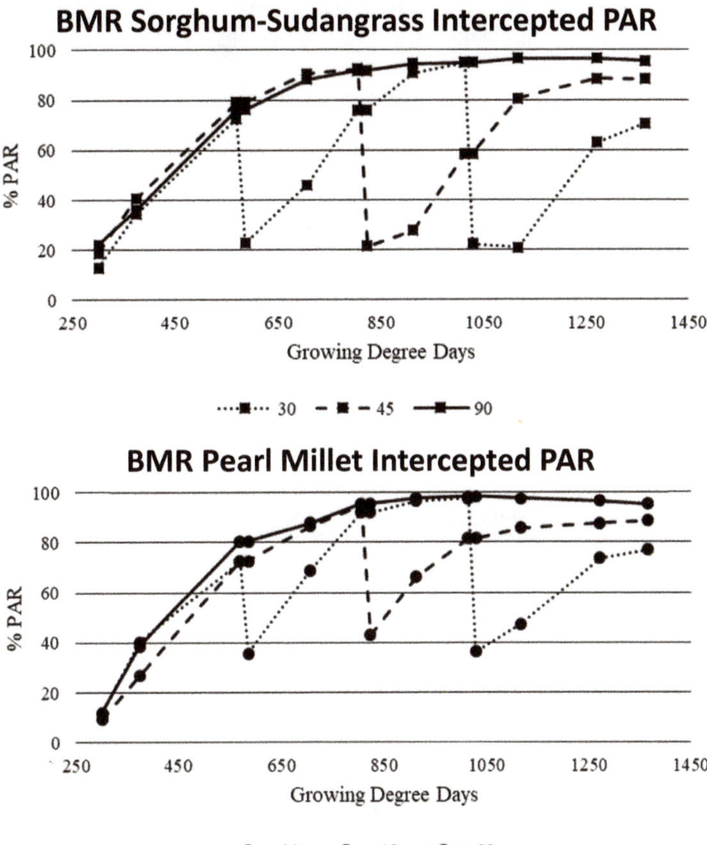

Figure 5. Brown midrib (BMR) sorghum-sudangrass and BMR pearl millet intercepted photosynthetically active radiation (PAR) for three different harvest intervals three 30-d, two 45-d, and one 90-d) in 2017 at Canyon, TX, USA.

3.5. Water Use Efficiency

Water use efficiency (WUE) was evaluated in 2017 only (Table 3). The 90-d harvest had higher WUE compared to other two harvest intervals in both SS and PM. When compared among the crop species, PM showed higher WUE than SS in 45 and 90-d harvest intervals. Overall, the PM WUEs in 30, 45, and 90-d were 10.9, 11.8, and 25.8 kg DM ha^{-1} mm^{-1} when compared to 10.2, 9.5, and 16.4 kg DM ha^{-1} mm^{-1} in SS, respectively. This concurs with Singh [26] that found PM to have a higher WUE than sorghum (17.9 vs. 14.4 kg DM ha^{-1} mm^{-1}).

Table 3. Water use efficiency (WUE), dry matter (DM), and total water use means of brown midrib (BMR) sorghum-sudangrass and BMR pearl millet under three different harvest intervals (three 30-d, two 45-d, and one 90-d) near Canyon, TX in 2017.

Crop	Interval	DM	Total Water Used	WUE
		kg ha^{-1} †	mm	kg ha^{-1} mm^{-1}
Sorghum-sudangrass	30	6528*bA*	599	10.2*bA*
	45	7051*bA*	596	9.5*bB*
	90	15,513*aA*	602	16.4*aB*
Pearl Millet	30	5996*bA*	591	10.9*bA*
	45	5637*bB*	591	11.8*bA*
	90	9873*aB*	601	25.8*aA*
	SE	417.8	2.2	0.701

† Columns with same lowercase letter are not different between harvest intervals within crop and columns with same uppercase letter are not different between crops within harvest interval.

4. Conclusions

The maximum amount of forage was produced in the single 90-d harvest for both crops. It was also concluded that rapid growth occurred between 30 and 60 d after emergence. Sorghum-sudangrass out-yielded PM; however, PM still may be a viable forage option for producers in the region. As the crop matured, forage quality decreased and forage DM production increased; however, some forage quality attributes can be retained with more frequent harvests. Although the 90-d harvest regime maximized forage DM production, if higher forage quality is desired, shorter cutting intervals are recommended. Frequent harvests reduce DM production potential while retaining high quality potential. When water is a limiting factor, a PM, 90-d harvest interval, production system is desirable due to higher WUE. Further research needs to be conducted to understand PM crop establishment and production for the Texas High Plains.

Author Contributions: Conceptualization, J.A.M. and B.C.B.; methodology, J.A.M. and B.C.B.; software, J.A.M., M.D., and M.B.R.; formal analysis, J.A.M., M.D., and M.B.R.; investigation, J.A.M. and B.C.B.; resources, B.C.B.; writing—original draft preparation, J.A.M. and B.C.B.; writing—review and editing, J.A.M., B.C.B., M.D., and M.B.R.; visualization, J.A.M., B.C.B., and M.D.; supervision, B.C.B.; project administration, J.A.M. and B.C.B.; funding acquisition, J.A.M. and B.C.B.

Funding: This research received no external funding.

Acknowledgments: The authors acknowledge partial funding support from the Killgore Research Center, West Texas A&M University, and seed donations from Richardson Seeds, Inc. (Vega, TX, USA) and Winfield United (Shoreview, MN, USA). The authors also gratefully acknowledge the technical and data analysis support of Bradley S. Crookston.

Conflicts of Interest: The authors declare no conflict of interest.

References

1. Terrell, B.L.; Johnson, P.N.; Segarra, E. Ogallala aquifer depletion: Economic impact on the Texas high plains. *Water Policy* **2002**, *4*, 33–46. [CrossRef]
2. McGuire, V.L. *Water-Level Changes in the High Plains Aquifer, Predevelopment to 2001, 1999 to 2000, and 2000 to 2001*; USGS Fact Sheet: FS-078-03; U.S. Geological Survey: Denver, CO, USA, 2003.
3. Colaizzi, P.D.; Gowda, P.H.; Marek, T.H.; Porter, D.O. Irrigation in the Texas High Plains: A brief history and potential reductions in demand. *Irrig. Drain.* **2009**, *58*, 257–274. [CrossRef]
4. Ejeta, G.; Knoll, J.E. Marker-Assisted selection in sorghum. In *Genomic-Assisted Crop Improvement: Vol 2: Genomics Applications in Crops*; Varshney, R.K., Tuberosa, R., Eds.; Springer Publications: Dordrecht, The Netherlands, 2007; pp. 187–205.
5. Lauriault, L.M.; Marsalis, M.A.; VanLeeuwen, D.M. Selecting sorghum forages for limited and full irrigation and rainfed conditions in semiarid, subtropical environments. *Forage Grassl.* **2011**, *9*. [CrossRef]
6. Marsalis, M. Sorghum Forage Production in New Mexico. 2011. Available online: https://aces.nmsu.edu/pubs/_a/A332/welcome.html (accessed on 10 August 2018).

7. Baltensperger, D.D. Progress with proso, pearl and other millets. In *Trends in New Crops and New Uses*; Janick, J., Whipkey, A., Eds.; ASHS Press: Alexandria, VA, USA, 2002; pp. 100–103.

8. Bishnoi, U.R.; Oka, G.M.; Fearon, A.L. Quantity and quality of forage and silage of pearl millet in comparison to Sudax, grain, and forage sorghum harvested at different growth stages. *J. Trop. Agric.* **1993**, *70*, 98–102.

9. Marsalis, M.A.; Lauriault, L.M.; Trostle, C. *Millets for Forage and Grain in New Mexico and West Texas*; NMSU Cooperative Extension Service Publication: Las Cruces, NM, USA, 2012; Available online: http://aces.nmsu.edu/pubs/_a/A417/welcome.html (accessed on 10 August 2018).

10. Andrews, D.J.; Kumar, K.A. Pearl millet for food, feed and forage. *Adv. Agron.* **1992**, *48*, 89–139.

11. Stephenson, R.J.; Posler, G.L. Forage yield and regrowth of pearl Millet. *Trans. Kansas Acad. Sci.* **1984**, *87*, 91–97. [CrossRef]

12. Burger, A.W.; Hittle, C.N. Yield, protein, nitrate, and prussic acid content of sudangrass, sudangrass hybrids, and pearl millets harvested at two cutting frequencies and two stubble heights. *Agron. J.* **1967**, *59*, 259–262. [CrossRef]

13. Staggenborg, S. Forage and renewable sorghum end uses. In *Sorghum: State of the Art and Future Perspectives*; Ciampitti, I., Prasad, V., Eds.; Agron Monogr: Madison, WI, USA, 2016.

14. AOAC International. *AOAC Method 990.03. Protein (Crude) in Animal Feed. Combustion Method. Official Methods of Analysis of AOAC International*, 19th ed.; Latimer, G.W., Jr., Ed.; AOAC International: Gaithersburg, MD, USA, 2012.

15. Ankom. *Ankom Technology Method 5. Acid Detergent Fiber in Feeds. Filter Bag Technique. Ankom Technology. Macedon, New York. This Method Is a Modification of AOAC Official Method 973.18. Fiber (Acid Detergent) and Lignin (H_2SO_4) in Animal Feed. Official Methods of Analysis of AOAC International*, 19th ed.; Latimer, G.W., Jr., Ed.; Ankom: Gaithersburg, MD, USA, 2006.

16. Ankom. *Ankom. Ankom Technology Method 6. Neutral Detergent Fiber in Feeds. Filter Bag Technique. Ankom Technology. Macedon, New York. This Method Is a Modification of AOAC Official Method 2002.04. Amylase-Treated Neutral Detergent Fiber in Feeds. Using Refluxing in Beakers or Crucibles. In Official Methods of Analysis of AOAC International*, 19th ed.; Latimer, G.W., Jr., Ed.; Ankom: Gaithersburg, MD, USA, 2006.

17. Venuto, B.; Kindiger, B. Forage and biomass feedstock production from hybrid forage sorghum and sorghum–sudangrass hybrids. *Grassl. Sci.* **2008**, *54*, 189–196. [CrossRef]

18. Virmani, S.M.; Tandon, H.L.S.; Alagarswamy, G. *Modeling the Growth and Development of Sorghum and Pearl Millet*; International Crops Research Institute for the Semi-Arid Tropics: Patancheru, India, 1989.

19. Maughan, M.; Voigt, T.; Parrish, A.; Bollero, G.; Rooney, W.; Lee, D.K. Forage and energy sorghum responses to nitrogen fertilization in central and southern Illinois. *Agron. J.* **2012**, *104*, 1032–1040. [CrossRef]

20. Duncan, R.R.; Gardner, W.A. The influence of ratoon cropping on sweet sorghum yield, sugar production, and insect damage. *Can. J. Plant Sci.* **1984**, *64*, 261–274. [CrossRef]

21. Sanderson, M.A.; Ali, G.; Hussey, M.A.; Miller, F.R. *Forage Quality and Agronomic Traits of Sorghum-Sudangrass Hybrids*; Texas Agricultural Experiment Station MP-1765: College Station, TX, USA, 1995; Available online: https://www.researchgate.net/publication/292175586_Forage_quality_and_agronomic_traits_of_sorghum-sudangrass_hybrids (accessed on 24 July 2019).

22. El-Lattief, E.A. Growth and fodder yield of forage pearl millet in newly cultivated land as affected by date of planting and integrated use of mineral and organic fertilizers. *Asian J. Crop Sci.* **2011**, *3*, 35–42.

23. Rostamza, M.; Chaichi, M.R.; Jahansouz, M.R.; Alimadadi, A. Forage quality, water use and nitrogen utilization efficiencies of pearl millet (*Pennisetum americanum* L.) grown under different soil moisture and nitrogen levels. *Agric. Water Manag.* **2011**, *98*, 1607–1614. [CrossRef]

24. Nasiyev, B.N. Selection of high-yielding agrophytocenoses of annual crops for fodder lands of frontier zone. *Life Sci. J.* **2013**, *10*, 267–271.

25. Cherney, D.J.; Patterson, J.A.; Johnson, K.D. Digestibility and feeding value of pearl millet as influenced by the brown-midrib, low-lignin trait. *J. Anim. Sci.* **1990**, *68*, 4345–4351. [CrossRef] [PubMed]

26. Singh, B.R.; Singh, D.P. Agronomic and physiological responses of sorghum, maize and pearl millet to irrigation. *Field Crops Res.* **1995**, *42*, 57–67. [CrossRef]

MDPI
St. Alban-Anlage 66
4052 Basel
Switzerland
Tel. +41 61 683 77 34
Fax +41 61 302 89 18
www.mdpi.com

Agronomy Editorial Office
E-mail: agronomy@mdpi.com
www.mdpi.com/journal/agronomy